Conformational Analysis of Molecules in Excited States

Methods in Stereochemical Analysis

Other Books in the Series:

Eiji Ōsawa and Osamu Yonemitsu
Carbocyclic Cage Compounds: Chemistry and Applications

Janet S. Splitter and Frantisek Turecek (editors)
Applications of Mass Spectrometry to Organic Stereochemistry

William R. Croasmun and Robert M. K. Carlson (editors)
Two-Dimensional NMR Spectroscopy: Applications for Chemists and Biochemists. Second Edition

Jenny P. Glusker with Mitchell Lewis and Miriam Rossi
Crystal Structure Analysis for Chemists and Biologists

Kalevi Pihlaja and Erich Kleinpeter
Carbon-13 NMR Chemical Shifts in Structural and Stereochemical Analysis

Louis D. Quin and John G. Verkade (editors)
Phosphorus-31 NMR Spectral Properties in Compound Characterization and Structural Analysis

Eusebio Juaristio (editor)
Conformational Behavior of Six-Membered Rings: Analysis, Dynamics and Stereo-electronic Effects

Motohiro Nishio, Minoru Hirota, and Yoji Umezawa
The CH/π Interaction: Evidence, Nature, and Consequences

David A. Lightner and Jerome E. Gurst
Organic Conformational Analysis and Stereochemistry from Circular Dichroism Spectroscopy

David Neuhaus and Michael Williamson
The Nuclear Overhauser Effect in Structural and Conformational Analysis Second Edition

Conformational Analysis of Molecules in Excited States

Edited by

Jacek Waluk

WILEY-VCH

New York • Chichester • Weinheim • Brisbane • Singapore • Toronto

This book is printed on acid-free paper. ♾

Published simultaneously in Canada.

For ordering and customer service, call 1-800-CALL-WILEY.

Library of Congress Cataloging-in-Publication Data:

Conformational analysis of molecules in excited states / edited by Jacek Waluk.
p. cm.– (Methods in stereochemical analysis)
Includes bibliographical references and index.
ISBN 0-471-29707-0 (alk. paper)
1. Excited state chemistry. 2. Molecular structure. I. Waluk, Jacek, 1952- II. Series.

QD461.5.C66 2000
541.2'2 – dc21

99-049632

Printed in the United States of America.

10 9 8 7 6 5 4 3 2 1

Contents

Preface

The last few decades have witnessed an enormous development in all branches of science devoted to studying molecules in excited states. Huge progress in laser technology has opened up completely new possibilities in both temporal and spectral domains. The achievement of femtosecond time resolution makes it possible to monitor the elementary processes that occur immediately after absorption of a photon. The use of tunable, narrowband lasers provides a means to selectively excite and follow the fate of a single molecule. Traditional divisions into separate areas of research, such as photochemistry and photophysics, are no longer valid. The same becomes true for the experimental techniques, which are getting more and more "hyphenated."

The progress in computational methods is equally impressive. Excited states of what the quantum chemists used to call "large" molecules (the size of naphthalene) can now be calculated by very sophisticated *ab initio* procedures. The calculations for yet larger molecules are limited to semiempirical methods. However, the software required for such computations is now available in user-friendly versions and can easily be installed on an average personal computer. Moreover, the development of molecular dynamics techniques makes it possible to study conformational changes in really huge molecules, such as proteins.

This stunning speed of the outgrowth of experiment and theory makes it very difficult, even for a person directly working with excited states, to keep track of all important publications, novelties and trends. Therefore, the idea of this book is to present an overview of a variety of important works performed in this field. The focus is on the conformational aspects of molecules in excited states. Obviously, a subject like that cannot be exhausted in just one volume. The intention was rather to expose the reader to the multitude of topics covered by this area. In preparing particular chapters, the authors were requested to follow two guiding principles: (i) to present the main issues related to their field of investigation and (ii) to describe the methods used to detect and study the excited-state species.

In choosing the topics, the idea was to present problems that are being heavily discussed and for which definitive solutions are still in the making. A typical example of such an issue, which is discussed in Chapter 1, is provided by various models that are put forward to explain the nature of the dual fluorescence observed in dimethylaminobenzonitrile and derivatives.

In the vast literature dealing with excited states, one can distinguish many papers that revolve about common subjects, such as (i) conformation changes accompanying photoinduced electron transfer; (ii) photochromism; (iii) photoisomerization, often involving a large-amplitude motion; (iv) intramolecular and intermolecular aspects of excited-state proton transfer; (v) photodissociation; and (vi) elementary photophysical and photochemical processes in living systems. Most of these subjects have been covered in the present volume, with the exception of photochromism (mentioned briefly in Chapter 2), on which separate books have been published recently. Photobiology was also not included explicitly, although many of the systems discussed in this volume serve as models in the studies of important photoprocesses occurring in living organisms.

The first chapter describes excited-state conformational changes that occur in molecules consisting of an electron donor and acceptor, linked by a single bond. Photoexcitation of such systems may lead to a charge transfer state for which the angle between the planes of the donor and the acceptor can be very different than in the ground state. This is a result of the competition between two factors, acting in opposite directions. On one hand, flattening of the molecule can be energetically advantageous because of increasing conjugation. On the other hand, twisting towards a fully perpendicular conformation leads to an increase of the dipole moment and is therefore helped by a polar solvent, which can further stabilize such a structure. As a result, the final conformation may be very sensitive to the environment. For the same molecule, flattening may occur in nonpolar solvents, while twisting may prevail in polar ones.

In Chapter 2, conformational aspects of excited-state proton transfer are discussed. These are of particular importance in systems in which not one, but two or more, protons are moving within the short lifetime of an excited state. In such cases, the reaction coordinate is often dominated by terms not necessarily related to proton motion, but rather by those that depict the spatial rearrangement (e.g., motion of an alcohol molecule around the excited chromophore, flattening of a dimeric structure, or a mutual rotation of proton donor and acceptor groups).

Chapter 3 describes an important, yet still relatively little-known, problem: the role played by friction between the solute and solvent molecules in chemical reactions, in particular those involving large-amplitude motion. Various theoretical models of friction are presented and compared with the experimental data. Modeling the friction, especially its frequency and spatial dependence, is a formidable task. Fortunately, it becomes feasible with the development of ultrafast spectral techniques and computational procedures.

Chapter 4 is devoted to exciplexes and other excited-state complexes. Mechanisms of their formation and decay are discussed, as well as the geometric and electronic structure. This chapter also contains a brief introduction to the Marcus theory of electron transfer—in particular, to the notion of "inverted region," a concept that is able to explain a phenomenon rather unusual for an ordinary chemist—a decrease of the reaction rate with the increasing driving force for the process. Another uncommon process presented in this chapter is the "harpooning mechanism," in which the photoinduced electron transfer is a prerequisite for a conformation change. Interestingly, this represents, in a way, a negative of the mechanism described in Chapter 2, by which a conformational process prepares the stage for a proton transfer reaction.

Chapter 5 is devoted to one of the basic problems in organic photochemistry: To what extent does the ground-state distribution of different conformers influence the outcome of a photochemical reaction? Experimental and theoretical procedures are described that allow one to determine the characteristics of each of the rotamers present both in the ground and excited states. The authors start from the classical "NEER principle," which postulates nonequilibration of excited rotamers. A direct consequence of this rule is that the products of a photochemical transformation are determined by the ground-state equilibria. Systems that obey the NEER principle are presented along with those that do not.

The NEER principle is also critically discussed, extended, and reformulated in Chapter 6, which is devoted to theoretical studies of excited-state conformational properties. The authors show that the ground-state distribution of conformers determines the accessibility of excited-state decay channels, which, in turn, is crucial for the fate of an excited molecule. A fundamental role in the photophysics is played by the geometry of conical intersections—extremely efficient channels for photochemistry and radiationless deactivation. It is remarkable how well the position of the conical intersection can be calculated, nowadays even for large molecules.

The structural and kinetic aspects of the excited state are closely related. In this volume, the emphasis is rather on the former, although in many places the latter also receive due attention. The book is not intended to describe in detail the experimental and theoretical techniques used for conformational studies. Still, the reader will find descriptions of many experimental approaches in Chapters 1–5, whereas Chapter 6 presents the principles underlying calculations of ground- and excited-state minima, transition states, and conical intersections.

The resulting volume is dedicated to audiences consisting of physical and organic chemists, photobiologists, and material science specialists. It is the hope of the editor that the ideas presented in this book may stimulate further activity in the fascinating area of conformational studies of excited molecules.

Jacek Waluk

Contributors

GIAMPIERO BARTOCCI. Dipartimento di Chimica, Università di Perugia, 06123 Perugia, ITALY

FERNANDO BERNARDI. Dipartimento di Chimica, Università di Bologna via Selmi n. 2, Bologna I-40126, ITALY

FRED BROUWER. Institute of Molecular Chemistry, Nieuwe Achtergracht 129, 1018 WS Amsterdam, NETHERLANDS

MICHAEL MAUS. Institute for Physical and Theoretical Chemistry, Humboldt University Berlin, Bunsenstr. 1, D–10117 Berlin, GERMANY

UGO MAZZUCATO. Dipartimento di Chimica, Università di Perugia, 06123 Perugia, ITALY

MASSIMO OLIVUCCI. Istituto di Chimica Organica, Pian dei Mantellini, Università degli Studi di Siena, I-53100 Siena, ITALY

WOLFGANG RETTIG. Institute for Physical and Theoretical Chemistry, Humboldt University Berlin, Bunsenstr. 1, D–10117 Berlin, GERMANY

MICHAEL A. ROBB. Dept. of Chemistry, King's College, London Strand, London WC2R 2LS, UK

ANNA SPALLETTI. Dipartimento di Chimica, Università di Perugia, 06123 Perugia, ITALY

DAVID H. WALDECK. Department of Chemistry, University of Pittsburgh, Pittsburgh, PA 15260

JACEK WALUK. Institute of Physical Chemistry, Polish Academy of Sciences, Kasprzaka 44, 01-224 Warsaw, POLAND

Conformational Analysis of Molecules in Excited States

CHAPTER

1

Conformational Changes Accompanying Intramolecular Excited State Electron Transfer

by Wolfgang Rettig and Michael Maus

Institute for Physical and Theoretical Chemistry, Humboldt-University Berlin.

1 Introduction

Molecular ionization often leads to large structural changes. For example, the pyramidal nitrogen of ammonia is transformed into the planar trigonal hybrid upon removal of a lone pair electron. This can also happen intramolecularly if the electron is captured by a nearby electron acceptor (e.g., in charge transfer complexes). In this case, the full ionization energy IP(D) does not have to be delivered (Eq. 1) (e.g., by a photon of short wavelength [vacuum uv]), but the photon may have reduced energy (Eq. 2)

$$h\nu > IP(\mathrm{D}). \tag{1}$$

$$h\nu > IP(\mathrm{D}) - EA(\mathrm{A}) + C. \tag{2}$$

where EA(A) is the electron affinity of the acceptor and C is the (negative) Coulomb energy, representing the mutual attraction of the donor (D) radical cation and acceptor (A) radical anion. This leads to the possibility that photon energies in the visible region are often sufficient to ionize the donor, which is especially likely to happen if the corresponding electron transfer (^{1}ET) state of the decoupled system is the lowest excited state of the supramolecular system consisting of D and A.

Conformational Analysis of Molecules in Excited States, Edited by Jacek Waluk
ISBN 0-471-29707-0 © 2000 Wiley-VCH

In addition to the structural changes within the individual donor and acceptor moieties, rearrangement of the complex D–A system occurs after intramolecular ET. If D and A are flexibly linked, the Coulomb energy, becoming more negative with shorter distances, will drive the D and A moieties to approach each other in what is known as the harpooning effect[1]. But not only the distances may change, but also angles and relative orientations of D and A, and the problem soon becomes too complicated for analysis.

If the dimensionality of the problem is reduced by rigidly linking D and A by using a single bond, then a certain bond shortening due to the harpooning effect can still be expected. However, the main geometrical effects of intramolecular relaxation will be related to intramolecular twisting of D with respect to A, and to the relaxations within D and A. If some molecular properties depend on the twist angle between D and A (φ_{DA}), their observation can be used for conformational analysis. This is possible both for ground states (before intramolecular ET) and for relaxed, excited ET states, and hence, the resulting structures can be compared.

A typical situation is encountered when the ET state is not pure, but couples strongly with other states of the molecule. The resulting ET character is then reduced with respect to full intramolecular ionization, but on the other hand, this partial ET or, synonymously, this charge-transfer (CT) transition from ground to excited or from excited to ground state is not as strongly forbidden as in the pure ET case. As a consequence, CT states can be reached more easily by photon absorption, and their relaxation path may also be different.

In the present review, we will focus on the angular relaxation of directly linked D–A systems in the excited state. Many of these systems are biaromatic compounds and are related to the family of twisted intramolecular charge transfer (TICT) compounds[2,3] where charge transfer leads to excited-state twisting towards an orbitally decoupled geometry (perpendicular arrangement). In other compounds, excited-state twisting occurs towards planarity with even stronger coupling between the states involved. Also, some aromatic compounds with small π substituents as donors belong to the class of TICT compounds. The most well-known examples, dialkylamino groups linked to various acceptor-substituted benzenes, even represent the prototype TICT systems such as dimethylaminobenzonitrile (**DMABN**). These systems have recently been reviewed extensively[4] and are not reviewed here. We will restrict the scope of this review by giving an introductory account relating to the TICT-forming anilines and focus on larger biaromatic systems, showing photo-induced torsional relaxations both towards planarity and perpendicularity, depending on intramolecular and environmental parameters. The various conformational changes in the excited state are reflected in the photophysical properties, which can therefore be used to determine excited-state conformations.

For this purpose, it is helpful to classify the various compounds into different families.

1.1 Twisting of Amino Groups

We may call the family of compounds with a flexible dialkylamino group that show excited-state twisting the *C–N-type compounds*. This is to indicate that the atoms at the flexible bond are carbon and nitrogen, with somewhat different properties. With a full electron pair in the lone pair orbital, the amino group tends to be pyramidal. This tendency is reduced if conjugation allows partial ET towards an acceptor, and planarity at the nitrogen can be expected for a full, intramolecular ET, as with the radical cation of ammonia or other amines. The carbon atom, on the other hand, has little tendency to become pyramidal. Hence, for the reference system **DMABN** and other derived systems, the excited-state relaxation process involves the twisting of the π-planes and the pyramidalization coordinate at the nitrogen, and both effects have to be considered simultaneously.

1.2 Biaromatic Compounds

The family of biaromatic compounds Ar–Ar′ can involve both pure aromatic hydrocarbons with no heteroatoms, or heteroatoms may be present in one or both of the aromatic systems. This can induce a donor or an acceptor character. If no heteroatoms are present in these compounds, pyramidalization effects can usually be neglected such that the torsional coordinate alone is sufficient to describe the conformationally relevant potential with respect to large-amplitude motions. In addition to the large-amplitude motions, small-amplitude relaxations like bond-length changes, will, of course, always be present, too. But these will be hindered very little by bridging substituents or by rigid surroundings, each of which can suppress the large-amplitude twisting motion.

1.2.1 Biaromatic Compounds Containing Donor- or Acceptor-Type Nitrogen

The lesser importance of pyramidalization also holds for biaromatic compounds containing nitrogen within an aromatic ring, because the pyramidal tendency of nitrogen is strongly reduced if this atom is part of an aromatic nucleus. Donor aromatics like pyrrole, as well as acceptor aromatics like pyridine or pyridinium cation, are known to be planar. The same arguments hold as for the dialkylamino donor group. If nitrogen is in the donor part, then partial or full ionization will favor the trigonal arrangement. If, however, nitrogen is situated in the acceptor moiety, the CT will lead to charge accumulation there, and the tendency for pyramidalization at N is increased. In either case, the two subsystems can be directly connected at C atoms only (C–C linkage) or involve a C and an N atom (C–N-type linkage).

1.2.2 Biaromatic Compounds without Nitrogen

There exist a few biaromatic compounds where the aromatic nuclei do not contain heteroatoms, or where any donor or acceptor substituents are not directly engaged in the twisting process, such as methoxy or cyano groups. In this case, the flexible bond connects two carbon atoms, and we can classify these compounds as C–C type. We then expect the smallest contribution of pyramidal components to the relaxation coordinate.

1.3 Compounds with Several Flexible bonds: Stilbenes, Polyenes, Cyanines

These classes of compounds, which are partially the subject of other chapters in this book,[5–7] generally involve several different twisting possibilities, and one may be preferred over the others. In many of these cases, simultaneous twisting of several bonds has to be considered as important for the reaction mechanism. This opens many new, and possibly complex dimensions for the relaxation coordinate. Nevertheless, the simple rules, as derived for the compounds with only one flexible bond, can also advantageously be applied if the kinetic effect of several competing processes is considered.[8–11] The approach can be especially fruitful if qualitatively opposite behavior is involved (e.g., fluorescence quantum yields that can either decrease or increase with solvent polarity[10]).

2 Basic Principles

Several principal factors govern the conformational behavior of D–A systems linked by one flexible bond. They are often of opposing influence, and the result depends on the relative strength of these influences. The general features of these factors will be outlined next, and specific examples will be given in Section 4.

2.1 Mesomeric Interaction

If two π-systems are linked by one single bond, they have a nonnegligible π–π interaction that strongly depends on the twist angle φ. The interaction is zero for perpendicular arrangements, and is strongest for the coplanar arrangement, which it stabilizes. This is generally called the mesomeric interaction (MI), leading to the preferred coplanar arrangement and a rotation barrier for 90° twist. Depending on the bond order of the linking bond, this barrier can be rather small in the ground state (ca. 0.1 eV). It increases with D–A substitution, because the MI increases. For **DMABN**-like compounds, the rotation barrier of the dimethylamino group has been determined by nuclear magnetic resonance (NMR) methods as ca. 0.4 eV.[12] MI usually increases in the excited state, and this is generally reflected by a tendency to flatten the D–A systems after light excitation ($\varphi \rightarrow 0$). In many cases, such as in biphenyls, the systems become coplanar in the excited state ($\varphi = 0$). In other cases with strong sterical hindrance, such as 9,9′-bianthryl, the increased MI leads to equilibrium geometries of the excited state that are nonplanar, but significantly closer to the planar situation than in the ground state (Fig. 1.1a).

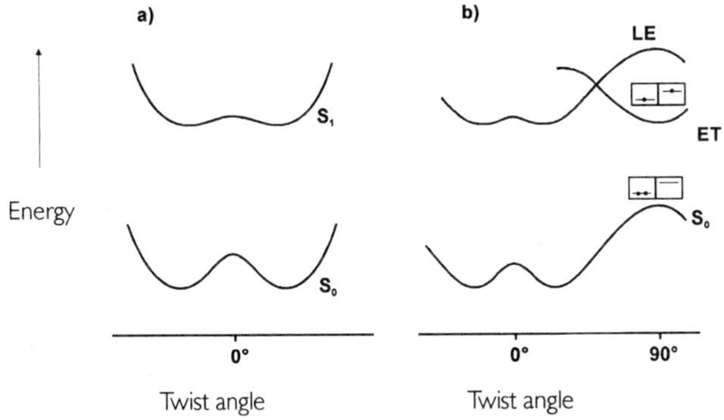

Figure 1.1. (a) General effect of increased mesomeric interaction in the singlet excited state. This effect causes a relative flattering of the biaromatic molecule with respect to the interannular twist angle φ. The energy minima can be far apart from 90°; (b) In D–A molecules, two general types of excited singlet states can be distinguished: locally excited (LE) states and charge transfer states of which the limiting form is electron transfer (ET) states. The mutual interaction of these states, as well as that with the ground state (S_0), determines the shape of the resulting lowest excited singlet state (S_1). The insets depict the electron distribution pattern in the frontier orbitals of D and A for the ground and ET states in the limiting perpendicular conformation. S_0–ET interaction for nonperpendicular twist angles is an important cause of the ground-state energy maximum and the excited-state energy minimum.

2.2 Self-Decoupling Biradicaloid Interaction with the Ground State

The explanation of the biradicaloid interaction is best introduced for noninteracting systems (e.g., a perpendicular D–A conformation). In this case, the orbitals on donor and acceptor are decoupled (i.e., they are localized on the respective moieties). For sufficiently good donor and acceptor properties, the excitation of an electron in the highest-lying donor orbital (HOMO(D)) to the lowest lying acceptor orbital (LUMO(A)) necessitates relatively little energy, which can be smaller than the excitation energy of either donor or acceptor alone or than that of the combined planar system (^{1}LE (locally excited) energy at 90° or planarity). In this case, the first excited state of the decoupled conformation is of pure ^{1}ET type, and the orbital occupation pattern is that of a biradicaloid species (i.e., it corresponds to one odd electron on each moiety (Fig. 1.1b), which can be called a "dot–dot" situation).[13] The electron configuration in the ground state is the complement, i.e., with the HOMO of the donor doubly occupied and the LUMO of the acceptor empty ("hole–pair" situation). The electronic and photophysical properties of ground and excited states are expected to be complementary, too.[11]

For angular deviations from 90°, ^{1}ET and ground state interact (due to the orbital overlap) and therefore energetically repel each other. This is a simple way to understand why a pure ^{1}ET system (without interaction with the ^{1}LE states) should possess an energetic minimum for decoupled geometries (like 90° twisting) where the ground state possesses an energy maximum. Due to the localization of orbitals, transition dipole moments are expected to be very small ("forbidden transitions"). The energetic maximum and minimum are obviously related to each other, and the size of the minimum will depend on the size of the interaction. In this simple picture, excitation into the ^{1}ET state will lead to angular relaxation towards perpendicularity (i.e. to self-decoupling).

2.3 Interaction with Locally Excited States

Except for very small model systems like amino-borane,[13] low-lying ^{1}LE states are also present, as depicted in Figure 1.1b. Depending on the twist angle φ_{DA} between the D and A moieties, on the symmetry of the system, and on the orbitals involved, ^{1}LE and the zero-order electron transfer (^{1}ET) states can strongly interact. This normally leads to avoided crossings between ^{1}LE and ^{1}ET states and may even introduce new minima on the excited-state surface, such that the simple picture of a 90° ^{1}ET minimum need not hold for strong ^{1}LE–^{1}ET interaction.[11] The interaction will also change the transition moments, and the general theory of this "three-state model"[14–16] will be applied in the examples that follow.

2.4 Steric Interaction

A further important factor that can be decisive for the photophysical behavior is the introduction of steric energy barriers to influence the shape of the potential-energy surface in the excited state. This is mostly done by introducing alkyl substituents close to the twisting bond, which causes an energetic disadvantage of the more planar conformations, without strongly affecting the electronic properties. This in turn leads to a shifting of both the ground- and excited-state rotational distribution function towards perpendicularity (Fig. 1.2a) and can help to overcome energy barriers for excited-state reactions (Fig. 1.2b). Moreover, any relaxation towards perpendicularity will be strongly accelerated by this effect, which also helps to identify the active reaction coordinate.

In some biaryls, like 9,9′-bianthryl, steric interaction through perihydrogen repulsion causes such a strong steric potential that even the resulting ground state is of perpendicular structure (Fig. 1.2c). As the MI is generally much stronger in the excited state, the relative contribution of the steric interaction will diminish in the excited state. This explains not only the behavior of biphenyls, which become planar in the excited state (Fig. 1.2a), but also the relaxation towards a nonperpendicular structure in 9,9′-bianthryl (Fig. 1.2c).

2.5 Pyramidalization at the Amino Nitrogen

In many D–A systems, the donor is represented by an aliphatic or aromatic amine, where the nitrogen atom is directly linked to the acceptor. Due to its surplus of electron density compared to carbon, this nitrogen has the tendency for sp^3

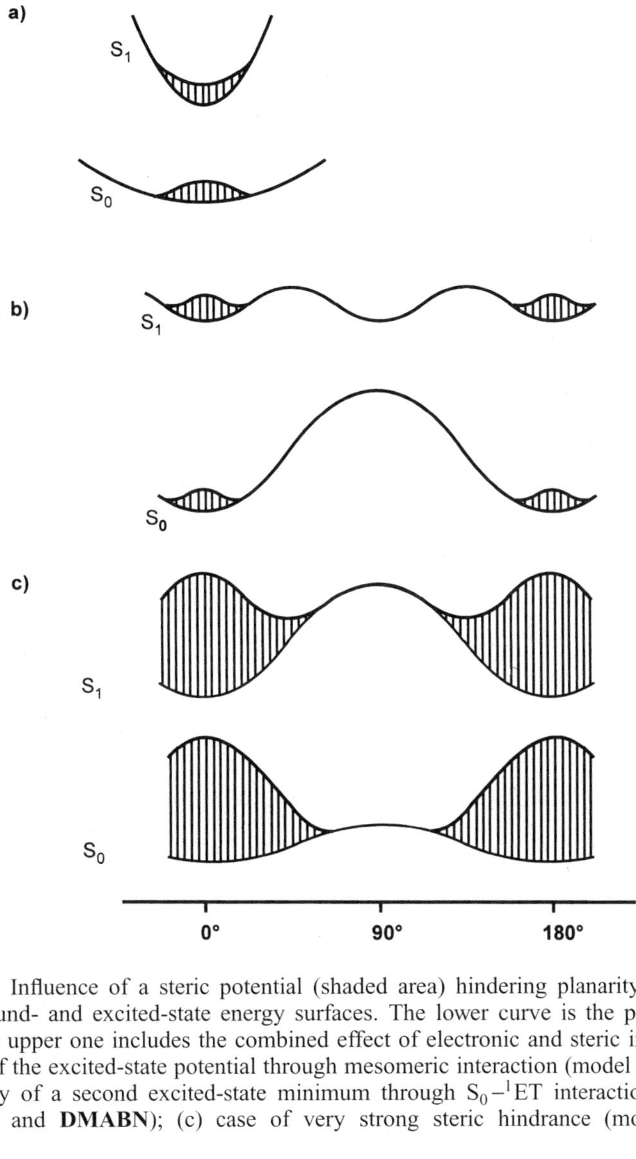

Figure 1.2. Influence of a steric potential (shaded area) hindering planarity for different types of ground- and excited-state energy surfaces. The lower curve is the pure electronic potential; the upper one includes the combined effect of electronic and steric interaction. (a) Steepening of the excited-state potential through mesomeric interaction (model for biphenyl); (b) possibility of a second excited-state minimum through $S_0-{}^1ET$ interaction (model for aminoborane and **DMABN**); (c) case of very strong steric hindrance (model for 9,9′-bianthryl).

hybridization (i.e., pyramidalization). The most well known system is the pyramidal structure of ammonia and its umbrella-type inversion coordinate. For alkyl- or especially aryl-substituted nitrogen, the pyramidalization tendency is reduced, but in many of these donor–acceptor amine derivatives, a residual pyramidalization remains in the ground state. Upon excitation, pyramidalization is also generally reduced, especially for the D–A systems, because the electron density on the nitrogen diminishes upon excitation (Fig. 1.3a, b). We can therefore state, as a

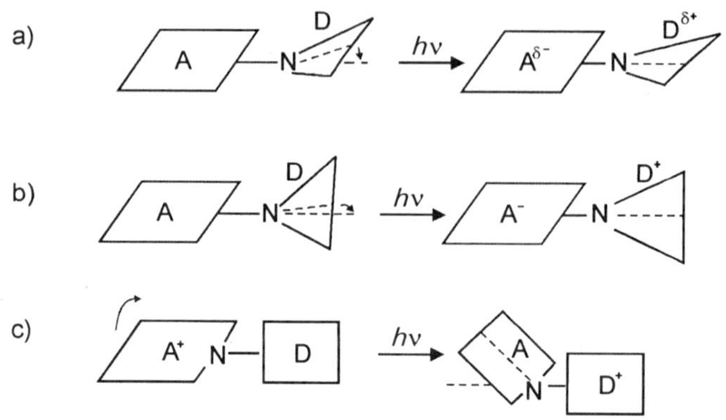

Figure 1.3 Excited-state rehybridization of nitrogen-containing D–A systems. (a) and (b), possessing amino donors with either a planar (a) or a perpendicular (b) twist angle tend to become nonpyramidal (with sp^2 hybridization) in the excited state; (c) twisted *N*-linked systems with strong CT show a bending relaxation due to excited-state pyramidalization of the nitrogen within the acceptor subsystem.

general tendency, that amine donor systems tend to lose their pyramidal character in the excited state (excited-state planarization) regardless of their twist angle, which can be controlled by steric interaction (see Section 3). This can be seen especially well in theoretical studies on the prototype system **DMABN**.[17,18]

2.6 Rehybridization in Nitrogen-containing Acceptor Groups

If a nitrogen is contained in the aromatic system of the acceptor, light-induced intramolecular charge transfer will increase the electron density on the nitrogen and therefore increase its pyramidalization tendency. Examples of very widespread acceptors of this type include pyridine and the pyridinium group. The consequences of this rehybridization are especially pronounced if the nitrogen is engaged in the bond linking D and A. Examples can be found in the pyridinium family (see Section 4.3.4.3). In this case, even if steric interaction causes a perpendicular D–A arrangement in the ground state, rehybridization leads to a bending of the acceptor in the excited state, and the resulting π-planes of donor and acceptor are no longer orthogonal (Fig. 1.3c). As a consequence, the forbidden character of the emissive transition is strongly reduced.

3 Acceptor-substituted Anilines: Evidence for Excited-State Twisting in DMABN and Derivatives

The best-studied example of an aniline-type D–A system, and one that has provoked a tremendous number of studies, is **DMABN** with its dual fluorescence. Because of its importance as a reference system, the main features will be reviewed here. The interested reader will find many more details in other more specialized reviews.[2–4,11,19–21]

Figure 1.4 depicts the dual fluorescence of **DMABN** and of the ester derivative, **DMABEE**. It can be seen that the relative intensity of the long-wavelength ("anomalous") emission band (the A band) increases with solvent polarity and is always larger for the ester than for the nitrile derivative. From fluorescence polarization studies,[22] it is known that these two derivatives differ in the ordering of their close-lying lowest singlet excited state: ^1L_a above ^1L_b for the nitrile **DMABN**, but ^1L_a below ^1L_b for the ester **DMABEE**. The strongly polar nature of the state emitting the A band can be inferred from its strong solvatochromic red shift. The short-wavelength fluorescence ("normal" fluorescence or *B* band) also shows a sizeable solvatochromic red shift, but considerably smaller than that of the A band.

The original structural hypothesis,[23] known later as the TICT model,[24,25] invokes a twisting motion of the dimethylamino group, coupled with charge transfer. About a

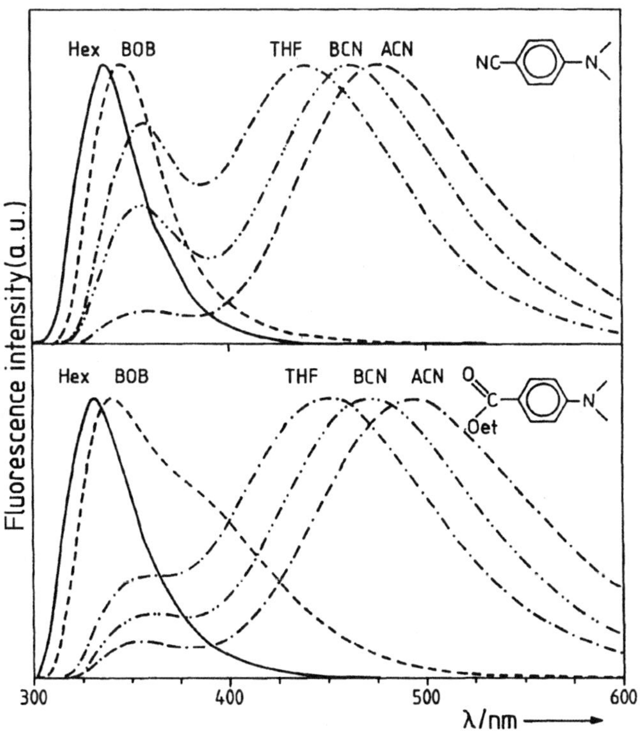

Figure 1.4. Dual fluorescence of **DMABN** and the corresponding ethyl ester in solvents of different polarity at room temperature: *n*-hexane (Hex), di-*n*-butyl ether (BOB), tetrahydrofuran (THF), butyronitrile (BCN), and acetonitrile (ACN).

decade ago, after the first general overview in the field,[3] the rapidly growing literature was extensively reviewed, focusing on the diversity of the TICT-related compounds[2] and on the physical phenomena involved.[19] Some further reviews were published,[11,20,21] covering various aspects of the phenomena, such as kinetic and thermodynamic implications. In the last few years, new experimental evidence and several mutually controversial mechanistic hypotheses[26–28] were published, which invoke other coordinates than twisting of the dimethylamino group. Nevertheless, the TICT hypothesis is still the most useful (especially for predictions) and the most widespread model. For a clear-cut distinction concerning the alternative models, it is therefore important to summarize the experimental evidence for excited-state twisting, to assess possible alternative explanations and to compare the predictions resulting from the alternative models.

In the following, the most important mechanisms explaining the "anomalous" A fluorescence band of **DMABN** and some of their implications and shortcomings are briefly described:

1. Twisted Intramolecular Charge Transfer (*TICT*) mechanism.[3] This model necessitates twisting of the dimethylamino group. It can be tested by investigating bridged and pretwisted model compounds: The A band should be absent for rigidly bridged planar compounds. No counterexample has been found yet.
2. Solute–solvent *exciplex* mechanism[26] Dual fluorescence should correlate with electron-donor–acceptor properties of the solvent and should be completely absent in alkane solvents. (Several counterexamples are given in References 2, 29, and 30.)
3. State switching model by *Lippert*.[31] This model presupposes an S_2 state more polar than S_1 and their reversal induced by solvent polarity. The model cannot explain cases of dual fluorescence, where S_1 is more polar than S_2.
4. State interaction or pseudo-Jahn–Teller (*psJT*) model.[27, 32] The importance of the A band, which, as a rule, increases with solvent polarity in all known cases, is inversely related to the S_1–S_2 energy gap between the benzenic ^1L_b- and ^1L_a-type states. This model requires an S_2 state (^1L_a) more polar than $S_1(^1L_b)$, such that the energy gap narrows with increasing solvent polarity. The mechanism cannot explain the forbidden nature of the A fluorescence and cases of dual fluorescence (ester-, aldehyde-, and keto-derivatives of **DMABN**, see Figure 1.4) where S_1 is more polar than S_2, such that the energy gap widens with increasing solvent polarity.
5. Rehybridization mechanism of the acceptor substituent (*RICT*).[28] The ^{1}CT state in **DMABN** is reached by an in-plane bending of the cyano group. This mechanism cannot explain the observed bridging and pretwisting effects at the amino substituent and cannot explain cases without separate acceptor substituent, such as the pyridine and pyrimidine derivatives.
6. Amino *pyramidalization* mechanism.[30] The ^{1}CT state in **DMABN** is believed to correspond to a highly pyramidalized, but untwisted, geometry. This mechanism has been tested by several theoretical approaches,[18,33] and the resulting ^{1}CT state has been found to be energetically very high lying.

3.1 Pretwisted and Planarized Model Compounds

The model compound approach has been especially fruitful for the class of **DMABN** derivatives, due to the comparatively good synthetic variability of the amino group, the benzene nucleus, and the acceptor substituent. Strictly speaking, every model compound should be treated as a separate case, but if the electronic changes are weak (which can be verified by inspecting absorption spectra, transition moments, ionization energies, etc.), then the model compound can be directly used in a comparative study referring to the original compound. Ideally, only one property should be changed in a model compound, and then the comparison will give the results of this specific change. In general, however, several factors are changed at the same time, and the answer is never completely clear-cut, but can be regarded as a good qualitative approach if the electronic changes are small enough. The strong potential of the model compound approach lies in the possibility of testing various mechanistic hypotheses linked with excited-state conformational changes. In the case of **DMABN**, the product of the conformational change is itself emissive, and a precursor–successor kinetic relationship between two species is observed, where, for historical reasons,[19] the precursor species is called B^* and the product A^*.

The twisting hypothesis contained within the TICT model can be tested by model compounds in two ways: either the precursor species is energetically less available, for example, by a steric potential mainly affecting the planar conformation (see Figure 1.2b), which leads to a "pretwisted" ground and Franck–Condon excited-state precursor geometry, as in **TMABN** (Scheme **1**); or the final product is unavailable by reducing the molecular flexibility away from the planar conformation (planarized model compounds like **MIN**). In the former case, the forward rate constant will increase and the backward one decrease, with the overall effect of shifting the excited-state equilibrium towards the product species A^*. In the latter case, the emission from A^* should be absent if this type of molecular twisting is really part of the reaction coordinate. The validity of this approach has been demonstrated for many pretwisted model compounds,[2,3] some of which are contained in Scheme **1**. The predictive strength of the TICT model goes so far, that for any given dual-fluorescent TICT system, a model compound can easily be devised where the A emission should be reduced or enhanced, and this can be tested by experiment. For the fully twisted model compound **CBQ**, the precursor species is completely absent, but the spectrum closely resembles that of the A^* species of **DMABN**, demonstrating the close relationship of the rigid perpendicular geometry of **CBQ** and of the geometrically relaxed species A^* in **DMABN**. At this stage it can be asked what the properties of an alternative mechanism would be that, without molecular twisting, could equally well predict the enhanced A band for **TMABN** and **CBQ** and its absence in **MIN**. Even for stronger changes of the electronic properties, the predictive power of the TICT model still remains applicable, such as for the ester and keto derivatives of **DMABN, DMABEE**, and **DMABK**. In both cases, bridging the amino group suppresses the A fluorescence band, but a bridging of the keto group in **DMABK–BA** has no effect. This latter model compound shows that the active relaxation coordinate is not situated at the acceptor substituent, as is assumed in the RICT model, but rather is located at the dimethylamino donor substituent.

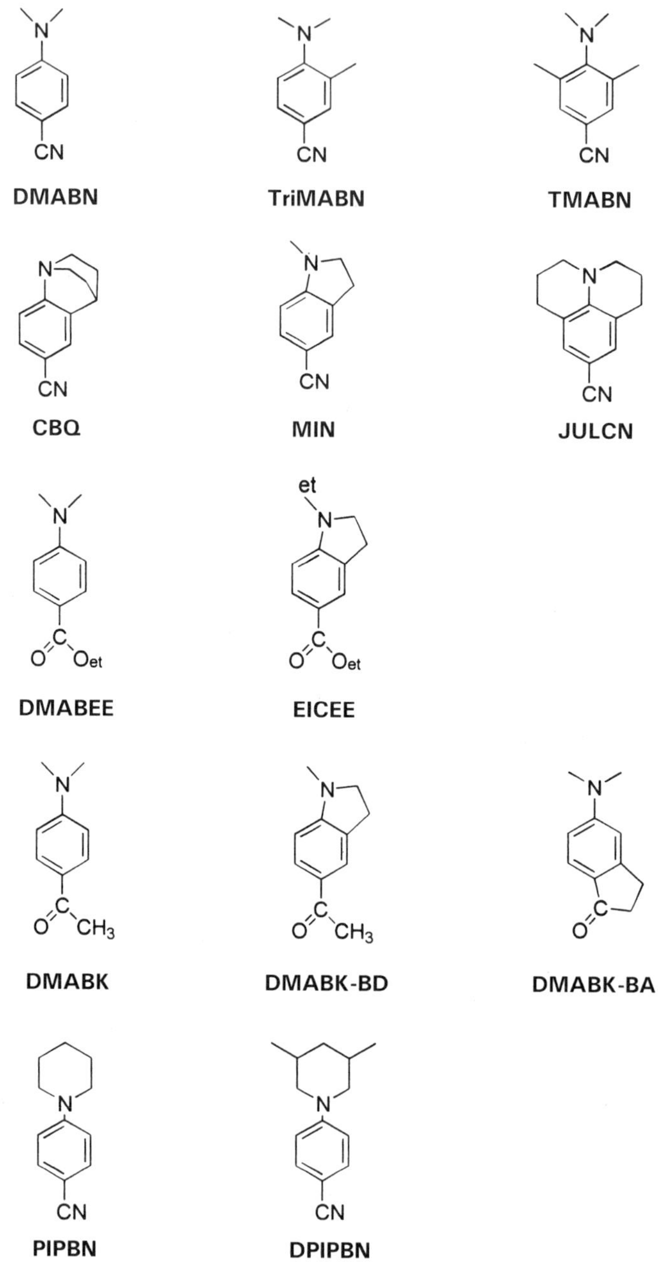

Scheme 1 Model compounds investigated to test the TICT hypothesis in **DMABN**.

We can also inspect the various predictions that would be made for these model compounds by the other mechanisms.

We already mentioned the RICT model, which would predict a stronger effect for the bridged-acceptor compound **DMABK–BA** than for the bridged-donor one, **DMABK–BD**. Experimentally, only the bridged-donor compound **DMABK–BD** shows a net effect, that is, the complete absence of A fluorescence and of the corresponding transient absorption band (Fig. 1.5).

Also, the ps-JT model, in which the importance of the A^* state is linked to the energy gap (ΔE) of the benzenic ^1L_a- and ^1L_b-type states, would not expect any direct correlation with molecular pretwisting or molecular bridging, except through ΔE. It can be shown that for **DMABN** and for some of the pretwisted compounds, ΔE in fact diminishes with increasing solvent polarity and can explain the enhanced A bands.[27,32] For the planarized compounds with small ΔE, the A band should,

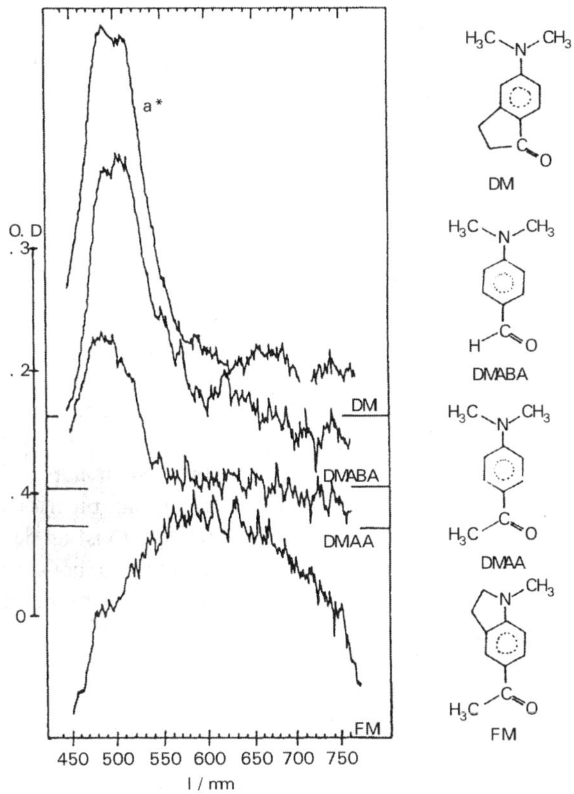

Figure 1.5. Transient absorption spectra (in acetonitrile at 290 K) of keto derivatives of **DMABN** with selectively bridged bonds. All compounds with a flexible dimethylamino group show the absorption band at around 500 nm, which is characteristic for the absorption of the benzaldehyde and acetophenone radical anion. But this band is absent in the compound with a bridged dimethylamino group. (Reproduced with permission from Ref. 34. Copyright Elsevier 1987.)

however, be equally prominent, and all the available experimental evidence speaks against this prediction. Not only does **MIN** possess two close-lying ^1L_a- and ^1L_b-type states, but so does **EICEE**, and photophysical studies, including fluorescence polarization, could even establish that ΔE formally goes through zero as the solvent polarity increases.[35] Some interesting effects are indeed observable and can be explained by the interaction of the two states,[35,36] but the long-wavelength A fluorescence is clearly absent, pointing to its different origin (e.g., in **DMABEE** and **DMABN**).

Other alternative models, like the Lippert or the pyramidalization model, are likewise unconnected to molecular pretwisting, and the same arguments hold. Within the exciplex model, some differences can be expected between the complexation of planar and of pretwisted molecules, but the model is unable to explain why the importance of the A^* band correlates with solvent polarity and not with the availability of lone-pair electrons of the solvent needed for the formation of the exciplex.[2]

In summary, the model compound studies represent a valuable approach to solving mechanistic questions, but they are not always sufficient, and for a clear-cut answer they should always be combined within the context of as broad an experimental basis as possible.

3.2 The Dynamic Approach

In the preceding section, static model compounds were considered wherein the equilibrium structure was modulated by alkane substituents or bridges and led to the exclusion or enhancement of certain reaction possibilities. A dynamic approach is also possible in which equilibrium structures and relative energies remain unaffected and only rate constants are influenced. This effect is obtained by introducing alkyl substituents that do not affect equilibrium conformations, but act as molecular "paddles," which slow down relaxational motions. Such a pair is **PIPBN** and **DPIPBN** (Scheme **1**). The methyl substituents on the piperidino ring are far removed from the amino nitrogen such that the average twist angle is not affected. Therefore, the thermodynamic properties are expected to be similar in the two compounds. The experimental dual-fluorescence spectra at room temperature (i.e., under equilibrated conditions with thermodynamic control) are indeed identical.[37] Only at low temperatures, where the back reaction is too slow to have a sizeable effect, does the slowed-down forward reaction for **DPIPBN** show up as a reduction in the relative importance of the A band (Fig. 1.6). In this case, the methyl substitution pattern induces a reduction of the reaction rate constant by a factor of around 1.8.[37] The result is consistent with the expectations from the TICT model, in which two molecular moieties twist against each other, and the rate constant will be related to the rotational volume for this movement. If the reaction coordinate were unconnected with a large-amplitude movement involving also the methyl substituents, much smaller differences would be expected. On this basis, the RICT model can be refuted, because it involves the relaxation of a small cyano group with respect to a large molecular assembly more or less unaffected by the methyl substitution.

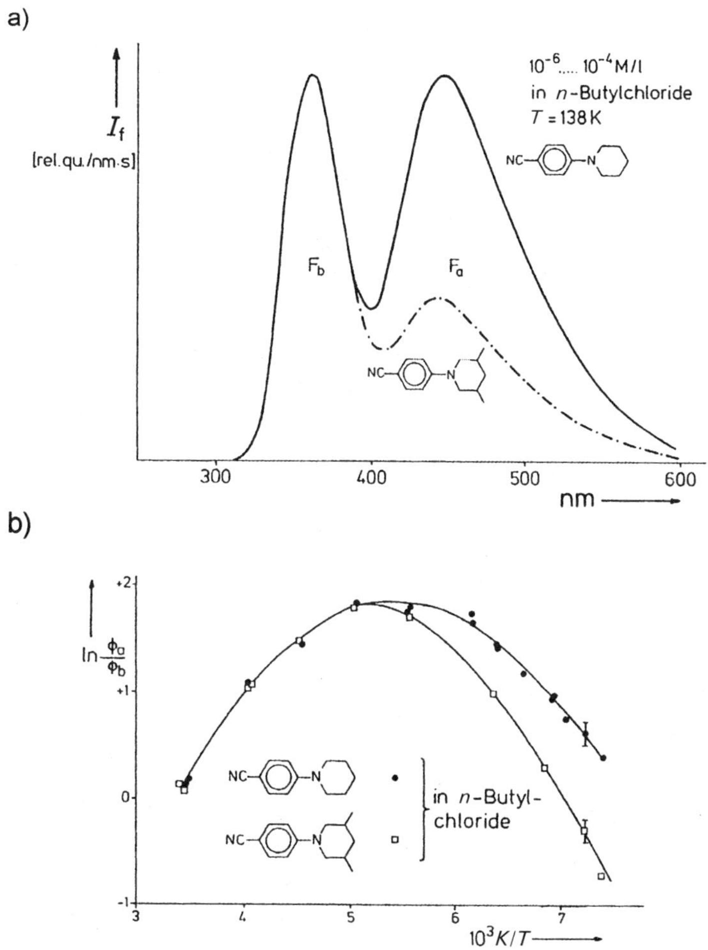

Figure 1.6. (a) Normalized fluorescence spectra of a pair of dynamically hindered **DMABN** derivatives in *n*-butyl chloride at 138 K. The CT band is considerably reduced for the compound with the larger rotational volume. (*b*) Ratio of CT to LE fluorescence quantum yield as a function of temperature. Only at low temperatures, in the kinetic regime, the dual fluorescence spectra differ. They are identical in the high-temperature thermodynamic range.

The same holds for mechanistic models that postulate only bond-length changes as the most important part of the reaction coordinate.

3.3 The Emission Transition Moment

As will be shown later, in the section on donor–acceptor biaryls, the comparison of absorption and emission transition moments can serve as a valuable tool for assessing molecular geometry changes after photoexcitation. For a quantitative treatment, this approach needs reference compounds with a known geometry. But even if these are not available, the size of the transition moment itself is a valuable indicator of the reaction coordinate and can often discriminate between alternative reaction mechanisms.

Emission processes with a very small transition moment are called "forbidden transitions." This term relates to the radiative rate constant k_f (proportional to the square of the transition moment), which is small compared to the "normal" value for allowed π–π^* transitions of around $10^9\ s^{-1}$. The inverse of k_f is called the radiative lifetime τ_{rf} and is equal to the fluorescence lifetime, which would be measured if all nonradiative processes were absent. An allowed transition thus corresponds to $\tau_{rf} = 1$ ns, and various degrees of "forbiddenness" are possible. For **DMABN** and derivatives, the radiative lifetime for emission from the ^1L_a-type state is approximately 2–5 ns, and that from the ^1L_b-type state, 15–20 ns.[22,35] Thus, the ^1L_b emission is considerably more forbidden, commonly interpreted by a near cancellation of larger transition moments of opposite direction.[38] If ^1L_a- and ^1L_b-type states come very close together, energetically, under the influence of the polar solvent or when induced by substituents, the states can mix, and the radiative lifetime will show intermediate values between those of the pure states.[35] The same applies to the fluorescence polarization, which also becomes mixed. This will be true as well for the qualitative ps-JT model.[27,32] The quantitative treatment goes back to Herzberg and Teller.[39]

The radiative lifetime observed for the A fluorescence of **DMABN** (160 ns)[40] is much longer than that of the less allowed of the two states, which are assumed to mix in the ps-JT model. This is a further indication that a two-states mixing model alone (Lippert and ps-JT models) is not sufficient to characterize the A^* state, which is of a fundamentally different nature from the ^1L_a- and ^1L_b-π–π^* states.

In the TICT model, the small transition moment (long radiative lifetime) was taken as an indication of a near perpendicular arrangement of donor and acceptor in the relaxed ^{1}CT state A^{*3}, because this leads to a localization of orbitals and hence, at the same time, to a maximization of charge separation and a minimization of the transition moment of the charge-transfer state. In principle, other arrangements are possible that lead to the same effect:[40] spatial separation of donor and acceptor, or an orthogonal arrangement, for example, in T-shape, or even the planar conformation if the orbitals involved are orthogonal for symmetry reasons.[41] (see Section 4.3.4.2, on phenyl pyrroles).

In view of the geometrical relaxation possibilities open for **DMABN**, the twisting relaxation seems the obvious choice for explaining the long radiative lifetime. This is, moreover, quantitatively supported by the twisted model compounds like **CBQ**, for which an even longer value of the radiative lifetime is measured than for **DMABN**.[42] The modeling of the reaction coordinate of **DMABN** by a twisted arrangement therefore leads to a quantitative agreement, not only of spectra (A^* band), but also of radiative lifetimes in a very unusual forbidden region.

3.4 Open Questions

Many arguments have been given in the preceding sections that support the twisting motion as being the important relaxation coordinate leading to the long-wavelength-emitting A^* state. However, none of these arguments can be counted as direct structural evidence, because they rely largely on the comparison with model compounds. From this point of view, a method that is able to identify directly the reactive coordinate in a given molecule, such as **DMABN**, without necessitating the access to model compounds, would be highly welcome. It may, for example, turn out that the twisting motion is coupled to a shortening of certain bonds and to the planarization of the pyramidal geometry of the amino nitrogen. This is certainly a field that deserves a concerted effort from the scientific community. The quantitative analysis of femtosecond time-resolved or ultrahigh-resolution energy-resolved spectra may lead in that direction.

Other open questions are less directly connected with conformational changes:

- Is the reaction leading from *B** to *A** of a barrierless nature[19,43] and under which conditions?
- What is the correct explanation for the strong retardation factor that slows down the reaction rate constant in the case of **DMABN** compared with compounds with a stronger acceptor (e.g, **DMABEE**) by a factor that can reach two orders of magnitude?[44] This could be the effect of hypersurface topology, with a conical intersection between the states A* and B* in the case of **DMABN**.[2,44] The conical intersection requires a further reaction coordinate other than the twisting; hence, it needs a description that is at least two dimensional. The involvement of a further reaction coordinate can slow down the reaction by entropic effects.
- Further open questions concern the way the alternative models "cooperate" with the twisting model. Earlier, we saw that only the twisting model is consistent with the full width of experimental evidence concerning the A band in **DMABN** and its model compounds. But the fact that the alternative mechanisms cannot explain some features concerning the A band does not preclude their importance for understanding the problem in its entirety. For example, the exciplex model, although not explaining the presence of the A band and its strong red shift, can account for additional, much smaller solvent-specific shifts of the A band, and supersonic jet spectroscopy with mass selection seems to be an interesting approach in this direction.[45] Furthermore, as the state interaction has been shown to be important for the understanding of the properties of the "normal" *B* band in compounds not exhibiting the A band,[35] the same will be true for compounds like **DMABN** with dual fluorescence, where a correct description of the *B* band will need similar concepts to those included in the ps-JT model, in addition to the TICT model (with a possible exciplex contribution) as necessary to understand the A fluorescence.

4 Aryl–Aryl Systems

The molecular conformation and, more specifically in this chapter, the interannular twist angle φ of biaryl compounds, can be well studied by conventional spectroscopic techniques, such as NMR, X-ray, electron or neutron diffraction, IR, Raman, etc., in the electronic ground state (S_0). In addition, quantum chemical calculations on an ab initio, semiempirical, or even force-field level are capable of modeling the twist potential curves for the gas-phase situation in S_0 to a sufficient precision. However, the conformational characterization in the excited state (S_1) is considerably more difficult, because of the following two features:

1. A more complicated electronic structure, consisting of different excited electronic configurations, that usually necessitates configuration interaction (CI) for a proper description.
2. A short transient lifetime of usually 10^{-12} to 10^{-7} s.

Furthermore, the analysis becomes even more complex if a photoinduced electron transfer is involved, because then the S_1 electronic structure strongly varies with the substituents and solvent, and, on the other hand, the conformational relaxations are superimposed by solvation processes. To address these problems, Section 4.1 outlines how electron transfer interactions can influence the excited-state geometry. Section 4.2 introduces possible optical methods for interrogating the S_1 conformation, and some examples are presented in Section 4.3.

4.1 From Biphenyl to Donor–Acceptor Biaryls

The potential energy of a molecular system is the sum of the electronic energy (E_{el}), the core repulsion energy (E_c), the steric interaction energy (E_{ster}), and the intermolecular stabilization energy by the solvent (E_{solv}):

$$E_{pot} = E_{el} + E_c + E_{ster} + E_{solv}. \tag{3}$$

Obviously, the twist potential function in the gas phase is determined by the counteracting interplay of E_{el} (energy minimum at $\varphi = 0°$ or $180°$) and E_{ster} (energy minimum at $\varphi = 90°$), leading to a single rotameric form, with φ between $0°$ and $90°$ for highly symmetric biaryls belonging to the C_{2v} or D_{2h} molecular-point group at $\varphi = 0°$ and $\varphi = 90°$. Two different energy minima result with φ between $0°$ and $90°$ and φ between $90°$ and $180°$ for less symmetric biaryls. Accordingly, the S_0 twist angle of biphenyl (**BP**) and 9, 9′-bianthryl (**BA**) in the gaseous state is about 45°[46] and 90°,[47] respectively, while for the less symmetric biaryl 1,1′-binaphthyl (**BN**), two stable rotamers with equilibrium angles of about 69°[48] for the *cis* conformation and 103°[49] for the *trans* conformation are reported for the crystalline forms.

The excited-state twist angle in the gas phase has been probed with the use of multiphoton ionization (MPI) or fluorescence excitation in a supersonic jet technique. The values of $\varphi(S_1)$ thus determined were found to be 0° for **BP**[49,50] and 67–78°[47,51,52] for **BA**. Such experiments prove that, in the absence of solvent and intramolecular electron transfer interactions, biaryls tend to relax towards a more planar structure in S_1. It is mainly the electronic energy term E_{el}, as determined by the electronic wave function in S_1 that is responsible for the change in the potential-energy curve. Due to the pairwise interaction of the orbitals on the submoieties, the π-atomic orbital coefficients at the linking bond of a symmetric biaryl system are usually of the same sign in the LUMO (bonding interaction), whereas in the HOMO they are of opposite sign (antibonding interaction). As a consequence, the resonance interaction between the aromatic units induced by the promotion of electrons from

HOMO to LUMO (antibonding to bonding) orbitals shifts the potential-energy minimum to more planar twist angles φ (as well as to shorter interannular bond lengths).

The electronic wave functions Ψ for the ground and excited states of a biaryl system can be described by a linear combination of electron configurations (i.e., the zero-order wave functions Φ for the ground state (S_0), the lowest-lying locally excited (^1L_b, ^1L_a in Platt's[53] perimeter notation), as well as by the electron transfer (^{1}ET) and higher excited configurations) according to the following equations:

$$^1\Psi(S_0) = g_0\,{}^1\Phi(S_0) + g_1\,{}^1\Phi(\mathrm{ET}) + g_2\,{}^1\Phi(L_a) + g_3\,{}^1\Phi(L_b) + \sum_i g_i\Phi_i \tag{4a}$$

$$^1\Psi(S_1) = e_0\,{}^1\Phi(S_0) + e_1\,{}^1\Phi(\mathrm{ET}) + e_2\,{}^1\Phi(L_a) + e_3\,{}^1\Phi(L_b) + \sum_i e_i\Phi_i \tag{4b}$$

In the absence of mutual electronic interactions, the electron configurations of S_0, ^1L_b, ^1L_a, and ^{1}ET can be termed as pure, or zero-order states.

Upon coupling the two aromatic moieties by a single bond in a symmetric biaromatic system, each of the locally excited states splits into two excitonic states, and the ^{1}ET transfer states can involve a transfer of one or more electrons from one to the other unit in both directions. For a given molecule, the CI coefficients depend on the nuclear coordinates, such as the interannular twist angle φ, but also on the solvent polarization z. The excited-state wave function of **BP** and **BN** (Scheme **2**) is dominated by the ^1L_b configurations ($e_3 \approx 1$) polarized perpendicularly to the linking bond, and that of **BA** and **BPery** is dominated by the ^1L_a configurations ($e_2 \approx 1$) polarized along the linking bond. Electron transfer interactions cannot be observed in the bare symmetric biaryls. However, if the biaryls are surrounded by solvent molecules, either in a liquid solvent or under jet conditions with solvent clusters, the symmetry of the molecule can be broken and ET becomes feasible, introducing some ^{1}ET character ($e_1 > 0$) to $\Psi(S_1)$. The increase in the ^{1}ET weight e_1 into $\Psi(S_1)$ can similarly be achieved by substituents like the cyano group in one of the moieties. In case of a dominant ^{1}ET character ($e_1 > e_2, e_3, e_0, e_i$), the S_1 state is generally called a charge transfer (^{1}CT) state, which means that a partial charge separation between the aryl entities occurs. Such a ^{1}CT state can be revealed by a red-shifted and broadened emission band compared with the more excitonic state with a dominant ^{1}LE character. No ^{1}CT characteristics were observed either under jet conditions or in the polar solution for **BP**[50,54,55] and **BN**[56,57] (low-lying ^1L_b with $e_2 \ll e_3$), while a ^{1}CT behavior is well documented for [58–61] **BPery**[62] (low-lying ^1L_a

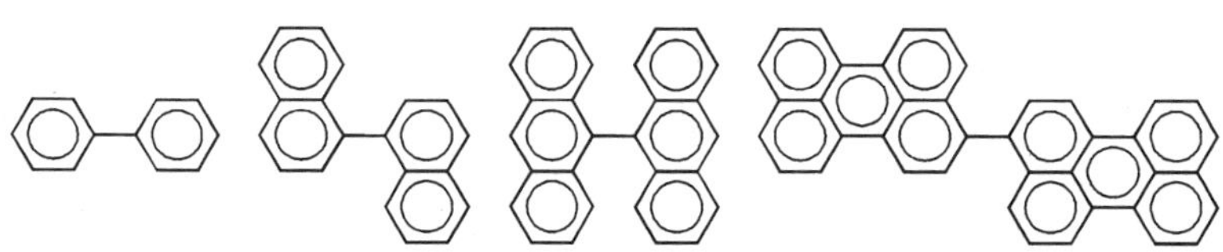

Scheme 2 Molecular structures of the biaryls biphenyl (**BP**), 1,1′-binaphthyl (**BN**), 9,9′-bianthryl (**BA**), and 3,3′-biperylenyl (**BPery**).

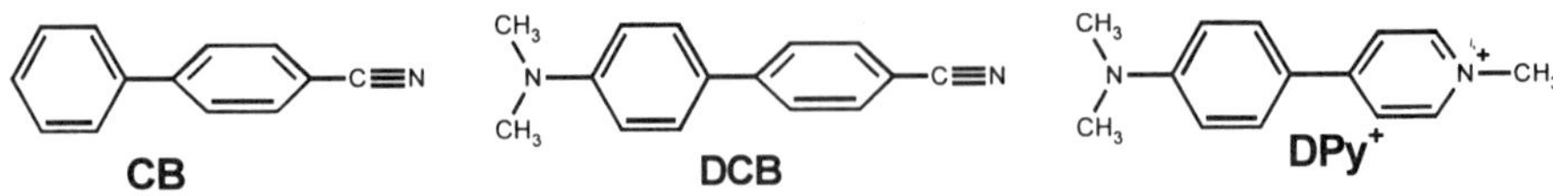

Scheme 3 Molecular structures of the biphenyls 4-cyano-biphenyl (**CB**), 4-dimethylamino-4′-cyanobiphenyl (**DCB**), 4-dimethylanilino-1-methyl-pyridinium ($\mathbf{DPy^+}$).

with $e_2 \gg e_3$) in polar solvents and even for supersonic jet conditions in the case of **BA**–acetone or **BA**–water clusters.[63, 64] A changeover occurs for 4-cyano-biphenyl (**CB**, Scheme **3**). In the gas phase, the electronic spectrum and the vibrational progression of the torsional mode are similar to that of **BP**, pointing to the ^1L_b character of S_1.[50] In solution phase, however, shorter fluorescence lifetimes and a solvent polarity-induced red shift and broadening of the fluorescence spectrum indicate the change of S_1 from ^1L_b in **BP** to the ^1L_a-type ^{1}CT state ($e_2, e_1 > e_3$) in **CB**.[65]

The electronic energy of the lowest-lying pure and **decoupled** ^{1}ET state ($c_1 = 1$, achievable through zero molecular orbital overlap between the linked aryl units acting as electron donor D and acceptor A) can be calculated by

$$E_{el}(\mathrm{ET}) = IP(\mathrm{D}) - EA(\mathrm{A}) + C + \Delta E_{solv}, \tag{5}$$

where IP(D) is the ionization potential of the electron donor aryl unit (D), EA(A) is the electron affinity of the acceptor aryl unit (A), C is the (negative) Coulomb interaction energy of the separated charges in the D–A biaryl, and ΔE_{solv} is the stabilization energy gained through solvation of the D^+–A^- dipole. From this equation, it can simply be anticipated that with an increase in the donor strength (D aryl unit with lower IP) or the acceptor strength (A aryl unit with larger EA), the state $E_{el}(^1\mathrm{ET})$ is lowered, leading to an enhanced ^{1}ET weight e_1 in the wave function $\Phi(S_1)$ of a coupled system. Such a case is given by 4-dimethylamino-4′-cyano-biphenyl (**DCB**, Scheme **3**),[66–68] where the smaller IP(D) of the dimethylaniline unit compared with the benzene donor unit in **CB** leads to a stronger ^{1}ET character of S_1, as is attested to by smaller fluorescence energies and lifetimes, but by a dipole moment twice as large in S_1 than in **CB**.[65] The ^{1}ET weight e_1 of planar **DCB** in the gas phase has been calculated by the CNDOS/CI method to be about 40%.[67,69] A further increase in the ^{1}ET weight e_1 is obtained by replacing the benzonitrile unit by the more electron-attracting pyridinium group-[lower EA(A)], as in 4-dimethylani-lino-1-methyl-pyridinium ($\mathbf{DPy^+}$). The zero–zero S_0–S_1 transition energy in nonpolar solvents is at 7000 cm^{-1} lower energy for $\mathbf{DPy^+}$,[70–72] compared with that of **DCB**, indicating the increased stabilization of S_1 by the enhanced ^{1}ET interactions in $\mathbf{DPy^+}$. It has to be noted that due to the charged nature of $\mathbf{DPy^+}$, an electron transfer transition leads to a charge shift and prevents the formation of a D^+–A^- dipole, resulting in only a relatively small solvent polarity-induced fluorescence red shift (a smaller ΔE_{solv} effect in Eq. 5). The observed (weakly radiative) S_1 state should therefore better be termed a *charge shift* (CSh) state, to discriminate the shift of the positive charge from the pyridinium unit towards the dimethylaniline unit from the *charge separation* taking place in the ^{1}CT state of **DCB**.

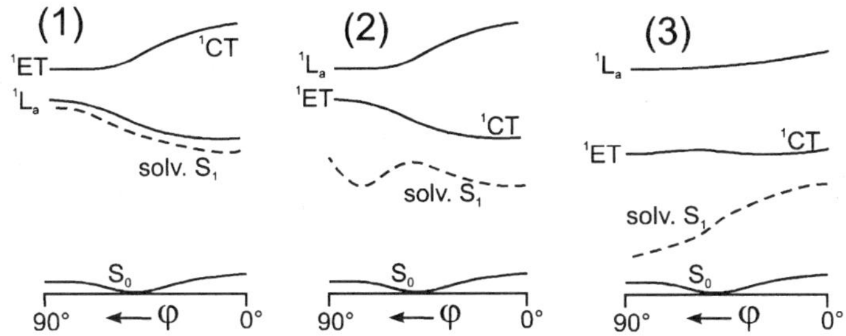

Figure 1.7. Schematic twist potential-energy curves of S_0, S_1, and solvated S_1 for three borderline cases. (See text.) The energy minima for S_2 at 90° are related to the S_1/S_2 interaction.

Three borderline cases for the shape of the twist potential-energy curves (Fig. 1.7) can be expected regarding the conformational structure, which depends on the strength of the ^{1}ET interactions in S_1:

1. *Case of negligible ^{1}ET interactions in S_1 (e.g.,* **BP***)*
   The lowest-lying excited singlet state S_1 is a locally excited or excitonic ^1L_b or ^1L_a state with a single minimum at a conformation more planar than in S_0. Solvation with dipolar solvents has minor effects on the twist potential $E_{pot}(\varphi)$ in S_0 and S_1.
2. *Case of strong ^{1}LE–^{1}ET interactions in S_1 (e.g.,* ***DCB****)*
   Due to the energetic repulsion between the close-lying and similarly polarized (along the linking axis) ^1L_a and ^{1}ET states, S_1 becomes the ^1L_a-type ^{1}CT state with a potential that has a deeper minimum towards the planar geometry than in Case 1. If the fully charge-separated ^{1}ET state is energetically below the ^1L_a state (cf. Eq. 5), S_0–^{1}ET repulsive interactions become more important, such that a second energy minimum can emerge for twist angles φ close to perpendicularity, where the ^{1}CT state develops into a state with almost a pure ^{1}ET character connected with a larger dipole moment.[2,11] Such a situation with an almost pure ^{1}ET being the lowest excited state can also be induced by a higher solvent polarity through the increase of the solvation energy term in Eq. 5. Hence, the counteracting resonance interaction energy (E_{el} in Eq. 3) and the solvation energy (ΔE_{solv} in Eq. 3) can lead either to a double minimum potential connected with an intrinsic energy barrier or to a single minimum in S_1 at higher twist angles than in S_0. Both situations can be viewed as being associated with an adiabatic[73] conformational photoreaction (conformer relaxation or equilibration in the excited state), which is viscosity dependent because of the large-amplitude twisting motion. The S_1 potential-energy surface may even become more complex if a third energy minimum appears for a ^1L_b-type ^{1}CT state, due to additional electronic interaction of ^{1}ET with ^1L_b induced by a small ^{1}ET–^1L_b energy gap or reduced symmetry.

3. *Case of leading ^{1}ET character in S_1 (e.g.,* ***DPy$^+$****)*
   A further lowering of the ^{1}ET energy results in a deeper energy minimum for the decoupled ^{1}ET state at twisted geometries connected with a possible vanishing of the energy barrier from the initially populated Franck–Condon state (close to planarity in **DPy$^+$**) to the relaxed state. Since the relaxed S_1 state contains a strong ^{1}ET character, radiative rate constants are expected to be very small, and this state might be observed as a weakly emissive one with a strong non-radiative channel (a "dark state").

Before we present some experimental details in Section 4.3, the basic principles for a conformational treatment of excited states are given in the next section.

4.2 Tools for a Conformational Analysis Using Optical Spectroscopy

It needs to be stressed that optical (absorption, fluorescence) spectroscopic methods can less directly extract information about molecular structures compared with methods such as IR, Raman, NMR, electron or neutron scattering, X-ray spectroscopy, and the like. Nevertheless, it will be shown that the combination of different electronic properties can indeed yield information about the molecular structure in the relaxed S_1, which is mostly not accessible with the aforementioned methods. This chapter is restricted to UV/vis absorption and emission spectroscopy. It is, however, necessary to mention that in the past decade time-resolved IR, Raman, and coherent anti-stokes Raman spectroscopy have been developed on timescales fast enough to yield information on short-lived transient species such as excited states.

4.2.1 Model Compound Study

For the elucidation of the excited-state molecular structure of a certain molecule, so-called "model compounds" can be investigated by using the same type of experiment. The basic idea is that a model compound is more or less restricted to a subspace of the full conformational freedom of the unhindered excited molecule under study. The results for the unhindered molecule are compared with those of the model compounds, and the model compound whose excited-state properties are in best agreement with the unhindered molecule is supposed to represent the most probable relaxed excited-state conformation. Some possible fluorescence properties that can be employed for such a comparative study are the fluorescence energy at the band maximum ν_f, static dipole moments μ_f, transition dipole moments M_f, lifetimes τ_f and quantum yields Φ_f. In addition, the determination of these quantities as a function of temperature T, pressure p, viscosity η, or the medium dielectric constant ε_r gives information about the factors that influence the excited-state conformational relaxation and can therefore lead to a consistent idea of the excited-state molecular structure. The comparison of model compounds is a powerful tool even though perfect model compounds cannot be designed. Slight electronic differences and the conformational flexibility of the model for the molecule under study always have to be taken into account. These drawbacks stress the necessity of a multidimensional analysis of fluorescence parameters, which takes into account the functional dependence of fluorescence properties on several independent relaxation coordinates.

4.2.2 Ratio of Absorption Versus Fluorescence Transition Moment

Generally, a difference between the transition moment experimentally derived from the lowest energy absorption band (M_a) and that derived from fluorescence (M_f) indicates a photoinduced change of the S_1 electronic structure as a result of a photoinduced rearrangement of nuclear coordinates, either those of the interacting solvents (reaction coordinate z) or those of intramolecular motions such as twisting (reaction coordinate q_i or, in particular, φ).

The determination of the absorption dipole moment M_a of the $S_0 \rightarrow S_1$ electronic transition requires a decomposition of the lowest-lying absorption band, the absorbance data of which can then be used to compute M_a from Eq 6:[74,75]

$$M_a^2 = f(n)\frac{3hc2303}{8\pi^3 Nn}\int\frac{\varepsilon(\nu_a)d\nu_a}{\nu_a} \tag{6}$$

where h is Planck constant, c is the velocity of light in vacuum, N is Avogadro's number, n is the refractive index of the medium, ν_a is the wave number of absorption, and $\varepsilon(\nu_a)$ is the decadic molar extinction coefficient. The function $f(n)$ is a correction term for refractive index effects, for which various expressions have been derived by different authors. While $f(n) = 1$ is given by Birks,[74] other forms, such as

$$f(n) = \frac{9n^2}{(n^2 - 2)^2}$$

and

$$f(n) = \frac{(2n^2 + 1)^2}{9n^2}, \tag{7}$$

are suggested by Chako and Bakshiev, respectively.[75] The fluorescence transition dipole moment M_f of the $S_1 \rightarrow S_0$ electronic transition can be calculated by[74]

$$M_f^2 = f(n)\frac{3h}{64\pi^4 n^3}\frac{\Phi_f}{\tau_f\langle\nu_f^3\rangle^3}, \tag{8}$$

where Φ_f and τ_f are the measured fluorescence quantum yield and lifetime, respectively. The mean cubic wave number $\langle\nu_f^3\rangle$ is obtained from the fluorescence spectrum $I_f(\nu_f)$ according to the equation[74]

$$\langle\nu_f^3\rangle = \frac{\int I_f(\nu_f)\nu_f^{-3}d\nu_f}{\int I_f(\nu_f)d\nu_f} \tag{9}$$

It can be seen that the ratio M_f^2/M_a^2 is independent of the choice of $f(n)$ and hence can be employed as an indicator of excited-state relaxations. If a photoinduced state reversal can be excluded, a value for M_f^2/M_a^2 larger than unity points to a better π-electron conjugation induced by a flattening of the molecular structure, whereas a value of M_f^2/M_a^2 smaller than unity indicates a reduced transition moment to the ground state, which can be the consequence of enhanced electron transfer interactions in S_1. The latter can be the result either of a rearranged and more strongly interacting solvent shell or of a D–A decoupling by a conformational change, such as the twisting of the interannular bond towards a more twisted geometry.

4.2.3 Calculation of Coupling Matrix Elements V_{S_0-ET} and V_{L_a-ET}

The calculation of the coupling matrix elements rests on the determination of the transition dipole moments and allows one to split the total transition moment M into contributions of S_0–ET and ^1L_a–^{1}ET electronic interactions according to the equation

$$M_{S_1} = g_0 e_1 M_{ET} + g_0 e_2 M_{L_a} + g_0 \sum_i e_i M_i, \tag{10}$$

where M_{S_1} is the transition moment of the spectroscopic S_0–S_1 absorption or emission transition and M_{ET}, M_{L_a}, and M_i denote the ground-state transition moments of the zero-order states ^{1}ET, ^1L_a, and the rest of configurations i.[16] Due to the large energy gap between the ground state and the excited configurations, the ground-state CI coefficient g_0 (see Eq. 4a) is usually very close to unity. A further simplification assumes either that the contributions e_i of configurations other than ^{1}ET and ^1L_a are small or that the transition moments M_i are negligible. Equation 10 then reduces to

$$M_{S_1} = e_1 M_{ET} + e_2 M_{L_a}, \tag{11}$$

The transition dipole moment of the zero-order ^{1}ET transition M_{ET} is connected to the change in the permanent dipole moments $\Delta\mu_{ET-S_0}$, the energy gap ΔE_{ET-S_0}, and the electronic matrix element V_{ET-S_0} between the pure ground and electron transfer state, respectively:[14,76]

$$M_{ET} = V_{ET-S_0} \frac{\Delta\mu_{S_0-ET}}{\Delta E_{S_0-ET}}. \tag{12}$$

M_{S_1} can therefore be expressed as

$$M_1 = e_1 V_{ET-S_0}(\mu_{ET} - \mu_{S_0})/\Delta E_{ET-S_0} + e_2 M_{L_a}. \tag{13}$$

The conformational information is hidden in the electron transfer coupling matrix element V_{ET-S_0} of Eq. 13. Dogonadze et al.[77] have proposed to approximate the electronic resonance integral V_{DA} between D(onor) and A(cceptor) complexes of transition metal ions contacted by the atoms d and a according to the equation

$$V_{DA} = S_{da}\beta_{da}c_d c_a, \tag{14}$$

where S_{da} is the overlap integral, β_{da} is the interatomic resonance integral, and c_d and c_a are the atomic orbital coefficients, each parameter corresponding to the connecting atoms d and a, respectively. In the case of D–A biaryls, the electron transfer occurs across the interannular bond linking the D and A aryl units by the atoms d and a. The interatomic normalized overlap integral is then

$$S_{da}(\varphi_{DA} r_{da}) = \cos(\varphi_{DA}) k(r_{da}), \tag{15}$$

with the dihedral angle φ_{DA} between the aryl moieties and with the bond-length-dependent resonance parameter $k(r_{da})$.[78] According to the formalism of Dogonadze,[77] the coupling element V_{ET-S_0} for the intramolecular electron transfer is then

$$V_{ET-S_0}(\varphi_{DA}, r_{da}) = \cos(\varphi_{DA}) k(r_{da}) \beta_0 c_{h(d)} c_{l(a)}, \tag{16}$$

where β_0 denotes the standard resonance integral between the neighboring d and a atoms (e.g., $\beta_{C=O} = 2.4\,\text{eV}$, $\beta_{C=N} = 2.0\,\text{eV}$) and $c_{h(d)}$ and $c_{l(a)}$ are the $p\pi_z$ atomic orbital coefficients of the atoms d and a connecting D and A in the HOMO(D) and LUMO(A) orbitals. If we assume that V_{ET-S_0} is determined only by the interactions of the $p\pi_z$ orbitals on atoms d and a (i.e., neglecting contributions from π orbitals between atoms different from d and a, as well as from σ orbitals), Eq. 16 relates V_{ET-S_0} to the desired structural parameters φ_{DA} and r_{da}.

In order to determine V_{ET-S_0} from Eq. 13, a procedure is needed to derive the wave-function coefficients e_1 and e_2. This can be achieved by using a reference model molecule in which the spectroscopic properties of the ^1L_a state are representative for the pure ^1L_a state of the D–A biaryl usually use the lowest energy transition. In this case, the transition moment of $S_0 \rightarrow {}^1L_a$ in the D–A biaryl can be expressed as

$$M_{L_a}(\text{D–A}) = c_1 M_{ET} + c_2 M_{ref}, \tag{17}$$

with M_{ref} being the transition moment of S_0–^1L_a in the reference system and with $c_1^2 + c_2^2 \approx e_1^2 + e_2^2 \approx 1$. Since the ^1L_a state of the reference system is distributed mainly between ^1L_a(D–A) and S′(D–A), $c_2^2 + e_2^2 \approx 1$ and by assuming that $\nu_a(\text{ref}) = \nu_a(\text{D–A})$ and $M_{ET} \approx 0$, the combination of Eqs. 6 and 17 yields

$$\frac{M_{L_a}(\text{D–A})^2}{M_{L_a}(\text{ref})^2} = \frac{\int \varepsilon_{L_a}(\text{D–A})}{\int \varepsilon_{L_a}(\text{ref})} d\nu_a \approx \frac{\varepsilon_{L_a}^{max}(\text{D–A})}{\varepsilon_{L_a}^{max}(\text{ref})} = 1 - e_1^2 = e_2^2, \tag{18}$$

where $\varepsilon_{L_a}^{max}$ represents the molar absorption coefficient at the absorption maximum of the decomposed $S_0 \rightarrow {}^1L_a$ absorption band. Furthermore, using the derived wave-function coefficients e_1 and e_2, we find that the secular equations of the ^{1}ET–^1L_a two-states interaction model (assuming negligible D and A molecular overlap, or $e_0 \ll e_1, e_2$) allow the estimation

$$V_{ET-L_a} \approx -\frac{e_2}{e_1}(E_{L_a} - E_{S_1}), \tag{19}$$

which, by substitution of $e_2{}^2 = 1 - e_1{}^2$, becomes

$$V^2{}_{ET-L_a} \approx \left(\frac{1}{e_1{}^2} - 1\right)(E_{L_a} - E_{S_1})^2. \tag{20}$$

Although the electronic parameters of the locally excited ^1L_a state (e.g., $M_{L_a} \approx M_{ref}$ in Eq. 13) are available from the spectroscopic measurements on the corresponding reference aryl system, a drawback of this approach is that the electronic properties of the pure (zero-order) ^{1}ET states (i.e., the dipole moment μ_{ET}, the energy E_{ET}, and the transition moment M_{ET} in Eq. 13) are experimentally not available from the D–A composite molecule, because of the mixed nature of S_1 and the higher excited states. In a first approximation, the spectroscopic properties of the observed strongly solvatochromic ^{1}CT state may be used. However, this simplification is applicable only to a limited number of D–A molecules wherein S_1(CT) is essentially of ^{1}ET character (back ^{1}ET and other ^{1}ET interactions are neglected) and the configuration interaction giving rise to the intensity-borrowing mechanism can mainly be attributed to some mixing with the nonpolar ^1L_a.

A good test whether the ^1L_a contribution e_2 (^{1}ET contribution e_1) is sufficiently small (large) is provided by applying Eq. 18, which yields the percentage of ^{1}ET contribution, e_1^2, for the equilibrium ground-state geometry. For the twisted acridine–anilines discussed in Section 4.3.1, this formalism could successfully be applied,[16,80] because the weight of ^1L_a in $S_1(^1\text{CT})$ is only 25–40%. However, the method fails for the D–A biphenyls, since the ^1L_a interaction with ^{1}ET is too strong.[67] The value $e_2{}^2$ should therefore be less than 40% and e_1^2 more than 60%, respectively, to justify the preceding approach.

If the contribution of a locally excited state, such as the ^1L_a state in biaryls, is even less than 10%, the coupling matrix elements for the three-state (S_0–^{1}ET–^1L_a) interaction case can be experimentally derived from plots of the fluorescence rate constants versus the average fluorescence energy. Such procedures have independently been developed by two groups[14,15] and employed for intermolecular exciplexes[14] and σ-bridged D–A systems.[15,81] However, the latter methods assume a solvent-polarity independent molecular structure, which prevents their application to flexible D–A biaryls.

4.2.4 Time-Resolved Transient Absorption and Emission Spectra

The methods previously mentioned are based on steady-state experiments (except the determination of the fluorescence lifetime τ_f). Consequently, they can only detect average S_1 parameters obtained by emission during the S_1 lifetime. Time-resolved techniques, however, allow one to follow the evolution towards the final electronic structure of the excited state or towards the equilibrium of a different molecular structure. In combination with model compound studies, this time-resolved approach is extremely powerful regarding the elucidation of conformational changes. Frequently, the transient absorption spectra (i.e., S_1 transmission spectra with contributions of $S_1 \rightarrow S_n$ absorption and $S_0 \leftarrow S_1$ stimulated emission) of donor–acceptor linked molecules are compared with the absorption spectra of the "model compounds" of the donor radical cation ($D^{+\cdot}$) and acceptor radical anion ($A^{-\cdot}$) to evaluate the extent of photoinduced intramolecular charge separation. If the transient spectrum of the D–A molecule is equal to a superposition of the D^+ and A^- electronic spectra, it can be concluded that a full electron transfer takes place in S_1 that requires electronically decoupled donor and acceptor moieties. This method is frequently applied to molecules in which D and A are separated by a long σ-bridge (e.g., for porphyrin–quinone D–A molecules).[82,83] The similarity of the transient absorption spectra of such D–A compounds to those of the cation radical of the porphyrin donor is then taken as evidence for the electron transfer from the porphyrin to the quinone moiety. In biaryls, the decoupling can be achieved by a perpendicular twist of the interannular bond, because the π-electron orbitals are then also spatially separated.

4.3 Analysis of Different Charge Transfer Biaryl Compounds

4.3.1 Acridine–Anilines

Acridine–anilines (Scheme **4**) belong to a family of D–A biaryl compounds that is well suited to an experimental analysis of the S_1 electronic structure and the associated coupling matrix elements within a three-state ($S_0/^1L_a/^1ET$) interaction model, outlined in Section 4.2.3. These matrix elements are experimentally accessible and allow one to draw conclusions on photoinduced conformational relaxations without the aid of quantum chemical calculations or extensive model compound studies.[16, 80] The advantageous electronic properties are due to the high symmetry of the chromophoric aryl backbone, which results in a parallel polarization of the $S_0 \rightarrow {}^1L_a$ and $S_0 \rightarrow {}^1CT$ transitions along the interannular bond, while the $S_0 \rightarrow {}^1L_b$ transitions are polarized perpendicularly to this axis, which can be viewed as the axis of approximate C_2 symmetry for all compounds and twist angles. The molecular systems considered are further simplified by the fact that the lowest lying LE electronic states are localized on the acridine electron acceptor unit. Moreover, the $S_0 \rightarrow {}^1L_b$ transition is energetically significantly above the $S_0 \rightarrow {}^1L_a$ transition ($1000–2000\,cm^{-1}$), and the latter is located at about $26,000\,cm^{-1}$ in the parent acridine compound. All these electronic properties permit one to neglect the influence of the ^1L_b state on the lowest lying singlet state S_1 and to describe the electronic structure of S_1 by an interaction of the acridine ^1L_a state and an aniline-to-

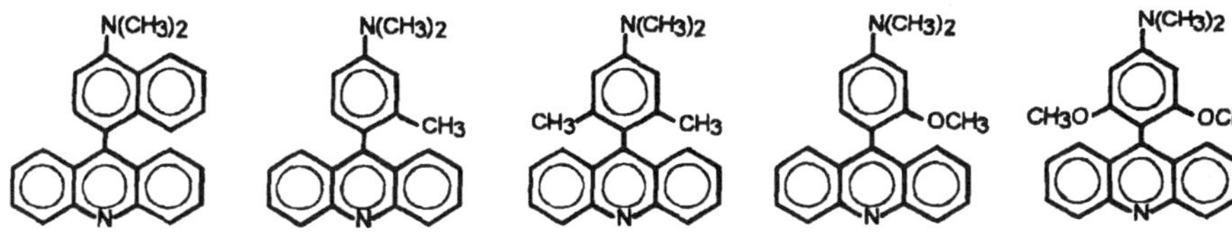

BDMN **BDMT** **BTMA** **BTMP** **BDMR**

Scheme 4 Molecular structures of acridine–anilines and an acridine–naphtylamine. The parent compound (not shown) is denoted **BDMA** in the tables (**BDMT** without the *ortho* methyl group), and the corresponding julolidino compound (**BDMA** with the methyl groups joined to the benzene ring) is denoted **BJ**.

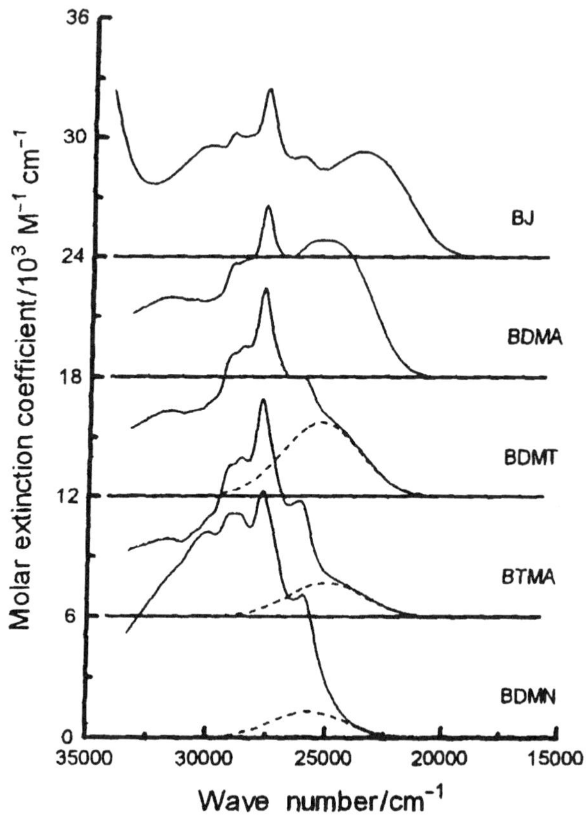

Figure 1.8. Absorption spectra of acridine–anilines and an acridine–naphthylamine. (Reproduced with permission from Ref. 16. Copyright American Chemical Society 1998.)

acridine electron transfer state ET according to the formalism in Section 4.2.3 and Eq. 11.

The decomposition of the absorption spectra of the acridine–anilines shown in Figure 1.8 reveals that the $S_0 \rightarrow S_1$ transition is predominantly of an ET nature and the $S_0 \rightarrow S_2$ transition of a ^1L_a nature.[16,80] The vibronic structure around 26,000 cm^{-1} corresponds exactly to the typical feature of the ^1L_a band in acridine, while the broad band on the low-energy side is characteristic of a CT band. For the D–A biphenyls described in Section 4.3.2, it will be shown that with increasing interannular twist, the electronic decoupling of the ^{1}ET and ^1L_a states leads to a stronger intensity of the ^1L_a band at the expense of the CT band.[67,69] Comparing the acridine series from **BDMA** to **BDMT** to **BTMA**, we observe the same behavior. In addition, the increase in the ^1L_a intensity is accompanied by a better resolved vibrational structure, because the contribution of the broad CT band decreases. These characteristic features undoubtedly show the electronic interaction between the two (zero-order) states ^1L_a and ^{1}ET. On the basis of Eq. 18, the ^{1}ET wave-

Table 1 Optical properties of acridine–anilines and an acridine naphtylamine[16]

	$\sqrt{\frac{\varepsilon_{LE}}{\varepsilon_{ref}}} \approx e_1$	ν_f in MCH[a] [cm^{-1}]	ν_f in BCN[b] [cm^{-1}]	M_f in MCH[a] [D]	M_f in BCN[b] [D]
BJ	0.82	21675	14200	3.5	1.9
BDMA	0.74	21900	15650	3.4	2.5
BDMT	0.86	22000	14850	3.0	1.7
BTMA	0.92	21775	14400	2.0	1.3
BDMN	0.95	21300	14150	2.9	1.5

[a] Methylcyclohexane; [b] *n*-butyronitrile.

function coefficient e_1 in the S_1 wave function for the ground-state equilibrium conformation is quantified in Table 1 to be in the range from 0.74 to 0.95.

Information on the relaxed geometry of the excited state can be gained from a quantification of the fluorescence properties. The solvatochromic red shift of the fluorescence spectra is similarly strong for all compounds, as indicated in Table 1 by the drop in the fluorescence energies ν_f from the nonpolar solvent methylcyclohexane (MCH) to the polar solvent butyronitrile (BCN). From an analysis of these solvatochromic data, the excited-state dipole moment is determined to be about 26–30 D, which corresponds to a charge separation of the core-to-core distance (ca. 4.5Å) between the aniline and acridine aryl units.[16] This large dipole moment points to an extremely large ^{1}ET character (e_1^2) of the conformationally relaxed emitting S_1 state. The ^{1}ET character of S_1, as estimated from absorption properties (Eq. 18) is indeed substantial (see Table 1), even though for the ground state Franck–Condon geometry, solvent stabilization of the ^{1}ET state is absent. If we compare the M_{S_1} transition moments obtained for a given molecule for different conditions (i) from the absorption spectrum M_{abs} and (ii) from the fluorescence spectrum in nonpolar, and (iii) polar solvents, M_{flu}(MCH) and M_{flu}(BCN), the considerable difference between these values indicates solvent-polarity dependent excited-state relaxations and molecular or electronic structural changes. Since the S_1 state is predominantly of ^{1}ET character, the coupling matrix elements V_{L_a-ET} and V_{S_0-ET} can be obtained from Eqs. 10–20. This allows us to decide whether the solvent effects on the transition moments are due to electronic changes alone (i.e., solvent-induced decoupling of ^{1}ET from ^1L_a), by an increased energy gap ΔE_{L_a-ET} and without any change in V_{L_a-ET} and V_{S_0-ET} or whether conformational changes are predominant, which will affect the value of the electron transfer matrix elements.

The coupling matrix elements V_{S_0-ET} and V_{L_a-ET} derived in Table 2 for the ground-state situation from absorption are determined from Eqs. 6, 13, and 17–20 by using $M_{ref} = 2.4$ D for the transition moment $S_0 \rightarrow {}^1L_a$ of the reference compound acridine and by using the difference of the dipole moments $\mu_{ET} - \mu_{S_0} = 22.5$ D.[16] The decrease in the values of V_{S_0-ET} and V_{L_a-ET} from **BJ** to **BDMN** reflects the increase in the twist angle φ_{DA} between the dimethylaniline and acridine moieties in the ground state.

Table 2 Coupling matrix elements and calculated (AM1) equilibrium angles in S_0 (from Reference 16)

	Absorption			Fluorescence[a]	
	$\varphi_{DA}(S_0)$ [deg]	V_{ET-L_a} [eV]	V_{S_0-ET} [eV]	V_{S_0-ET} in MCH[b] [eV]	V_{S_0-ET} in BCN[c] [eV]
BJ	64	0.21	0.18	**0.31**–0.42	0.05–**0.15**
BDMA	68	0.12	0.20	**0.29**–0.41	0.10–**0.22**
BDMT	80	0.06	0.12	**0.25**–0.36	0.05–**0.14**
BTMA	82	0.06	0.06	**0.14**–0.24	0.03–**0.10**
BDMN	84	0.02	0.08	**0.27**–0.34	0.06–**0.12**

[a] The range of the coupling matrix elements corresponds to the borderline cases, where e_1 for the ground state situation, as listed in Table 1, is used (left number) and where $e_1 = 1$ for the pure ^{1}ET character of S_1 is employed (right number). The more probable value is printed in bold. (See text.)

[b] Methylcyclohexane.

[c] *n*-Butyronitrile.

The corresponding coupling matrix elements for the relaxed excited-state geometry can be estimated from the fluorescence data. However, they cannot directly be determined from fluorescence, because spectral information about the ^1L_a transition is not available to evaluate e_1 from Eq. 18. This is possible only for the borderline cases in which either e_1(flu) for the excited-state situation is equal to that for the ground-state situation probed by absorption through Eq. 18 (Table 1) or e_1(flu) is close to unity (pure ^{1}ET-type S_1 state). Even though e_1(flu) cannot directly be obtained, the respective coupling matrix elements V_{S_0-ET} in the nonpolar solvent MCH are more reasonable with $e_1(\text{flu}) = e_1(\text{abs})$ (the lower limit of V_{S_0-ET}) due to the similar absorption and fluorescence energy, while in the polar solvent BCN, the use of $e_1(\text{flu}) = 1$ (the upper limit of V_{S_0-ET}) is more probable due to the strong solvent stabilization of ^{1}ET and hence of S_1.

Table 2 reveals that in MCH the values of V_{S_0-ET} from absorption are clearly smaller for all acridine–anilines than the range of values for V_{S_0-ET} obtained from fluorescence. This indicates that in nonpolar solvents, a considerable structural relaxation takes place that results in an improved π-electronic molecular orbital overlap. According to Eq. 16, this is most efficiently achieved by a geometrical relaxation towards planarity and, additionally, by a bond shortening especially of the interannular bond. Such conformational effects can indeed be reproduced by quantum chemical calculations for gas-phase conditions.[67] On the other hand, the range of the V_{S_0-ET} values derived from fluorescence in a polar solvent like BCN is similar or even smaller than that for the ground-state situation. This points either to negligible structural relaxation in S_1 or to a torsional relaxation towards perpendicularity, particularly if the same excited-state bond-length shortening as in MCH is assumed in BCN. The latter process is then reminiscent of the TICT concept described in Sections 2 and 3. Furthermore, the ranges of the ET coupling matrix elements V_{S_0-ET} obtained for the excited-state situation show no overlap between MCH and BCN, which undoubtedly demonstrates solvent-polarity dependent equilibrium conformations in the excited state, with a much smaller dependence in the ground state (e.g., V_{S_0-ET} of **BJ** from absorption is independent of the solvent polarity). With the reasonable assumption that the bond lengths of the relaxed excited states are equal in MCH and BCN and do not induce differences in V_{S_0-ET}, it can be concluded that the conformation of the D–A biaryls is more twisted in polar than in nonpolar solvents.

4.3.2 Donor–Acceptor Biphenyls

A typical model compound study has been performed for the donor–acceptor biphenyl **DCB** by investigating simultaneously the planar D–A biphenyl (fluorene) model **DCF** and the more twisted model D–A biphenyl **DMDCB** (see Scheme **5**a).[67–69,84] The difference in the absorption spectra of these three D–A biphenyls and, especially, the decrease in the extinction coefficient of the CT band (Fig. 1.9a) can readily be assigned to the increase in the S_0 interannular twist angle φ_{D-A} from **DCF** (the model for $\varphi_{D-A} = 0°$) to **DCB** and to **DMDCB**. The analysis of the electronic transitions reveals that the lowest lying intense absorption band is due to the ^1L_a-type $S_0 \rightarrow {}^1CT$ transition, where the contribution of $e_1^2 + e_2^2$ in Eq. 4b is larger than 90% and where the ^1ET contribution e_1^2 results from an electron transfer from the D (dimethylaniline) unit to the A (benzonitrile unit).[67,69] Due to the ^1L_a

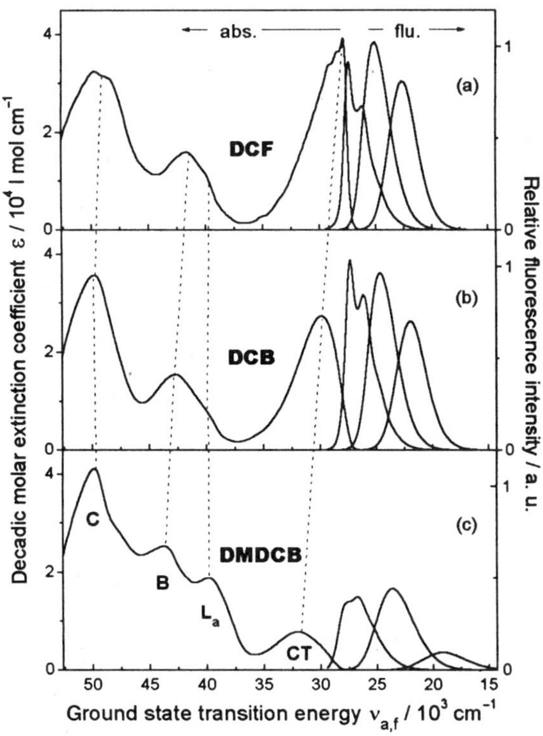

Figure 1.9. Absorption and fluorescence spectra of the donor–acceptor biphenyls **DCF**, **DCB**, and **DMDCB**.

character, the increase in transition energy and the decrease in absorbance of the $S_0 \rightarrow {}^1CT$ absorption bands reflects the increase in the twist angle φ_{D-A} between the D and A. Unfortunately, the procedure for deriving the coupling matrix elements V_{S_0-ET} and V_{L_a-ET} in Section 4.2.3 cannot be applied here because the zero-order ^{1}ET and ^1L_a states are energetically quasi-degenerate such that strong electronic configuration interaction occurs for twist angles differing from perpendicularity ($\varphi_{D-A} < 90^\circ$). For example, a quantum chemical study[67,69] reveals that for twist angles smaller than the experimental twist angle of unsubstituted biphenyl (ca. $\varphi_{D-A} = 40^\circ$), the ratio of e_1^2/e_2^2 is about unity.

The comparative investigation of the fluorescence properties (Table 3) clearly reveals similarities between the flexible biphenyl **DCB** and the planar model compound **DCF** only, whereas **DMDCB** behaves differently. This indicates that a photoinduced relaxation towards a more twisted geometry for **DCB** does not take place, but points to a planar structure in the relaxed S_1 state, as is usually observed for biphenyls. (See Section 4.1.) The planar conformational distribution of **DCB** in S_1 is further attested to by the (dis-)agreement between **DCB** and (**DMDCB**) **DCF** with respect to the static dipole moments μ_e and the transition moments M_f characterizing the emitting $S_1(^1CT)$ state. The occurrence of a sizeable photoinduced conformational relaxation from a medium twisted to planar structure in **DCB** is also demonstrated by the fact that, in contrast to **DCF**, the ratio M_f^2/M_a^2 is larger than unity, consistent with a geometry relaxation towards planarity for **DCB**, but not for **DCF**. This intramolecular rotational process in **DCB** is not affected by the solvent polarity, since M_f and M_f^2/M_a^2 remain constant in solvents with different dielectric constants.

A completely different and more complex photophysical behavior is observed for the twisted model compound **DMDCB** [(e.g., regarding the solvent dependence of the fluorescence properties (Table 3)]. In Table 4, the larger excited-state dipole moment μ_e and the considerably smaller transition moments indicate a stronger ^{1}ET character in **DMDCB** compared to **DCB** and **DCF**, and the strong decrease in the ratio M_f^2/M_a^2 with solvent polarity is consistent with an angular relaxation towards a more planar geometry in nonpolar solvents in S_1, but towards a more twisted structure in highly polar solvents. In medium polar solvents, the fluorescence decays of **DMDCB** become biexponential, with the shorter lifetime appearing as a rise time in the low-energy region of the fluorescence band.[68,69] In particular, the short lifetime increases from the picosecond into the nanosecond timescale with a lowering of the temperature. These two lifetimes can be associated with an adiabatic photoreaction between two excited-state conformers differing in the twist angle and separated by an energy barrier.[68,69,86] At room temperature, a photodynamic equilibrium between the two conformer species within a double minimum excited-state potential is experimentally observed for **DMDCB** in the medium polar solvent diethyl ether.[68,69] By analyzing the temperature dependence of the time-resolved fluorescence and by comparing it with semiempirical quantum chemical calculations, one of the conformers could be assigned to a rotamer more planar than the S_0 conformer, while the second one is associated with a rotamer more twisted than in S_0.[68,69,86,87]

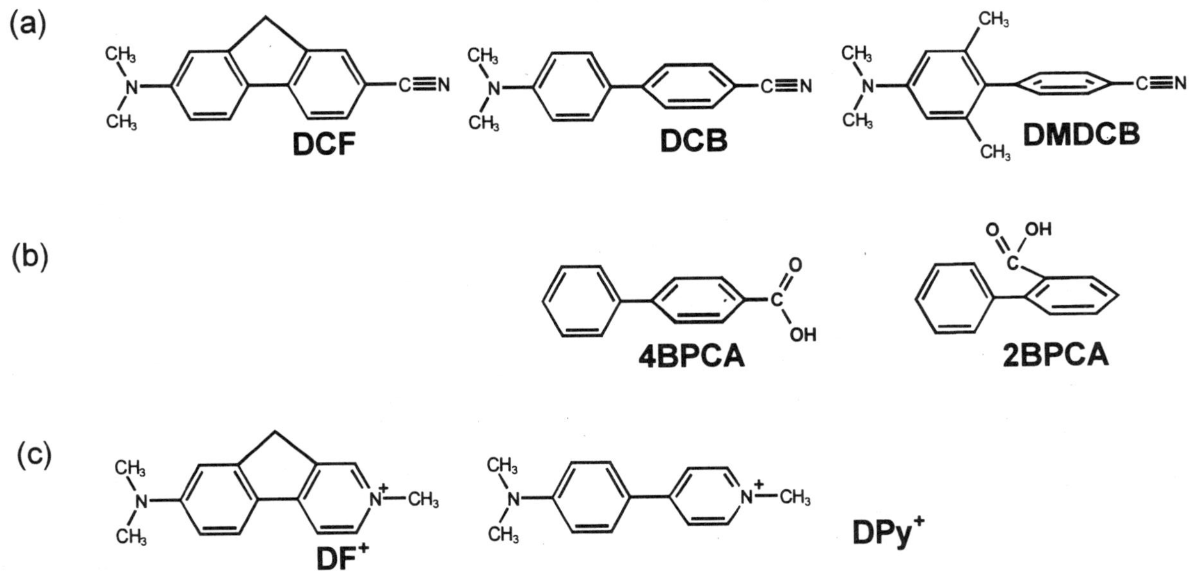

Scheme 5 Molecular structure of donor–acceptor biphenyls with different interannular twist angles φ: (a) 4-dimethylamino-cyanofluorene (**DCF**), 4-dimethylamino-cyanobiphenyl (**DCB**), and dimethyl-dimethylamino-cyanobiphenyl (**DMDCB**); (b) 2- and 4-biphenylcarboxylic acid (**2BPCA** and **4BPCA**); (c) dimethylamino-azafluorenium ($\mathbf{DF^+}$), and 4-dimethylanilino-pyridinium ($\mathbf{DPy^+}$). Semiempirical calculations yield the equilibrium ground-state angles $\varphi = 0$, 39, and 78° (AM1); 39° (MM2) and 54°; 0 and 20° (AM1) respectively.

Table 3 Fluorescence properties (band maxima, full width at half maximum, quantum yields and lifetimes) of the donor–acceptor biphenyls **DCF (I)**, **DCB (II)**, and **DMDCB (III)** (from references 68 and 69)

	$\nu_f^{max}[10^3\ cm^{-1}]$			$fwhm_f[10^3\ cm^{-1}]$			Φ_f			$\tau_f[10^{-9}\ s]$		
Solvent	**I**	**II**	**III**	**I**	**II**	**III**	**I**	**II**	**III**	**I**	**II**	**III**
n-Hexane	26.7	26.6	27.0	2.5	2.7	3.6	0.57	0.71	0.35	1.4	1.2	1.2
Diethyl ether	25.1	24.5	23.4	2.8	3.0	4.0	0.83	0.84	0.55	1.7	1.7	4.6
Acetonitrile	22.5	21.8	18.7	3.0	3.1	4.6	0.81	0.79	0.21	2.1	2.2	7.6

The transient absorption spectra shown in Figure 1.10 correspond to the relaxed S_1 situation and therefore reflect the conformational structure in S_1. In *n*-hexane, they are quite similar for the three D–A biphenyls, especially for **DCB** and **DCF**, consistent with a planar or, in the case of **DMDCB**, a slightly twisted excited-state conformation. In acetonitrile, the excited-state spectra of **DCF** and **DCB** are quite similar, too, because of their solvent-independent planar molecular structure in S_1. **DMDCB**, however, exhibits a totally different spectrum. The good agreement of the absorption maxima at 19,000 cm^{-1} and 26, 000 cm^{-1} with that of the radical cation spectrum of dimethylaniline[88,89] **PhDMA**$^{+\cdot}$ at 21,000 cm^{-1} and that of the radical anion spectrum of benzonitrile[88,90] **PhCN**$^{-\cdot}$ at 24,700 cm^{-1} results from an almost complete intramolecular electron transfer between the donor aniline and the acceptor benzonitrile units, which is achieved by the highly twisted conformation of **DMDCB** only in very polar solvents.

The effect that a pretwisted (sterically hindered) biphenyl structure can suppress the usual excited-state relaxation towards planarity, especially in polar solvents, has also been found in photophysical studies of the biphenylcarboxylic acids **2BPCA** and **4BPCA** (Scheme **5**b).[91–93] While the sterically unhindered derivative **4BPCA** displays a single fluorescence band in aprotic solvents, the fluorescence of the pretwisted biphenyl **2BPCA** shows a shoulder at around 400 nm, the intensity of

Table 4 Excited state dipole moments, transition dipole moments, and ratio of the squared fluorescence to absorption transition dipole moments for the fluorescent ^{1}CT state of the donor–acceptor biphenyls **DCF (I)**, **DCB (II)**, and **DMDCB (III)**. (From References 68 and 69)

	μ_e[D]			M_f[D]			M_f^2/M_a^2		
Solvent	**I**	**II**	**III**	**I**	**II**	**III**	**I**	**II**	**III**
n-Hexane				5.1[b]	6.2	4.3	0.71[b]	1.33	2.28
Diethyl ether	21.3[a]	22.0[a]	27.2[a]	6.3	6.4	3.4	1.03	1.32	1.40
Acetonitrile				6.6	6.6	2.3	1.13	1.33	0.60

[a] Average excited-state dipole moment obtained from the Lippert–Mataga solvatochromic plot. (See ref. 85.)
[b] Relatively small value because of additional mixing with ^1L_b (sizeable CI coefficient e_3 for DCF only in *n*-hexane).

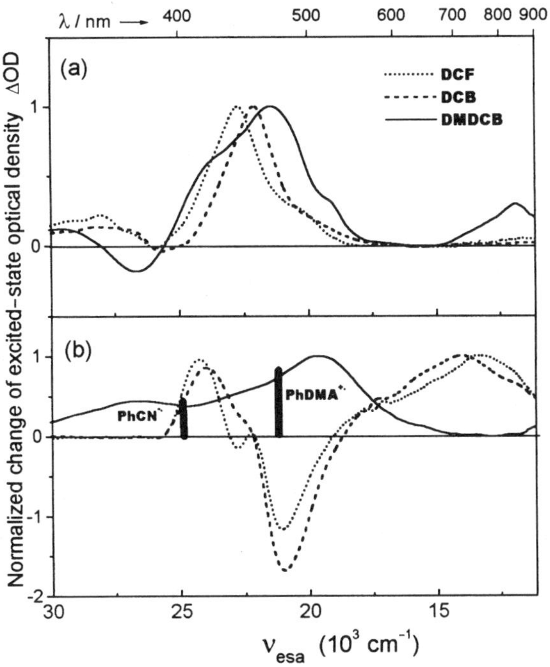

Figure 1.10. Transient absorption spectra of the D–A biphenyls **DCF, DCB**, and **DMDCB** in (a) *n*-hexane and (b) acetonitrile in the equilibrated ^{1}CT state after 200 ps. The lines in (b) represent the main experimentally observed absorption bands of the dimethylaniline radical cation **PhDMA**$^{+\cdot}$ and the benzonitrile radical anion **PhCN**$^{-\cdot}$. (Transition energies for radical anion spectra from Refs. 88–90.)

which increases with solvent polarity (Fig. 1.11). This can be explained as the contribution from a second anomalously red-shifted fluorescence band that derives from highly twisted rotamers possessing stronger ^{1}ET character than the more planar excited-state conformers responsible for the "normal" emission band at 350–370 nm in **2BPCA** and **4BPCA**. Protic solvents and, most favorably, water lead to a further enhancement of the anomalous emission from twisted and highly dipolar rotamers, such that even for **4BPCA**, the anomalous band appears as a shoulder. These observations are reasonably explainable by hydrogen bonding of the carboxyl group with the protic solvents. In **4BPCA**, the specific solvent interaction increases the electron-withdrawing capability of the carboxyl group, introducing ^{1}ET character to the hydrogen-bonded complex, which emits at longer wavelengths than the uncomplexed molecule. This process may also hinder the relaxation towards planarity. On the other hand, the hydrogen-bonded complex with the ortho-substituted biphenyl **2BPCA** additionally enhances the pretwist, such that the energy minimum for highly twisted rotamers can be reached much more easily than that for the usually observed minimum at more planar geometries. A switching between the different excited-state biphenyl rotamers can be achieved upon adding cyclodextrins into aqueous solutions

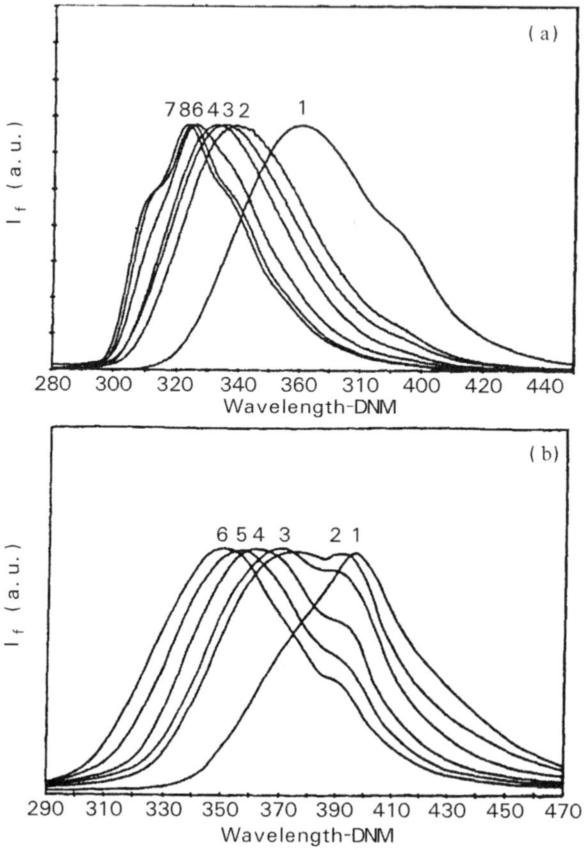

Figure 1.11. Normalized fluorescence spectra of (a) **4BPCA** and (b) the sterically hindered **2BPCA** in different solvents: 1 = H_2O (pH 3.0), 2 = methanol, 3 = ethanol, 4 = acetonitrile, 5 = dioxane, 6 = cyclohexane, 7 = hexane. (Reproduced with permission from Ref. 91. Copyright Korean Chemical Society 1992.) For molecular structures, see Scheme **5**.

of **2BPCA**[93] or **4BPCA**.[92] If the cyclodextrin cavities are voluminuous enough to accommodate the biphenyl molecules and thereby protect them from the water molecules, the photoinduced intramolecular rotation towards planarity is no longer inhibited, as is attested to by a quenching of the CT band at 400 nm and simultaneous enhancement of the locally excited fluorescence band at around 330 nm. By this procedure, dual fluorescence with separate band maxima is observed for **2BPCA**.[93]

An alternative way to induce excited-state relaxation towards perpendicularity is to increase the electron transfer character in S_1 by smaller *IP*(D) or larger *EA*(A) values according to Eq. 5 and by using charged compounds characterized by a reversed bond-order alternation in S_0 and S_1, compared with neutral biphenyls.[72] Both conditions are accomplished in **DPy**$^+$ and its planar model compound **DF**$^+$

(Scheme **5c**). The larger *EA*(A) of the acceptor pyridinium unit compared with the benzonitrile fragment in **DCB** is responsible for a 7000 cm^{-1} more red-shifted fluorescence band in nonpolar solvents, and the larger π-bond order of the interannular bond in S_0 results in a more planar conformation of **DPy**$^+$ than for **DCB**.[72] The photoinduced electron transfer shifts the positive charge from the acceptor pyridinium unit to the donor dimethylaniline unit for both **DPy**$^+$ and **DF**$^+$. However, only the intramolecularly flexible biphenyl **DPy**$^+$ shows a strong decrease in fluorescence quantum yields with increasing solvent polarity (Fig. 1.12), because it can perform a geometric relaxation to nonradiative highly twisted rotamers.[71,72] The involvement of a large-amplitude motion like the twisting is confirmed by the solvent viscosity influence on the nonradiative rate constants.[70] This twisting relaxation is facilitated in **DPy**$^+$ compared with **DCB** because (a) the S_1 wave function possesses larger ^{1}ET character for all twist angles and leads to a smaller energy barrier to reach the highly twisted conformer with almost complete ET; (b) the final state with twisted geometry is of considerably lower energy in **DPy**$^+$ than in **DMDCB**, where it is endothermic in the gas phase; and (c) **DPy**$^+$ possesses considerable polymethinic character[94,95] and leads to a nearly single-bond character of the interannular bond in S_1, which no longer favors a planar conformation. The fast nonradiative decay of **DPy**$^+$, which is not observed even for the highly twisted **DMDCB** in polar solvents, may also be connected with the facile access for ionic compounds to conical intersections with the ground state that quench fluorescence.[96]

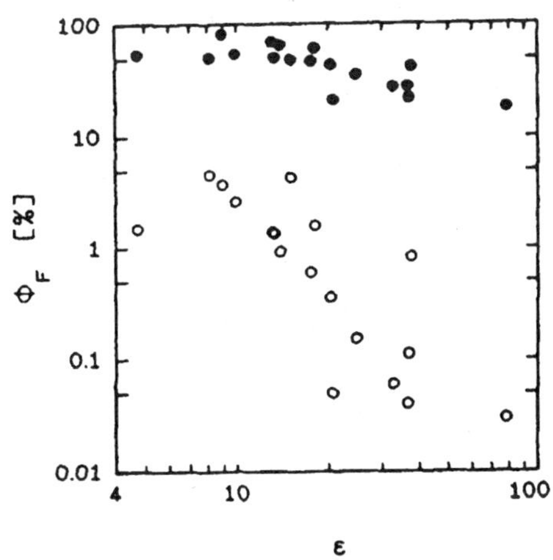

Figure 1.12. Logarithm of the fluorescence quantum yields Φ_f versus the logarithm of the solvent dielectric constant ε for analogues of **DF**$^+$ (●) and **DPY**$^+$ (○). (Reproduced with permission from Ref. 71. Copyright Chemical Society 1992.)

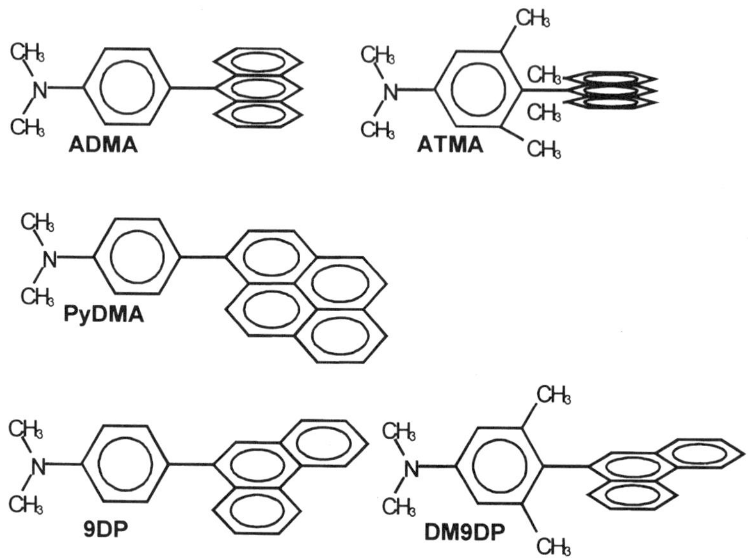

Scheme 6 Structures of aniline–anthracenes **ADMA** and **ATMA**, aniline–pyrene **PyDMA**, and aniline–phenanthrenes **9DP** and **DM9DP**. Their calculated equilibrium twist angles φ in S_0 are 84 and 87° (Ref. 105), 50° (Ref. 106), and 62 and 82° (Ref. 110), respectively.

4.3.3 Other Aniline–Aryl Systems

Next to **DMABN** and 9,9′-bianthryl, the aniline–anthracene biaryl **ADMA** is one of the most extensively studied molecules showing photoinduced intramolecular CT in which single-bond twisting relaxations are involved.[16,24,80,97–107] Solvent-dependent fluorescence spectra of **ADMA**, were first reported in 1972 and established the idea that in polar solvents an intramolecular CT state becomes slightly lower lying than the localized ^1L_a state of anthracene.[97] Later on, both the observation of strongly decreasing radiative rate constants with solvent polarity[98,103] and model compound studies[99,102] led to the suggestion of an excited-state adiabatic photoreaction from a nonpolar locally excited (^1L_a) state to a more twisted-CT state, describable within the TICT model, (see Section 3). Further time- and frequency-resolved fluorescence and transient absorption[100–102] studies suggested that the excited-state behavior could hardly be explained with a simple two-state equilibrium. A multiple-species description, resulting from various twist-angle and solvent configurations, which in turn introduce different degrees of ET electronic character to S_1, seems to be more appropriate to account for the multiexponential photokinetics and transient absorption spectra differing from a simple superposition of two (LE and CT) excited-state spectra. An alternative approach, without the necessity of twisting relaxations, suggests that the electron transfer reaction to the relaxed ^{1}CT state takes place in two steps:

1. A primary ultrafast diabatic transition from the $^1L_a(S_2)$ to the $^1CT(S_1)$, with little geometrical change and influence of the solvent coordination. This process resembles the internal conversion in polyaromatic molecules, but it is surprisingly fast with respect to the large energy gap between S_2 and S_1 of about $1400\,cm^{-1}$.
2. An adiabatic transition on the S_1 surface from the "L_a-like" to the relaxed ^1CT state. This transition occurs on the slower time scale of the reorientation dynamics of the solvent molecules, which thereby accommodate to the enhanced dipole moment of **ADMA** in the ^1CT with respect to the weakly polar ^1L_a state.[103,104]

The latter model simulates the electron transfer kinetics, free-energy potentials, and the time-dependent emission spectra on the basis of an LE–ET mixed excited-state wave function, which is changing with the solvent motions that obey the generalized Smoluchowski equation. There, the solvent coordinate uniquely determines the adiabatic LE–ET interaction and is therefore responsible for the observed solvent polarity and time-dependent fluorescence rate constants. A similar conclusion has been reached elsewhere[105] by fluorescence band shape analysis related to the Marcus ET concept.[108] The inner reorganization energy of various aryl–anilines, which is due to the low-frequency intramolecular motions (e.g., internal rotations of submoieties) needed to achieve the final conformation of the fluorescent ^1CT state, is found to increase with the ground-state twist angle. This behavior is inconsistent with an intramolecular twisting rotation towards perpendicularity, but it indicates large-amplitude motions towards a more planar structure, facilitated by a less restrictive steric twist potential. However, the fact that the transition moments of the localized $S_0 \rightarrow {}^1L_a$ transition are smaller than those of the observed $^1CT(S_1) \rightarrow S_0$ fluorescence transition in different aryl–anilines stresses the additional importance of S_0–ET interactions.

More recently, **ADMA** has been compared with the closely related pyrene–aniline derivative **PyDMA** by time-resolved transient absorption and emission in alcohols.[106] In spite of the fact that for both compounds, the electron transfer reaction is believed to correspond to an $S_2(^1L_a) \rightarrow S_1(^1CT)$ transition, a quite different behavior is found. The following experimental findings indicate an additional or different intramolecular relaxation accompanying the electron transfer in **ADMA:** (1) Whereas the inner reorganization energy in methyltetrahydrofuran is zero for **PyDMA**, it is about 0.15 eV in **ADMA;** (2) the fluorescence transition moments of **PyDMA** remain constant in solvents of different polarity, but those of **ADMA** decrease with the solvent dielectric constant; (3) the solvation correlation times obtained from exponential fitting of the time-resolved fluorescence maxima for **ADMA** are twice as large as those for **PyDMA**; (4) the decay of the transition moment (proportional to the integrated and ν_f^3 corrected fluorescence band intensity) in hexanol equals the monoexponential fluorescence decay time obtained by the time-correlated single-photon counting method for **PyDMA** ($\tau_f = 3.8$ ns), but it is biexponential for **ADMA**, with a second decay component of 870 ps beside the fluorescence lifetime of 15 ns. These effects were attributed to a more extensive excited-state intramolecular rotational relaxation in **ADMA** than in **PyDMA**. Even though most authors believe that the photoinduced structural relaxation in **ADMA** occurs towards planarity, a narrowing of the rotational distribution around perpendicularity could not yet be excluded. It is one of the challenging questions whether the decrease of the fluorescence transition moments with increasing solvent polarity of pretwisted D–A biaryls is due only to the solvent-induced increased energy gap of the ET state to the higher lying and emissive LE states inducing a decreased ^1L_a–^{1}ET coupling (i.e., smaller e_2 values in Eq. 4) or whether intramolecular twisting relaxations are additionally involved to reduce M_f. Quantum chemical treatments, including bulk and specific solvent interactions, may serve as helpful tools for the elucidation of this question, as well as detailed ultrafast time-resolved spectroscopic studies of D–A biaryls in solvents of well-characterized relaxation behavior.[109]

Usually, the discussion of the transition moments is based (i) on the ^1L_a–^{1}ET interactions, which lead to a reduction in the transition moments with increasing energy gap between ^1L_a and ^{1}ET by solvent stabilization of ^{1}ET and (ii) on the S_0–^{1}ET interactions, which lead to an enhancement of the transition moments by a molecular flattening. Very recently, however, it was demonstrated for phenanthrene derivatives[110,111] and D–A biphenyl,[67,84] that the interactions with a less allowed ^1L_b transition should also be taken into account to understand the CT dynamics. For the D–A biphenyls, the observation of dual or, in the case of the pretwisted derivative **DMDCB**, triple stimulated emission bands has been explained with three excited-state minima (i.e., a primarily populated state with mixed ^1L_b/^1L_a/^{1}ET character, a planar ^1L_a-type ^{1}CT state, and a more relaxed ^{1}CT state with enhanced ^{1}ET character due to increased interannular twist and dipolar solvation). A strongly solvent-polarity dependent S_1 wave function has also been suggested for phenanthrene–anilines. In contrast to other aryl–anilines, the phenanthrene–anilines **9DP** and **DM9DP** at first exhibit an increase in M_f, with increasing solvent polarity, as shown in Figure 1.13.[110] This behavior is due to the very low-lying ^1L_b-type locally excited state in phenanthrene, which mainly constitutes S_1 in weakly polar solvents and the influence of which on the S_1 wave function decreases as the solvent polarity increases and introduces a more ^1L_a-type and electron transfer character. Since the ratio M_f^2/M_a^2 exceeds unity in medium polar solvents, it can additionally be concluded that the geometric relaxation occurs towards planarity. In highly polar solvents, the transition moment M_f decreases, which can again be explained either by an inhibition of the planarization relaxation with a possible change to a relaxation into the opposite direction towards higher twist angles or by a decoupling of ^{1}ET from the more emissive ^1L_a state through the polarity-induced enlargement of the ^{1}ET–^1L_a energy gap. The molecular flattening in less polar solvents and the preservation of a twisted structure in more polar solvents seems to be a general feature of photoinduced intramolecular charge transfer in pretwisted donor–acceptor biaryls. This means that a typical TICT photoreaction leading from a planar precursor to a near-perpendicular product (see Section 3.1) is less probable in the larger D–A biaryls compared with small-sized D–A compounds like **DMABN**

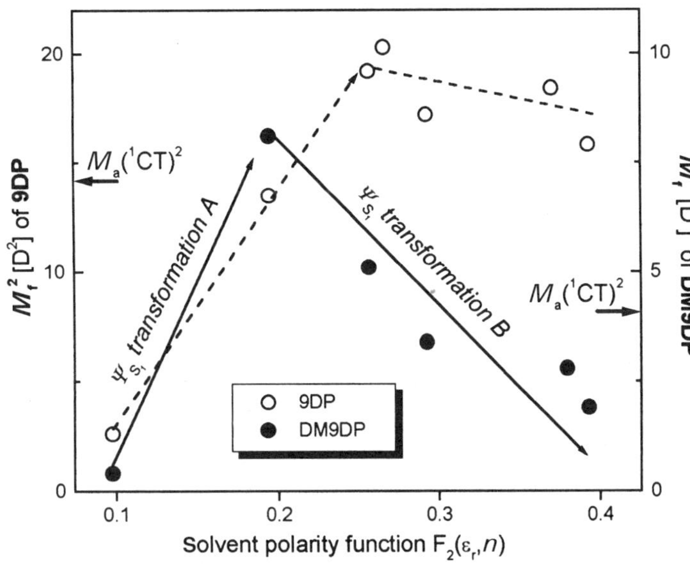

Figure 1.13. Plot of the squared fluorescence transition moments M_f^2 vs. the solvent polarity parameter $F_2(\varepsilon_r, n) = (\varepsilon_r - 1)/(2\varepsilon_r + 1) - 0.5(n^2 - 1)/(2n^2 + 1)$ for **DM9DP** (•) and **9DP** (○). The values of the squared absorption transition moments $M_a(\mathrm{CT})^2$ of the CT band in isooctane are indicated by arrows for **DM9DP** on the right and for **9DP** on the left axis. Transformation *A* indicates the change in the wave function of $S_1(\Psi_{S_1})$ from ^1L_b to ^1L_a/ET character, and transformation *B* indicates the change from ^1L_a to ^{1}ET.

derivatives or amino–boranes. Four principal reasons can account for such a general behavior: (1) In the biaryls, the LE states are much lower lying, such that ^1L_a–^{1}ET state repulsions, which shift the S_1 energy minimum towards planarity, are more dominant than S_0–^{1}ET state repulsions, which shift S_1 minima towards perpendicularity; (2) biaromatic systems prefer a mutually planar arrangement, due to the increased electron delocalization (mesomeric interaction); (3) Coulomb stabilization (see Eq. 5) between the donor and acceptor part is less strong for larger aromatic units because the average distance between the separated charges is larger; (4) according to Onsager's reaction field theory,[112] solute–solvent stabilization of polar species increases with the ratio of the squared dipole moment to the third power of the Onsager radius (μ_e^2/a^3). This ratio is smaller in large D–A systems, and hence, solvent stabilization is reduced. In any case, the current results for large D–A systems speak in favor of a more twisted excited-state conformational distribution in polar solvents than in nonpolar or weakly polar ones. In accordance with this view, the coupling matrix elements $V_{\mathrm{ET-S_0}}$ reflecting the twist-angle-dependent aryl–aryl electronic interaction (Eq. 16) are determined for **ADMA** by the procedure described in Section 4.2.3 to amount to 0.31 eV in nonpolar methylcyclohexane, but to only 0.16 eV in polar acetonitrile.[16]

4.3.4 Biaryls of the C–N type

4.3.4.1 Aryl–Carbazoles

Carbazole is isoelectronic with anthracene, and therefore, anthrylcarbazoles can directly be compared with the corresponding bianthryls. 9,9′-bianthryl (**BA**) is the prototype bianthryl and has been studied extensively since the discovery of its dual fluorescence in polar solvents.[113] The second fluorescence band is identified as a CT band from its solvatochromic red shift,[60,113,114] and corresponds to a solvent-polarity-induced charge separation creating a symmetry-broken pair of anthracene cation–anion radicals. This effect has been demonstrated by transient absorption measurements.[115] As carbazole is a better electron donor than anthracene, the carbazole derivative corresponding to 9,9′-bianthryl, 9-(N-carbazolyl)anthracene (**C9A**) is expected to exhibit an increased tendency towards CT formation, leading to a decreased weight of the locally excited state (LE) in the excited-state equilibrium between LE and CT. This tendency can be seen in the steady-state fluorescence spectra in polar solvents, in which different from **BA**, the LE band is missing for **C9A** and only the CT band is present,[60] as well as in the transient absorption results.[115] Both **BA** and **C9A** show ultrafast formation of the CT state in the picosecond range.[116–118]

Other carbazole compounds with acceptors different from anthracene have also been studied in solution.[119,120] The data can be treated in the manner previously described. However, for the system **C9A**, supersonic jet studies are available as well, and the enormously enhanced spectral resolution enables a completely new picture of the ongoing electron transfer reaction. In this section, we will therefore focus on the jet results rather than on the solution studies. These supersonic jet studies have revealed a tremendous amount of detailed information and have uncovered an outstanding example of a photoreaction with resonance behavior (i.e., a slowing down of the reaction occurs if a certain energetic threshold is passed, contrary to the usual Arrhenius-type barrier behavior, where increasing thermal energy always leads to an acceleration of the reaction). **BA**, its cyano derivative **BACN**, and **C9A** have been studied under jet conditions,[47,51,52,121] and the gas-phase twist potentials of S_0 and S_1 can be accurately determined from fluorescence excitation and dispersed fluorescence spectra as a function of excess energy even if the 0–0 band is not observable.[52,121] **C9A** possesses energy minima that are further distant from 90° (Fig. 1.14). This is explained by the reduced steric hindrance for less twisted conformations in **C9A**, because in the carbazole nucleus, the hydrogen atoms on the two benzene rings fused to the central five-membered ring are, on average, at a larger distance from the hydrogen atoms on the anthracene. The result of the reduced steric strain is a considerably higher energy barrier for **C9A** than for **BA** at 90° in the fluorescing excited state and a weak double-minimum potential in the ground state compared with the single perpendicular minimum determined for **BA** (Fig. 1.14).

The minima of the ground-state double-minimum potential of **C9A** are situated below a steeply rising potential region of S_1, and many levels of the S_1 torsional vibration can be excited optically. This leads to a very rich fluorescence excitation

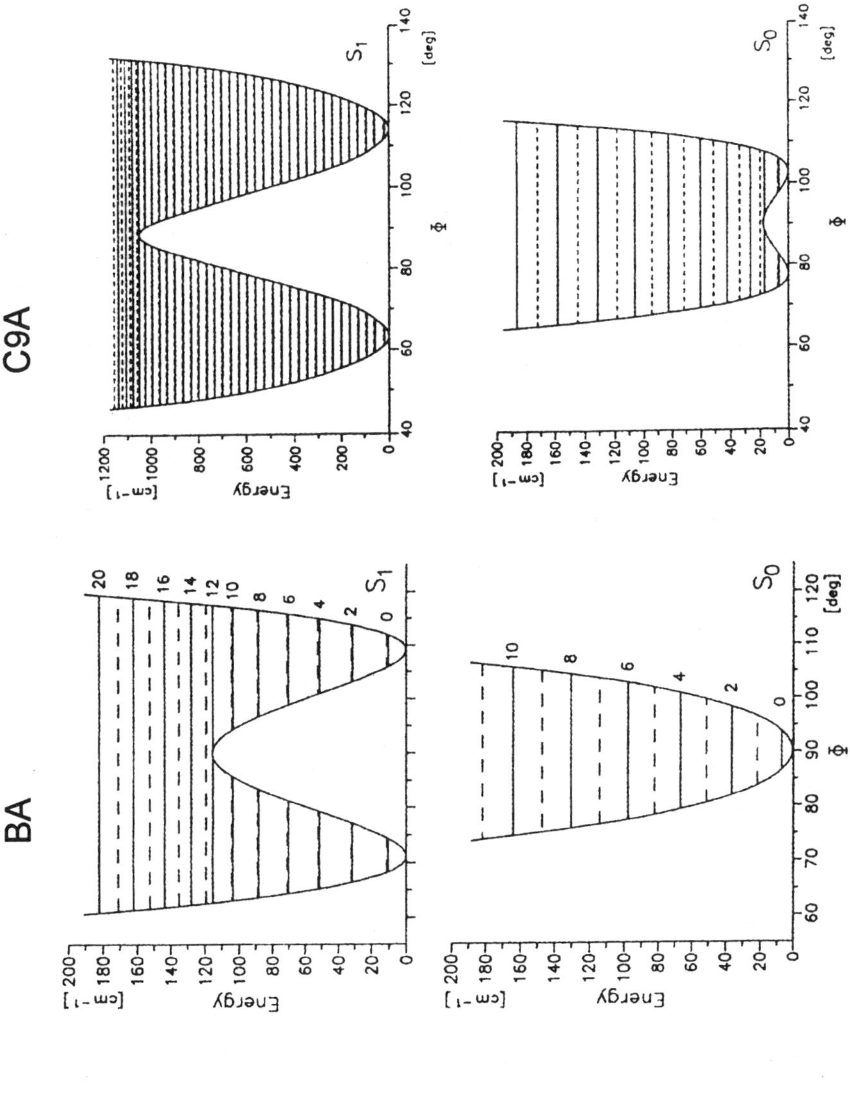

Figure 1.14. Twist potentials for **BA** and **C9A**, as determined from supersonic jet studies. Note the difference in the ordinates. (Taken from Refs. 121 and 52.)

spectrum consisting of a multitude of lines spaced approximately 37 cm^{-1} apart (Fig. 1.15) with an unobservable 0–0 transition. From the detailed knowledge of S_0 and S_1 potentials, the Franck–Condon factors (intensity distribution) can be accurately calculated and compared with the experimental values. Most remarkably, the envelope of the spectrum presents a pronounced minimum of the intensity, which is not explainable from the potential shape. It is, however, comprehensible if a process competing with fluorescence leads to fluorescence quenching at this amount of excess energy (ca. 470 cm^{-1}). The assumption has been verified by separately measuring the fluorescence lifetimes, which also show a pronounced shortening at this excess-energy amount (Fig. 1.15), corresponding to a photoprocess that is fastest at that amount, but that slows down if the energy is larger or smaller. Since this work has been published, a further independent confirmation of the resonance-type photoprocess has been gained by measuring the absorption spectrum for these jet conditions[122] with the cavity-ring-down method. A comparison with the fluorescence excitation spectrum clearly identifies the lines with reduced intensity. Most remarkably, this photoprocess is highly mode specific (i.e., two peaks of nearly equal energy can show a strongly different behavior with respect to the photoprocess that quenches the fluorescence).

A possible explanation for this unusual behavior is the crossing of an emissive state S_1, the potential of which is known very accurately from the fluorescence measurements, with a nonemissive "dark" state with a minimum at strongly different twist angles. Since planar twist angles are energetically strongly disfavored, a potential with a minimum at 90° twist was proposed for the "dark" state S_X (Fig. 1.16). Although many arguments speak in favor of this state being of a CT nature, an experimental proof is still lacking.

At this point, theory can be helpful in unravelling additional details of the ongoing photoprocess. The twist dynamics of **C9A** has been modeled on the femtosecond and picosecond timescale based on the time-dependent density matrix formalism, and the spectra have been calculated therefrom and compared with experimental values.[123,124] These studies show that the assumption of a dark state with a potential minimum at 90° is a reasonable approach that satisfactorily reproduces the observed spectra. Moreover, the model calculations even allow one to estimate the relative energy of the minima of S_1 and S_x. If the excitation energy E_{ex} is too small to reach the crossing of the two states at the critical energy E_c, the wave packet that is generated oscillates rather undisturbed (i.e., coherently around the minima of S_1, with a characteristic torsional period of about 1 ps). For excitation energies that reach the crossing exactly—that is, where the wave-packet motion has its classical turning point there (Fig. 1.16)—efficient transfer of the population to S_x occurs, whereupon the population is nearly completely transferred back from S_x to S_1 after half an oscillation period (Fig. 1.17). Some of this population can, however, remain trapped in S_x and accumulate there if intramolecular vibrational redistribution (IVR) from the torsional to the vibrational modes is faster in this state than in S_1. More efficient IVR in S_x than in S_1 points to stronger coupling of the torsional and background vibrations in S_x than in S_1, and this could be a consequence of the possible CT nature of the S_x state, in comparison with the non-CT nature of S_1. As a consequence, rapid IVR in S_X induces energy relaxation down to the zero level of the

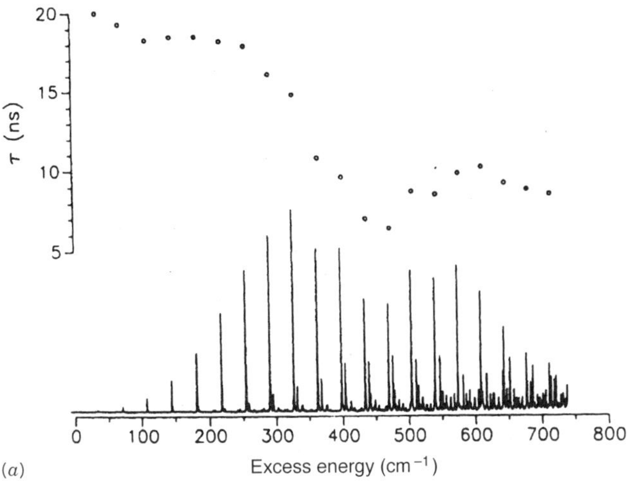

(*a*)

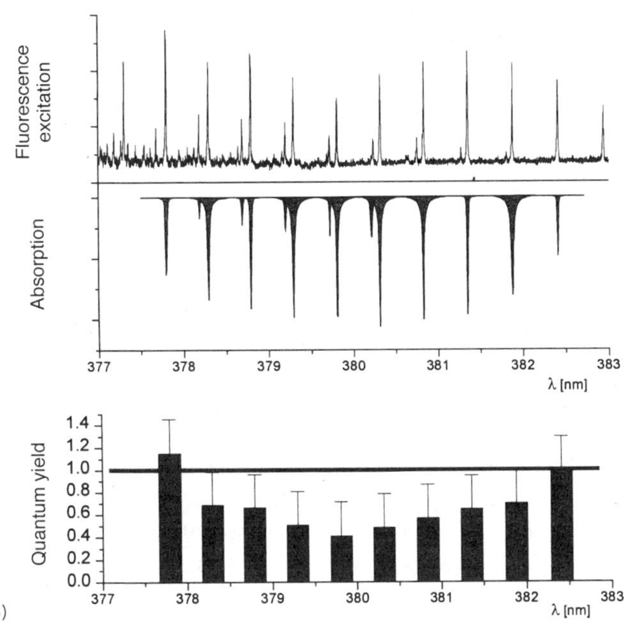

(*b*)

Figure 1.15. (a) Fluorescence lifetimes following selective band excitation and the corresponding part of the excitation spectrum (first strong progression). In addition to the general decrease in the lifetimes with increasing excitation energy, a pronounced lifetime minimum at about 470 cm^{-1} correlates with the intensity minimum of the excitation spectrum. (Taken with permission from Ref. 52.) (b) Comparison of the part of the fluorescence excitation spectrum showing the intensity minimum and the absorption spectrum measured by the cavity-ring-down method. The absorption spectrum does not show the characteristic minimum. Moreover, only the main torsional progression—the lines with higher intensity—is affected. The combination progression with lower intensity does not show the intensity minimum in this energy region, but at significantly higher values. This is the signature of a mode-specific reaction. The lowest part of the figure shows the fluorescence quantum yields derived from absorption and excitation spectra. (Taken from Ref. 122.)

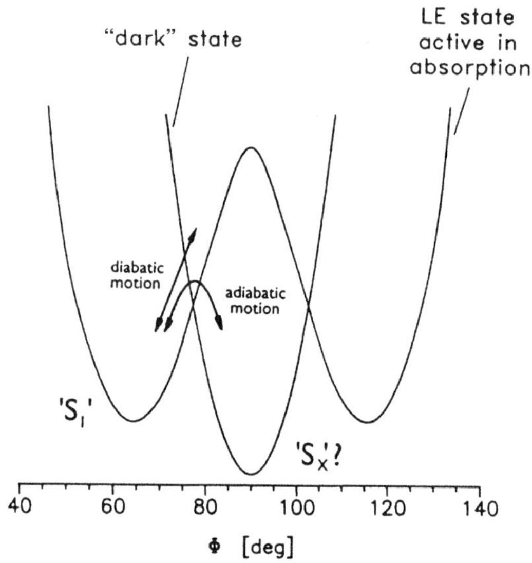

Figure 1.16. Model of the diabatic surface crossing of the lowest electronically excited states of **C9A**. The diabatic crossing takes place along the torsional coordinate between the optically active excited-state S_1 and the dark background state S_x, with no significant oscillator strength to S_0. (Taken from Ref. 52.)

torsional motion within S_x, with the accepting vibrational modes playing the role of a thermal heat bath introducing irreversibility. The measured fluorescence decay time is in the nanosecond range, and its shortening at the critical energy E_c corresponds to the irreversible trapping of the population in S_x. These types of calculations also allow one to describe femtosecond time-resolved pump-probe experiments that have not yet been conducted, but that are predicted to show a stepwise diminution of intensity on the ps timescale, with pairs of signals of similar intensity for the case $E_{ex} = E_c$ and a very long train of signals with slow decay for $E_{ex} < E_c$[123,124] corresponding to nearly coherent torsional motions of **C9A**(S_1) on the timescale of ca. 1 ps to 100 ps and longer. This type of quasi-coherent vibrational motion on a long (1 ps–1 ns) timescale is quite surprising for a large organic molecule such as **C9A**(S_1), implying that IVR for this torsional mode is on the very slow (nanosecond) timescale, presumably a consequence of the rather weak coupling of the torsional motion to the other vibrations. This detailed picture of ET, which is possible due to the high resolution of the jet experiment, thus shows that

1. the CT state is confined to more strongly twisted geometries than is the LE state
2. the actual step occurs at a specific geometry and can be treated as a nonadiabatic surface-crossing process in this case,
3. the mode specificity suggests that a one-dimensional treatment using the torsional coordinate may be sufficient, and

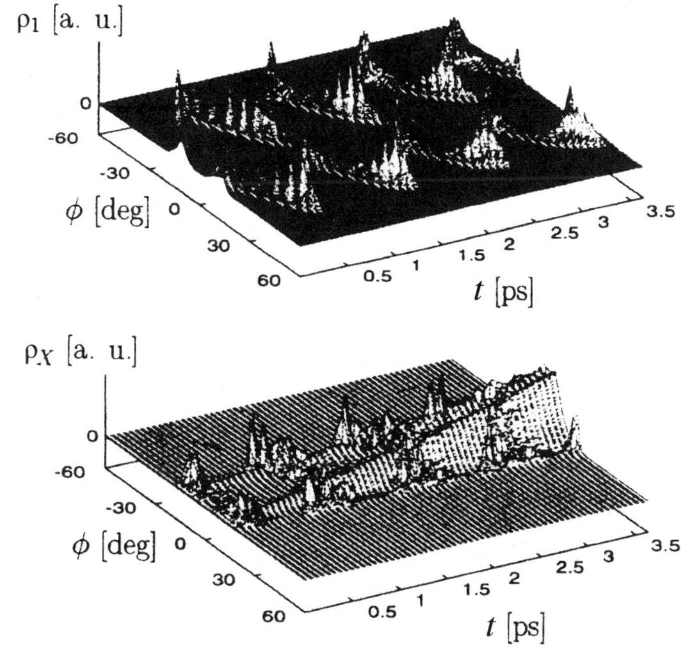

Figure 1.17. Calculated twist dynamics of **C9A** on the picosecond and femtosecond (ps/fs) timescale, as represented by the time-dependent densities in the bright state S_1 ($\rho_1(\phi, t)$, upper panel) and the dark state S_x ($\rho_x(\phi, t)$, lower panel). To make the effect of IVR clearer, especially regarding the trapping of the population in S_x, IVR has been artificially enhanced one-hundredfold for S_x. The population accumulation seen within picoseconds in this model simulation at $\phi = 0$ (which corresponds to the perpendicular potential minimum of S_x in this representation) thus mimics the accumulation within nanoseconds in the real system. (Reproduced with permission from Ref. 124. Copyright Springer-Verlag 1997.)

4. in the neighborhood of the state crossing, however, which can be viewed as a conical intersection,[125] a second coordinate—possibly the inversion mode—could be necessary.

Recently, high-level quantum chemical calculations (e.g., within the density functional theory for excited states) have been attempted for systems of the size of **C9A**[122,126] in order to reproduce the experimental potentials of S_1 and S_0. Such calculations would also give information on the S_x potential and its electronic nature. An experimental approach to verify the presumed CT character of S_x is given by the investigation of related molecules in which a CT state should be considerably lowered in energy with respect to **C9A**.

4.3.4.2 Orbital Symmetry Effects in N-Phenyl Pyrroles

Orbital localization and forbidden CT transitions can also result from pure symmetry effects because orbitals belonging to different irreducible representations do not couple with each other. For example, all donor (D) and acceptor (A) π orbitals of 90° twisted **DMABN** are orthogonal to each other, preventing orbital interaction. Hence, since the HOMO is localized on the donor and the LUMO is localized on the acceptor, HOMO–LUMO excitation leads to the transfer of a full electron from donor to acceptor. However, for the planar conformation of **DMABN**, D and A orbitals of the same symmetry type exist such that orbital mixing leads to delocalized HOMO and LUMO. The resulting HOMO–LUMO excitation therefore has only partial ^{1}ET character.

The situation in pyrroles is quite different, because the highest donor orbital on the pyrrole moiety is antisymmetric with respect to the symmetry plane normal to the pyrrole ring, and coupling with other orbitals is negligible even for the planar conformation. This is the reason that HOMO and LUMO in pyrrolobenzonitriles remain localized on the donor and acceptor subsystems, irrespective of the twist angle, and the corresponding excited state is expected to be of full ^{1}ET character and to possess a small transition moment for both planar and perpendicular (and any intermediate) conformations.

Such a situation leads to a new pattern with respect to excited-state angular relaxation: either planarity in sterically unhindered systems or perpendicularity in pretwisted systems. This conclusion could be drawn from a recent study of pyrrolobenzonitriles (Scheme **7**).[127] In the unhindered derivative **PBN**, the CT emission is found to be strongly forbidden (τ_{rf} = 230 ns in acetonitrile),[127] but in the pretwisted compound **DPBN**, the forbidden character is even increased by a factor of around 3(τ_{rf} = 770 ns), consistent with a more perpendicular emissive conformation. It is not clear whether the emissive conformation differs from the Franck–Condon conformation, but the strong excitation wavelength dependence of the dual fluorescence observed in rigid low-temperature glasses[127] points to the twisting coordinate as the major controlling factor. In unsubstituted *N*–phenylpyrrol **PB**, the dual fluorescence is also observed in liquid solvents at room temperature, and the strong enhancement of the CT band for pretwisted compounds in this case seems to indicate a preferred perpendicular conformation of the relaxed excited state.[128]

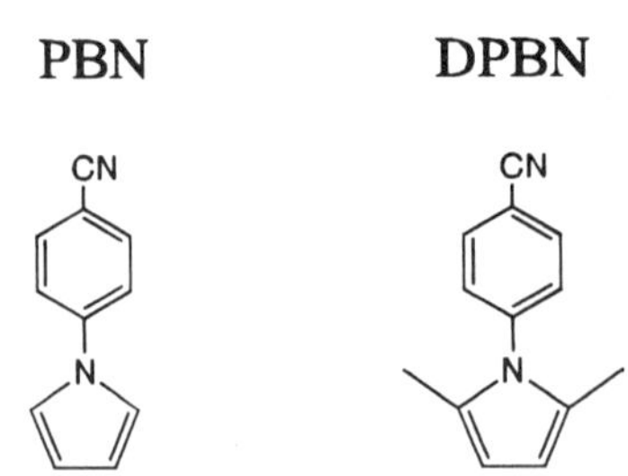

Scheme 7 Molecular structures of **PBN** and **DPBN**.

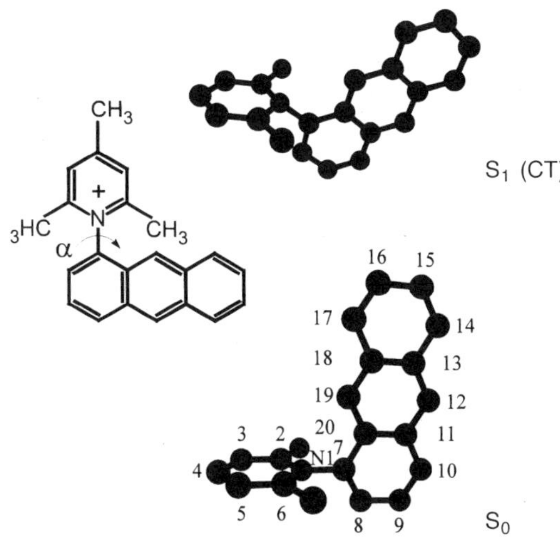

Scheme 8 Anthracene–pyridinium with perpendicular structure of the ground state and bent structure of the (CT) excited state (from Ref. 130).

4.3.4.3 Pyridinium as Acceptor: Pyramidalization as Coupling Coordinate

Anthryl–pyridinium is an interesting CT system that behaves differently from the previously described donor–acceptor systems like **DCB** or **DMDCB**.[129,130] It possesses an anthracene donor group and an *N*-linked pyridinium acceptor held perpendicular by ortho-methyl groups (Scheme **8**). In the ^{1}CT state, an electron is transferred from anthracene to the pyridinium group, and the latter becomes bent due to pyramidalization at the nitrogen atom (see Section 2.6). As a result, the transition moment does not decrease as in **DMDCB**, but increases because this relaxation leads away from the orthogonal situation. The corresponding radiative lifetime is strongly forbidden for CT absorption (>4000 ns) and becomes less forbidden for CT emission (ca. 80 ns).[129,130] At low temperatures, a dual fluorescence from the CT and the anthracene-type LE state is observable,[130] and the reaction coordinate in this case has to be ascribed to molecular bending rather than to twisting.

5 Concluding Remarks

In reviewing the possibilities for elucidating excited-state relaxational pathways, we have focused on the twisting relaxation of dimethylanilines acting as an electron donor linked by a single bond either to nonaromatic acceptor substituents (Section 3) or to an aromatic acceptor moiety (Section 4). The importance of differentiating between C–C bonded systems and C–N bonded systems has been emphasized. In the latter systems, twisting and bending (i.e., pyramidalization and rehybridization) can become important. If a nitrogen atom is in the donor moiety, the relaxation occurs towards less bending (pyramidalization) upon formation of the ^{1}CT state, irrespective of the twisted or planar overall structure, but when N is part of the acceptor system, the relaxation tends to occur towards a more pyramidal system. On the other hand, the C–C bonded systems, which make up the larger part of the biaromatic systems, possess much less complications from pyramidalization effects.

We have outlined two different experimental approaches, one based on sterically controlled model compounds and the other relying mainly on radiative transition moments. In favorable cases, it is possible to extract quantitative conformational information from the dependence of coupling matrix elements on molecular geometry. By comparing these values for the ground state and the relaxed excited state, it can be determined whether the relaxation pathway is towards a more perpendicular or a more planar conformation.

Acknowledgment:

We are grateful to the Deutsche Forschungsgemeinschaft for its generous support and wish to thank Professors Zbigniew R. Grabowski, Krystyna Rotkiewicz, Jan Verhoeven, and Jörn Manz, as well as Dr. Jerzy Herbich, for their illuminating discussions.

References

1. (a) Wegewijs, B.; Hermant, R.M.; Verhoeven, J.W.; Kunst, A.G.M.; Warman, J.M. *Chem. Phys. Lett.*, **1990**, *168*, 185; (b) Verhoeven, J.W. *'From Close-Contact to Long-Range Intramolecular Electron Transfer.'* In Jortner J. and Bixon M. eds. *Electron Transfer—From Isolated Molecules to Biomolecules*. Part I. (Series editors: I. Prigogine and S.A. Rice, *Advances in Chemical Physics*, Vol. 106), John Wiley and Sons, Inc.: New York, **1999**; (c) Wegewijs, B.; Verhoeven, J.W. In Jortner, J. and Bixon, M. eds. *Electron Transfer—from Isolated Molecules to Biomolecules*. Part 1 *Advances in Chemical Physics*, Vol. 106 Series Editors Prigogine, I.; Rice S.A.; John Wiley and Sons, Inc.: New York, **1999**.
2. Rettig, W. *Angew. Chem., Int. Ed. Engl.*, **1986**, *25*, 971.
3. Grabowski, Z.R.; Rotkiewicz, K.; Siemiarczuk, A.; Cowley, D.J.; Baumann, W. *Nouv. J. Chim.*, **1979**, *3*, 443.
4. Grabowski, Z.R.; Rotkiewicz K.; Rettig, W. *Chem. Rev.*, submitted.
5. Waldeck, D. (Chapter 3 of this book.)
6. Bartocci, G.; Spaletti, A.; Mazzucato, U. (Chapter 5 of this book.)
7. Olivucci, M.; Robb, M.A.; Bernardi, F. (Chapter 6 of this book.)
8. Lapouyade, R.; Czeschka, K.; Majenz, W.; Rettig, W.; Gilabert, E.; Rullière, C. *J. Phys. Chem.*, **1992**, *96*, 9643.
9. Rettig, W.; Majenz, W.; Lapouyade, R.: Haucke, G. *J. Photochem. Photobiol. A: Chem.*, **1992**, *62*, 415.
10. Rettig, W.; Majenz, W.; Herter, R.; Létard, J.F.; Lapouyade, R. *Pure and Applied Chem.*, **1993**, *65*, 1699.

11. Rettig, W. *Topics in Curent Chem.*, **1994**, *169*, 253.

12. Mackenzie, R.M.; MacNicol, D.C. *J. Chem. Soc., Chem. Commun.*, **1970**, *3*, 1299.

13. Michl, J.; Bonačić-Koutecký, V. *Electronic Aspects of Organic Photochemistry*, Wiley and Sons, Inc: New York, **1990**.

14. Gould, I.R.; Young, R.H.; Mueller, L.J.; Albrecht, A.C.; Farid, S. *J. Am. Chem. Soc.*, **1994**, *116*, 8188.

15. Bixon, M.; Jortner, J.; Verhoeven, J.W. *J. Am. Chem. Soc.*, **1994**, *116*, 7349.

16. Herbich, J.; Kapturkiewicz, A. *J. Am. Chem. Soc.*, **1998**, *120*, 1014.

17. Kato, S.; Amatatsu, Y. *J. Chem. Phys.*, **1990**, *92*, 7241.

18. Gorse, A.D.; Pesquer, M. *J. Phys. Chem.*, **1995**, *99*, 4039.

19. Lippert, E.; Rettig, W.; Bonačić-Koutecký, V.; Heisel, F.; Miehé, J.A. *Adv. Chem. Phys.*, **1987**, *68*, 1.

20. Rettig, W. In *Modern Models of Bonding and Delocalization*, eds. J.E. Liebman, A. Greenberg, VCH Publ.: New York, **1988**, 229.

21. Rettig, W.; Baumann, W. In *Progress in Photochemistry and Photophysics*, Vol. VI (J.F. Rabek, ed.) CRC Press: Boca Raton, **1990**, 79.

22. Rettig, W.; Wermuth, G.; Lippert, E. *Ber. Bunsenges. Phys. Chem.*, **1979**, *83*, 692.

23. Rotkiewicz, K.; Grellmann, K.H.; Grabowski, Z.R. *Chem. Phys. Lett.*, **1973**, *19*, 315; *Erratum*, ibid, **1973**, *21*, 212.

24. Siemiarczuk, A.; Grabowski, Z.R.; Krówczyński, A.; Asher, M.; Ottolenghi, M. *Chem. Phys. Lett.*, **1977**, *51*, 315.

25. Grabowski, Z.R.; Rotkiewicz, K.; Siemiarczuk, A. *J. Lumin.*, **1979**, *18/19*, 420.

26. Visser, R.J.; Varma, C.A.G.O. *J. Chem. Soc., Faraday Trans.II*, **1980**, *76*, 453.

27. Zachariasse, K.A.; von der Haar, T.; Hebecker, A.; Leinhos, U.; Kühnle, W. *Pure and Appl. Chem.*, **1993**, *65*, 1745.

28. Sobolewski, A.L.; Domcke, W. *Chem. Phys. Lett.*, **1996**, *250*, 428; ibid. *259*, 119.

29. Wermuth, G.; Rettig, W.; Lippert, E. *Ber. Bunsenges. Phys. Chem.*, **1981**, *85*, 64.

30. Schuddeboom, W.; Jonker, S.A.; Warman, J.M.; Leinhos, U.; Kühnle, W.; Zachariasse, K.A. *J. Phys. Chem.*, **1992**, *96*, 10809.

31. Lippert, E.; Lüder, W.; Boos, H. In *Advances in Molecular Spectroscopy*, Mangini, A. ed., Pergamon Press: Oxford, **1962**, 443.

32. von der Haar, T.; Hebecker, A.; Il'ichev, Y.; Jiang, Y.B.; Kühnle, W.; Zachariasse, K.A. *Recl. Trav. Chim. Pays-Bas.*, **1995**, *114*, 430.

33. (a) Gorse, A.-D.; Pesquer, M. *J. Mol Struct. (THEOCHEM)*, **1993**, *281*, 21; (b) Serrano-Andrés, L.; Merchán, M.; Roos B.O.; Lindh, R. *J. Am. Chem. Soc.*, **1995**, *117*, 3189.

34. Rullière, C.; Grabowski, Z.R.; Dobkowski, J. *Chem. Phys. Lett.*, **1987**, *137*, 408.

35. Wermuth, G.; Rettig, W. *J. Phys. Chem.*, **1984**, *88*, 2729.

36. Heisel, F.; Miehé, J.A.; Eckert, C.; Rettig, W. *Chem. Phys. Lett.*, **1991**, *187*, 45.

37. Rettig, W. *J. Phys. Chem.*, **1982**, *86*, 1970.

38. L. Salem, *The Molecular Orbital Theory of Conjugated Systems*, W.A. Benjamin, Inc.: Reading, MA, **1966**.

39. Herzberg, G.; Teller, E. *Z. Phys. Chem., Abt. B*, **1933**, *21*, 410.

40. Van der Auweraer, M.; Grabowski, Z.R.; Rettig, W. *J. Phys. Chem.*, **1991**, *95*, 2083.

41. Rettig, W.; Marschner, F. *New J. Chem.*, **1990**, *14*, 819

42. Köhler, G.; Rechthaler, K.; Luboradzki, R.; Suwińska, K.; Rotkiewicz, K. *J. Phys. Chem.*, **1997**, *A 101*, 8518.

43. Rettig, W.; Fritz, R.; Braun, D. *J. Phys. Chem.*, **1997;** *A 101*, 6830.

44. Rettig, W. In Mataga, N.; Okada, T.; Masuhara, H.; eds. *Dynamics and Mechanisms of Photoinduced Electron Transfer and Related Phenomena*, Elsevier Science Publ. B.V.: Austerdam **1992**, 57.

45. Shang, Q.; Bernstein, E.R. *J. Chem. Phys.*, **1992**, *97*, 60.

46. Almennigen, A.; Bastiansen, O.; Fernholt, L.; Cyvin, B.N.; Cyvin, S.J.; Samdal, S. *J. Mol. Struct.*, **1985**, *128*, 59.

47. Khundkar, L.R.; Zewail, A.H. *J. Chem. Phys.*, **1986**, *84*, 1302.

48. Kerr, K.A.; Robertson, J.M. *J. Chem. Soc., B*, **1969**, *7*, 1146.

49. Kress, R.B.; Duesler, E.N.; Etter, M.C.; Paul, I.C.; Curtin, D.Y. *J. Am. Chem. Soc.*, **1980**, *102*, 7709.

50. (a) Murakami, J.; Ito, M.; Kaya, K. *J. Chem. Phys.*, **1981**, *74*, 6505. (b) Takei, Y; Yamaguchi, T.; Osamura, Y.; Fuke, K.; Kaya, K. *J. Phys. Chem.*, **1988**, *92*, 577.

51. Yamasaki, K.; Arita, K.; Kajimoto, O. *Chem. Phys. Lett.*, **1986**, *123*, 24.

52. Monte, C.; Roggan, A.; Subaric-Leitis, A.; Rettig, W.; Zimmermann, P. *J. Chem. Phys.*, **1993**, *98*, 2580.

53. Platt, J.R. *J. Chem. Phys.*, **1949**, *17*, 484.

54. Berlmann, I.B.; Steingraber, O.J. *J. Chem. Phys.*, **1965**, *43*, 2140.

55. Berlmann, I.B. *Fluorescence Spectra of Aromatic Molecules*, Academic Press: New York, **1971**.

56. Hochstrasser, R.M. *Can. J. Chem.*, **1961**, *39*, 459.

57. Shank, C.V.; Ippen, E.P.; Teschke, O.; Eisenthal, K.B. *J. Chem. Phys.*, **1977**, *67*, 5547.

58. Schneider, F.; Lippert, E. *Ber. Bunsenges. Phys. Chem.*, **1968**, *72*, 1155.

59. Nakashima, N.; Murakawa, M.; Mataga, N. *Bull. Chem. Soc. Jpn.*, **1976**, *49*, 854.

60. Rettig, W.; Zander, M. *Ber. Bunsenges. Phys. Chem.*, **1983**, *87*, 1143.

61. Grabner, G.; Rechthaler, K.; Köhler, G. *J. Phys. Chem., A*, **1998**, *102*, 689 and references therein.

62. Dobkowski, J.; Grabowski, Z.R.; Paeplow, B.; Rettig, W.; Koch, K.H.; Müllen, K.; Lapouyade, R. *New J. Chem.*, **1994**, *18*, 525.

63. Fujiwara, T.; Fujimura, Y.; Kajimito, O. *Chem. Phys. Lett.*, **1996**, *26*, 201.

64. Honma, K.; Kajimoto, O. *J. Chem. Phys.*, **1994**, *101*, 1752.

65. David, C.; Baeyens-Volant, D. *Mol. Liq. Cryst.*, **1980**, *59*, 181.

66. Lahmani, F.; Breheret, E.; Zehnacker-Rentien, A.; Amatore, C.; Jutand, A. *J. Photochem. Photobiol. A: Chem.*, **1993**, *70, 39.*

67. Maus, M. *Photoinduced Intramolecular Charge Transfer in Donor–Acceptor Biaryls and Resulting Applicational Aspects*, PhD. dissertation.com, Parkland, **1998**.

68. Maus, M.; Rettig, W.; Bonafoux, D.; Lapouyade, R. *J. Phys. Chem., A*, **1999**, *103*, 3388.

69. Maus, M.; Rettig, W. *Chem. Phys.*, **1997**, *218*, 151.

70. Ephardt, H.; Fromherz, P. *J. Phys. Chem.*, **1991**, *95*, 6792.

71. Fromherz, P.; Heilemann, A. *J. Phys. Chem.*, **1992**, *96*, 6864.

72. Rettig, W.; Kharlanov, V.A.; Maus, M. *Chem. Phys. Lett.*, in press (**2000**).

73. Turro, N.J.; McVey, J.; Ramamurthy, V.; Lechtken, P. *Ang. Chem.*, **1979**, *91*, 597.

74. Birks, J.B. *Photophysics of Aromatic Molecules*, Wiley-Interscience: New York, **1979**.

75. Hirayama, S.; Phillips, J. *J. Photochem.*, **1980**, *12*, 139 and references therein: Chako, N.Q. *J. Chem. Phys.*, **1934**, *2*, 644; Bakshiev, N.G.; Girin, O.P.; Libov, V.S. *Opt. Spectrosc. (U.S.S.R)* **1963**, *14*, 255395.

76. Cannon, R.D. *Electron Transfer Reactions*, Section 8.3, Butterworths: Boston, **1980**.

77. Dogonadze, R.R.; Kutznetsov, A.M.; Marsagishvili, T.A. *Electrochim. Acta*, **1980**, *25*, 1.

78. Suzuki, H. *Electronic Absorption Spectra and Geometry of Organic Molecules*, Academic Press: New York, **1967**.

79. Beens, H.; Weller, A. In *Organic Molecular Photophysics*, Vol. 2, Chapter 4, Birks, J.B., ed.; Wiley: New York, **1975**.

80. Herbich, J.; Kapturkiewicz, A. *Chem. Phys. Lett.*, **1997**, *273*, 8.

81. Verhoeven, J.W.; Scherer, T.; Wegewijs, B.; Hermant, R.M.; Jortner, J.; Bixon, M.; Depaemelare, S.; De Schryver, F.C. *Recl. Trav. Chim. Pays-Bas*, **1995**, *114*, 443.

82. Wasielewski, M.R.; Niemczyk, M.P. In *Porphyrins Excited States and Dynamics*, Gouterman, M.; Rentzepis, P.M.; Straub, K.D., eds., American Chemical Society: Washington, DC, **1986**, 154.

83. Wasielewski, M.R.; Johnson, D.G.; Svec, W.A. In *Supramolecular Photochemistry*, Balzani, V., ed., D. Reidel Publishing Company: Dordrecht, **1987**, 255.

84. Maus, M.; Rettig, W.; Jonusauskas, G.; Lapouyade, R.; Rullière, C. *J. Phys. Chem.*, **1998**, *102*, 7393.

85. Mataga, N.; Kaifu, Y.; Koizumi, M. *Bull. Chem. Soc. Jpn.*, **1956**, *29*, 465.

86. (a) Maus, M.; Rettig *J. Inf. Rec.* **1998**, *22*, 451. (b) Maus, M.; Rettig, W.; Lapouyade, R. *J. Inf. Rec.*, **1996**, *22*, 451.

87. Rettig, W.; Maus, M.; Lapouyade, R. *Ber. Bunsenges. Phys. Chem.* **1996**, *100*, 2091.

88. Shida, T. *Electronic absorption spectra of radical ions*, Elsevier Science Publishers B.V.: Amsterdam, **1988**.

89. (a) Shida, T.; Iwata, S. *J. Am. Chem. Soc.*, **1973**, *95*, 3473. (b) Yoshihara, K.; Yartsev, A.; Nagasawa, Y.; Kandori, H.; Douhal, A.; Kemnitz, K. *Ultrafast Phenomena VIII*, Springer Series in Chemical Physics, Vol. 55, Martin, J.-L.; Migus, A.; Zewail, A.H. eds., Springer-Verlag: Berlin, Heidelberg, **1993**, 571.

90. (a) Shida, T.; Haselbach, E.; Bally, T. *Acc. Chem. Res.*, **1984**, *17*, 180. (b) Chutny, B.; Swallow, A. *Trans. Faraday Soc.*, **1970**, *66*, 2847.

91. Yoon, M.; Cho, D.W.; Lee, J.Y. *Bull. Korean Chem. Soc.*, **1992**, *13*, 613.

92. Cho, D.W.; Kim, Y.H.; Kang, S.G.; Yoon, M. *J. Phys. Chem.*, **1994**, *98*, 558.

93. Cho, D.W.; Kim, Y.H.; Kang, S.G.; Yoon, M.; Kim, D. *J. Chem. Far. Trans.*, **1996**, *92*, 29.

94. Dähne, S. *Science (Washington)*, **1978**, *199*, 1163.

95. Gorman, C.B.; Marder, S.R. *Proc. Natl. Acad. Sci. U.S.A.*, **1993**, *90*, 11297.

96. Jurczok, M.; Plaza, P.; Martin, M.M.; Rettig, W. *J. Phys. Chem.: A*, **1999**, *103*,

97. Okada, T.; Fjita, T.; Kubota, M.; Masaki, S.; Mataga, N.; Ide, R.; Sakata, Y.; Misumi, S. *Chem. Phys. Lett.*, **1972**, *14*, 563.

98. Okada, T.; Fujita, T.; Mataga, N. *Z. Phys. Chem.*, **1976**, *54*, 3304.

99. Okada, T. *Proc. Indian Acad. Sci. (Chem. Sci.)*, **1992**, *104*, 173.

100. Okada, T.; Kawai, M.; Ikemachi, T.; Mataga, N.; Sakata, Y.; Misumi, S.; Shionoya, S. *J. Phys. Chem.*, **1984**, *88*, 1976.

101. Okada, T.; Mataga, N.; Baumann, W.; Siemarczuk, A. *J. Phys. Chem.*, **1987**, *91*, 4490.

102. Siemarczuk, A.; Ware, W.R. *J. Phys. Chem.*, **1987**, *91*, 3677.

103. Tominaga, K.; Walker, G.C.; Jarzeba, W.; Barbara, P.F. *J. Phys. Chem.*, **1991**, *95*, 10475.

104. Tominaga, K.; Walker, G.C.; Kang, T.J.; Barbara, P.F.; Fonseca, T. *J. Phys. Chem.*, **1991**, *95*, 10485.

105. Herbich, J.; Kapturkiewicz, A. *Chem. Phys.*, **1993**, *170*, 221.

106. Wiessner, A.; Hüttmann, G.; Kühnle, W.; Staerk, H. *J. Phys. Chem.*, **1995**, *99*, 14923.

107. Herbich, J.; Kapturkiewicz, A. *Chem. Phys.*, **1991**, *158*, 143.

108. Marcus, R.A. *J. Phys. Chem.*, **1989**, *93*, 3078.

109. Horng, M.L.; Gardecki, J.A.; Papazyan, A.; Maroncelli, M. *J. Phys. Chem.*, **1995**, *99*, 17311.

110. Maus, M.; Rettig, W.; Depaemelaere, S.; Onkelinx, A.; De Schryver, F.C.; Iwai, K. *Chem. Phys. Lett.*, **1998**, *292*, 115.

111. Onkelinx, A.; De Schryver, F.C.; Viane, L.; Van der Auweraer, M.; Iwai, K.; Yamamoto, M.; Ichikawa, M.; Masuhara, H.; Maus, M.; Rettig, W. *J. Am. Chem. Soc.*, **1996**, *118*, 2892.

112. (a) Onsager, L. *J. Am. Chem. Soc.*, **1936**, *58*, 1486. (b) Böttcher, C.J.F. *Theory of Electronic Polarisation*, Elsevier: Amsterdam, **1973**.

113. Schneider, F.; Lippert, E. *Ber. Bunsenges. Phys. Chem.*, **1968**, *72*, 1155; ibid., **1970**, *74*, 624.

114. Nakashima, N.; Murakawa, M.; Mataga, N. *Bull. Chem. Soc. Jpn.*, **1976**, *49*, 854.

115. Mataga, N.; Yao, H.; Okada, T.; Rettig, W. *J. Phys. Chem.*, **1989**, *93*, 3383.

116. Nagarajan, V.; Brearley, A.M.; Kang, T.J.; Barbara, P.F. *J. Chem. Phys.*, **1987**, *86*, 3183.

117. Barbara, P.F.; Kang, T.J.; Jarzeba, W.; Fonseca, T. In *Perspectives in Photosynthesis*, Jortner J. and Pullman, B. eds., Kluwer Acad. Publ: New York, **1990**, 273; Barbara, P.F.; Jarzeba, W. *Adv. Photochem.*, **1990**, *15*, 1.

118. Kang, T.J.; Jarzeba, W.; Barbara, P.F.; Fonseca, T. *Chem. Phys.*, **1990**, *149*, 81.

119. Rettig, W.; Zander, M. *Chem. Phys. Lett.*, **1982**, *87*, 229.

120. Kapturkiewicz, A.; Herbich, J.; Karpiuk, J.; Nowacki, J. *J. Phys. Chem.: A*, **1997**, *101*, 2332.

121. Subaric-Leitis, A.; Monte, C.; Roggan, A.; Rettig, W.; Zimmermann, P.; Heinze, J. *J. Chem. Phys.*, **1990**, *93*, 4543.

122. Evers, F.; Giraud-Girard, J.; Grimme, S. Manz, J.; Monte, C.; Oppelt, M.; Rettig, W.; Saalfrank, P.; Zimmermann, P. *J. Chem. Phys.*, in preparation.

123. Manz, J.; Proppe, B.; Schmidt, B. *Z. Phys. D*, **1995**, *34*, 111.

124. Giraud-Girard, J.; Manz, J.; Scheurer, C. *Z. Phys. D*, **1997**, *39*, 291.

125. Zilberg, S.; Haas, Y.; *Chemistry—A European Journal*, **1999**, *5*, 1755.

126. Grimme, S. *Chem. Phys. Lett.*, **1996**, *259*, 128.

127. Cornelissen-Gude, C.; Rettig, W. *J. Phys. Chem.: A*, **1998**, *102*, 7754.

128. Rettig, W.; Marschner, F. *Nouv. J. de Chimie*, **1983**, *7*, 425.

129. Kharlanov, V.A.; Knyazhansky, M.I.; Rettig, W. *J. Mol. Structure*, **1996**, *380*, 113.

130. Kharlanov, V.A.; Rettig, W.; Knyazhansky, M.I.; Makarova, N. *J. Photochem. Photobiol. A: Chem.*, **1997**, *103*, 45.

CHAPTER

2

Conformational Aspects of Intra- and Intermolecular Excited-State Proton Transfer

by Jacek Waluk

Institute of Physical Chemistry, Polish Academy of Sciences

1 Introduction

Nearly half a century ago, Albert Weller observed that photoexcited 1- and 2-naphthols emit dual fluorescence, even though only one form is present in the ground state.[1] The higher energy emission was assigned to the initially excited neutral species and the lower energy fluorescence to the monoanion, created in the lowest excited singlet state as a result of deprotonation (Fig. 2.1). This finding provided the experimental proof of an earlier prediction by Theodor Förster[2] that acid–base properties of photoexcited molecules can be quite different from those in the ground state, and thus, proton transfer can occur upon excitation. Förster proposed a procedure for the determination of the excited state pK_a values,[3] also valid in cases when the equilibrium is not established. The method is based on the experimental determination of $\tilde{\nu}_{00}^A$ and $\tilde{\nu}_{00}^B$, the energies of the electronic transition between the ground state and the excited state under consideration, in the acidic and basic form, respectively (Fig. 2.2). Assuming that the entropy changes are the same in both states, the shift of chemical equilibrium upon excitation can be described by the formula

$$\Delta pK_a = pK_a^* - pK_a^G \approx -0.0021(\tilde{\nu}_{00}^A - \tilde{\nu}_{00}^B),$$

where pK_a^* and pK_a^G denote the values in the excited and ground state, respectively, the transition energies are expressed in wave numbers.

Conformational Analysis of Molecules in Excited States, Edited by Jacek Waluk
ISBN 0-471-29707-0

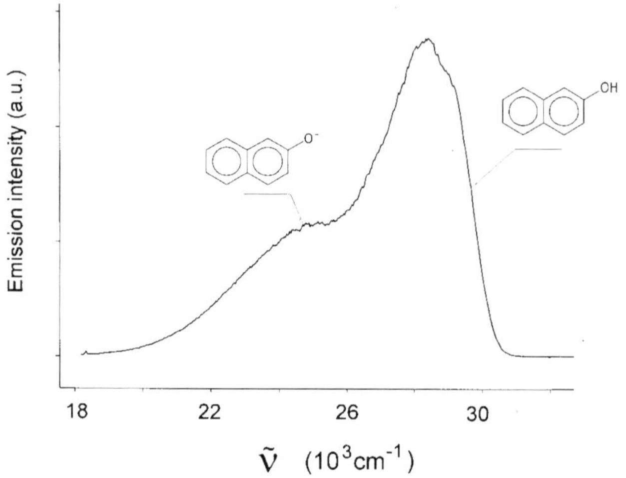

Figure 2.1. Fluorescence of 2-naphthol in water (293 K, pH = 4.62), revealing bands corresponding to the initially excited neutral form and to the anion produced in the excited state.

Using the "Förster cycle," numerous authors have reported dramatic changes of pK_a values of molecules in electronically excited states.[4] The source of this effect is a large electronic density redistribution occurring after excitation. Many azaaromatic systems become strong bases in the excited state, revealing an increase in pK_a, which often exceeds even 10 units. For instance, for the first protonation of 1,4-diazatriphenylene, $pK_a = -0.36$ and 10.1 in the ground and lowest excited singlet state, respectively.[5] Thus, $\Delta pK_a = 10.5$ after passing from the ground state (S_0) into the lowest excited singlet state (S_1). A large increase in basicity in 1,4-diazatriphenylene is also observed in the lowest triplet state: $\Delta pK_a = 6.7$. Hence, a driving force for proton attachment is created upon photoexcitation.

The opposite trend—an increase in the ability to detach a proton—occurs in molecules showing negative values of ΔpK_a. Such behavior is typical, for instance, of hydroxy derivatives of many aromatic hydrocarbons, which become much more

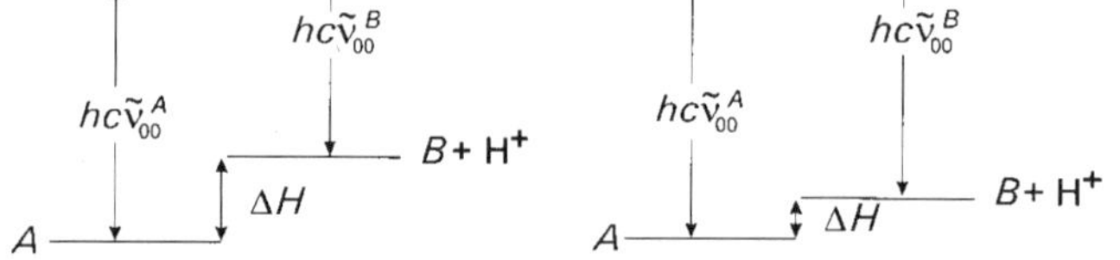

Figure 2.2. Principle of the Förster cycle for the reaction $A \rightarrow B + H^+$. The left part corresponds to the enhancement of acidity in the excited state, the right part to the enhancement of basicity.

acidic when excited. For instance, a large decrease in pK_a in the lowest excited singlet state, about 7 units, is observed for both 1- and 2-naphthols.[6]

Chemistry textbooks often depict proton transfer as not only the simplest, but also a ubiquitous, reaction. While ubiquitous it certainly is, that it is the simplest reaction might be debated. First, proton motion can be very fast, occurring on a timescale comparable to the period of vibration. On the other hand, the time necessary for the encounter between reaction partners, a proton donor and an acceptor, can be quite long, and is dictated, for example, by the diffusion coefficients. Opposed to the latter situation is the case when the reaction occurs along the preformed hydrogen bond. Finally, the small mass of hydrogen compared to all other atoms is often a source of a significant role played by tunneling in the proton transfer mechanism. The involvement of all these factors makes it sometimes quite difficult to define a proper coordinate of the proton transfer reaction and to find the rate-determining step, which is, of course, necessary for understanding the kinetics of the reaction. In many systems, the reaction can be controlled by conformational characteristics. Such a situation is encountered, for instance, in molecules that can exist in two or more rotameric forms, of which only one is capable of forming an intramolecular hydrogen bond—a prerequisite for proton transfer. In this case, it is the interconversion between rotamers, not the proton movement itself, that may turn out to be the rate-limiting factor for the proton transfer process.

For intermolecular excited-state proton transfer reactions, the situation is even more complex. Since the reaction partners must encounter each other prior to the proton transfer step, many processes that are thermodynamically possible do not actually occur. This is due to kinetic limitations, as the diffusion time is longer than the excited-state lifetime. Therefore, excited-state intermolecular proton transfer is usually observed for systems that are linked already in the ground state (e.g., by hydrogen bonding). However, even in such cases, the proton transfer step must frequently be preceded by the achievement of a proper structure, which requires mutual reorientation of the partners. This conformational relaxation often controls the reaction rate.

Acid–base properties of molecules in electronically excited states, as well as multiple examples of intramolecular and intermolecular photoinduced proton transfer reactions, have been the subjects of numerous papers and reviews.[6–24] In this chapter, we will focus on those aspects of photoinduced proton transfer reactions that relate to conformational problems. The idea is to expose to the reader the current state of knowledge in the field by presenting results of investigations of several most representative and most intensely studied systems. We start in Section 2 by reviewing ample work that has been done on 7-azaindole dimers and alcohol complexes, long known to undergo photoinduced double proton transfer. The studies of 7-azaindole have touched upon virtually every problem related to the tautomerization process: proper definition of the reaction path, the characterization of reactive and nonreactive forms, the role of the solvent, the determination of reaction rates, isotope effects, the contribution of tunneling, the consideration of synchronous and stepwise mechanisms, and more. Closely related to the problems discussed for 7-azaindole are the investigations of solvent-assisted photoinduced intermolecular proton transfer processes in 1-azacarbazole, 2-(2′-pyridyl)indoles, and 1H-pyrrolo[3,2-*h*]quinoline and derivatives, presented in Section 2.4. A special case of solvent-assisted reactions involves tautomerization that occurs via proton relay mechanism, which will be described for 7-hydroxyquinoline (Section 3).

Section 4 is devoted to conformational aspects of intramolecular excited-state proton transfer that are related to rotational isomerism. A separate part (Section 5) will discuss proton transfer in photochromic and thermochromic substances. It will be shown that even in the case of a formally intramolecular reaction, the role of the environment can be crucial. Therefore, Section 6 will address the issues concerning specific and nonspecific effects related to solvent control of the proton transfer processes.

Since we are not going to describe separately the experimental techniques used in the investigation of excited-state proton transfer, some general remarks seem pertinent here. The most important tools for studying photochemical reactions and proton transfer in particular are provided by luminescence measurements. Emission studies can directly prove the occurrence of the reaction. Such measurements can be complemented by transient absorption methods, which are extremely advantageous if one or both reaction partners are not luminescent. Moreover, using time-resolved techniques, one is able to extract the rates of the proton transfer processes from the analysis of temporal profiles of luminescence or transient absorption. The enormous development of femtosecond laser spectroscopy, coupled with supersonic beam techniques and mass spectrometry, made it possible to fulfill a chemists dream: to follow the proton transfer reaction in real time.

2 The Never-Ending Story of 7-Azaindole and Its Relatives

2.1 Excited-State Double Proton Transfer (ESPDT) in 7-Azaindole Dimers

7-Azaindole (7AI, **1**) forms dimers in nonpolar solvents. The association constant, $1.8 \times 10^3\ M^{-1}$ in 3-methylpentane at 293 K, ensures that 85% of the molecules are dimerized at the concentration of 10^{-2} M.[25] The large negative enthalpy of the process, (−9.6 kcal/mol) means that at low temperatures, practically no monomers are present even at much lower concentrations. Fluorescence spectra of 7AI, taken at high concentration (10^{-2} M) at room temperature and at low concentration (10^{-5} M) at 77 K and below, reveal the presence of two emissions: Apart from a "normal" violet fluorescence (F_1), not strongly shifted from the absorption, another (green) fluorescence (F_2) is observed (Fig. 2.3). The F_2 band was assigned to arise from a product of the biprotonic phototautomerization occurring in the excited singlet state of cyclic, doubly hydrogen-bonded dimers[26] (Fig. 2.4). Other possibilities for the assignment of the F_2 emission, such as excimer formation, excited-state single-proton transfer, and the interchange of the two lowest excited singlet states, the ^1L_a and the ^1L_b, were excluded. The strongest and most direct argument for the assignment of the F_2 band to a phototautomer was provided by spectral measure-

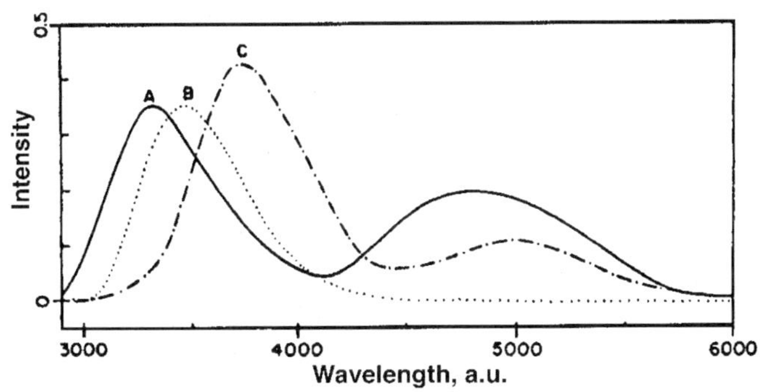

Figure 2.3. Room-temperature fluorescence of 7AI in 3-methylpentane (A), ethyl ether (B), and ethanol (C). The ratio of the two bands is strongly dependent on the concentration in 3-methylpentane, whereas the shape of curves (B) and (C) does not depend on concentration. (Reprinted from Ref. 30. with permission from Academic Press.)

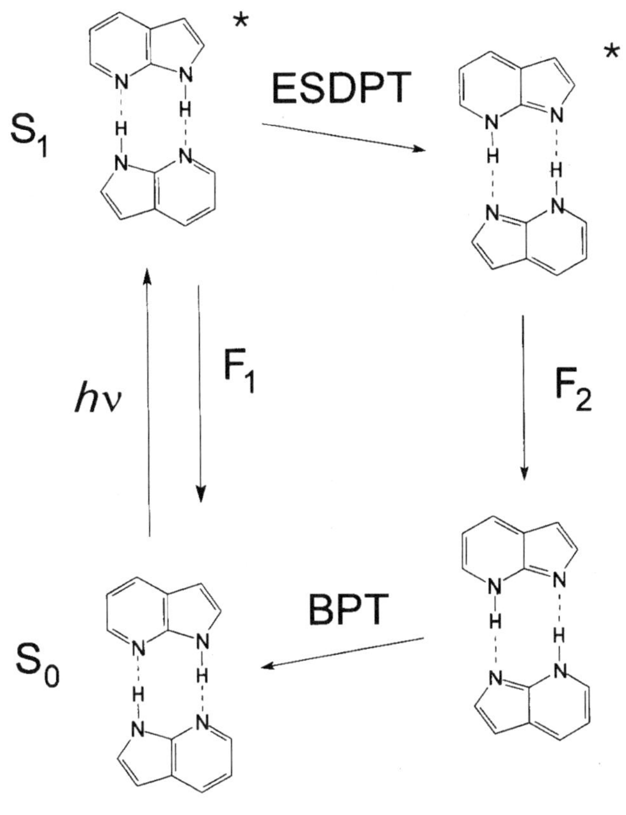

Figure 2.4. A scheme of the ESDPT reaction in 7AI and the back proton transfer (BPT) in the ground state.

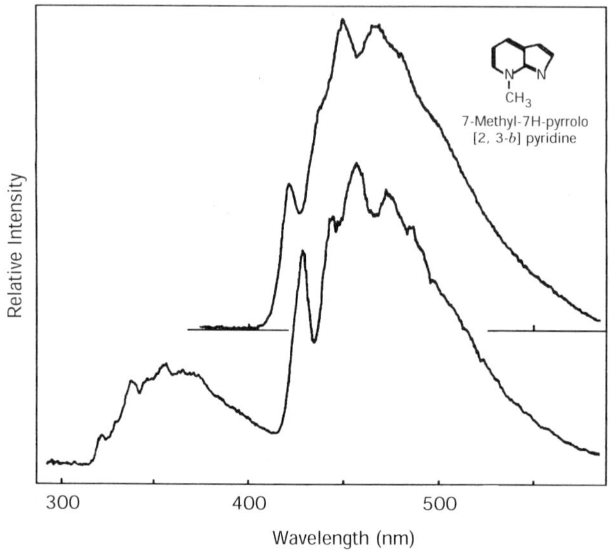

Figure 2.5. Comparison of the F_2 emission of 7AI (bottom) with the fluorescence of 7-methyl-7H-pyrrolo[2,3-*b*]pyridine, a chemical model for the 7AI tautomer (top). Both spectra were recorded in 3-methylpentane at 77 K. (Reprinted with permission from Ref. 27. Copyright, 1971 American Chemical Society.)

ments performed for 7-methyl-7H-pyrrolo[2,3-b]pyridine, a chemical model of the 7AI tautomer[27] (Fig. 2.5). The fluorescence of this compound turned out to be similar to the F_2 band observed for 7AI.

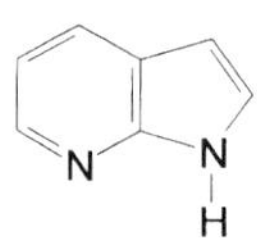

1

Since 7AI dimers may be considered models of DNA base pairs, the finding of photoinduced tautomerization raised much interest. A mechanism of mutagenesis involving double proton transfer in DNA and the possibility of light-induced mutation through phototautomerization was discussed by various authors.[28–31]

In order to quantitatively characterize the double proton phototautomerization, relative intensities of both fluorescence bands were studied as a function of temperature. The value of the F_2/F_1 intensity ratio was found to increase upon cooling, to pass through a maximum near 200 K, and then to level off to a nonzero value, obtained even at liquid helium temperatures.[25] The proposed mechanism of the reaction assumed a double minimum on the excited-singlet state potential-energy surface, with a lower limit of 1.4 kcal/mol for the tautomerization barrier. From the fact that the F_2 emission persisted even at 4 K, it was inferred that proton tunneling was the dominant mechanism of the reaction at low temperatures. Consistent with this hypothesis were the changes in the F_2/F_1 intensity ratio observed in 7AI samples in which the hydrogen of the NH group was replaced by deuterium.

Time-resolved room-temperature fluorescence studies revealed a very rapid rise time of the F_2 band, faster than 5 ps.[32] (In a later work, this rise time was determined more exactly to be 1.4 ps.[33]) At 77 K, on the other hand, two channels of tautomerization were detected. In addition to a fast component of less than 5 ps, a relatively slow rise time of the tautomer fluorescence, about 1 ns, was observed.[32] The relative contributions of the two components were found to be dependent on the excitation energy. The authors interpreted their results as a manifestation of two pathways for proton transfer: (i) thermally activated crossing of the barrier and (ii) direct tautomerization from the initially prepared, nonrelaxed state. While both processes were postulated to be rapid at room temperature, the former was considered to be much slower at 77 K.

The preceding results and conclusions were obtained by using a simple one-dimensional model of the potential-energy surface governing the proton movement. In the studies of the F_2/F_1 intensity ratio, as well as in the kinetic fluorescence measurements, it was tacitly assumed that the F_1 band is homogeneous and corresponds only to the primary fluorescence from the planar, doubly hydrogen-bonded cyclic 7AI dimers. Similarly, the F_2 band was treated as corresponding to the phototautomer only. However, subsequent findings showed that the situation is more complicated. Studies of the triplet state of 7AI using a combination of ESR and luminescence techniques demonstrated that at low temperatures a large fraction of the F_2 band is due to phosphorescence.[34] Detailed investigations of the F_2/F_1 ratio as a function of concentration and temperature also revealed that the F_1 emission is not homogeneous and contains contributions from both dimers and higher oligomers. A possibility of nonplanar dimeric species was also suggested.[35] Tautomerization in such forms would require energy in order to attain a planar conformation. This would lead to slowing down the process at low temperatures. Such nonplanar dimers could be assigned to structures responsible for the slow component observed in the rise time of the tautomeric emission.

A possibility of higher oligomers of 7AI in solutions was also suggested on the basis of the NMR investigation.[36] The determination of the crystal structure of 7AI strongly supported this possibility. It was found that in the crystalline phase 7AI forms basketlike tetramers, with each of the four molecules engaged in two hydrogen bondings with two different partners (Fig. 2.6).[37] Therefore, no possibility of double proton transfer within a dimeric unit exists. These results explain why only "normal" fluorescence was observed for oligomeric structures and why no tautomeric emission could be detected in the 7AI crystals.

In view of the diversity of possible structures in the condensed phase, an investigation of isolated 7AI dimers seemed highly desirable. The development of supersonic beam techniques made such studies possible. The combination of laser spectroscopy in supersonic jets and mass selection resulted in the observation of two conformational isomers of the 7AI dimer, of which only one was able to undergo tautomerization in the cold molecular beam.[38] Studies of this reactive species by femtosecond techniques, using time-resolved mass spectrometry, brought new,

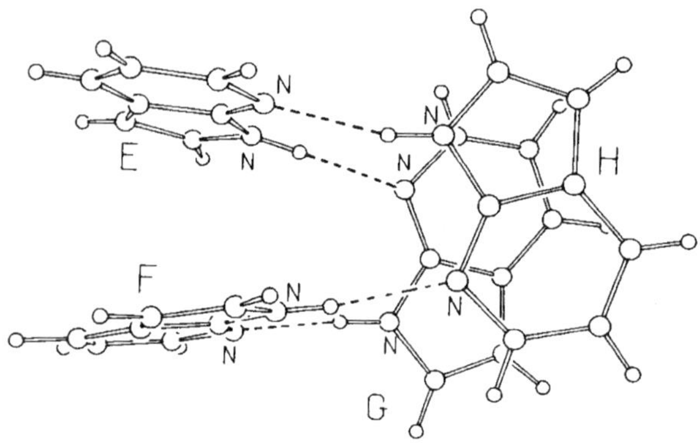

Figure 2.6. Crystal structure of 7AI. (Reprinted from Ref. 37 with permission from National Research Council of Canada.)

fascinating results concerning the reaction mechanism. It was found that the phototautomerization occurs in a stepwise fashion. The transfer of the first proton takes several hundred femtoseconds, while the second proton shift requires a somewhat longer time, several picoseconds.[39] The exact values depend on the initially excited vibronic level. Proton transfer is observed even for excitation near the zero-point energy. A large isotope effect was observed, pointing to tunneling. Thus, a double-minimum potential initially proposed as a model of the potential-energy surface for the excited-state tautomerization had to be replaced by a triple-minimum curve (Fig. 2.7). In addition to the minima corresponding to the initially excited dimer and to the product of the transfer of two protons, an intermediate structure involving a transfer of only one proton has to be included as a separate minimum. Calculations indicate that the excitation is localized on one-half of the dimer and that the first proton that moves is the one located on the excited moiety.[40]

These findings were confirmed by a beautiful experiment that showed an ingenious procedure for arresting a single proton transfer intermediate state on a femtosecond timescale.[41] The technique used the phenomenon of Coulomb explosion. Following laser excitation, a second high-power laser pulse was applied, causing multiple ionization resulting in the dimer "explosion" (i.e., breakup into monomeric units that were subsequently detected by a mass spectrometer). With a suitable choice of time delay between the two laser pulses (about 600 fs), the signal due to the monomeric unit with one extra proton attached was dominant (Fig. 2.8). Its contribution was decreased at larger delays between the pump and probe pulses, when the species with a mass equal to one-half that of the dimer took over. The timescales for the first and second proton transfer steps, 650 fs and 5 ps, respectively, were in excellent agreement with previous findings.

It may be expected that knowledge of the reaction mechanism and kinetics in an isolated dimeric 7AI can be used to reach a deeper understanding of the nature of tautomerization in the condensed phases. A recent work performed in 3-methylpentane and hexadecane, two nonpolar solvents of different viscosity, again revealed two

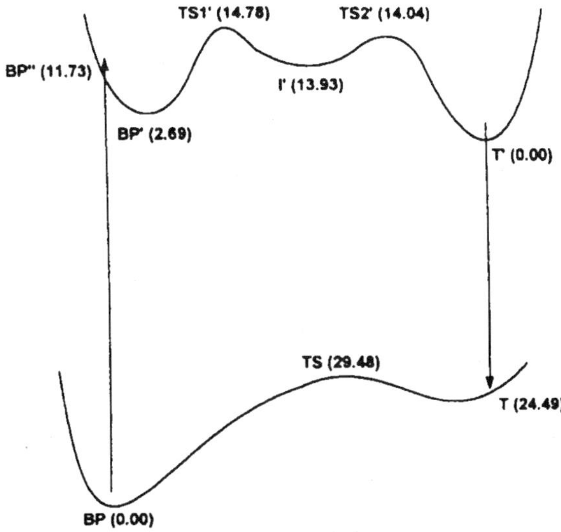

Figure 2.7. Potential-energy curves for the ESDPT and BPT processes proposed on the basis of femtosecond experiments and calculations. In parentheses are the relative energies (kcal/mol) calculated with respect to S_0 and S_1 lowest energy minima. Note that the calculations incorrectly predict that the transfer of the first proton involves a larger barrier than the transfer of the second proton. (Reprinted from Ref. 40 with permission from Elsevier Science.)

channels of tautomer formation: a faster one, of about 1 ps, and a slower one, of about 12 ps.[42] The authors proposed a model involving direct and indirect reaction pathways for the double proton transfer on a global potential-energy surface. However, in another work,[43] the slow rise time was attributed to vibrational cooling, not to proton transfer. It was also claimed that in solution, a transfer of both protons may occur simultaneously, not in a stepwise fashion. On the contrary, recent studies

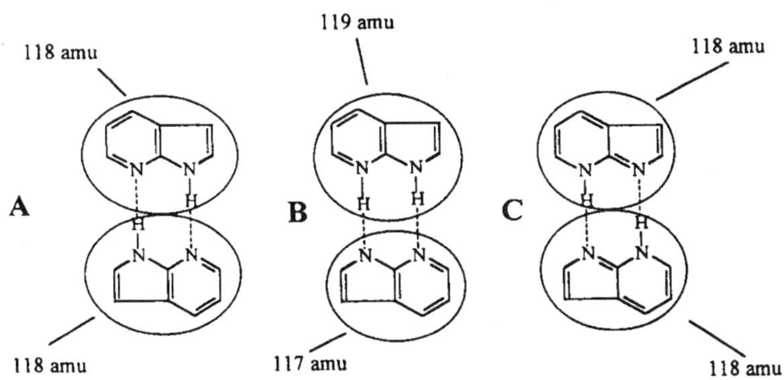

Figure 2.8. Products of "Coulomb explosion" obtained at different stages of the stepwise ESDPT in 7AI dimers. (Reprinted from Reference 41 with permission from Elsevier Science.)

in solution[45] have led to the conclusion that a stepwise mechanism is also possible for low internal energy of the excited dimer. Evidently, many questions concerning the reactivity of 7AI dimers have yet to be answered. Further work on 7AI is being carried out in many laboratories.[43–47] The central issues relate to conformational problems having to do with the nature of the species that reacts slowly in solution and the characterization of the dimeric structure of the species which does not form tautomers in supersonic jets. It remains to be seen whether both forms turn out to be the same species.

2.2 7-Azaindole and 1-Azacarbazole: Similarities and Differences

Parallel to the work on 7-azaindole, several research groups have investigated a closely related compound: 1-azacarbazole (1AC, **2**). As in 7AI, dual fluorescence was detected in solutions containing dimers of 1AC.[48–49] Again, the lower energy band was assigned to the product of photoinduced two-proton tautomerization. Similar behavior was also observed for the 1:1 "heterodimers" formed between 7AI and 1AC.[50–51] In all these dimers, two conformational isomers were detected for both solutions[49] and supersonic beams.[38] Only one of the forms was able to undergo phototautomerization. Selective excitation of the other conformer, feasible in supersonic jets, produced the UV fluorescence exclusively, analogous to the F_1 band of 7AI. Studies of 1AC in solutions revealed only the F_2 band at 293 K and only the F_1 emission at low temperatures. In the intermediate temperature range, where both bands were observed, it was found that the lifetime of F_2 is shorter than the decay of F_1. This finding excluded the latter as a kinetic precursor of the tautomer emission. Interestingly, the range of F_1 and F_2 coexistence turned out to be strongly dependent on solvent viscosity, shifting to much higher temperatures in more viscous solvents. These findings were interpreted as evidence of two conformers: a planar structure prone to tautomerization and a nonplanar one in which the process is inhibited. Strong support of this hypothesis came after the X-ray structure showed that 1AC, contrary to 7AI, forms quasi-planar doubly hydrogen-bonded dimers in the crystalline phase[52] (Fig. 2.9). A luminescence study of crystalline 1AC revealed the tautomeric emission even at temperatures below 2 K.[53] Thus, once the right conformation for proton transfer was achieved, the photoinduced reaction was very efficient.

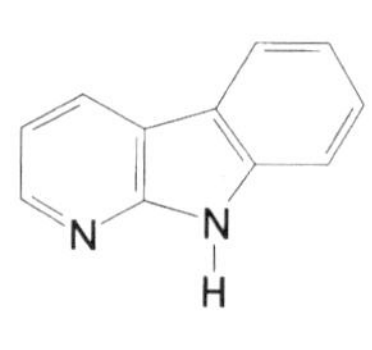

2

2.3 Excited-State Double Proton Transfer in Alcohol and Water Complexes of 7-Azaindole

The phenomenon of dual fluorescence of 7AI is not limited to its dimeric form: It was also observed in alcohol solutions[26] (Fig. 2.3). Contrary to the case of the

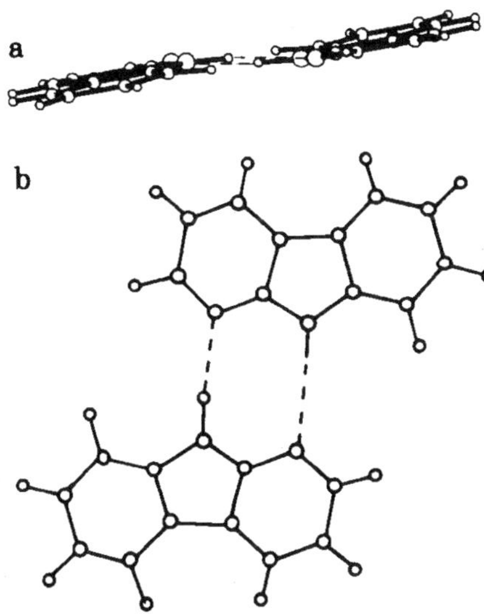

Figure 2.9. Crystal structure of 1AC. (Reprinted with permission from Reference 53. Copyright 1986 American Chemical Society.)

dimers, the intensity ratio of the F_1 and F_2 bands is independent of concentration and excitation wavelength.[54] Moreover, the F_2 emission disappears at low temperatures, which indicates that the reaction involves a barrier.

There is a general agreement that the double proton transfer reaction occurs in a cyclic complex formed by one molecule of the solvent with one 7AI molecule. Thus, the tautomerization must be preceded by the formation of such a cyclic structure if it does not correspond to the most stable form. Proton transfer would thus be a two-step process (Fig. 2.10), of which the first step involves solvent relaxation, but no proton movement coordinate whatsoever. A lot of controversies remain concerning the mechanism of formation of such species, the rate-determining step of tautomerization, and the factors that have most influence on this process.[55]

The kinetics of the excited-state reaction was studied by following the decay of the F_1 band and the rise time of the tautomeric emission, F_2. Various models of tautomer formation, sometimes contradicting each other, have been proposed. The first[56] considered two channels of tautomerization, due to the presence of two different ground-state conformations of alcohol complexes. A fast channel involved cyclic structures, giving rise to instantaneous proton transfer upon excitation. The slow channel was due to noncyclic complexes, in which solvent reorientation is required in order to form a cyclic hydrogen bond. The possibility of the slow channel was contested in a paper which claimed that there is no evidence for the excited-state formation of cyclic complexes that would then undergo tautomerization.[57] In yet another work, only the slow channel was observed.[58] The single exponential decay time of the primary fluorescence, 194 ± 5 ps in ethanol, was the same as the rise time of the tautomeric emission (Fig. 2.11). The time profile of F_2 was satisfactorily described by a combination of a single exponential growth and a single exponential decay. It was concluded that only one type of species was producing the F_1 emission,

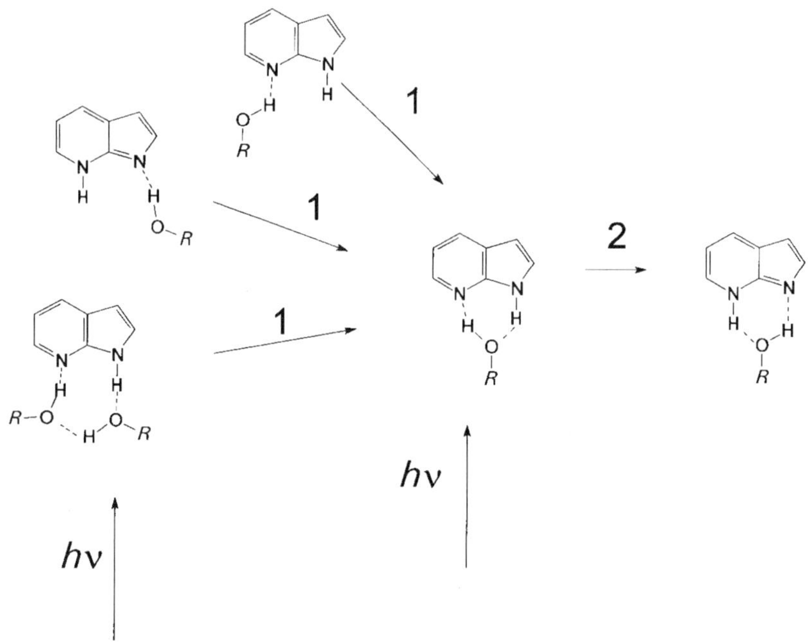

Figure 2.10. Solvent reorientation (1) and double proton transfer (2) in excited 7AI : alcohol complexes.

being the precursor of the tautomeric structure. These results were independently confirmed.[59]

It has been observed that the F_1/F_2 intensity ratio varies in different alcohols and that the activation energy for the tautomer formation can be correlated with the activation energy of viscous flow of the alcoholic solvent.[60] In another paper, alcohol acidity, not viscosity, was considered as the main factor controlling the rate of the process.[58] It was also observed that the reaction rate shows a logarithmic dependence on the solvent polarity, expressed by the $E_T(30)$ values.[59,61]

It now seems generally accepted that the proper form for proton transfer in alcohol is a 1:1 cyclic complex with an alcohol solvent molecule and that in bulk alcohols this form is very rare. The latter observation was recently confirmed by molecular dynamics and Monte Carlo calculations.[62] Therefore, a two-step mechanism seems justified. Still, there is no agreement as to the rate-controlling step. Some authors use the expression $k = k_{PT} \exp(-\Delta G/RT)$, where k_{PT} is the (fast) rate of proton transfer in the system that has acquired the proper conformation for the reaction and ΔG is the free energy of solvent rearrangement required to achieve such a form. Given that in properly arranged dimers of 7AI the proton transfer is very fast, it would seem reasonable to assume that this may also be the case for alcohol complexes, and thus, the rate-limiting step involves the realization of the correct solvation by the alcohol. However, based on the isotopic substitution experiments, it was claimed that the proton transfer itself might be the rate-determining step.[63–64]

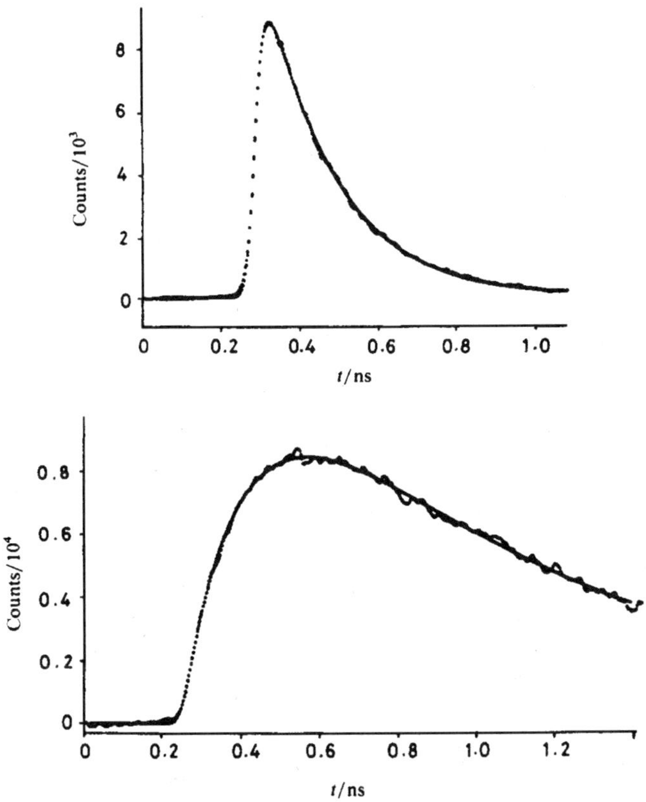

Figure 2.11. Kinetic profiles of the intensity of fluorescence of 7AI in ethanol. Top: intensity measured at 370 nm ("normal" band). Bottom: Intensity measured at 500 nm (tautomeric emission). (Reprinted from Reference 58 with permission from the Royal Society of Chemistry.)

The situation in water solutions is even less clear. Contrary to the observation of dual emission in dimers and alcohol solutions, stationary fluorescence measurements of 7AI in water reveal the presence of only one band, assigned to the "normal" form.[54] It was, however, possible to induce a second, tautomeric emission by saturating ethyl ether or *p*-dioxane solutions with water[65] (Fig. 2.12). The authors concluded that the 7AI monohydrate that is present under such conditions can undergo excited-state proton transfer, but the reaction is not possible for polyhydrates, which are prevalent in pure water.

Time-resolved studies, on the other hand, reveal double exponential decays and growths even in pure water. The interpretation of this phenomenon varies between two extremes. Maroncelli and co-workers claim that 7AI in water undergoes the same solvent-catalyzed tautomerization as in alcohols. The process is slow, about 900 ps, compared to that in alcohols. Similarly, a low rate of reaction is postulated for polyhydroxy solvents such as ethylene glycol and glycerol.[61] On the contrary, the works from the group of Petrich postulate that in water, only a small fraction, less

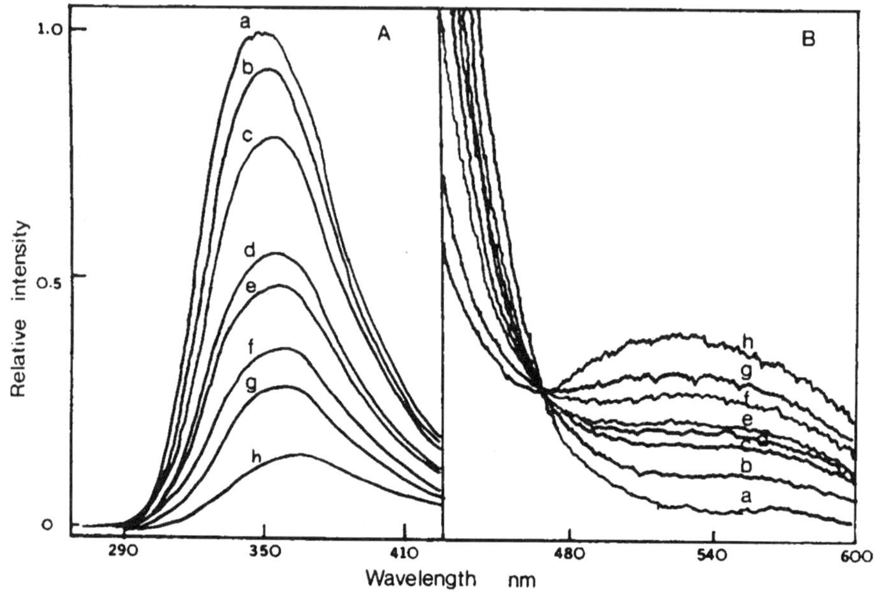

Figure 2.12. Room-temperature fluorescence of 7AI in ethyl ether containing increasing concentrations of water: (*a*) no water; (*h*) water saturated. (Reprinted with permission from Reference 65. Copyright 1992 American Chemical Society.)

than 20%, is able to tautomerize (in about 70 ps). The dominant structures correspond to "blocked" complexes for which solvent rearrangement leading to the tautomerizing form would take more than 10 ns.[66] The same authors claim that also in alcohol solvents, a fraction of complexes unable to tautomerize can be detected.[64]

Using both experiment and theory, Chou and co-workers found that the rate of excited-state proton transfer can be greatly accelerated with increasing energies of hydrogen bonding between 7AI and the complexing partner.[67–68] For instance, the ESDPT rate in 1:1 complexes between 7AI and acetic acid was so fast that only tautomeric emission was observed.

It is clear from these studies that the central issues of the proton transfer mechanism in alcohol and water solvates of 7AI demand solving conformational problems. In this respect, recent results obtained by supersonic jet spectroscopy are very promising.[47] Analysis of the rotational structure of fluorescence excitation spectra of 7AI:water complexes revealed a cyclic, doubly hydrogen-bonded structure of a 1:1 solvate. It also showed that both the second and third water molecule in 1:2 and 1:3 complexes are located in the molecular plane of 7AI.

It may be quite instructive to study the photophysical characteristics of systems that are structurally related to 7AI. Alcohol complexes of 1-azacarbazole do not seem to differ qualitatively from their 7AI counterparts.[62,69] Large differences, however, are observed for a series of azaheterocyclic compounds based on pyrrole, pyridine, indole, and carbazole units, discussed separately in the next section.

2.4 Excited State Double-Proton Transfer in 2-(2′-Pyridyl)Indoles, 1H-Pyrrolo[3,2-*h*]Quinoline, and Related Structures

It has been observed that room-temperature fluorescence of several 2-(2′-pyridyl)-indoles (**3–7**) is strongly quenched in alcohol solutions: Quantum yields become about 50 times lower than in aprotic polar and nonpolar solvents.[70–73] The quenching is observed only when the chromophore possesses both the hydrogen bond donor—the NH group—and the hydrogen bond acceptor—the pyridine-type nitrogen atom. In compounds **8–9**, in which one of these groups is blocked by methylation on the pyrrolic nitrogen or by replacing pyridyl by a phenyl group, fluorescence becomes as intense in alcohols as in other solvents. The quenching is much weaker in isomeric 2-(3′-pyridyl)indole (**10**) and is not observed at all in 2-(4′-pyridyl)-indole (**11**). The fluorescence ability of alcohol solutions of 2-(2′-pyridyl)-indoles can also be restored by lowering the temperature. The activation energy of the quenching process, determined from the temperature dependence of the fluorescence intensity is clearly correlated with the activation energy of solvent viscosity.[72]

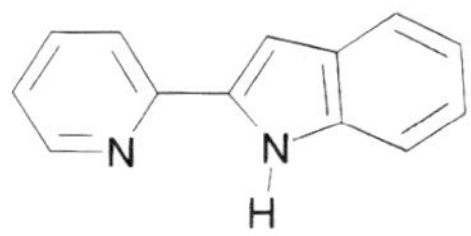

3

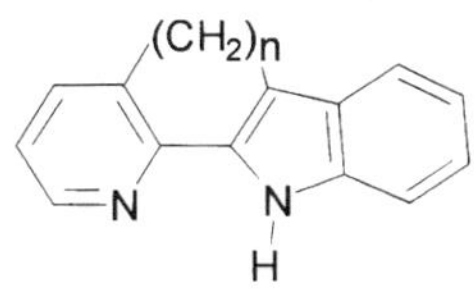

n=1,4

4-7

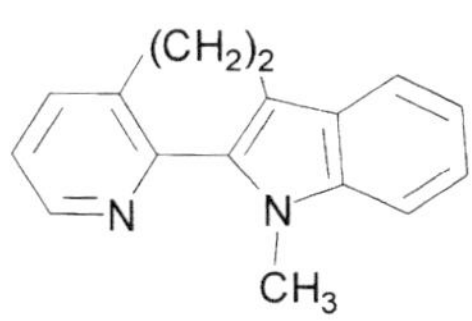

8

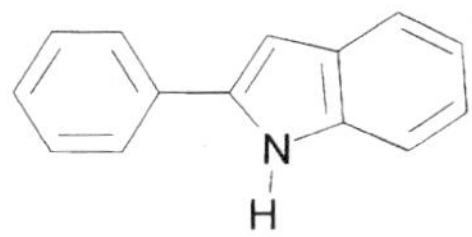

9

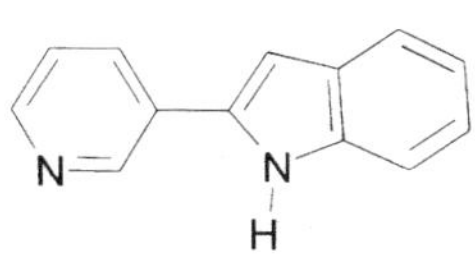

10

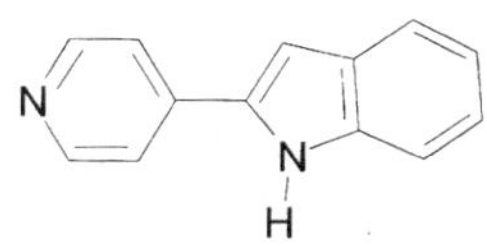

11

However, several other structurally related compounds behave quite differently. For instance, in alcohol solutions of dipyrido[2,3-*a*:3′,2′-*i*]carbazole (DPC, **12**), the fluorescence intensity remains quite weak even at low temperatures. In alcohol glasses, triple luminescence is observed (Fig. 2.13): Two fluorescence emissions are separated by a phosphorescence band. The excitation spectra (Fig. 2.14) show that the phosphorescence has the same precursor as the higher energy fluorescence. However, the low-energy tautomeric fluorescence must originate from a different ground-state species. Such an unusual luminescence pattern has been attributed to the presence of two different conformeric forms of alcohol complexes: one of them can readily undergo excited-state tautomerization, while in the other species an activation energy for the reaction is required.[74] Therefore, at low temperatures, photoinduced proton transfer leading to the appearance of the low-energy fluorescence band F_2, is possible only in the former structure. In the latter form, "blocked" complexes emit a "normal" F_1 fluorescence and a "normal" phosphorescence.

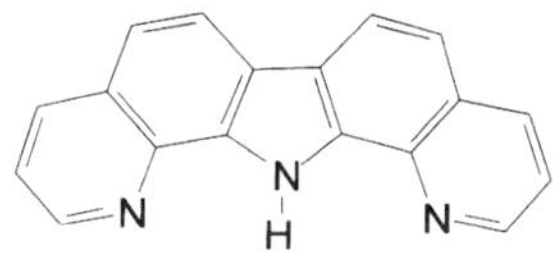

12

Femto- and picosecond fluorescence measurements revealed a biexponential decay of the primary emission.[75] The faster decay can be correlated with the rise time of the tautomeric fluorescence. On the other hand, the slower decay component of F_1 has no counterpart in the F_2 rise. Thus, there is evidence for two different structures emitting the "normal" F_1 fluorescence. One of them is able to undergo tautomerization, while the other is deactivated via a different channel: internal conversion to the ground state. Even though the deactivation does not lead to a

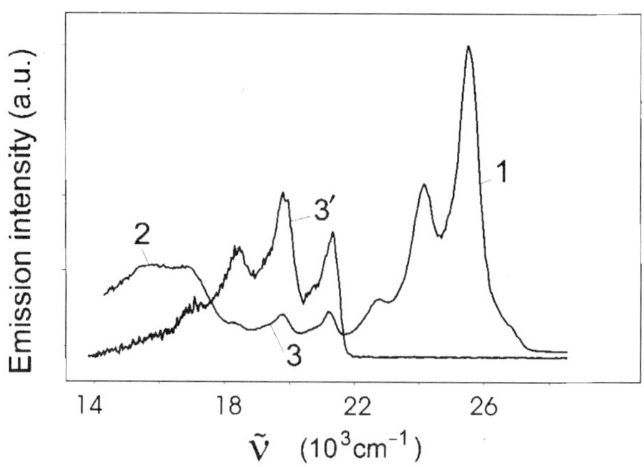

Figure 2.13. Triple luminescence of DPC (**12**) in *n*-propyl alcohol at 123 K: 1, "normal" fluorescence; 2, tautomeric fluorescence; 3, phosphorescence; 3′, phosphorescence separated from fluorescence by using choppers.

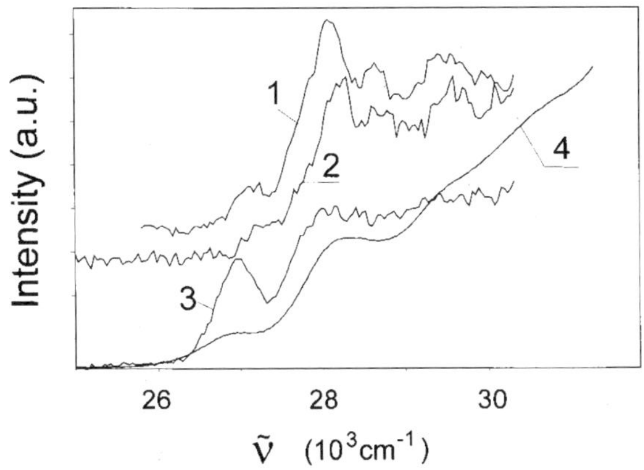

Figure 2.14. Luminescence excitation spectra for DPC (**12**) in *n*-propyl alcohol at 77 K: fluorescence monitored at 23,800 cm^{-1} (1) and at 16,000 cm^{-1} (3); phosphorescence monitored at 19,800 cm^{-1} (2); room-temperature absorption (4).

phototautomer, it must be related to hydrogen bonding with the alcohol, since in aprotic solvents, both polar and nonpolar, the monoexponential decay is nearly an order of magnitude slower.

It is natural to identify the candidate for a fast phototautomerization as a cyclic, doubly hydrogen-bonded complex with alcohol. The form responsible for the slow component in the F_1 band decay profile would correspond to noncyclic $1:n$ complexes, where n is the number of alcohol molecules attached to the substrate.

Detailed studies of **4–7**, **12**, 1H-pyrrolo[3,2-*h*]quinoline (**13**), 7,8,9,10-pyrido[2,3-*a*]carbazole (**14**), and tetrahydropyrido-[2,3-*a*]carbazole (**15**) showed that all these compounds exhibit the same qualitative behavior: Dual fluorescence is observed in alcohol solutions at room temperature.[74c] Quantitative differences, however, are quite large. In pyridylindoles **4–7** and in **14**, the tautomeric emission is barely detectable, whereas in **12**, **13**, and **15**, it is comparable in intensity to, or even stronger than, the "normal" fluorescence. The fraction of tautomers that are able to undergo photoinduced tautomerization depends on the particular compound and may be either very small or very large, close to unity.

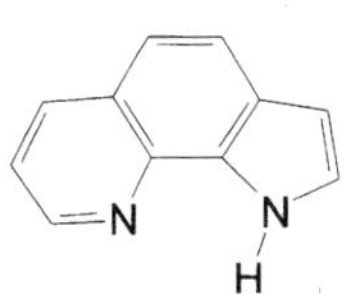

13

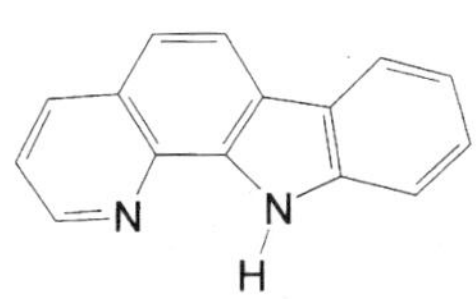

14

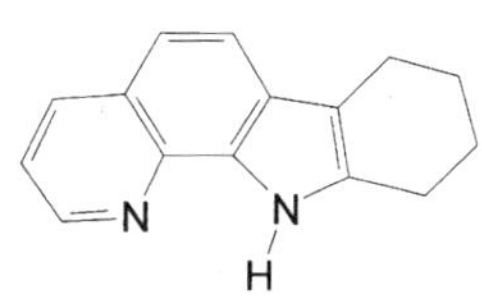

15

A general picture that emerges from the study of these structurally similar molecules is that of a ground-state equilibrium between cyclic and noncyclic alcohol complexes.[74c] If the former type prevails, photoinduced tautomerization cannot be stopped even at low temperatures. If the latter is in excess, an energy barrier exists, associated with solvent reorganization; thus, phototautomerization can be blocked by increasing the solvent viscosity or lowering the temperature. Moreover, in the case when the cyclic structure can be attained only after excitation, $S_0 \leftarrow S_1$ internal conversion effectively competes with tautomerization.

It remains to be clarified what the factors are that lead to shifts in the conformational equilibria between various types of alcohol complexes in a series of structurally similar compounds. Some help may come from calculations. Monte Carlo and molecular dynamics methods were recently used to obtain the structures of 7AI, 1AC, and DPC alcohol complexes.[62] Simulations were performed for various alcohol concentrations in a nonpolar solvent. At low alcohol content, the main fraction of the complexes was predicted to consist of cyclic, double hydrogen-bonded structures. However, in bulk alcohol, the cyclic 1:1 structures were dominant only for DPC. In contrast, preferred solvation for both 7AI and 1AC involved more than one alcohol molecule (Fig. 2.15). These results correlate nicely with the experimental findings that show that DPC, contrary to 7AI and 1AC, is able to undergo ESDPT in alcohols at low temperatures.[74a] At room temperature, the reaction proceeds in all three compounds; however, its rate is much faster in DPC than in the other two molecules, since no major solvent reorganization is required for the cyclic complex of the former.

3 7-Hydroxyquinoline: The Case of Proton Relay

Kasha has divided intramolecular proton transfer processes into four classes:[17a] (i) intrinsic intramolecular proton transfer, (ii) concerted biprotonic transfer, (iii) static and dynamic catalysis of proton transfer, and (iv) proton-relay tautomerization. In fact, only process (i) can be considered truly intramolecular. The other types of proton transfer occur with the participation of at least one "partner" molecule. When more than one such partner is required, the mechanism of proton relay transfer is realized. It may happen in the case of bifunctional molecules having proton donor and acceptor groups apart from each other. For 7-hydroxyquinoline (7HQ, **16**), two methanol molecules can act as a bridge between the hydrogen-bonding donor and acceptor centers, providing an example of a three-proton relay. In 3-hydroxy-xanthone (**17**), three alcohol molecules and four protons could take part in the reaction (Fig. 2.16). Such a mechanism of proton transfer seems to play a role in biological systems, in which proton transfer has to occur across membranes.[76]

Excited-state double proton transfer in 7HQ has been studied by fluorescence and transient absorption measurements.[77–89] Solutions of 7HQ reveal dual emission when the solvent contains traces of protic impurities such as water or alcohol. The low-energy fluorescence was assigned to a product of an ESPT reaction. Two mechanisms of this process can be envisaged: (i) a two-step process, involving separate deprotonation of the OH group and protonation of the nitrogen atom, and (ii) a concerted proton transfer, occurring via a hydrogen-bonded bridge. The former mechanism probably operates in water, while the latter occurs in alcohols. The high-

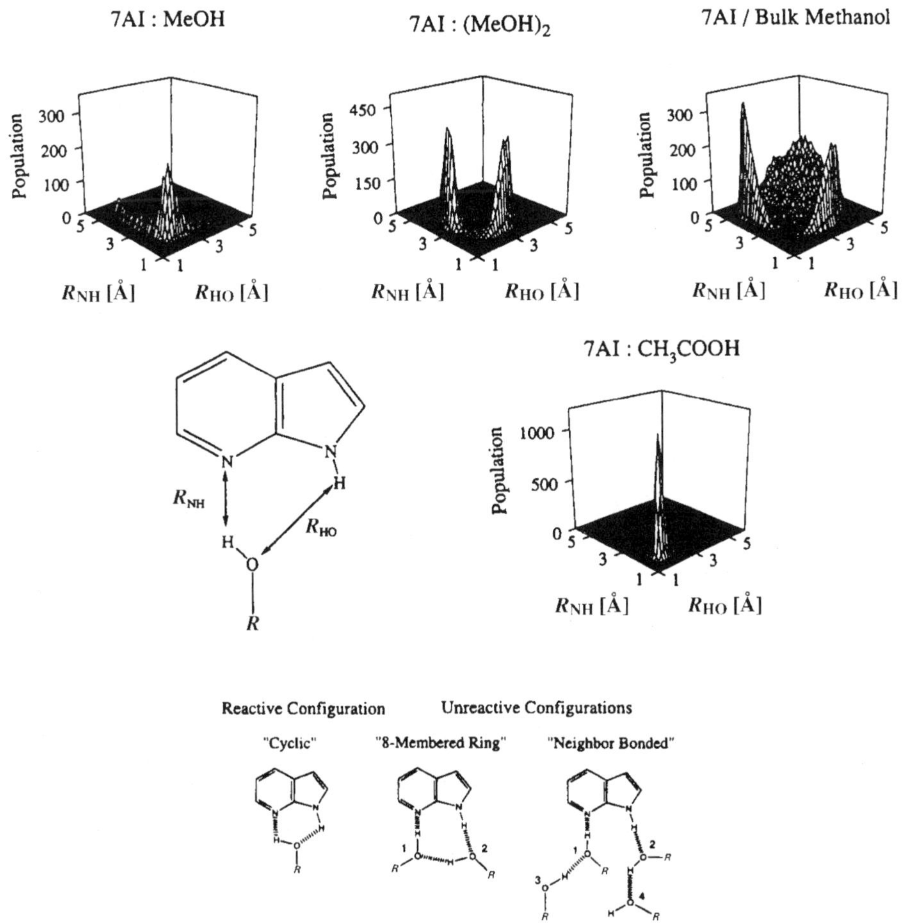

Figure 2.15. Results of simulations of the structure of alcohol complexes: hydrogen-bonding distance distributions for (a) 7AI in cyclohexane containing either one or two molecules of methanol or one molecule of acetic acid, and 7AI in bulk methanol solvent. (Reprinted with permission from Reference 62. Copyright, 1998 American Chemical Society.)

energy emission exhibits double exponential decay in pure alcohols; the low-energy fluorescence is characterized by a biexponential rise time.[85] The faster component (about 50 ps) has been attributed to the 1:2 complex with alcohol, in which solvent reorientation is not required before proton transfer can take place.[79–85] On the other hand, such rearrangement is necessary for other types of complexes—for instance, those containing more than two alcohol molecules. The slower component in the fluorescence kinetic curves (200–500 ps, depending on the alcohol) was attributed to this kind of structure. Interestingly, in a nonpolar aprotic solvent containing small amounts of alcohol, only the faster component was detected.

A convincing confirmation of the cyclic structure of the conformer taking part in the reaction was provided by a comparison of fluorescence measurements performed

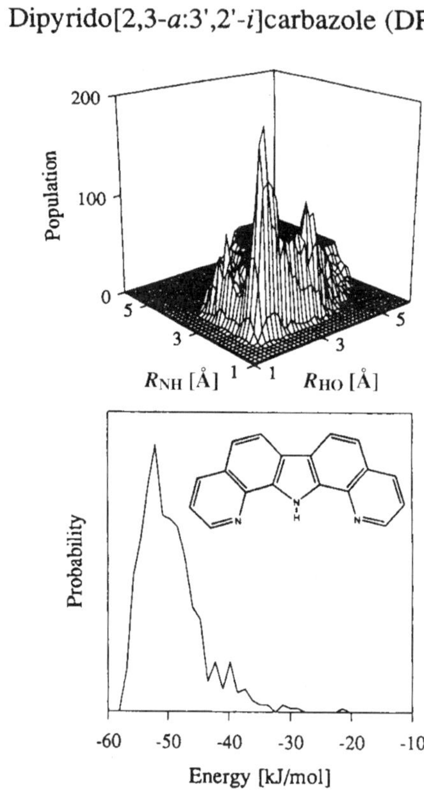

Figure 2.15. (b) DPC in bulk methanol. (Reprinted with permission from Reference 62. Copyright, 1998 American Chemical Society.)

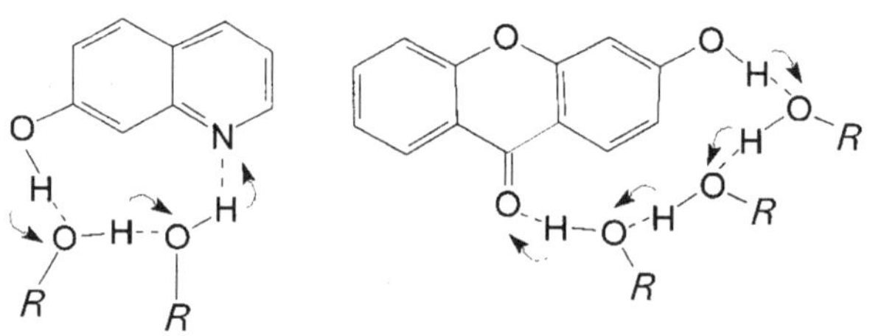

16 **17**

Figure 2.16. Proton relay mechanisms in 7-hydroxyquinoline (**16**) and 3-hydroxyxanthone. (**17**).

in alcohol solutions of 6-methyl-7-hydroxyquinoline (6-Me-7HQ, **18**) and 8-methyl-7-hydroxyquinoline (8-Me-7HQ, **19**).[80] For steric reasons, only the former molecule is able to form a cyclic 1:2 complex with alcohol. In the latter, this is impossible, since the methyl substituent does not allow the *cis* conformation of the 7-OH group and the ring nitrogen atom. No tautomeric fluorescence was observed for 8-Me-7HQ, while the dual emission pattern, analogous to that of 7HQ, was observed for 6-Me-7HQ (Fig. 2.17).

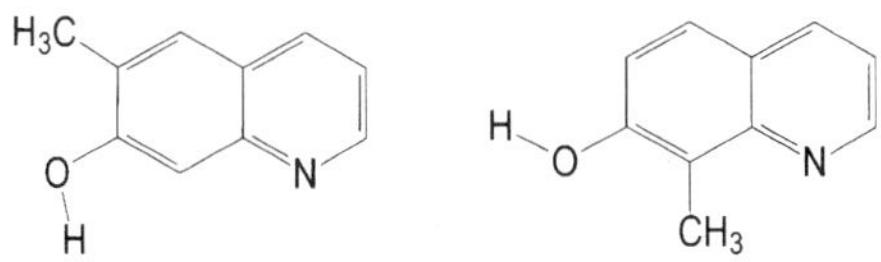

18 19

The presence of bridged alcohol complexes of 7HQ has been discussed with respect to rare-gas low-temperature matrices,[84] supersonic beams,[88] and polymeric matrices.[89a–b] The triple proton transfer reactions of the ground and excited states of 7HQ in methanol have been also studied by ab initio methods.[90]

Studies of 7HQ provided a basis for a design of a proton transfer molecular device.[86] An *N*-morpholinomethyl group was substituted at position 8 of 7HQ. This group was found to act as a "molecular crane," first accepting a proton from the OH site and then depositing it onto the nitrogen atom (Fig. 2.18). The first step was found to be very rapid (faster than 5 ps), due to the hydrogen bonding between the OH and the N atom on the morpholinomethyl moiety. The reaction was controlled

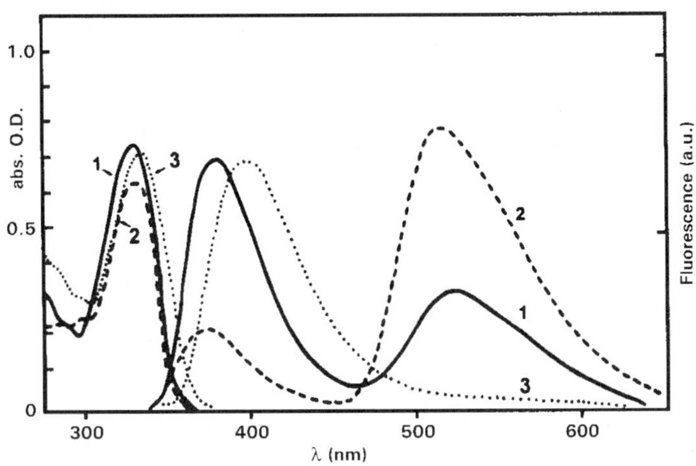

Figure 2.17. Absorption and fluorescence of 7HQ (1), 6-Me-7HQ (2), and 8-Me-7HQ (3) in methanol at room temperature. (Reprinted with permission from Reference 80. Copyright 1995 American Chemical Society.)

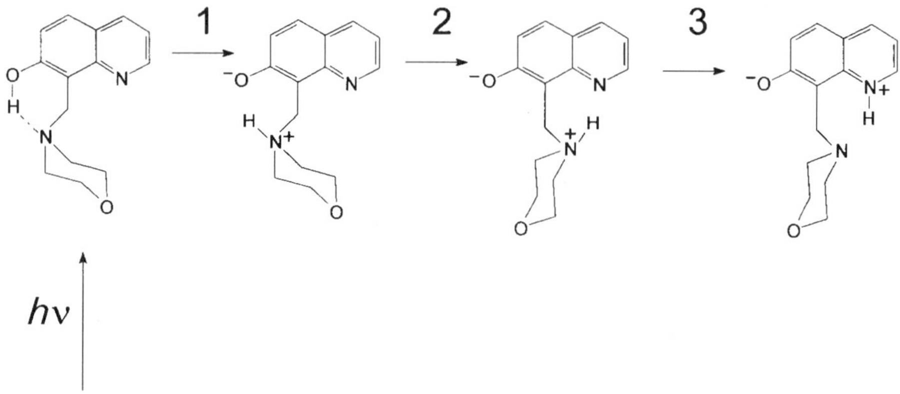

Figure 2.18. The action of a "molecular crane": 1,3-proton transfer steps, 2-internal rotation of the protonated N-morpholinomethyl moiety.

by the second step, involving rotation of the protonated *N*-morpholinomethyl moiety. The rate of this process was found to be correlated with solvent viscosity.

The results obtained so far for 3-hydroxyxanthone seem to indicate that in that molecule photoinduced tautomerization occurs in a stepwise fashion.[91] This is also the case with water solutions of other hydroxyquinolines—for example, 6-hydroxyquinoline (**20**)[92] and 8-hydroxyquinoline (**21**).[93] It is remarkable that the excited-state deprotonation of the former can occur even in 10 M $HClO_4$, and proton uptake by the heterocyclic nitrogen atom can take place in 12 M NaOH!

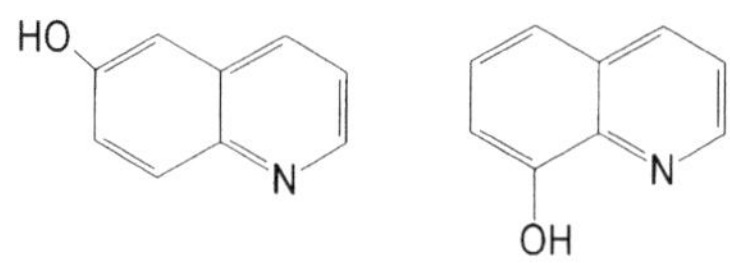

20 21

4 Rotational Isomerism and Proton Transfer

4.1 Salicylic Acid and Its Derivatives

Salicylic acid (**22**) and methyl salicylate (MS, **23**) were the first compounds in which the excited-state intramolecular proton transfer (ESIPT) was reported. Two fluorescence bands, lying in the UV and blue regions, were observed as early as 1924 for solutions of methyl salicylate.[94] Some 30 years later, Weller interpreted the two emissions as being due to the initially excited species and to the product of excited-state tautomerization, respectively. He also suggested that the ESIPT reaction

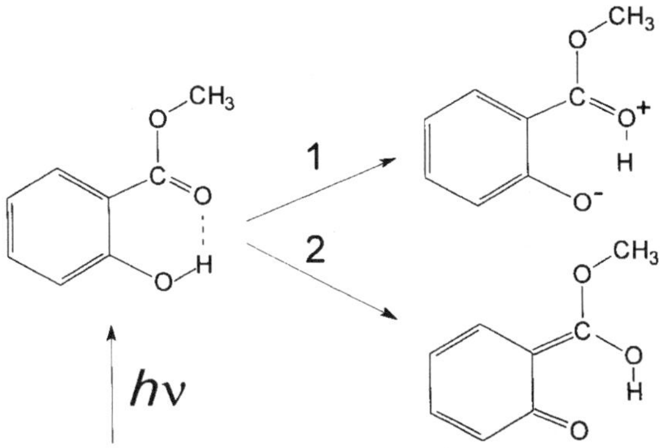

Figure 2.19. Two possible tautomerization pathways in methyl salicylate: proton transfer (1) and hydrogen transfer (2).

involved an activation energy of less than 10 kJ/mol and that the phototautomer could be represented as a zwitterionic structure (Fig. 2.19).[95]

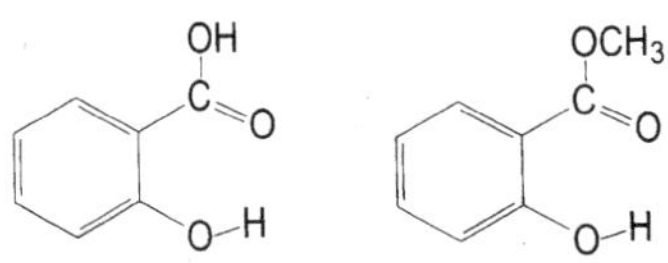

22 23

Further studies corrected and complicated this picture.[96–113] The tautomeric emission was observed at very low temperatures in supersonic beams,[96–102] as well as in a neon matrix[103] and in a durene host.[97] Even at 4 K, the proton transfer rate was found to be extremely rapid. Femtosecond fluorescence depletion studies at room temperature in the gas phase led to the conclusion that the transfer occurs within 60 fs and that the phototautomer is best described not by the zwitterionic form, but by the uncharged structure.[98] The two possible products of the tautomerization process presented in Fig. 2.19 correspond to the extreme cases of proton transfer (1) and hydrogen transfer (2). They should not be understood as two resonance forms of one chemical species. The geometry and charge distribution would strongly differ for the two forms, making them spectroscopically distinguishable. If the reaction can be described as a proton transfer, structural changes accompanying the process are localized in the vicinity of the reaction center. In contrast, during hydrogen transfer the whole molecular skeleton is modified, and the reaction path can be a complex function of many internal coordinates, not necessarily related to proton movement. Studies of MS as well as resonance Raman experiments for structurally similar *o*–hydroxyacetophenone (**24**) support the latter mechanism. Raman intensity analysis shows that excited *o*–hydroxyacetophenone undergoes a rapid relaxation involving a large number of low- and high-frequency coordinates. Somewhat surprisingly, they do not include proton motion.[105]

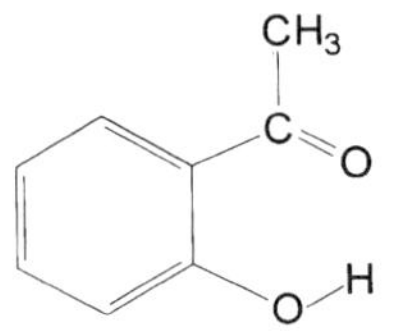

24

These findings, along with the long-known fact that the excitation spectra of the two emissions are not identical[96,106–108] (Fig. 2.20), are difficult to reconcile with the double minimum character of the potential for the proton motion. It is now generally accepted that several conformers are responsible for the emission bands in MS. In structure **23a**, which gives rise to the phototautomer, intramolecular hydrogen bonding occurs between the hydroxyl group and the carbonyl oxygen. The ESIPT in such species proceeds without a barrier (Fig. 2.21). This has been confirmed by the absence of any (OH/OD) isotope effect on the rise time of tautomeric fluorescence.[98,101] Another argument for the barrierless character of the reaction is the finding that in salicylic acid, dual fluorescence occurring after the excitation of just one rotamer was observed in supersonic jets for all excitation energies.[100] The intensity ratio of the blue to the UV emission was lower for higher excitation energies. Just the opposite behavior would be expected for the double minimum potential. For the single minimum case, low-energy excitation corresponds to a nonvertical transition, directly populating the hydrogen-transferred form. Increasing

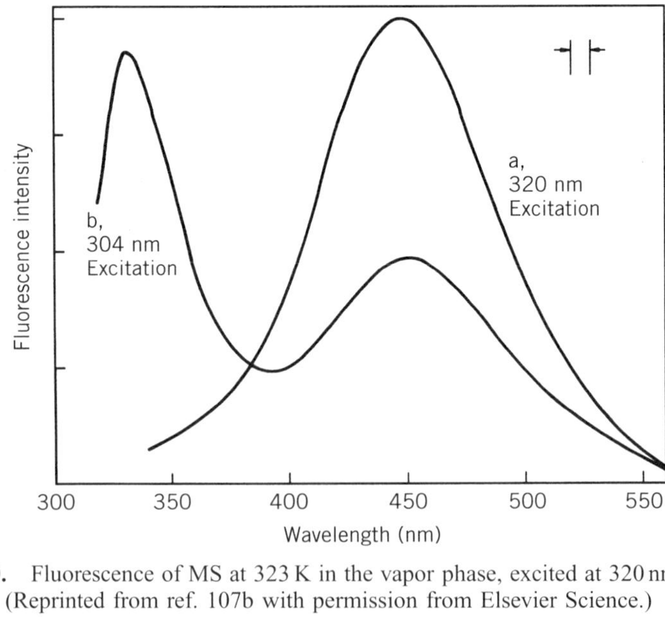

Figure 2.20. Fluorescence of MS at 323 K in the vapor phase, excited at 320 nm (*a*) and at 304 nm (*b*). (Reprinted from ref. 107b with permission from Elsevier Science.)

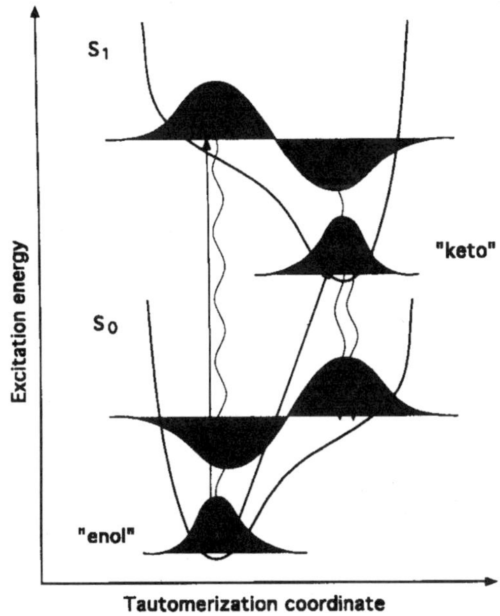

Figure 2.21. Potential-energy surface for phototautomerization of salicylic acid. (Reprinted with permission from Reference 100.) A similar surface has been proposed for methyl salicylate.[98a,101,103]

the excitation energy leads to more favorable Franck–Condon factors for the UV emission.

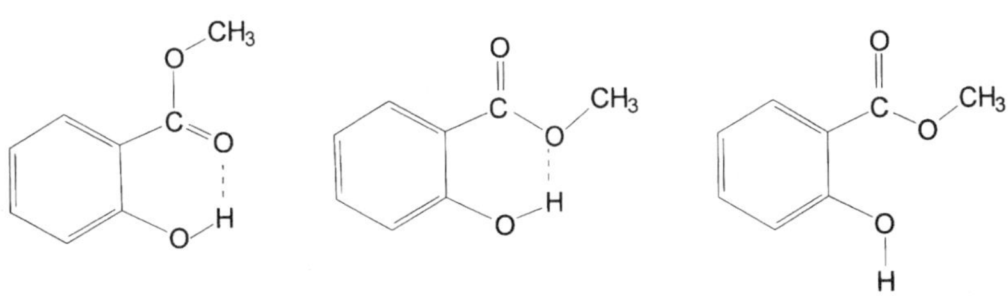

23a 23b 23c

The other intramolecularly hydrogen-bonded form (**23b**) does not undergo ESIPT and thus emits the "normal" UV fluorescence. The existence of both rotamers (**23a** and **23b**) was nicely confirmed by matrix isolation spectroscopy.[109] Yet another species, **23c**, was postulated to contribute to the short-wavelength emission.[107–108] It corresponds to an "open" structure, with no intramolecular hydrogen bond present.

The picture may become even more complex when the molecule is dissolved in a protic solvent. The formation of intermolecular hydrogen bonds may then block the ESIPT process (see Section 6.1). Moreover, salicylic acid can also form cyclic, doubly hydrogen-bonded dimers, consisting of moieties corresponding to both types of hydrogen-bonded rotamers (**25a**–**25b**). The two forms may be transformed into each other by intermolecular double proton transfer. Experimental evidence suggests the occurrence of this process.[100,102,112–113].

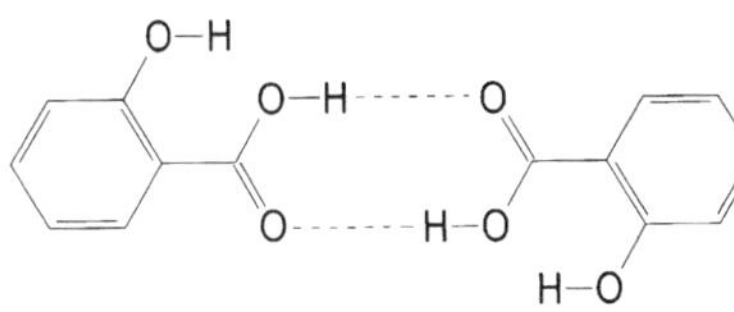

25a

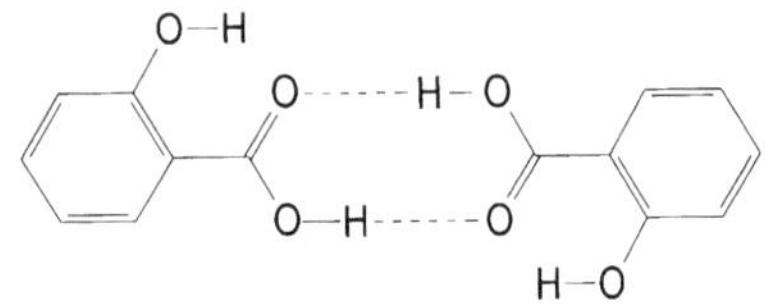

25b

Numerous derivatives of salicylic acid have been studied.[8,10] The interpretation of the fluorescence pattern is similar to that put forward for MS. Two different ground-state species are responsible for the UV and the blue emission, respectively. Their ratio may be changed by chemical modification of the ring (e.g., by chlorine substitution). In some compounds that can only form one "closed" conformer with an intramolecular hydrogen bond, such as *o*-hydroxyacetophenone (**24**), *o*-hydroxybenzaldehyde (**26**), or salicylamide (**27**), no UV fluorescence was detected in the solution.[104] This implies the absence of the "open" conformer, analogous to **23c**. However, the open conformers of **24**, **26** and **27** have been produced upon irradiation in low-temperature rare-gas matrices.[109–111] It is interesting that of the three different possible rotameric products **26b–26d**, the one that is formed (**26d**) involves a 180° rotation of both aldehyde and hydroxy groups.[110]

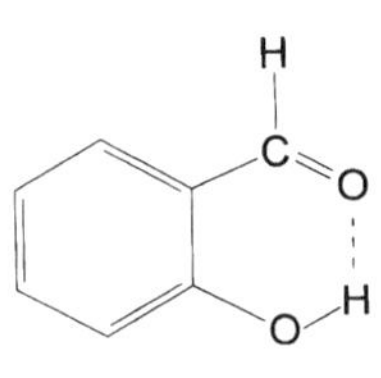

26a

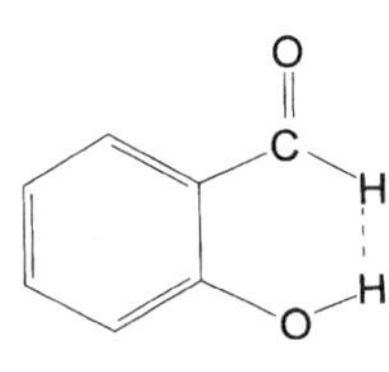

26b

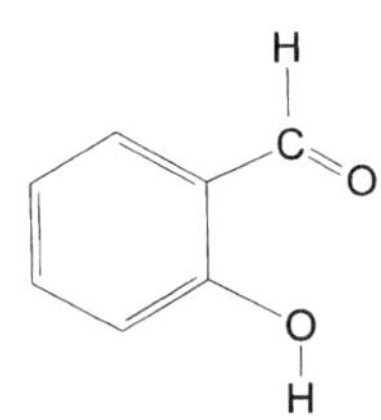

26c

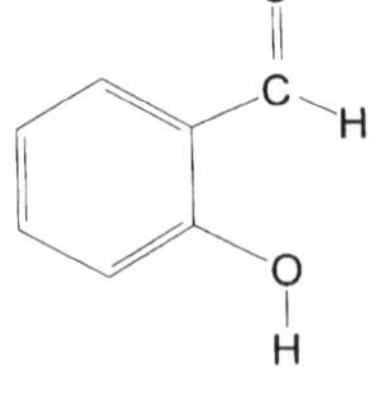

26d

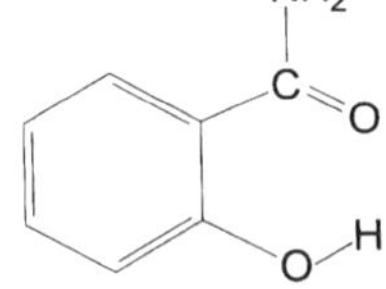

27

4.2 2-(2′-Hydroxyphenyl)benzazoles[114–180]

2-(2′-Hydroxyphenyl)benzothiazole (HBT, **28**),[15,114–140,158b] 2-(2′-hydroxyphenyl)-benzoxazole (HBO, **29**),[114,136–162] and 2-(2′-hydroxyphenyl)benzimidazole (HBI, **30**)[114,158b,162–167] show similar photophysical properties. An extremely rapid ESIPT reaction is observed, manifested in aprotic solvents by the lack of the primary emission and the appearance of a strongly Stokes-shifted low-energy tautomeric fluorescence (Fig. 2.22). In tetrachloroethylene solution of HBT at 293 K, the rate of ESIPT is 160–170 fs.[126–127] The same rate is observed in the deuterated analogue, which suggests a barrierless process.

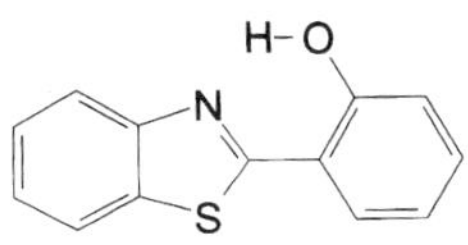

HBT

28

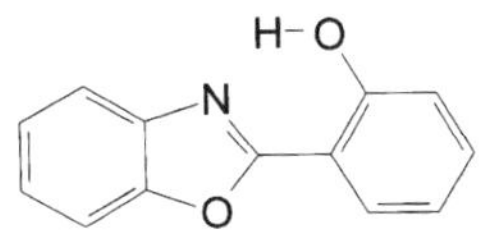

HBO

29

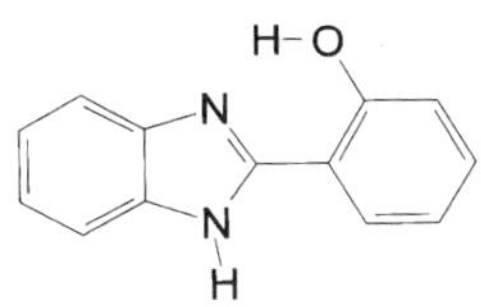

HBI

30

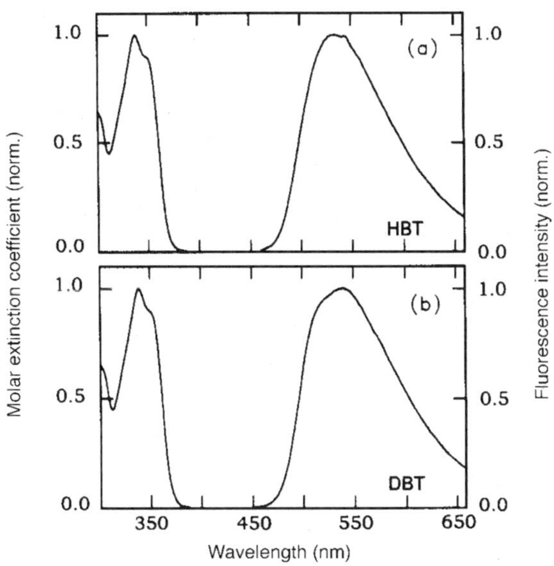

Figure 2.22. Absorption and fluorescence of HBT and its O-deuterated derivative (DBT) in C_2Cl_4. (Reprinted with permission from Reference 127. Copyright 1991 American Chemical Society.)

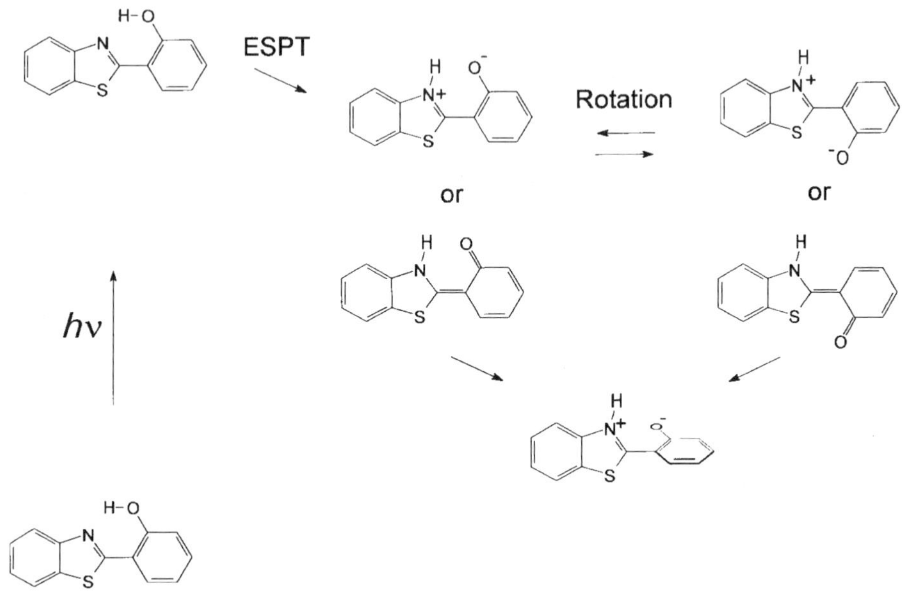

Figure 2.23. Possible conformers of the product of ESIPT reaction in HBT.

There is no doubt that the ESIPT occurs from a *cis* enol structure with an intramolecular hydrogen bonding between the hydroxyl group and the nitrogen atom. The product of the reaction, a zwitterionic or a neutral (keto) *cis* form, can undergo rotational isomerization, leading to the *trans* form. The presence of both *cis* and *trans* transient tautomers in the ground electronic state, with lifetimes in the microsecond range, was detected in solution.[121,131–133,136] Yet another form, a twisted intermediate between *cis* and *trans* structures, was also proposed to account for the efficient nonradiative deactivation of the phototautomer (Fig. 2.23).[125,131] Irradiation of HBT in a polystyrene matrix produced new absorption and emission bands, which disappeared after the sample was heated to 90°C. These new bands were attributed to the *trans* keto ground-state conformers, produced from the *trans* keto excited state.[118] Such interpretation requires that the *cis–trans* rotation barrier be much lower in the excited state.

Much discussion has arisen about assigning the structure of the phototautomer of HBT (and somewhat related salicylideneanilines; see Section 5.1) to either neutral or zwitterionic forms (cf. Fig. 2.23). Zwitterionic character has been postulated for the HBT tautomer in crystals.[114] On the other hand, picosecond infrared studies of the photoproduct in solution revealed bands at 1540 cm^{-1} and 3000 cm^{-1}, consistent with C=O and NH vibrations of the keto species.[128]

HBO is one of the few molecules for which ESPT occurring in the triplet state has been studied.[145–153] The molecule emits dual phosphorescence, due to both enol and keto tautomeric forms (Fig. 2.24). In contrast to the situation in the lowest excited singlet state, triplet levels of both tautomers are nearly isoenergetic and are separated by a barrier of about 30–50 kJ/mol.[148]

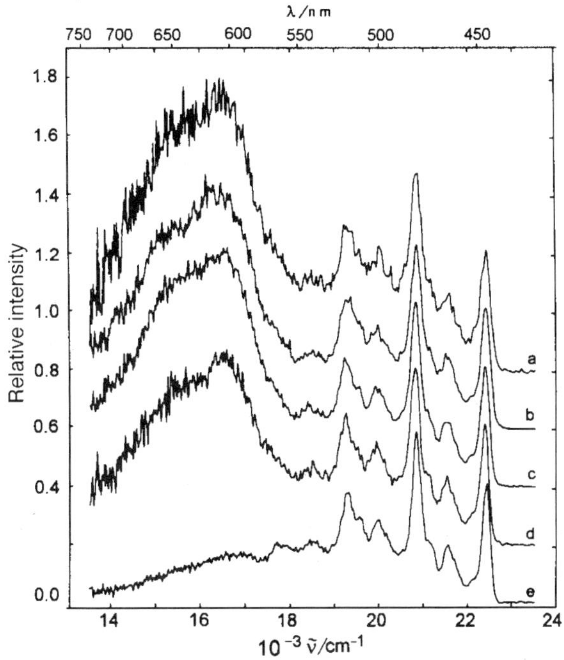

Figure 2.24. Phosphorescence spectra of solutions of HBO in 3-methylpentane at different temperatures: (a) 140 K; (b) 121 K; (c) 101 K; (d) 90 K; (e) 85 K. (Reprinted from Reference 146 with permission from Elsevier Science.)

Excited-state proton transfer has also been detected in the "double" benzoxazoles of **31** and **32**.[168–180] Recent work shows that only one proton is shifted upon excitation in both compounds.[177]

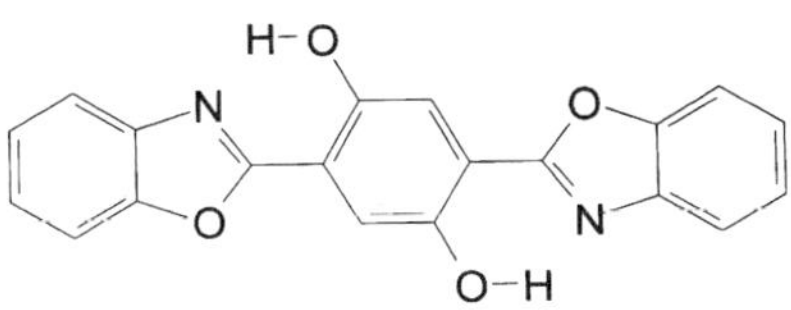

31

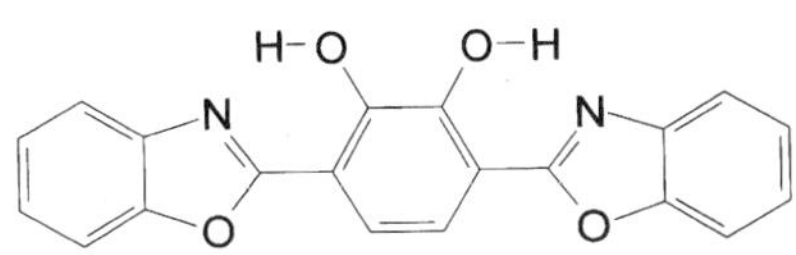

32

4.3 2-(2′-Hydroxyphenyl)benzotriazoles

Numerous studies have been devoted to 2-(2′-hydroxy-5′-methylphenyl)-benzotriazole (Tinuvin-P, **33**), a molecule used as an ultraviolet photostabilizer in polyolefin polymers.[181–207] The mechanism of its protective action is still under discussion. What seems important is that the molecule dissipates nearly all the absorbed energy as heat, as a consequence of the ultrafast ESIPT. A detailed femtosecond study of

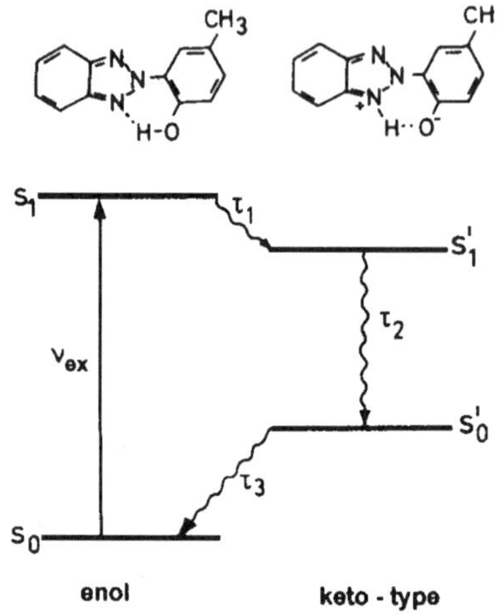

Figure 2.25. Scheme of the proton transfer cycle for Tinuvin-P dissolved in nonpolar solvents. See text for the values of $\tau_1 - \tau_3$. (Reprinted from Reference 200 with permission from Elsevier Science.)

Tinuvin-P led to the determination of the rate constants of the processes that form the proton transfer cycle (Fig. 2.25).[200] All three steps are extremely rapid. ESIPT occurs in about 100 fs (τ_1). The tautomeric form decays by internal conversion to its ground state in 150 fs (τ_2), which is followed by the reverse proton transfer reaction, occurring in 600 fs (τ_3).

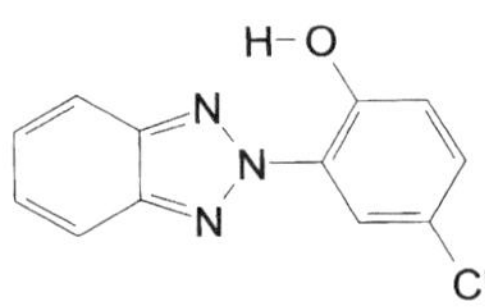

33

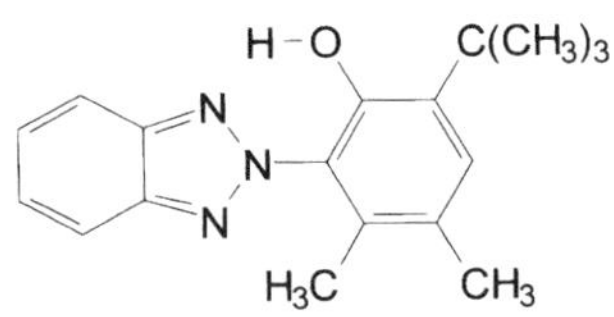

34

The phototautomerization process may be hampered by changing the conformation of the molecule from planar towards twisted. Putting a methyl group in the 6′ position of the phenyl ring (as in **34**) makes the two moieties practically perpendicular to each other. This leads to the breaking of the intramolecular hydrogen bond and to the recovery of "normal" luminescence properties.[184]

5 Photochromic and Thermochromic Substances

5.1 Salicylideneanilines

Crystals of N-salicylideneaniline (**35**) and its derivatives have long been known to be either thermochromic or photochromic.[208–216] In the latter case, a new species, differing from the substrate by the strongly red-shifted absorption spectrum, is produced upon irradiation. Some other derivatives of salicylideneaniline, however, reveal only thermochromism in the crystalline form. In both cases, the effect is due to enol → keto tautomerization caused by the intramolecular proton transfer between the hydroxyl group and the nitrogen atom. The factor that makes the substance either photochromic or thermochromic is the character of molecular packing in the lattice. Thermochromic salicylideneaniline crystals are tightly packed, with both rings practically coplanar. This makes it possible for the proton to move via thermal activation. On the other hand, photochromic derivatives reveal loose packing in the crystal. The dihedral angle between the two rings is about 40–50°. Upon excitation, such systems undergo proton transfer with a subsequent conformational change. *Cis* and *trans* keto forms are possible (Fig. 2.26); experimental evidence seems to favor the latter species.[215]

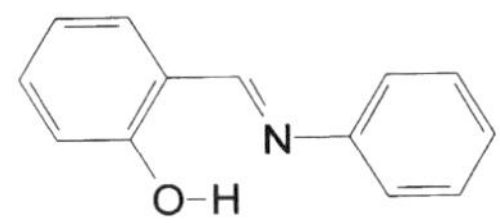

35

In solution, the constraints due to crystal forces are no longer present. Depending on the properties of the medium, such as viscosity, polarity, and temperature, it is possible to detect both photochromic and thermochromic properties in the same molecule.[217–231] Photochromism involves two steps of different time span: (i) excited-state intramolecular proton transfer, occurring in less than 10 ps, and

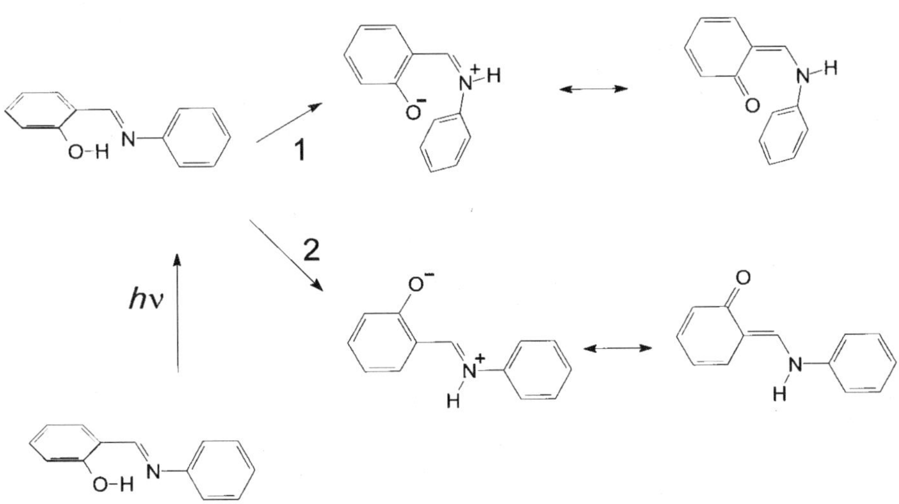

Figure 2.26. Possible *cis* (1) and *trans* (2) isomeric products of the photochromic reaction in *N*-salicylideneaniline.

(ii) conformational change, leading to the final species. The latter step takes longer, and its rate is strongly dependent on solvent properties. The exact structure of the photochromic product is still under debate, regarding both the spatial arrangement of atoms (*cis* versus *trans* keto isomers) and the electronic distribution ("normal" versus zwitterionic form).

Studies of salicylideneanilines have been extended to molecules possessing two identical proton transfer centers (e.g., **36** and **37**).[229–234] The experiments led to the conclusion that upon excitation of such systems, only half of the chromophore takes part in the photochromic process. The other fragment does not seem to be of much influence regarding the photochromic properties.

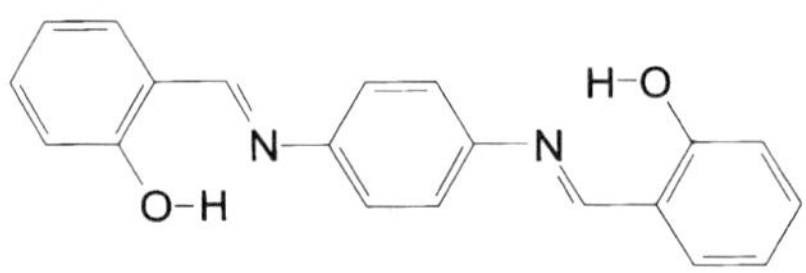

36

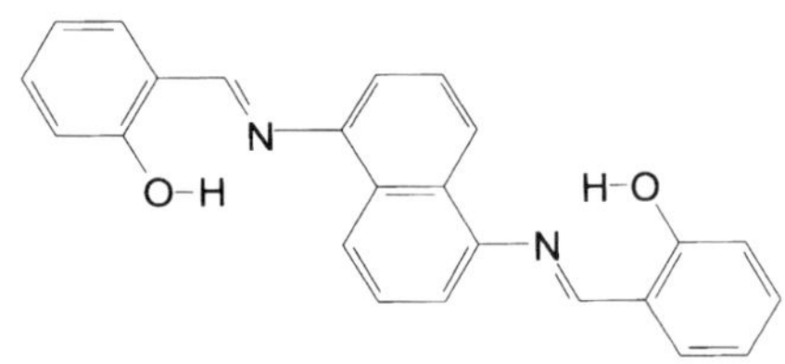

37

5.2 Dinitrobenzylpyridines and Derivatives

2-(2′,4′-Dinitrobenzyl)pyridine (α-DNBP, **38**) represents an interesting example of a photochromic compound that may find possible applications (e.g., in optical data storage). Although the molecule has been known since 1925,[235] the nature of the photochromic species is still under discussion.[236–242] Upon irradiation of α-DNBP, a photogenerated metastable tautomer is produced (Fig. 2.27) that lives for 4.7s in ethanol at 298 K[239] and about 4.6 hours in the crystal.[240–241] Various structures have

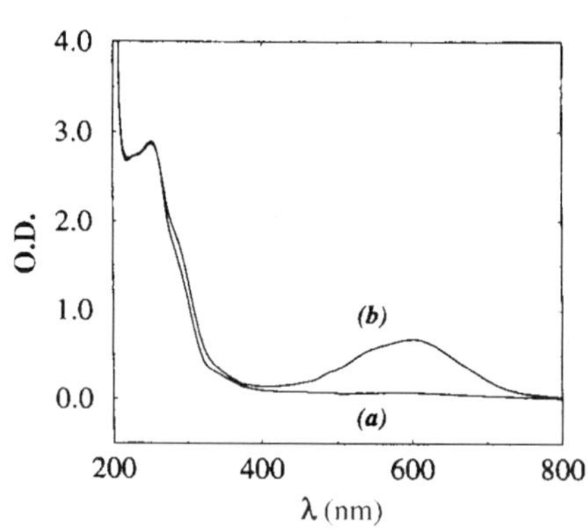

Figure 2.27. Room-temperature absorption spectra of polycrystalline α-DNBP (*a*) before and (*b*) after UV irradiation. (Reprinted with permission from Reference 240. Copyright 1996 American Chemical Society.)

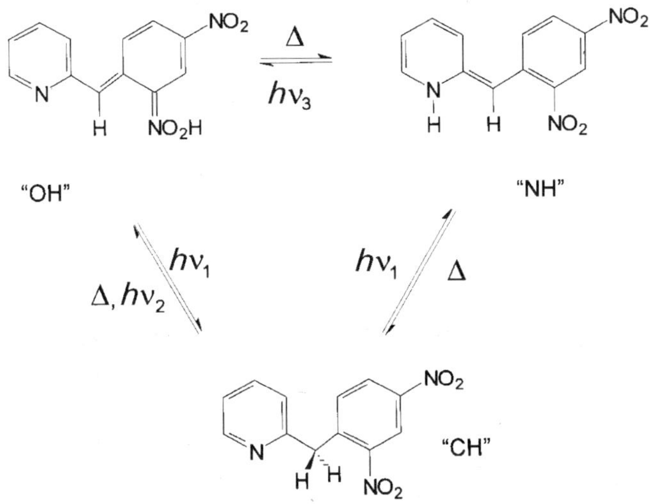

Figure 2.28. The scheme of possible thermal and photochromic rearrangements in α-DNBP.

been proposed for the photochromic species. The scheme of possible thermal and photochromic rearrangements is shown in Figure 2.28. Recent work assigned the tautomeric form of **38** to the enamine "NH" structure, predominantly in the *cis* configuration.[238] Detailed spectroscopic studies of crystalline α-DNBP and derivatives have revealed the role of supramolecular effects in the proton transfer mechanism.[239] The relative yields and rates of tautomerization were found to be controlled by the supramolecular arrangement of the chromophore and the closest layer of its neighbors. A clear manifestation of these effects is provided by the remarkable difference between two very similar molecules, **38** and **39**. In the latter, the "OH" form was found to be more stable than the "NH" tautomer; the opposite occurs in the former.

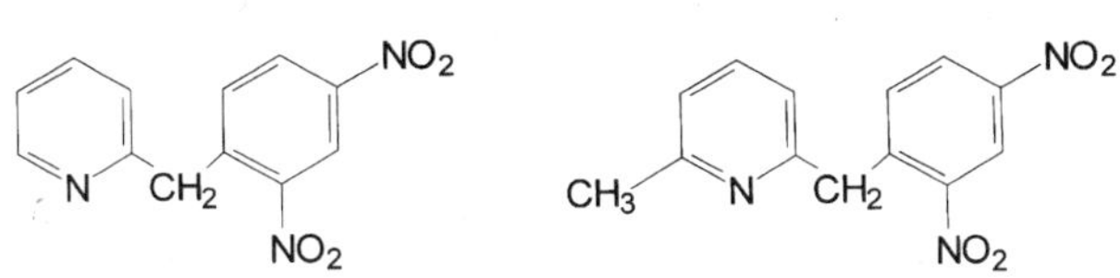

38 39

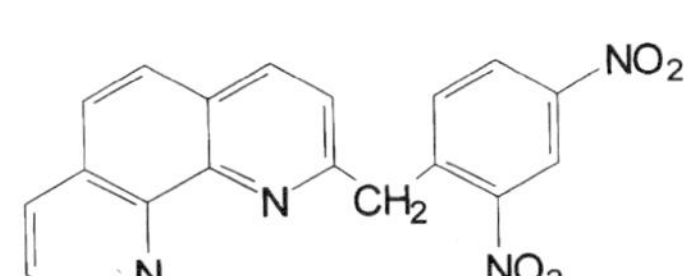

40

Chemical modification of α-DNBP resulted in a significant stabilization of the tautomer.[238] The ground-state energy of the "NH" form of 2-(2′,4′-dinitrobenzyl)-phenanthroline **(40)** is only 2.9 kcal/mol above the energy of the thermodynamically stable "CH" form. In α-DNBP, the corresponding energy difference amounts to more than 8 kcal/mol. The reason for the stabilization was attributed to the hydrogen bonding between the transferred proton and the additional pyridine nitrogen atom.

6 Solvent Control of the Excited-State Proton Transfer

6.1 Solvent Interference with the Excited-State Intramolecular Proton Transfer

As discussed in the previous sections, the formation of an intramolecular hydrogen bond is a prerequisite for a rapid ESIPT. If a molecule is dissolved in a protic solvent, competition arises between intramolecular bonding and the tendency to form the intermolecular bonds with the solvent molecules. It has already been shown that improper intramolecular hydrogen bonding or the absence of such bonding may block the proton transfer process. The same is true if intermolecular bonds to the solvent are formed.

Numerous works have demonstrated that in intermolecularly hydrogen-bonded species ESIPT cannot effectively occur. In methyl salicylate, for instance, two emission bands are observed in hydrogen-bonding solvents. Their intensity ratio is dependent on the excitation wavelength, and the excitation spectra are shifted by about 10 nm.[107] This indicates that the two emissions have different ground-state precursors. The intramolecularly hydrogen-bonded form (with the bonding to the carboxylic oxygen) gives rise to the phototautomer and thus to the visible fluorescence, while the molecules that are engaged in the intermolecular bonds to the solvent emit only the "normal" UV band. Since the absorption spectra practically do not depend on solvent, it was concluded that the fraction of the intermolecularly bonded species was dominant

Similar behavior has been observed in many other systems (e.g., various derivatives of salicylic acid)[121a,243] 2-(2′-hydroxyphenyl)benzazoles,[121a,122,139,161] and 2-(2′-hydroxyphenyl)benzotriazoles.[186,192] In most cases, the absorption spectra of the intra- and intermolecularly hydrogen-bonded forms are similar, and it is thus difficult to distinguish between the two species. This difficulty was the source of some confusion in the literature, because the traces of protic impurities, often present in nonpolar solvents, lead to the appearance of emission bands that were mistakenly assumed to belong to the kinetic precursors of the tautomeric species. Such was the case of 3-hydroxyflavone (3HF, **41**), one of the most thoroughly investigated phototautomerizing molecules.[244–293] Two emission bands, observed in solutions of 3HF, were initially attributed to the primarily excited form and the product of excited-state proton transfer, respectively (Fig. 2.29). However, it was subsequently shown that careful purification of a nonpolar solvent leads to the appearance of only one fluorescence band, that of the tautomer, even at 77 K.[247] Adding trace amounts

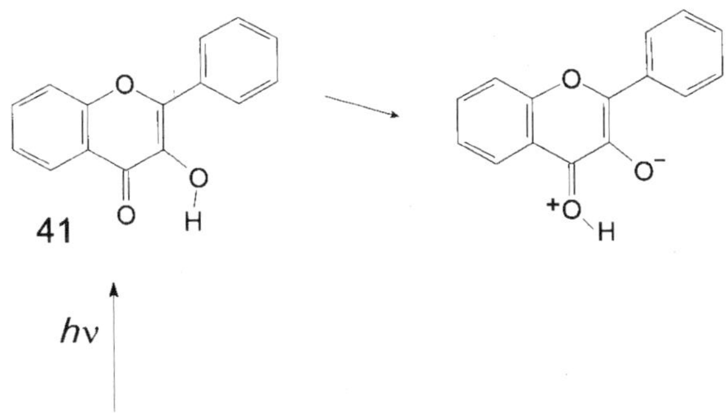

Figure 2.29. Excited-state proton transfer in 3-hydroxyflavone.

of water changes the emission spectrum drastically. In addition to the low-energy band, two other emissions are observed (Fig. 2.30). They have been assigned to the solvated species, which is unable to undergo ESIPT, and to the 3HF anion, respectively. The presence of the anion is an additional complication that is often observed in such solvents as alcohols or pyridine.

In the absence of hydrogen-bonding impurities, the rate of the intramolecular excited-state proton transfer in 3HF was found to be $4 \times 10^{12}\ s^{-1}$, which corresponds to 250 fs.[281] In various possible solvated species (**41a**–**41e**), the rate at which

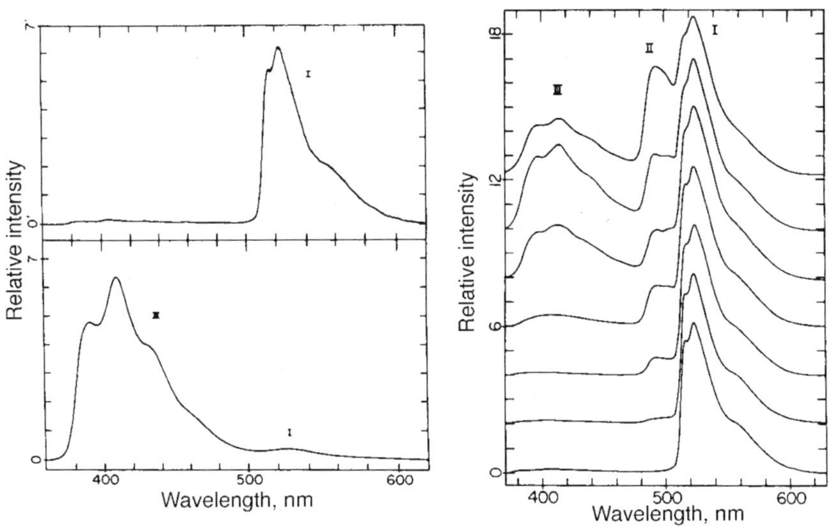

Figure 2.30. Left: fluorescence spectra of 3-hydroxyflavone at 77 K in highly purified dried methylcyclohexane (MCH) (top) and in commercial spectroquality MCH (bottom). Right: fluorescence in quick-frozen MCH glass at 77 K as a function of the addition of traces of water. The lowest curve shows only the tautomer fluorescence in anhydrous MCH; the highest curve is for solvent saturated by the addition of a drop of water. The curves are normalized with respect to intensity at 523 nm. (Reprinted with permission from Reference 247. Copyright 1984 American Chemical Society.)

this process takes place can be dramatically different, depending on their structure. Matrix isolation studies made it possible to distinguish between different complexes.[277–279] In doubly and polysolvated species, proton transfer is blocked at low temperatures. On the other hand, in 1:1 complexes of 3HF with alcohols, ammonia, and ethyl ether, the reaction was found to occur in Ar matrices at 10 K in less than 10 ps, about 10 ps, and 40 ps, respectively.[278]

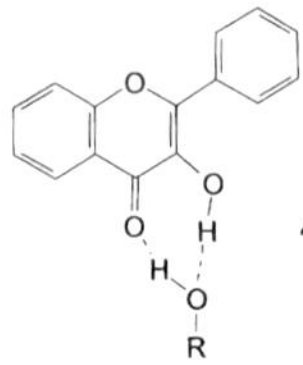

41a

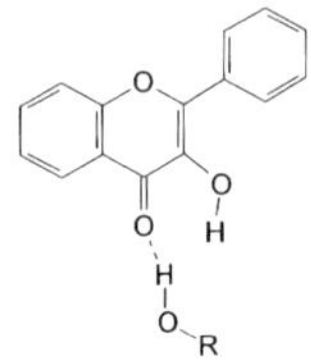

41b

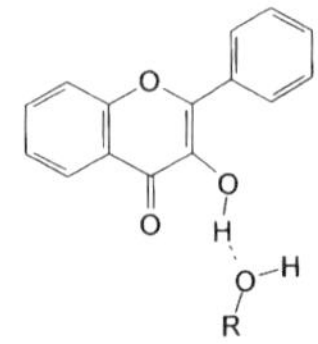

41c

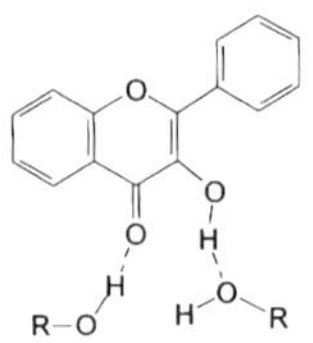

41d

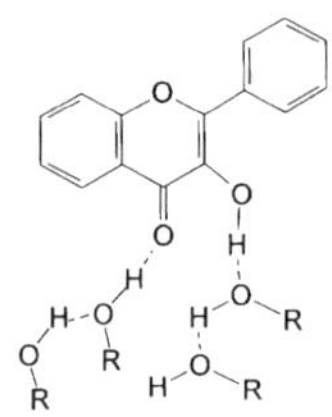

41e

Room-temperature solutions of 3HF in alcohols reveal a biexponential rise of the tautomer emission. An extremely fast component, shorter than 125 fs, was attributed to a cyclic monosolvate with methanol, while the slower process was associated with desolvation of the 1:2 complex.[281] However, such interpretation is at variance with the finding that the rate of the slow process does not correlate with solvent viscosity and is thus not dependent on solvent dynamics.[282] This rate exhibits an Arrhenius dependence, with the activation energy increasing with the strength of the intermolecular hydrogen bond. In order to explain such behavior, a model was proposed[282] that assumes the existence of two differently solvated structures: $\mathbf{N_x}$ and $\mathbf{N_y}$ (Fig. 2.31). In $\mathbf{N_x}$, the presence of a strong intermolecular hydrogen bond weakens the intramolecular bond and thus decreases a driving force for proton transfer. In $\mathbf{N_y}$, which can be thermally excited from $\mathbf{N_x}$, the intermolecular hydrogen bond is weakened, and the intramolecular phototautomerization can proceed very rapidly. Studies of methylated derivatives of 3HF have indeed confirmed that the energetics of solute : solvent complexes, not solvent relaxation, is responsible for the photoreaction mechanism.[283]

An elegant way of getting rid of external perturbations caused by intermolecular hydrogen bonding to solvent was demonstrated for 2-(2′-hydroxyphenyl)imidazo[1,2-*a*]pyridine (HPIP, **42**).[294–295] The behavior of HPIP resembles that of the structurally similar benzazoles **28**–**30**. Phototautomerization occurs in non-hydrogen-bonding solvents, manifested by the appearance of a red fluorescence. In protic solvents, such as water, the intramolecular hydrogen bond is broken, and only the

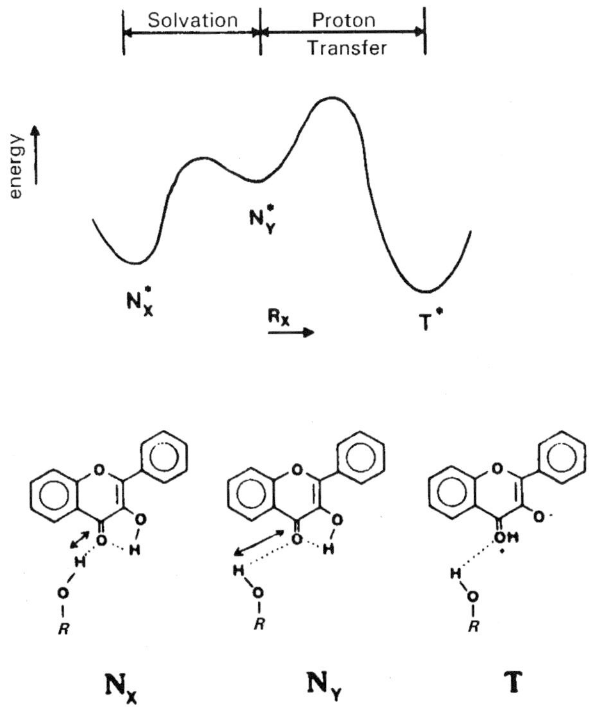

Figure 2.31. A model for ESPT in 3-hydroxyflavone. (Reprinted with permission from Reference 283. Copyright 1985 American Chemical Society.)

emission of a nontautomerizing species is observed in the UV region. Adding β-cyclodextrin to water solutions of HPIP leads to the appearance of the tautomeric emission and to a simultaneous decrease in the intensity of the UV band (Fig. 2.32). This clearly shows that by encapsulating HPIP in the hydrophobic cavity of β-cyclodextrin, it is possible to obtain "unperturbed" molecules that are able to undergo intramolecular proton transfer.

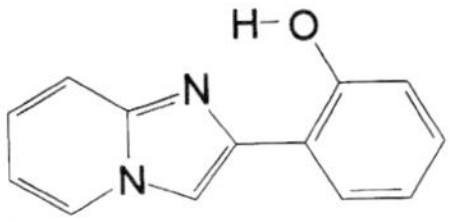

42

6.2 Solvent-Assisted Excited-State Proton Transfer

As previously shown, the formation of an intermolecular hydrogen bonding to the solvent can often block the ESIPT reaction. Sometimes, however, the process can

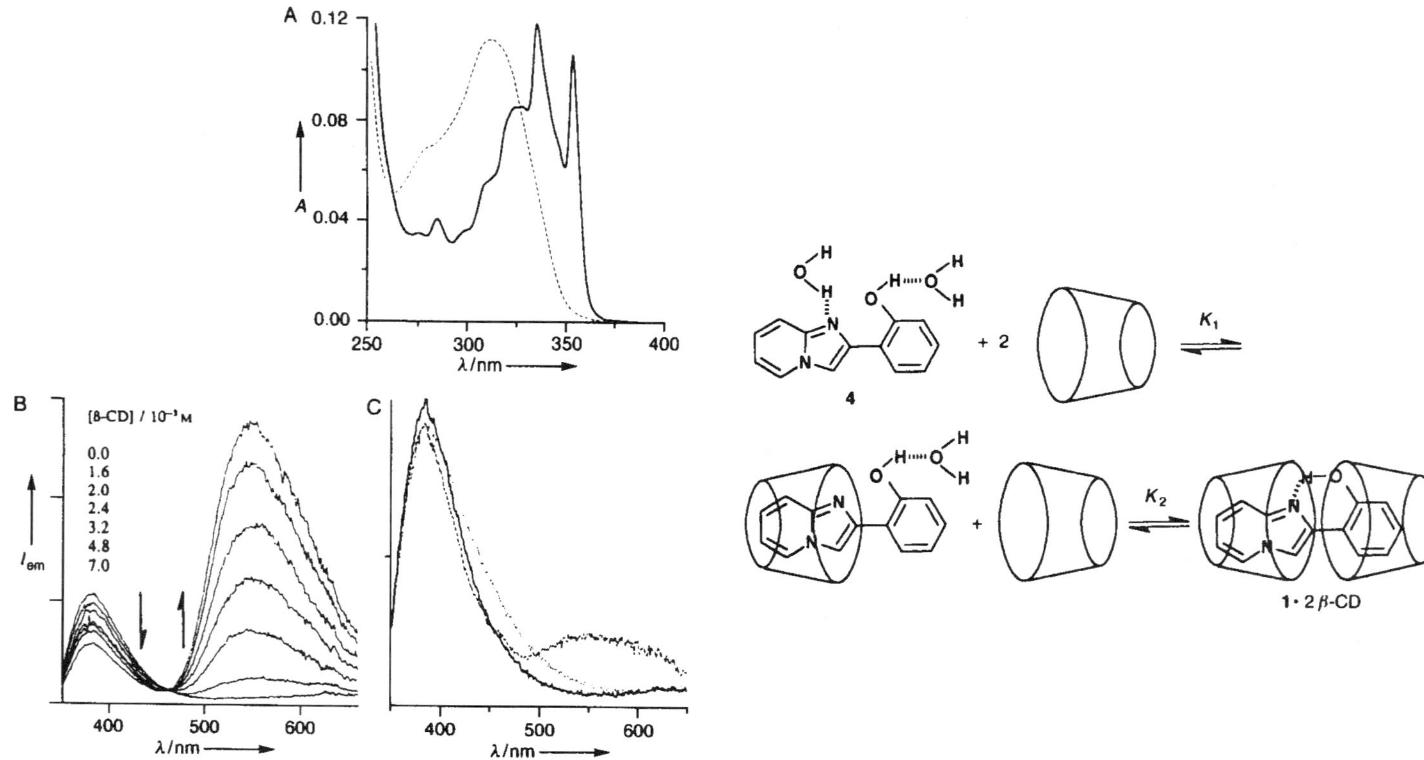

Figure 2.32. Left: (A) Normalized absorption spectra of HPIP in methylcyclohexane (solid line) and water (dashed line); (B) fluorescence spectra in water after the addition of β-cyclodextrin (β-CD); (C) fluorescence in water (solid line), water plus β-CD (dashed line), and water plus maltoheptose (linear oligosaccharide lacking the cavity of β-CD) (dotted line). Right: Scheme of the fitting of HPIP into the β-CD cavity. (Reprinted from Reference 294 with permission from Wiley-VCH.)

still occur, although on a much different timescale. A rate-determining step in this case involves solvent reorientation, associated with the breaking of hydrogen bonds or with the formation of a solvate in which ESPT can occur rapidly. The latter mechanism occurs in bifunctional molecules, that have their proton donor and acceptor groups too far from each other to undergo proton transfer without assistance from the solvent. The best known examples are provided by the tautomerization of 7AI and related structures in alcohols, discussed in Section 2, and by the proton-relay mechanism in 7HQ (Section 3). A similar mechanism takes place in 2-(2′-pyridyl)benzimidazole (**43**), which apparently undergoes ESPT assisted by a protic solvent.[296–297] As in 7AI, two protons are engaged in the reaction, one belonging to the chromophore itself and the other to the solvent molecule (Fig. 2.33).

A crucial role in the kinetics of the solvent-assisted ESPT is played by the ground-state structure of the complex with the solvent. It is generally agreed that, prior to the proton transfer step, a planar, cyclic, hydrogen-bonded structure must be attained. In most cases, the ground-state solvate does not correspond to such species. Therefore, the phototautomerization kinetics is controlled by the rearrangement of the solvent around the excited chromophore, leading to the formation of the planar structure. This process usually requires several tens or hundreds of picoseconds. However, when the solvate is preformed already in the ground state, the ESPT rate can be extremely high, of the same order as is observed in truly intramolecular processes. Recent kinetics studies of molecules discussed in Section 2.4, including dipyrido[2,3-*a*:3′,2′-*i*]carbazole (**12**), 1H-pyrrolo[3,2-*h*]quinoline (**13**), and 7,8,9,10-tetrahydropyrido-[2,3-*a*]carbazole (**15**) in alcohol solutions show that the ESPT in alcohol solvates occurs in several picoseconds[75] and that the process cannot be stopped by lowering the temperature.[74] These findings strongly indicate that if a cyclic hydrogen-bonded structure is thermodynamically favorable, ESPT occurs with a very small barrier or without any barrier at all.

A rather unique example of a solvent-assisted ESPT occurs in solutions of alloxazine (**44**) derivatives in the presence of pyridine.[298–301] Dual fluorescence is observed, indicating the presence of both initially excited alloxazine and an isoalloxazine phototautomer (**45**). Picosecond experiments suggest that an ion pair is formed upon excitation of the alloxazine, consisting of protonated pyridine and

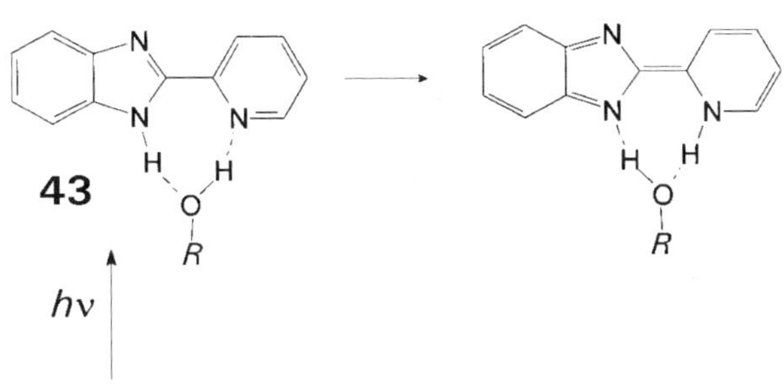

Figure 2.33. ESDPT in the alcohol complex of 2-(2′-pyridyl)benzimidazole.

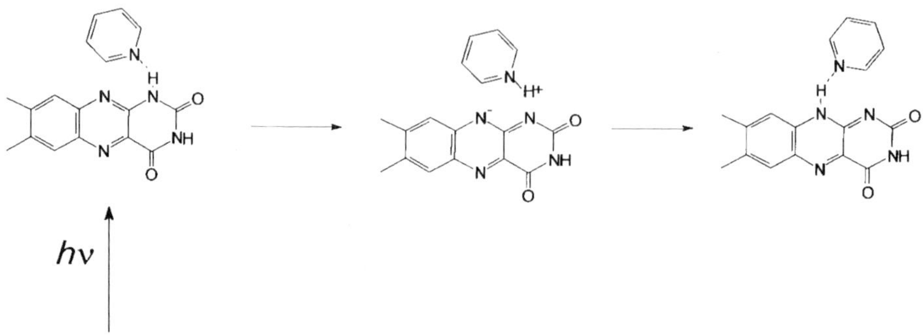

Figure 2.34. Pyridine-assisted ESPT in lumichrome (7,8-dimethylalloxazine).

the alloxazine anion.[17a] The mechanism would thus involve pyridine carrying a proton from the NH group onto a pyridine-type nitrogen atom of alloxazine (Fig. 2.34), with rotational relaxation of pyridinium cation as a rate-controlling step.

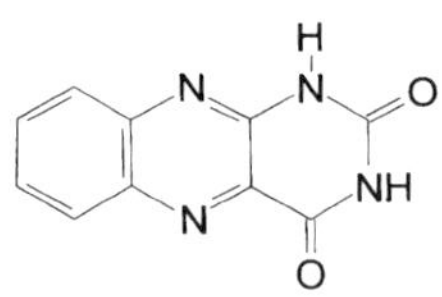

44

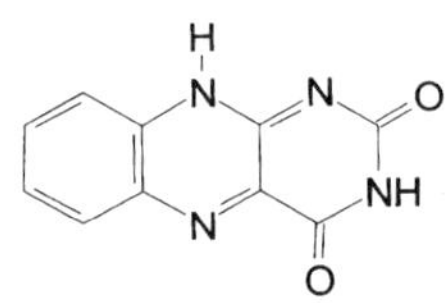

45

6.3 Nonspecific Solvent Effects

In the preceding examples of solvent participation in proton transfer reactions, only the specific interactions were discussed, such as those caused by hydrogen bonding between the solvent and the excited chromophore. However, the role of nonspecific effects due to solvent polarity should not be neglected. This has been demonstrated for 3HF.[274–276] The ESIPT dynamics in that molecule has been studied in different nitriles, as well as in acetonitrile/benzene and acetonitrile/CCl_4 solvent mixtures of varying polarity. The ESIPT rate was found to vary with solvent polarity, which was explained in terms of a simple dielectric continuum model (Fig. 2.35). Due to the interaction of the dipole moments of the normal and tautomeric species of 3HF with the polar solvents, a solvent-induced barrier for ESIPT can be created. This was found to be the case in acetonitrile, for which the measured phototautomerization time was 6 ± 2 ps. In less polar benzonitrile, ESIPT occurs practically without a barrier, and the observed ESIPT time was about 3.2 ps.[275] It was concluded that in the case of polar solvents the rate of proton transfer is limited by both barrier crossing and solvent reorganization, while in the limit of low polarity and no barrier this rate is limited by the rate of solvent reorganization.

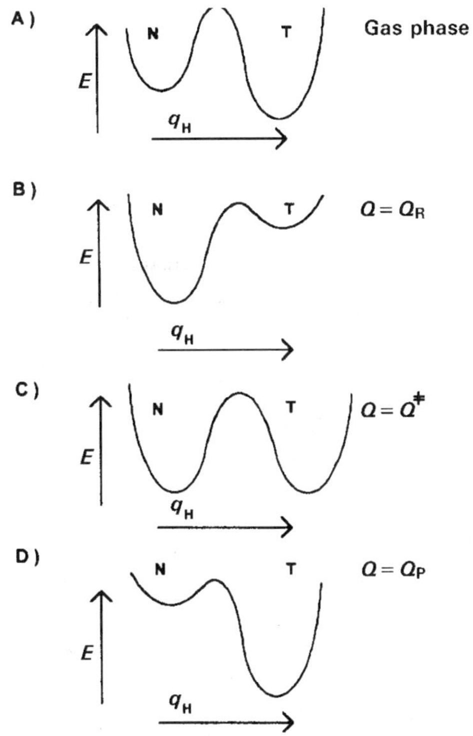

Figure 2.35. Schematic plots of the excited-state energy as a function of the proton motion coordinate q_H. *N* and *T* are normal and tautomeric products, respectively. Case (A) refers to the gas phase. Cases (B) and (D) correspond to the equilibrium polarization of the reactant and product, respectively. Case (C) corresponds to the situation in which the solvent polarization makes the reactant and product states isoenergetic. (Reprinted with permission from Reference 274. Copyright 1991 American Chemical Society.)

Although Figure 2.35 presents the case of nonspecific solvent effects related to the modulation of the barrier to proton transfer by solvent polarization, its meaning is much more general. Any kind of interaction with the solvent, be it specific or nonspecific, may be described in similar terms. In order for proton transfer to proceed, the system must be in an energetically favorable configuration. The factors that can change the shape of the potential-energy curve along the proton transfer coordinate may be due to solvent fluctuations or to intramolecular processes, such as modulating the distance between the heavy atoms involved in intramolecular hydrogen bonding.[302] In fact, femtosecond studies show that the latter process may be rate limiting in many different systems undergoing ESIPT. Tautomerization occurring in about 200 fs, which corresponds to a low-frequency vibration of about 200 cm^{-1}, was observed in various molecules.[127,281,314]

6.4 Excited-State Proton Transfer in Clusters

We started this chapter by discussing 1-naphthol, the first structure that was found to exhibit ESPT. Let us finish by describing studies of the same molecule, performed some 30–40 years later with completely different instrumentation. The rapid development of molecular beams, laser spectroscopy, and mass spectrometry techniques led to investigations of proton transfer reactions in clusters of precisely defined stoichiometry. It became possible to add "solvent" molecules to the species of interest, one by one, and to monitor the ensuing changes. Such experiments provide a link between the gas phase and the bulk liquid and enable one to understand both static and dynamic aspects of solvation phenomena on a molecular level. 1-Naphthol and other acid–base systems have been studied, probing reactivity changes as a function of cluster size.[303–311] For complexes consisting of a single acid molecule and n-base molecules, a threshold value of n was found above which the ESPT reaction could be observed. For 1-naphthol:$(NH_3)_n$ complexes, the reaction proceeds when $n \geq 4$, as is revealed by the appearance of the anionic emission (Fig. 2.36).[303–305] For an even stronger base, piperidine, $n = 2$.[307] With such small clusters, critical proton affinity of the base cluster is required to induce the reaction, the value of which was determined to be 243 kcal/mol.[304,307–308] In 1-naphthol:-water clusters, the situation is different: ESPT is observed only when the number of water molecules reaches about 25.[307–308] The process is controlled by the relaxation of water around the excited naphthol, which, in turn, leads to an inversion of the ordering of the two lowest excited electronic states.

These kinds of experiments can be extended to clusters containing hundreds of solvent molecules, providing invaluable help in studies of liquid solution—probably the most complex condensed phase.

7 Concluding Remarks

The examples presented in this chapter may give the reader an idea about the degree of complexity involved in the studies of photoinduced excited-state proton transfer. The issues closely associated with the subject of this book are related to the conformational aspects of that process. It turns out in many cases that the understanding of the proton transfer mechanism requires a careful analysis of conformational problems, such as the determination of the structure of possible rotamers or solvates. Only after one determines these structures can the "real" proton transfer be properly described.

In this context, a strong need arises for good model systems for studies of photoinduced proton transfer systems that are devoid of conformational complexities. The desired features of such model molecules include rigidity of the chromophore skeleton, a well-defined proton transfer path, and insensitivity to external solvent perturbation. A molecule that certainly deserves mention in this respect is [2,2′-bipyridyl]-3,3′-diol (**46**), in which a very rapid double proton transfer occurs upon excitation.[312–317] Both the synchronous and the stepwise transfer of two

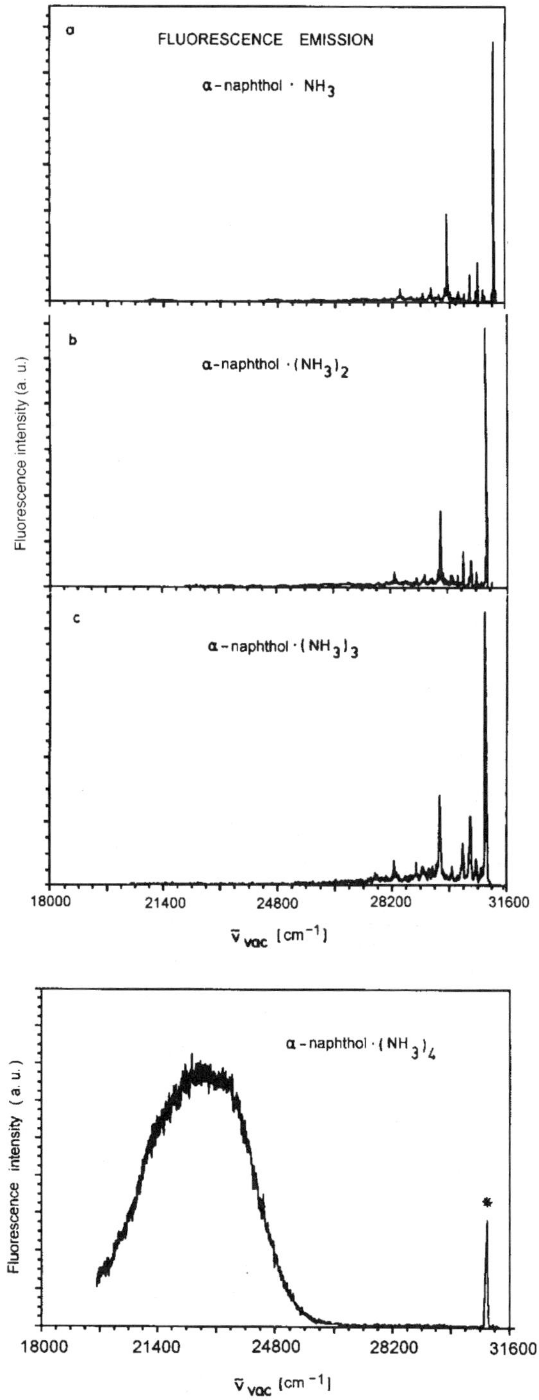

Figure 2.36. Fluorescence of 1-naphthol: $(NH_3)_n$ solvent cluster in a supersonic beam. Top, $n = 1$ (a), 2 (b), 3 (c); bottom, $n = 4$. (Reprinted with permission from Reference 304.)

protons from the oxygen to the nitrogen atoms in this compound have been detected.[314] Two strong intramolecular hydrogen bonds ensure rigidity of the system and are not easily broken by most solvents. Therefore, the proton transfer path seems to be well defined.

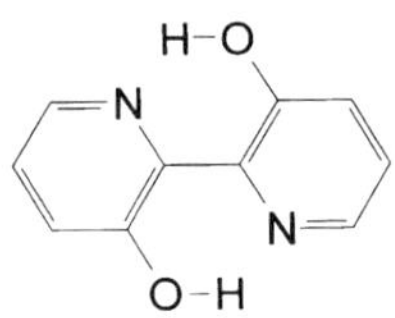

46

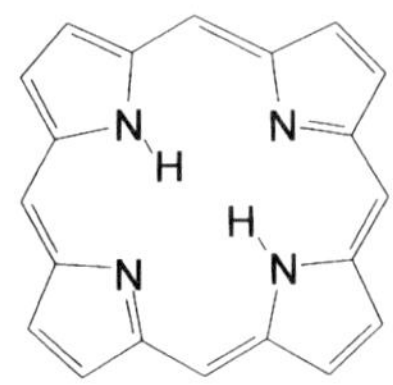

47

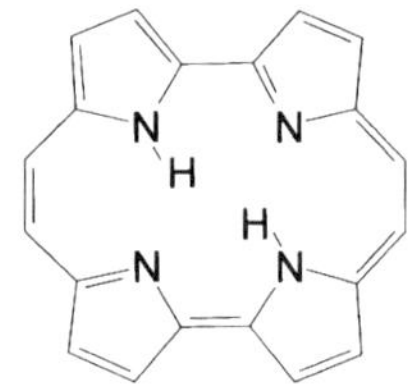

48

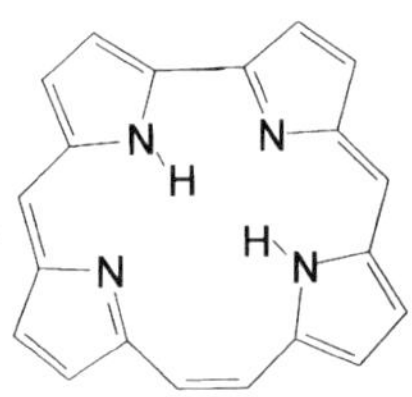

49

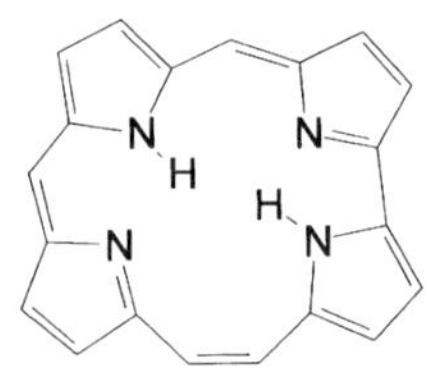

50

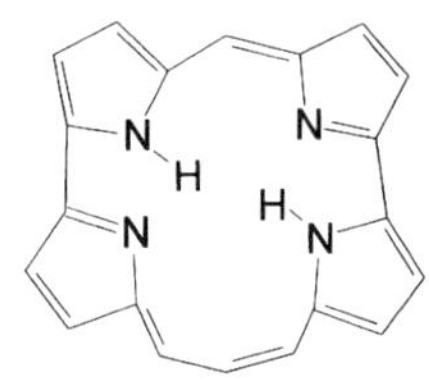

51

Another very promising class of compounds consists of porphyrin (**47**) and its recently synthesized constitutional isomers: porphycene[318] (**48**), corrphycene[319] (**49**), hemiporphycene[320] (**50**), and isoporphycene[321] (**51**). Photoinduced tautomerization in porphyrin, involving the shifting of two inner protons between the nitrogen atoms, has been the subject of numerous studies.[322–332] Fast proton transfer in the ground and lowest excited singlet state of porphycene and its alkyl derivatives was discovered not long after the synthesis of the compound.[333–336] Being located in the internal cavity of the macrocycle, the reaction center is well "insulated" from the outside perturbations. The proton transfer rate seems to be well correlated with the NH–N distance, which, in turn, is a measure of the hydrogen-bonding strength.[334]

References

1. Weller, A. *Z. Elektrochem. Ber. Bunsenges. Phys. Chem.*, **1952**, *56*, 662.
2. Förster, T. *Naturwissenschaften*, **1949**, *36*, 186; *Z. Elektrochem. Ber. Bunsenges. Phys. Chem.*, **1950**, *54*, 531.
3. Förster, T. *Z. Elektrochem. Ber. Bunsenges. Phys. Chem.*, **1950**, *54*, 42.
4. See, for example Grabowski, Z. R.; Grabowska, A. *Z. Phys. Chem. NF*, **1976**, *101*, 197, and references therein.

5. Bulska, H.; Chodkowska, A.; Grabowska, A.; Pakuła, B.; Slanina, Z. *J. Luminescence*, **1975**, *10*, 39.
6. Weller, A. *Prog. Reaction Kinetics*, **1961**, *1*, 189.
7. Ormson, S.M.; Brown, R.G. *Prog. Reaction Kinetics*, **1994**, *19*, 45.
8. Le Gourrierec, D.; Ormson, S.M.; Brown, R.G. *Prog. Reaction Kinetics*, **1994**, *19*, 211.
9. Arnaut, L.G.; Formosinho, S.J. *J. Photochem. Photobiol., A: Chem.*, **1993**, 75, 1.
10. Formosinho, S.J.; Arnaut, L.G. *J. Photochem. Photobiol., A: Chem.*, **1993**, 75, 21.
11. Wan, P.; Shukla, D. *Chem. Rev.*, **1993**, 93, 571.
12. *J. Phys. Chem.*, **1991**, *95(25)*, whole issue.
13. *Chem. Phys.*, **1989**, *136(2)*, whole issue.
14. *Ber. Bunsenges. Phys. Chem.*, **1998**, *102(3)*, whole issue.
15. Barbara, P.F.; Walsh, P.K.; Brus, L.E. *J. Phys. Chem.*, **1989**, *93*, 29.
16. Kelly, R.N.; Schulman, S.G. In *Molecular Luminescence Spectroscopy: Methods and Applications;* Schulman, S.G., ed.; Wiley-Interscience: New York, **1988**; Part 2, Chapter 6.
17. (a) Kasha, M. *J. Chem. Soc., Faraday Trans. 2*, **1986**, *82*, 2379; (b) *Acta Phys. Pol.*, **1986**, *A71*, 717.
18. Lahiri, S.C. *J. Sci. Ind. Res. (India)*, **1979**, *38*, 492.
19. Klöpfer, W. *Adv. Photochem.*, **1977**, *10*, 311.
20. Martynov, I.Y.; Demyashkevich, A.B.; Uzhinov, B.M.; Kuzmin, M.G. *Russ. Chem. Rev.*, **1977**, *46*, 1.
21. Ireland, J.F.; Wyatt, P.A.H. *Adv. Phys. Org. Chem.*, **1976**, *12*, 131.
22. Schulman, S.G. *Rev. Anal. Chem.*, **1971**, *1*, 85.
23. Schulman, S.G.; Winefordner, J.D. *Talanta*, **1970**, *17*, 607.
24. Vander Donckt, E. *Prog. Reaction Kinetics*, **1970**, *5*, 273.
25. Ingham, K.D.; El-Bayoumi, M.A. *J. Am. Chem. Soc.*, **1974**, *96*, 1674.
26. Taylor, C. A.; El-Bayoumi, M.A.; Kasha, M. *Proc. Natl. Acad Sci. U.S.A.*, **1969**, *63*, 253.
27. Ingham, K.C.; Abu-Elgheit, M.; El-Bayoumi, M.A. *J. Am. Chem. Soc.*, **1971**, *93*, 5023.
28. Watson, J.D.; Crick, F.H.C. *Nature (London)*, **1953**, *171*, 964; *Proc. Roy. Soc. Ser. A.*, **1954**, *233*, 80.
29. Löwdin, P.O. *Rev. Mod. Phys.*, **1963**, *35*, 724; *Biopolym. Symp.*, **1964**, *1*, 161.
30. Kasha, M.; Horowitz, P.; El-Bayoumi, M.A. *Molecular Spectroscopy: Modern Research*, Academic Press: New York and London, **1972**, 287.
31. Goodman, M.F. *Nature*, **1995**, *378*, 237.
32. Hetherington, W.M., III; Micheels, R.M.; Eisenthal, K.B. *Chem. Phys. Lett.*, **1979**, *66*, 230.
33. Share, P.E.; Sarisky, M.J.; Pereira, M.A.; Repinec, S.T.; Hochstrasser, R.M. *J. Lumin.*, **1991**, *48/49*, 204.
34. Bulska, H.; Chodkowska, A. *J. Am. Chem. Soc.*, **1980**, *102*, 3259.
35. Bulska, H.; Grabowska, A.; Pakuła, B.; Sepioł, J.; Waluk, J.; Wild, U.P. *J. Lumin.*, **1984**, *29*, 65.
36. Walmsley, J.A. *J. Phys. Chem.*, **1981**, *85*, 3181.
37. Dufour, P.; Dartiguenave, Y.; Dartiguenave, M.; Dufour, N.; Lebuis, A.M.; Bélanger-Gariépy, F.; Beauchamp, A.L. *Can. J. Chem.*, **1990**, 68, 193.

38. Fuke, K.; Kaya, K. *J. Phys. Chem.*, **1989**, *93*, 614.

39. Douhal, A.; Kim, S.K.; Zewail, A.H. *Nature*, **1995**, *378*, 260.

40. Douhal, A.; Guallar, V.; Moreno, M.; Lluch, J.M. *Chem. Phys. Lett.*, **1996**, *256*, 370.

41. Folmer, D.E.; Poth, L.; Wisniewski, E.S.; Castleman, A.W., Jr. *Chem. Phys. Lett.*, **1998**, *287*, 1.

42. Chachissvilis, M.; Fiebig, T.; Douhal, A.; Zewail, A.H. *J. Phys. Chem. A.*, **1998**, *102*, 669.

43. (Takeuchi, S.; Tahara, T. *J. Phys. Chem. A* **1998**, *102*, 7740.

44. Takeuchi, S.; Tahara, T. *Chem. Phys. Lett.* **1997**, *277*, 340.

45. Fiebig, T.; Chachissvilis, M.; Manger, M.; Zewail, A.H.; Douhal, A.; Garcia-Ochoa, I.; de La Hoz Ayuso, A. *J. Phys. Chem. A* **1999**, *103*, 7419.

46. Lopez-Martens, R.; Long, P.; Solgadi, D.; Soep, B.; Syage, J.; Millie, P.H. *Chem. Phys. Lett.*, **1997**, *273*, 219.

47. Nakajima, A.; Hirano, M.; Hasumi, R.; Kaya, K.; Watanabe, H.; Carter, C.C.; Williamson, J.M.; Miller, T.A. *J. Phys. Chem. A.*, **1997**, *101*, 392.

48. Chang, C.; Shabestary, N.; El-Bayoumi, M.A. *Chem. Phys. Letters*, **1980**, *75*, 107.

49. Waluk, J.; Grabowska, A.; Pakuła, B.; Sepioł, J. *J. Phys. Chem.*, **1984**, *88*, 1160.

50. Sepioł, J.; Wild, U.P. *Chem. Phys. Lett.*, **1982**, *93*, 204.

51. Fuke, K.; Yabe, T.; Chiba, N.; Kohida, T.; Kaya, K. *J. Phys. Chem.*, **1986**, *90*, 2309.

52. Suwińska, K. *Acta Cryst.*, **1985**, *C41*, 973.

53. Waluk, J.; Herbich, J.; Oelkrug, D.; Uhl, S. *J. Phys. Chem.*, **1986**, *90*, 3866.

54. Avouris, P.; Yang, L.L.; El-Bayoumi, M.A. *Photochem. Photobiol.*, **1976**, *24*, 211.

55. Smirnov, A.V.; English, D.S.; Rich, R.L.; Lane, J.; Teyton, L.; Schwabacher, A.W.; Luo, S.; Thornburg, R.W.; Petrich, J.W. *J. Phys. Chem. B*, **1997**, *101*, 2758.

56. McMorrow, D.; Aartsma, T.J. *Chem. Phys. Lett.*, **1986**, *125*, 581.

57. Moog, R.S.; Bovino, S.C.; Simon, J.D. *J. Phys. Chem.*, **1988**, *92*, 6545.

58. Konijnenberg, J.; Huizer, A.H.; Varma, C.A.G.O. *J. Chem. Soc. Faraday Trans., 2* **1988**, *84*, 1163.

59. Moog, R.S.; Maroncelli, M. *J. Phys. Chem.*, **1991**, *95*, 10359.

60. Herbich, J.; Sepioł, J.; Waluk, J. *J. Mol. Struct.*, **1984**, *114*, 329.

61. Chapman, C.F.; Maroncelli, M. *J. Phys. Chem.*, **1992**, *96*, 8430.

62. Mente, S.; Maroncelli, M. *J. Phys. Chem. A*, **1998**, *102*, 3860.

63. Chen, Y.; Gai, F.; Petrich, J.W. *J. Am. Chem. Soc.*, **1993**, *115*, 10158.

64. Chen, Y.; Gai, F.; Petrich, J.W. *Chem. Phys. Lett.*, **1994**, *222*, 329.

65. Chou, P.-T.; Martínez, M.L.; Cooper, W.C.; Collins, S.T.; McMorrow, D.P.; Kasha, M. *J. Phys. Chem.*, **1992**, *96*, 5203.

66. Chen, Y.; Rich, R.L.; Gai, F.; Petrich, J.W. *J. Phys. Chem.*, **1993**, *97*, 1770.

67. Chang, C.-P.; Wen-Chi, H.; Meng-Shin, K.; Chou, P.-T.; Clements, J.H. *J. Phys. Chem.*, **1994**, *98*, 8801.

68. Chou, P.-T.; Wei, C.-Y.; Chang, C.-P.; Meng-Shin, K. *J. Phys. Chem.*, **1995**, *99*, 11994.

69. Waluk, J.; Komorowski, S.J.; Herbich, J. *J. Phys. Chem.*, **1986**, *90*, 3868.

70. Herbich, J.; Rettig, W.; Thummel, R.P.; Waluk, J. *Chem. Phys. Lett.*, **1992**, *195*, 556.

71. Herbich, J.; Waluk, J.; Thummel, R.P.; Hung, C.Y. *J. Photochem. and Photobiol. A:* Chem., **1994**, *80*, 157.

72. Herbich, J.; Hung, C.-Y.; Thummel, R.P.; Waluk, J. *J. Am. Chem. Soc.*, **1996**, *118*, 3508.

73. Dobkowski, J.; Herbich, J.; Galievsky, V.; Thummel, R.P.; Waluk, J. *Ber. Bunsenges. Phys. Chem.*, **1998**, *3*, 469.

74. (a) Herbich, J.; Dobkowski, J.; Thummel, R. P.; Hegde, V.; Waluk, J. *J. Phys. Chem. A*, **1997**, *101*, 5839; (b) Kyrychenko, A.; Herbich, J.; Izydorzak, M.; Gil, M.; Dobkowski, J.; Wu, F.; Thummel, R.P.; Waluk, J. *Isr. J. Chem.*, **1999**, *39*, 309; (c) Kyrychenko, A.; Herbich, J.; Izydorzak, M.; Wu, F.; Thummel, R.P.; Waluk, J. *J. Am. Chem. Soc.*, **1999**, *121*, 11179.

75. Zhang, H.; Borowicz, P.; Marks, D.; Waluk, J.; Glasbeek, M. (In preparation.)

76. Lewis, A.; Marcus, M.A.; Ehrenberg, B.; Crespi, H. *Proc. Natl. Acad Sci. U.S.A.*, **1978**, *75*, 4642; Marvin, D.A.; FEBS, **1983**, *156*, 1; Ohno, K.; Govindjee, R.; Ebrey, T.G. *Biophys. J.*, **1983**, *43*, 251.

77. Mason, B.E.; Philp, J.; Smith, B.E. *J. Chem. Soc. A*, **1968**, 3051.

78. Thistlethwaite, P.J. *Chem. Phys. Lett.*, **1983**, *96*, 509; Thistlethwaite, P.J.; Corkill, P.J. *Chem. Phys. Lett.*, **1982**, *85*, 317.

79. Itoh, M.; Adachi, T.; Tokumura, K. *J. Am. Chem. Soc.*, **1983**, *105*, 4828; *ibid.*, **1984**, *106*, 850.

80. Nakagawa, T.; Kohtani, S.; Itoh, M. *J. Am. Chem. Soc.*, **1995**, *117*, 7952.

81. Tokumura, K.; Natsume, M.; Nakagawa, T.; Hashimoto, M.; Yuzawa, T.; Hamaguchi, H.; Itoh, M. *Chem. Phys. Lett.*, **1997**, *271*, 320.

82. Kang, W.K.; Cho, J.S.; Lee, M.; Kim, D.H.; Ryoo, R.; Jung, K.H.; Jang, D.J. *Bull. Korean Chem. Soc.*, **1992**, *13*, 140.

83. Lee, S.-I.; Jang, D.-J. *J. Phys. Chem.*, **1995**, *99*, 7537.

84. Lavin, A.; Collins, S. *Chem. Phys. Lett.*, **1993**, *204*, 96; ibid., **1993**, *207*, 513; Bohra, A.; Lavin, A.; Collins, S. *J. Phys. Chem.*, **1994**, *98*, 11427.

85. Konijnenberg, J.; Ekelmans, G.B.; Huizer, A.H.; Varma, C.A.G.O. *J. Chem. Soc. Faraday Trans. 2*, **1989**, *85*, 39.

86. Jalink, C.J.; Van Ingen, V.M.; Huizer, A.H.; Varma, C.A.G.O. *J. Chem. Soc. Faraday Trans.*, **1991**, *87*, 1103.

87. Chou, P.T.; Martinez, S.S. *Chem. Phys. Lett.*, **1995**, *235*, 463.

88. Lahmani, F.; Douhal, A.; Breheret, E.; Zehnacker-Rentien, A. *Chem. Phys. Lett.*, **1994**, *220*, 235; Bach, A.; Leutwyler, S. *Chem. Phys. Lett.*, **1999**, *299*, 381.

89. (a) Douhal, A.; Dabrio, J.; Sastre, R. *J. Phys. Chem.*, **1996**, *100*, 149; (b) Douhal, A.; Sastre, R. *Chem. Phys. Lett.*, **1994**, *219*, 91; (c) García-Ochoa, I.; Díez-López, M.-A.; Viñas, M.H.; Santos, L.; Martínez-Ataz, E.; Sánchez, F.; Douhal, A. *Chem. Phys. Lett.*, **1998**, *296*, 335.

90. Fang, W.-H. *J. Am. Chem. Soc.*, **1998**, *120*, 7568.

91. Mukaihata, H.; Nakagawa, H.; Kohtani, S.; Itoh, M. *J. Am. Chem. Soc.*, **1994**, *116*, 10612.

92. Bardez, E.; Chatelain, A.; Larrey, B.; Valeur, B. *J. Phys. Chem.*, **1994**, *98*, 2357.

93. Bardez, E.; Devol, I.; Larrey, B.; Valeur, B. *J. Phys. Chem. B*, **1997**, *101*, 7786.

94. Marsch, J. K. *J. Chem. Soc.*, **1924**, *125*, 418.

95. Weller, A. *Z. Elektrochem. Ber. Bunsenges. Phys. Chem.*, **1956**, *60*, 1144.

96. Heimbrook, L.; Kenny, J.E.; Kohler, B.E.; Scott, G.W. *J. Phys. Chem.*, **1983**, *87*, 280; *J. Chem. Phys.*, **1981**, *75*, 5201.

97. Nishiya, T.; Yamauchi, S.; Hirota, N.; Baba, M.; Hanazaki, I. *J. Phys. Chem.*, **1986**, *90*, 5730.

98. (a) Herek, J.L.; Pedersen, S.; Bañares, S.; Zewail, A.H. *J. Chem. Phys.*, **1992**, *97*, 9046; (b) Felker, P.M.; Lambert, W.R.; Zewail, A.H. ibid. **1982**, *77*, 1603.

99. Kuper, J.W.; Perry, D.S. *J. Chem. Phys.*, **1984**, *80*, 4640.

100. Bisht, P.B.; Petek, H.; Yoshihara, K.; Nagashima, U. *J. Chem. Phys.*, **1995**, *103*, 5290.

101. Douhal, A.; Lahmani, F.; Zewail, A.H. *Chem. Phys.*, **1996**, *207*, 477.

102. Lahmani, F.; Zehnacker-Rentien, A. *Chem. Phys. Lett.*, **1997**, *271*, 6; *J. Phys. Chem. A*, **1997**, *101*, 6141.

103. Goodman, J.; Brus, L.E. *J. Am. Chem. Soc.*, **1978**, *100*, 7472.

104. (a) Acuña, A.U.; Toribio, F.; Amat-Guerri, F.; Catalán, J. *J. Photochem.*, **1985**, *30*, 339; (b) Catalán, J.; Toribio, F.; Acuña, A.U. *J. Phys. Chem.*, **1982**, *86*, 303; (c) Acuña, A.U.; Catalán, J.; Toribio, F. *J. Phys. Chem.*, **1981**, *85*, 241; (d) Toribio, F.; Catalán, J.; Amat, F.; Acuña, A.U. *J. Phys. Chem.*, **1983**, *87*, 817.

105. Peteanu, L.A.; Mathies, R.A. *J Phys. Chem.*, **1992**, *96*, 6910.

106. López-Delgado, R.; Lazare, S. *J. Phys. Chem.*, **1981**, *85*, 763.

107. (a) Klöpffer, W.; Naundorf, G. *J. Lumin.*, **1974**, *8*, 457; (b) Klöpffer, W.; Kaufmann, G. *J. Lumin.*, **1979**, *20*, 283.

108. Sandros, K. *Act. Chem. Scand. Ser. A*, **1976**, *A30*, 761.

109. Orton, E.; Morgan, M.A.; Pimentel, G.C. *J. Phys. Chem.*, **1990**, *94*, 7936.

110. Morgan, M.A.; Orton, E.; Pimentel, G.C. *J. Phys. Chem.*, **1990**, *94*, 7927.

111. Gębicki, J.; Krantz, A. *J. Chem. Soc., Perkin Trans. 11*, **1984**, 1617.

112. Pant, D.D.; Joshi, H.C.; Bisht, P.B.; Tripathi, H.B.; *Chem. Phys.*, **1994**, *185*, 137.

113. Denisov, G.S.; Golubev, N.S.; Schreiber, V.M.; Shajakhmedow, S.S.; Shurukhina, A.V. *J. Mol. Struct.*, **1996**, *381*, 73.

114. Williams, D.L.; Heller, A. *J. Phys. Chem.*, **1970**, *74*, 4473.

115. Brown, R.G.; Hepworth, J.D.; Hodgson, K.W.; May, B.; West, M.A. *Chem. Ind.*, **1982**, 129.

116. Anthony, K.A.; Brown, R.G.; Hepworth, J.D.; Hodgson, K.W.; May, B.; West, M.A. *J. Chem. Soc. Perkin Trans. 11*, **1984**, 2111.

117. Kirkbright, G.F.; Spillane, D.E.M.; Anthony, K.; Brown, R.G.; Hepworth, J.D.; Hodgson, K.W.; West, M.A. *Anal. Chem.*, **1984**, *56*, 1644.

118. Bross, A.D.; Pla-Dalmau, A.; Spangler, C.W. *Radiat. Phys. Chem.*, **1993**, *41*, 379.

119. Orlando, C.M., Jr.; Wirth, J.G.; Heath, D.R. *J. Chem. Soc., Chem. Comm.*, **1971**, *23*, 1551.

120. Nakagaki, R.; Kobayashi, T.; Nagakura, S. *Bull. Chem. Soc. Jpn.*, **1978**, *51*, 1671.

121. (a) Becker, R.S.; Lenoble, C.; Zein, A. *J. Phys. Chem.*, **1987**, *91*, 3509; (b) ibid. **1987**, *91*, 3517.

122. Potter, C.A.S.; Brown, R.G. *Chem. Phys. Lett.*, **1988**, *153*, 7.

123. Barbara, P.F.; Brus, L.E.; Rentzepis, P.M. *J. Am. Chem. Soc.*, **1980**, *102*, 563.

124. Ding, K.; Courtney, S.J.; Strandjord, A.J.; Flom, S.; Friedrich, D.; Barbara, P.F. *J. Phys. Chem.*, **1983**, *87*, 1184.

125. Potter, C.A.S.; Brown, R.G.; Vollmer, F.; Rettig, W. *J. Chem. Soc., Faraday Trans.*, **1994**, *90*, 59.

126. Laermer, F.; Elsaesser, T.; Kaiser, W. *Chem. Phys. Lett.*, **1988**, *148*, 119.

127. Frey, W.; Laermer, F.; Elsaesser, T. *J. Phys. Chem.*, **1991**, *95*, 10391.

128. Elsaesser, T.; Kaiser, W. *Chem. Phys. Lett.*, **1986**, *128*, 231.

129. Elsaesser, T.; Schmetzer, B.; Lipp, M.; Bäuerle, R.J. *Chem. Phys. Lett.*, **1988**, *148*, 112.

130. Elsaesser, T.; Schmetzer, B. *Chem. Phys. Lett.*, **1987**, *140*, 293.

131. Brewer, W.E.; Martínez, M.L.; Chou, P.T. *J. Phys. Chem.*, **1990**, *94*, 1915.

132. Chou, P.T.; Studer, S.L.; Martínez, M.L. *Chem. Phys. Lett.*, **1991**, *178*, 393.

133. Chou, P.T.; Martínez, M.L.; Studer, S.L.; Shannon, L. *Chem. Phys. Lett.*, **1992**, *195*, 586.

134. Tero-Kubota, S.; Akiyama, K.; Shoji, F.; Ikegami, Y. *J. Chem. Soc., Chem. Comm.*, **1992**, 641.

135. Dey, J.K.; Dogra, S.K. *Bull. Chem. Soc. Jpn.*, **1991**, *64*, 3142.

136. Itoh, M.; Fujiwara, Y. *J. Am. Chem. Soc.*, **1985**, *107*, 1561.

137. Nagaoka, S.; Itoh, A.; Mukai, K.; Hoshimoto, E.; Hirota, N. *Chem. Phys. Lett.*, **1992**, *192*, 532.

138. Nagaoka, S.; Itoh, A.; Mukai, K.; Nagashima, U. *J. Phys. Chem.*, **1993**, *97*, 11385.

139. Roberts, E.L.; Dey, J.; Warner, I.M. *J. Phys. Chem.*, **1996**, *100*, 19681.

140. Premvardham, L.; Peteanu, L. *Chem. Phys. Lett.*, **1998**, *296*, 521.

141. Mordziński, A.; Grabowska, A. *Chem. Phys. Lett.*, **1982**, *90*, 122.

142. Woolfe, G.J.; Melzig, M.; Schneider, S.; Dörr, F. *Springer Ser. Chem. Phys.*, **1982**, *23*, 273.

143. Woolfe, G.J.; Melzig, M.; Schneider, S.; Dörr, F. *Chem. Phys.*, **1983**, *77*, 213.

144. Krishnamurthy, M.; Dogra, S.K. *J. Photochem.*, **1986**, *32*, 235.

145. Mordziński, A.; Grellmann, K.H. *J. Phys. Chem.*, **1986**, *90*, 5503.

146. Rodríguez-Prieto, M.F.; Nickel, B.; Grellmann, K.H.; Mordziński, A. *Chem. Phys. Lett.*, **1988**, *146*, 387.

147. Nickel, B.; Rodríguez Prieto, M.F. *Chem. Phys. Lett.*, **1988**, *146*, 393.

148. Grellmann, K.H.; Mordziński, A.; Heinrich, A. *J. Chem. Phys.*, **1989**, *136*, 201.

149. Al-Soufi, W.; Grellmann, K.H.; Nickel, B. *J. Phys. Chem.*, **1991**, *95*, 10503.

150. Eisenberger, H. Nickel, B.; Ruth, A.A.; Al-Soufi, W.; Grellmann, K.H.; Novo, M. *J. Phys. Chem.*, **1991**, *95*, 10509.

151. Stephan, J.S.; Grellmann, K.H. *J. Phys. Chem.*, **1995**, *99*, 10066.

152. Nickel, B.; Walla, P.J. *Chem. Phys. Lett.*, **1998**, *237*, 371.

153. Nakamura, H.; Terazima, M. Hirota, N. *J. Phys. Chem.*, **1993**, *97*, 8952.

154. Sepioł, J.; Mordziński, A. *Exp. Tech. Phys.*, **1991**, *39*, 371.

155. Arthen-Engeland, T.; Bultmann, T.; Ernsting, N.P.; Rodríguez, M.A.; Thiel, W. *Chem. Phys.*, **1992**, *163*, 43.

156. Lavtchieva, L.; Enchev, V.; Smedarchina, Z. *J. Phys. Chem.*, **1993**, *97*, 306.

157. Smedarchina, Z.; Enchev, V. *J. Photochem. Photobiol. A: Chem.*, **1994**, *80*, 135.

158. (a) Rios, M.A.; Rios, M.C. *J. Phys. Chem.*, **1995**, *99*, 12456; (b) *J. Phys. Chem. A*, **1998**, *102*, 1560.

159. Yang, G.; Morlet-Savary, F.; Peng, Z.; Wu, S.; Fouassier, J.-P. *Chem. Phys. Lett.*, **1996**, *256*, 536.

160. Guallar, V.; Moreno, M.; Lluch, J.M.; Amat-Guerri, F.; Douhal, A. *J. Phys. Chem.*, **1996**, *100*, 19789.

161. Yang, G.; Dreger, Z.A.; Li, Y.; Drickamer, H.G. *J. Phys. Chem. A*, **1997**, *101*, 7948.

162. Das, K.; Sarkar, S.; Ghosh, A.K.; Majumdar, D.; Nath, D.N.; Bhattacharya, K. *J. Phys. Chem.*, **1994**, *98*, 9126.

163. Rodríguez-Prieto, M.F.; Penedo, J.C.; Mosquera, M. *J. Chem. Soc., Faraday Trans.*, **1998**, *94*, 2775.

164. Mosquera, M.; Penedo, J.C.; Rios Rodríguez, M.C.; Rodríguez-Prieto, M.F.; *J. Phys. Chem.*, **1996**, *100*, 5398.

165. Das, K.; Sarkar, N.; Majumdar, D.; Bhattacharya, K. *Chem. Phys. Lett.*, **1992**, *198*, 443.

166. Sinha, K.; Dogra, S.K. *J. Chem. Phys.*, **1986**, *102*, 337.

167. (a) Acuña, A.U.; Amat, F.; Catalan, J.; Costela, A.; Figuera, J.M. Muñoz, J.M. *Chem. Phys. Lett.*, **1986**, *132*, 567; (b) Costela, A.; Amat, F.; Catalán, J.; Douhal, A.; Figuera, J.M.; Muñoz, J.M.; Acuña, A.U. *Opt. Commun.*, **1987**, *64*, 457.

168. Mordziński, A.; Grabowska, A.; Kühnle, W.; Krówczyński, A. *Chem. Phys. Lett.*, **1983**, *101*, 291.

169. Mordziński, A.; Grabowska, A.; Teuchner, K. *Chem. Phys. Lett.*, **1984**, *111*, 383.

170. Brackman, U.; Ernsting, N.P.; Ouw, D.; Schmitt, K. *Chem. Phys. Lett.*, **1984**, *110*, 319.

171. Ernsting, N.P. *J. Phys. Chem.*, **1985**, *89*, 4932.

172. Mordziński, A.; Kühnle, W. *J. Phys. Chem.*, **1986**, *90*, 1455.

173. Grabowska, A.; Mordziński, A.; Tamai, N.; Yoshihara, K. *Chem. Phys. Lett.*, **1990**, *169*, 450.

174. Sepioł, J. *Chem. Phys. Lett.*, **1990**, *175*, 419.

175. Grabowska, A.; Sepioł, J.; Rullière, C. *J. Phys. Chem.*, **1991**, *95*, 10493.

176. Grabowska, A.; Mordziński, A.; Kownacki, K.; Gilabert, E.; Rullière, C. *Chem. Phys. Lett.*, **1991**, *177*, 17.

177. Wortmann, R.; Lebus, S.; Reis, H.; Grabowska, A.; Kownacki, K.; Jarosz, S. *Chem. Phys.* **1999**, *243*, 295.

178. (a) Dick, B. *Ber. Bunsenges. Phys. Chem.* **1987**, *91*, 1205; (b) Dick, B. *Chem. Phys. Lett.* **1989**, *158*, 37.

179. Ernsting, N.P.; Arthen-Engeland, T.; Rodríguez, M.A.; Thiel, W. *J. Chem. Phys.*, **1992**, *97*, 3914.

180. Mühlpfordt, A.; Even, U.; Ernsting, N.P. *Chem. Phys. Lett.*, **1996**, *263*, 178.

181. Werner, T. *J. Phys. Chem.*, **1979**, *83*, 320.

182. Woessner, G.; Goeller, G.; Rieker, J.; Hoier, H.; Stezowski, J.J.; Daltrozzo, E.; Neureiter, M.; Kramer, H.E.A. *J. Phys. Chem.*, **1985**, *89*, 3629.

183. Goeller, G.; Rieker, J.; Maier, A.; Stezowski, J.J.; Daltrozzo, E.; Neureiter, M.; Port, H.; Wiechmann, M.; Kramer, H.E.A. *J. Phys. Chem.*, **1988**, *92*, 1452.

184. Rieker, J.; Lemmeri-Schmitt, E.; Goeller, G.; Roessle, M.; Stueber, G.J.; Schettler, H.; Kramer, H.E.A.; Stezowski, J.J.; Hoier, H.; Henkel, S.; Schmidt, A.; Port, H.; Wiechmann, M.; Rody, J.; Rytz, G. Slongo, M.; Birbaum, J.-L. *J. Phys. Chem.*, **1992**, *96*, 10225.

185. Ghiggino, K.P.; Scully, A.D.; Bigger, S.W. *ACS Symp. Ser.*, **1989**, *381*, 57.

186. Huston, A.L.; Scott, G.W.; Gupta, A. *J. Chem. Phys.*, **1982**, *76*, 4978.

187. Ghiggino, K.P.; Scully, A.D.; Leaver, J.H. *J. Phys. Chem.*, **1986**, *96*, 5089.

188. Wiechmann, M.; Port, H.; Laermer, F.; Frey, W.; Elsaesser, T. *Chem. Phys. Lett.*, **1990**, *165*, 28.

189. Wiechmann, M.; Port, H.; Frey, W.; Lärmer, F.; Elsaesser, T. *J. Phys. Chem.*, **1991**, *95*, 1918.

190. Catalán, J.; Fabero, F.; Guijaro, M.S.; Claramunt, R.M.; Santa Maria, M.D.; de la Concepción Foces-Foces, M.; Cano, F.H.; Elguero, J.; Sastre, R. *J. Am. Chem. Soc.*, **1990**, *112*, 747.

191. Huston, A.L.; Merrit, C.D.; Scott, G.W.; Gupta, A. *Springer Ser. Chem. Phys.*, **1980**, *14*, 232.

192. Huston, A.L.; Scott, G.W. *J. Phys. Chem.*, **1987**, *91*, 1408.

193. Woessner, G.; Goeller, G.; Kollat, P.; Stezowski, J.J.; Hauser, M.; Klein, U.K.A.; Kramer, H.E.A. *J. Phys. Chem.*, **1984**, *88*, 5544.

194. O'Connor, B.; Scott, G.W.; Coulter, D.R.; Gupta, A.; Webb, S.P.; Yeh, S.W.; Clark, J.H. *Chem. Phys. Lett.*, **1985**, *121*, 417.

195. O'Connor, B.; Scott, G.W.; Coulter, D.R.; Yavrouian, A. *J. Phys. Chem.*, **1991**, *95*, 10252.

196. Lee, M.; Yardley, J.T.; Hochstrasser, R.M. *J. Phys. Chem.*, **1987**, *91*, 4621.

197. Kim, Y.R.; Yardley, J.T.; Hochstrasser, R.M. *J. Chem. Phys.*, **1989**, *136*, 311.

198. Shizuka, H.; Machii, M.; Higaki, Y.; Tanaka, M.; Tanaka, I. *J. Phys. Chem.*, **1985**, *89*, 320.

199. Elsaesser, T.; Lärmer, F.; Frey, W. *Inst. Phys. Conf. Ser.*, **1972**, *126*, 543.

200. Frey, W.; Elsaesser, T. *Chem. Phys. Lett.*, **1992**, *189*, 565.

201. Flom, S.R.; Barbara, P.F. *Chem. Phys. Lett.*, **1983**, *94*, 488.

202. Wiechmann, M.; Port, H. *J. Lumin.*, **1991**, *48/49*, 217.

203. Scully, A.D.; Bigger, S.W.; Ghiggino, K.P.; Vogl, O. *J. Photochem. Photobiol A: Chem.*, **1991**, *55*, 387.

204. Ghiggino, K.P.; Scully, A.D.; Bigger, S.W.; Yandell, M.D.; Vogl, O. *J. Polym. Sci. C: Polym. Lett.*, **1988**, *26*, 505.

205. Chudoba, C.; Lutgen, S.; Jentzsch, T.; Riedle, E.; Woerner, M.; Elsaesser, T. *Chem. Phys. Lett.*, **1995**, *240*, 35.

206. Chudoba, C.; Riedle, E.; Pfeiffer, M.; Elsaesser, T. *Chem. Phys. Lett.*, **1996**, *263*, 622.

207. Hass, K.C.; Schneider, W.F.; Estévez, C.M.; Bach, R.D. *Chem. Phys. Lett.*, **1996**, *263*, 414.

208. Hadjoudis, E. In *Photochromism. Molecules and Systems*. Dürr, H. and Bouas-Laurent, H. eds., Oxford: Elsevier, **1990**, 685.

209. Hadjoudis, E. *J. Photochem.*, **1981**, *17*, 355.

210. Hadjoudis, E.; Moustakali-Mavridis, I.; Xexakis, J. *Isr. J. Chem.*, **1979**, *18*, 202.

211. Cohen, M.D.; Schmidt, G.M.J. *J. Phys. Chem.*, **1962**, *6*, 2442.

212. Cohen, M.D.; Schmidt, G.M.J.; Flavian, S. *J. Chem. Soc.*, **1964**, 204.

213. Cohen, M.D.; Flavian, S.; Leiserowitz, L. *J. Chem. Soc.(B)*, **1967**, 329.

214. Cohen, M.D.; Flavian, S. *J. Chem. Soc.(B)*, **1967**, 334.

215. Nakagaki, R.; Kobayashi, T.; Nakamura, J.; Nagakura, S. *Bull. Chem. Soc. Jpn.*, **1977**, *50*, 1909.

216. Higelin, D.; Sixl, H. *J. Chem. Phys.*, **1983**, *77*, 39.

217. Becker, R.S.; Richey, W.F. *J. Am. Chem. Soc.*, **1967**, *89*, 1298.

218. Richey, W.F.; Becker, R.S. *J. Chem. Phys.*, **1968**, *49*, 2092.

219. Becker, R.S.; Chagneau, F. *J. Am. Chem. Soc.*, **1992**, *114*, 1373.

220. Potashnik, P.; Ottolenghi, M. *J. Chem. Phys.*, **1969**, *51*, 3671.

221. Ottolenghi, M.; McClure, D.S. *J. Chem. Phys.*, **1967**, *46*, 4620.

222. Barbara, P.F.; Rentzepis, P.M.; Brus, L.E. *J. Am. Chem. Soc.*, **1980**, *102*, 2786.

223. Lewis, J.W.; Sandorfy, C. *Can. J. Chem.*, **1982**, *60*, 1738.

224. Turbeville, W.; Dutta, P.K. *J. Phys. Chem.*, **1990**, *94*, 4060.

225. Yuzawa, T.; Takahashi, H.; Hamaguchi, H. *Chem. Phys. Lett.*, **1993**, *202*, 221.

226. Nurmukhametov, R.N.; Shigorin, D.N.; Kozlov, Y.I.; Puchkov, V.A. *Opt. Spektrosk.*, **1961**, *11*, 327.

227. Gegiou, D.; Lambi, E.; Hadjoudis, E. *J. Phys. Chem.*, **1996**, *100*, 17762.

228. Ledbetter, J.W. *J. Chem. Phys.*, **1982**, *86*, 2449.

229. Kownacki, K.; Kaczmarek, L.; Grabowska, A. *Chem. Phys. Lett.*, **1993**, *210*, 373.

230. Grabowska, A.; Kownacki, K.; Kaczmarek, Ł. *Acta Phys. Pol. A*, **1995**, *88*, 1081; *J. Lumin.*, **1994**, *60/61*, 886.

231. Kownacki, K.; Mordziński, A.; Wilbrandt, R.; Grabowska, A. *Chem. Phys. Lett.*, **1994**, *227*, 270.

232. Mordziński, A.; Kownacki, K.; Wilbrandt, R.; Rios, C. *J. Inf. Rec. Mats.*, **1994**, *21*, 601.

233. Takeda, S.; Chichara, H.; Inabe, T.; Mitani, T.; Maruyama, Y. *Chem. Phys. Lett.*, **1992**, *189*, 13.

234. Hoshino, N.; Inabe, T.; Mitani, T.; Maruyama, Y. *Bull. Chem. Soc. Jpn.*, **1988**, *61*, 4207.

235. Chichibabin, A.E.; Kuindshi, B.; Benewolenskaja, S.V. *Ber. Dtsch. Chem. Ges.*, **1925**, *58B*, 1580.

236. Sixl, H.; Warta, R. *Chem. Phys.*, **1985**, *94*, 147.

237. Seff, K.; Trueblood, K.N. *Acta Cryst.*, **1968**, *B24* 1406.

238. Corval, A.; Kuldová, K.; Eichen, Y.; Pikramenou, Z.; Lehn, J.-M.; Trommsdorff, H.P. *J. Phys. Chem.*, **1996**, *100*, 19315, and references therein.

239. Eichen, Y.; Lehn, J.-M.; Scherl, M.; Haarer, D.; Casalegno, R.; Corval, A.; Kuldová, K.; Trommsdorff, H.P. *J. Chem. Soc., Chem. Commun.*, **1995**, 713.

240. Scherl, M.; Haarer, D.; Fischer, J.; DeCian, A.; Lehn, J.-M.; Eichen, Y. *J. Phys. Chem.*, **1996**, *100*, 16175.

241. Clark, W.; Lothian, G. *Trans. Faraday Soc.*, **1958**, *54*, 1790.

242. Frank, I.; Grimme, S.; Peyerimhoff, S.D. *J. Phys. Chem.*, **1996**, *100*, 16187.

243. Nagaoka, H.; Hirota, N.; Sumitami, M.; Yoshihara, K. *J. Am. Chem. Soc.*, **1983**, *105*, 4220.

244. Frolov, Y.L.; Sapozhnikov, Y.M.; Barer, S.S.; Pagodaeva, N.N.; Tyukavkina, N.A. *Izv. Akad. Nauk USSR Ser. Khim.*, **1974**, 10, 2364.

245. Sengupta, P.K.; Kasha, M. *Chem. Phys. Lett.*, **1979**, *68*, 382.

246. McMorrow, D.; Kasha, M. *J. Am. Chem. Soc.*, **1983**, *105*, 5133.

247. McMorrow, D.; Kasha, M. *J. Phys. Chem.*, **1984**, *88*, 2235.

248. McMorrow, D.; Kasha, M. *Proc. Natl. Acad. Sci. U.S.A.*, **1984**, *81*, 3375.

249. McMorrow, D.; Dzugan, T.P.; Aartsma, T.J. *Chem. Phys. Lett.*, **1984**, *103*, 492.

250. Dzugan, T.P.; Schmidt, J.; Aartsma, T.J. *Chem. Phys. Lett.*, **1986**, *127*, 336.

251. Chou, P.T.; McMorrow, D.; Aartsma, T.J.; Kasha, M. *J. Phys. Chem.*, **1984**, *88*, 4596.

252. Chou, P.; Aartsma, T.J. *J. Phys. Chem.*, **1986**, *90*, 721.

253. Brewer, W.E.; Studer, S.L.; Standiford, M.; Chou, P.T. *J. Phys. Chem.*, **1989**, *93*, 6088.

254. Brewer, W.E.; Studer, S.L.; Chou, P.T.; Orton, E. *Chem. Phys. Lett.*, **1989**, *158*, 345.

255. Martínez, M.L.; Studer, S.L.; Chou, P.T. *J. Am. Chem. Soc.*, **1990**, *112*, 2427.

256. Khan, A.U.; Kasha, M. *Proc. Natl. Acad. Sci. U.S.A.*, **1983**, *80*, 1767.

257. Kasha, M. *Environmental and Process Monitoring Technologies*, **1992**, *1637*, 2.

258. Parthenopoulos, D.A.; Kasha, M. *Chem. Phys. Lett.*, **1988**, *146*, 77.

259. Chou, P.T.; Martínez, M.L.; Clements, J.H. *J. Phys. Chem.*, **1993**, *97*, 2618.

260. Wolfbeis, O.S.; Knierzinger, A.; Schipfer, R. *J. Photochem.*, **1993**, *21*, 67.

261. Woolfe, G.J.; Thistlethwaite, P.J. *J. Am. Chem. Soc.*, **1981**, *103*, 6916.

262. Itoh, M.; Tokumura, K.; Tanimoto, Y.; Okada, Y.; Takeuchi, H.; Obi K.; Tanaka, I. *J. Am. Chem. Soc.*, **1982**, *104*, 4146.

263. Itoh, M.; Fujiwara, Y.; Sumitani, M.; Yoshihara, K. *J. Phys. Chem.*, **1986**, *90*, 5672.

264. Itoh, M.; Tanimoto, Y.; Tokumura, K. *J. Am. Chem. Soc.*, **1983**, *105*, 3339.

265. Itoh, M.; Fujiwara, Y. *J. Phys. Chem.*, **1983**, *87*, 4558.

266. Itoh, M.; Kurokawa, H. *Chem. Phys. Lett.*, **1982**, *91*, 487.

267. Itoh, M.; Fujiwara, Y. *Chem. Phys. Lett.*, **1986**, *130*, 365.

268. Tokumura, K.; Kurauchi, M.; Nagata, N.; Itoh, M. *Chem. Phys. Lett.*, **1996**, *258*, 495.

269. Dick, B.; Ernsting, N.P. *J. Phys. Chem.*, **1987**, *91*, 4261.

270. Ernsting, N.P.; Dick, B. *Chem. Phys.*, **1989**, *136*, 181.

271. Mühlpfordt A.; Bultmann, T.; Ernsting, N.P.; Dick, B. *Chem. Phys.*, **1994**, *181*, 447.

272. Dick, B. *J. Phys., Chem.* **1990**, *94*, 5752.

273. Peluso, A.; Adamo, C.; Del Re, G. *J. Math. Chem.*, **1992**, *10*, 249.

274. Brucker, G.A.; Swinney, T.C.; Kelley, D.F. *J. Phys. Chem.* **1991**, *95*, 3190.

275. Swinney, T.C.; Kelley, D.F. *J. Phys. Chem.*, **1991**, *95*, 10369.

276. Swinney, T.C.; Kelley, D.F. *J. Phys. Chem.*, **1993**, *99*, 211.

277. Brucker, G.A.; Kelley, D.F. *J. Phys. Chem.*, **1989**, *93*, 5179.

278. Brucker, G.A.; Kelley, D.F. *J. Phys. Chem.*, **1987**, *91*, 2856.

279. Brucker, G.A.; Kelley, D.F. *J. Phys. Chem.*, **1988**, *92*, 3805.

280. Parsapour, F.; Kelley, D.F. *J. Phys. Chem.*, **1996**, *100*, 2791.

281. Schwartz, B.J.; Peteanu, L.A.; Harris, C.B. *J. Phys. Chem.*, **1992**, *96*, 3591.

282. Strandjord, A.J.G.; Barbara, P.F. *J. Phys. Chem.*, **1985**, *89*, 2355.

283. Strandjord, A.J.G.; Smith, D.E.; Barbara, P.F. *J. Phys. Chem.*, **1985**, *89*, 2362.

284. Strandjord, A.J.G.; Courtney, S.H.; Friedrich, D.M.; Barbara, P.F., *J. Phys. Chem.* **1983**, *87*, 1125.

285. Strandjord, A.J.G.; Barbara, P.F. *Chem. Phys. Lett.*, **1983**, *98*, 21.

286. Salman, O.A.; Drickamer, H.G. *J. Chem. Phys.*, **1981**, *75*, 572.

287. Sepioł, J.; Kołos, R. *Chem. Phys. Lett.*, **1990**, *167*, 445.

288. Dawes, J.M.; Wallace, S.C. *Chem. Phys. Lett.*, **1993**, *208*, 335.

289. Ormson, S.M.; Brown, R.G.; Vollmer, F.; Rettig, W. *J. Photochem. Photobiol. A: Chem.*, **1994**, *81*, 65.

290. Ormson, S.M.; LeGourrierec, D.; Brown, R.B.; Foggi, P. *J. Chem. Soc., Chem. Commun.*, **1995**, 2133.

291. Dharia, J.R.; Johnson, K.; Schlenoff, J.B. *Macromolecules*, **1994**, *27*, 5167.

292. Sarkar, M.; Sengupta, P.K. *Chem. Phys. Lett.*, **1991**, *179*, 68.

293. Sarkar, M.; Ray, J.G.; Sengupta, P.K. *Spectrochim. Acta A*, **1996**, *52*, 275.

294. Douhal, A.; Amat-Guerri, F.; Acuña, A.U. *Angew. Chem. Int. Ed. Engl.*, **1997**, *36*, 1514.

295. Douhal, A. *Ber. Bunsenges. Phys. Chem.*, **1998**, *102*, 448.

296. Kondo, M. *Bull. Chem. Soc Jpn.*, **1978**, *51*, 3027.

297. Brown, R.G.; Entwistle, N.; Hepworth, J.D.; Hodgson, K.W.; May, B. *J. Phys. Chem.*, **1982**, *86*, 2418.

298. Song, P.-S.; Sun, M.; Koziołowa, A.; Kozioł, J. *J. Am. Chem. Soc.*, **1974**, *96*, 4319.

299. Fugate, R.D.; Song, P.-S. *Photochem. Photobiol.*, **1976**, *24*, 479.

300. Koziołowa, A. *Photochem. Photobiol.*, **1979**, *29*, 459.

301. Choi, J.D.; Fugate, R.D; Song, P.-S. *J. Am. Chem. Soc.* **1980**, *102*, 5293.

302. Borgis, D.; Hynes, J.T.; *Chem. Phys.*, **1993**, *170*, 315; Borgis, D.; Lee, S.; Hynes, J.T. *Chem. Phys. Lett.*, **1989**, *162*, 19.

303. Cheshnovsky, O.; Leutwyler, S. *Chem. Phys. Lett.*, **1985**, *121*, 1.

304. Cheshnovsky, O.; Leutwyler, S. *J. Chem. Phys.*, **1988**, *88*, 4127.

305. Knochenmuss, R. *Chem. Phys. Lett.*, **1998**, *293*, 191.

306. Knochenmuss, R.; Muiño, P.; Wickleder, C. *J. Phys. Chem.*, **1996**, *100*, 11218.

307. Knochenmuss, R.; Cheshnovsky, O.; Leutwyler, S. *Chem. Phys. Lett.*, **1988**, *144*, 317.

308. Knochenmuss, R.; Leutwyler, S. *Chem. Phys.*, **1989**, *92*, 1268.

309. Knochenmuss, R.; Holtom, G.; Ray, D. *Chem. Phys. Lett.*, **1993**, *215*, 188.

310. Knochenmuss, R.; Smith, D.E. *J. Chem. Phys.*, **1994**, *101*, 7327.

311. Knochenmuss, R.; *J. Chim. Phys.*, **1995**, *92*, 445.

312. Bulska, H. *Chem. Phys. Lett.*, **1998**, *83*, 398.

313. Bulska, H.; Grabowska, A.; Grabowski, Z.R. *J. Lumin.*, **1986**, *35*, 189.

314. Zhang, H.; van der Meulen, P.; Glasbeek, M. *Chem. Phys. Lett.*, **1996**, *253*, 97.

315. Marks, D.; Zhang, H.; Glasbeek, M.; Borowicz, P.; Grabowska, A. *Chem. Phys. Lett.*, **1997**, *275*, 370.

316. Glasbeek, M.; Marks, D.; Zhang, H. *J. Lumin.*, **1997**, *72–74*, 832.

317. Grabowska, A.; Kaczmarek, Ł. *Pol. J. Chem.*, **1992**, *60*, 715.

318. Vogel, E.; Kocher, M.; Schmickler, H.; Lex, J. *Angew. Chem. Int. Ed. Engl.*, **1987**, *25*, 257.

319. Sessler, J.L.; Brucker, E.A.; Weghorn, S.J.; Kisters, M.; Schäfer, M.; Lex, J.; Vogel, E. *Angew. Chem. Int. Ed. Engl.*, **1994**, *33*, 2308.

320. Callot, H.J.; Rohrer, A.; Tschamber, T.; Metz, B. *New. J. Chem.*, **1995**, *19*, 155.; Vogel, E.; Bröring, M.; Weghorn, S.J; Scholz, P.; Deponte, R.; Lex, J.; Schickler, H.; Schaffner, K.; Braslavsky, S.E.; Müller, M.; Pörting, S.; Fowler, C.J.; Sessler, J.L. *Angew. Chem. Int. Ed. Engl.*, **1997**, *36*, 1651.

321. Vogel, E.; Bröring, M.; Erben, C.; Demuth, R.; Lex, J.; Nendel, M.; Houk, K.N. *Angew. Chem. Int. Ed. Engl.* **1997**, *36*, 353; (b) Vogel, E.; Scholz, P.; Demuth, R.; Erben, C.; Bröring, M.; Schmickler, H.; Lex, J.; Hohlneicher, G.; Bremm, D.; Wu, Y.-D. *Angew. Chem. Int. Ed. Engl.* **1999**, *38*, 2919.

322. Solov'ev, K.N.; Zalesski, I.E.; Kotlo, V.N.; Shkirman, S.F. *Phys. Lett.*, **1973**, *17*, 332.

323. Zalesski, I.E.; Kotlo, V.N.; Sevchenko, A.N.; Solov'ev, K.N.; Shkirman, S.F. *Sov. Phys.-Dokl. (Eng. Transl.)*, **1973**, *17*, 1183.

324. Korotaev, O.N.; Personov, R.I. *Opt. Spectrosc. (Engl. Transl.)*, **1972**, *32*, 479.

325. Völker, S.; van der Waals, J.H. *Mol. Phys.*, **1976**, *32*, 1703.

326. Völker, S.; McFarlane, R.M.; Genack, A.Z.; Trommsdorff, H.P.; van der Waals, J.H. *J. Chem. Phys.*, **1977**, *67*, 1759.

327. Völker, S.; McFarlane, R.M. *IBM J. Res. Dev.*, **1979**, *23*, 547.

328. Butenhoff, T.J.; Chuck, R.S.; Limbach, H.-H.; Moore, C.B. *J. Phys. Chem.*, **1990**, *94*, 7847; Butenhoff, T.J.; Moore, C.B. *J. Am. Chem. Soc.*, **1988**, *110*, 8336.

329. Radziszewski, J.; Waluk, J.; Michl, J. *Chem. Phys.*, **1989**, *136*, 165.

330. Radziszewski, J.; Waluk, J.; Michl, J. *J. Mol. Spectrosc.*, **1990**, *140*, 373.

331. Radziszewski, J.; Waluk, J.; Nepraš, M.; Michl, J. *J. Phys. Chem.*, **1991**, *95*, 1963.

332. Radziszewski, J.; Nepraš, M.; Balaji, V.; Waluk, J.; Vogel, E.; Michl, J. *J. Phys. Chem.*, **1995**, *99*, 14254.

333. Waluk, J.; Müller, M.; Swiderek, P.; Köcher, M.; Vogel, E.; Hohlneicher, G.; Michl, J. *J. Am. Chem. Soc.*, **1991**, *113*, 5511.

334. Waluk, J.; Vogel, E. *J. Phys. Chem.*, **1994**, *88*, 4530.

335. Andersen, K.B.; Vogel, E.; Waluk, J. *Chem. Phys. Lett.*, **1997**, *271*, 341.

336. Sepioł, J.; Stepanenko, Y.; Vdovin, A.; Mordziński, A.; Vogel, E.; Waluk, J. *Chem. Phys. Lett.*, **1998**, *296*, 549.

CHAPTER

3

The Role of Solute–Solvent Friction in Large-Amplitude Motions

by David H. Waldeck

Department of Chemistry, University of Pittsburgh

1 Introduction

Intermolecular interactions play a central role in chemical reactions in solution, especially those involving large-amplitude conformational changes. The influence of solvent on reaction rates can be separated into two general categories. The first, and often most significant effect, is the change in the reactive system's energetics that is associated with the differential solvation of the reactant, product, and transition-state species. This effect can be very large because it modifies the activation energy for the reaction; consequently, it has been studied for many years.[1] The second category concerns dynamical features of the solute–solvent interaction and may be associated with changes in reactant geometry, solute–solvent frictional coupling, steric constraints, and more. These effects are less clearly understood, but they also affect the reaction in significant ways, and there are cases in which the effects control the reaction rate or modify the reaction coordinate.

This chapter describes current developments in the understanding of solute–solvent frictional coupling and how it affects chemical reaction dynamics, with particular regard to reactions that involve large-amplitude motions. The quantitative description of solute–solvent frictional coupling in typical organic systems remains at a crude level; nevertheless, the underlying physical interactions that are important have been defined, and the development of modern computer technology promises an adequate numerical description of them. The discussion proceeds in three stages. The first stage describes recent studies of the overall rotational motion of single

Conformational Analysis of Molecules in Excited States, Edited by Jacek Waluk
ISBN 0-471-29707-0

molecules in solution that investigate the nature of the solute–solvent interaction and its role in determining the solute–solvent friction. The second stage describes the internal torsional dynamics of bimolecular complexes of organic molecules. The third and final stage discusses the photoisomerization of organic molecules in solution, with particular regard to the solute–solvent friction dependence.

This order of presentation of topics is chosen to reflect the increasing complexity of the physical description that is required in their study. In the first stage, the overall rotational motion of the solute is that of a rigid body moving in the intermolecular solute–solvent potential. This situation is the simplest one, because the internal degrees of freedom of the solute are separated from the dynamics being studied, and one is more clearly able to focus on the solute–solvent coupling and friction. In the second stage, the complexity increases because one must now consider two rigid bodies moving relative to each other in a torsional potential. The nature of the motion and the solute–solvent interactions will depend strongly on the structure of the complex and on the properties of the potential. Lastly, the photoisomerization studies involve two parts of a molecule moving along a reactive trajectory on an intramolecular potential. In this case, the dynamics of the motion and the nature of the solute–solvent friction are strongly influenced by the potential surface. The quantitative study of solute–solvent friction in these three situations should help provide a comprehensive description of how the potential surface modifies the solute–solvent interaction.

The measurement of the overall rotational relaxation of medium-size (100 to 500 $Å^3$) solutes provides an important avenue for probing molecular dynamics in solution. The timescales investigated for overall rotational relaxation range from 10 ps to 1 ns, but can be the order of 1 ps for smaller solutes and many nanoseconds for macromolecules. This time range complements the 100-fs to 10-ps time range investigated by dynamical Stokes shift studies, another powerful probe of polar solvent dynamics. It seems clear that the longer timescale solute rotational motion in liquids may be described by diffusion models.[2–3] In this case, the measurement of one or more rotational relaxation times is connected to the solute–solvent frictional coupling through the diffusion constant and the Einstein relation. These measurements provide a method to investigate, on a short timescale, the influence of solute–solvent interactions on a transport property of the solute. Rotational relaxation studies are also distinguished from solvation studies because they can probe the anisotropic nature of the polarization field in the solvent, whereas the solvation dynamics studies probe an energy, which is scalar. The rotational motion about the different principal axes of the diffusion tensor depends on different parts of the solute–solvent interaction potential. These effects are clearly evident in molecular dynamics simulations and in some experimental systems.

The second part of this discussion describes experimental studies of bimolecular (or host–guest) complexes in liquid solution. Less work has been performed in this area because the model host–guest systems and ultrafast methods have only recently become available. The development of a quantitative understanding of the dynamics of host–guest complexes, both reactive and nonreactive, is important to a number of long-standing issues in chemistry. For example, homogeneous catalysis and enzymolysis often proceeds by bimolecular association and subsequent geometric changes; however, our understanding of the nuclear dynamics in these systems is mostly from a theoretical perspective, rather than an experimental one. Secondly, a bimolecular host–guest complex may be viewed as a bimolecular encounter pair whose reactive channel is blocked. An adequate understanding of the temporal behavior of reactive encounter pairs in liquid solution is still lacking. From another perspective, host–guest systems provide the ability to create solvation environments whose properties can be tuned synthetically (e.g., the case of an anthracene encapsulated in a carcerand[4] or bound to a guanidinium group[5]). The ability to tune the interactions in "designer" host–guest systems and probe their internal motion provides an avenue for developing a proper description of their dynamics and, ultimately, controlling it.

The third part of this discussion describes photochemical isomerization reactions that involve large-amplitude motion and significant solute–solvent frictional coupling. The ability to predict and control the outcome of photochemical isomerization reactions may be important to a number of nanotechnologies, such as "molecular motors," "molecular turnstiles," and photochemical switches.[6–7] Understanding and quantifying the solvent effects on such reactions is very challenging because one requires detailed features of the excited-state potential-energy surfaces in solution and a realistic model to describe the frictional coupling between the solute and solvent. In particular, the timescale of the motion through the transition state is largely determined by the curvature of the potential surface in this region. In addition, the solute–solvent friction will depend on the timescale of that motion because the medium's response is dependent upon frequency. Despite these challenges, significant progress has been made in this area. The description here aims to discuss a few representative systems in depth, rather than providing a survey of all the work in the area.

2 Rotational Relaxation

The mechanism for the rotational relaxation of organic molecules that are "medium sized" or larger in simple liquids is taken to be small-step diffusion.[2] This mechanism is consistent with the exponential character of the orientational correlation function and the observation that the solvent dynamics [as measured by dynamic Stokes-shift and optically heterodyned Raman-induced Kerr-effect spectroscopy (OHRIKES) studies][8] occurs on significantly shorter timescales than the solute rotational relaxation. In this case, the diffusion coefficient D may be directly related to the solute–solvent frictional coupling by the Einstein relation,

$$D = \frac{kT}{\zeta} \tag{1}$$

where k is Boltzmann's constant and ζ is the friction coefficient.[9] It is this relationship that provides the fundamental connection between the measured relaxation time(s)—(or diffusion coefficient(s))—and the solute–solvent frictional coupling. In turn, the friction coefficient is determined by features of the fluid's interparticle interaction potential. To date, most experimental studies use spectroscopic measurements to determine rotational relaxation times, which can be represented as simple combinations of the diffusion coefficients about each principal axis of the solute molecule's diffusion tensor. The evaluation of models for the solute–solvent friction proceeds by comparing the measured relaxation times with those predicted by different models.

The relationship between the observed rotational relaxation time and the diffusion coefficients depends on both the molecular geometry and the experimental method that is used to determine them. Most of the measurements described here use transient electronic spectroscopy and determine the anisotropy

$$r(t) = \frac{2}{5}\langle P_2(\boldsymbol{\mu}(t) \cdot \boldsymbol{\mu}(0))\rangle, \tag{2}$$

where $P_2(x) = \frac{1}{2}(3x^2 - 1)$ and $\boldsymbol{\mu}(t)$ is the transition dipole moment vector at time t. The brackets indicate an average over an equilibrium ensemble of solute molecules. The anisotropy is the second-order orientational correlation function, and it provides a direct measure of how the transition dipole's orientation relaxes in time. If the transition dipole is fixed in the molecular frame, a reasonable assumption for many systems, then the relaxation function provides the orientational relaxation of the entire solute molecule with time. Other techniques probe different moments of the orientational correlation function.[10] If the molecule is spherical, then the anisotropy decay is single exponential with a relaxation time

$$\tau = \frac{1}{6D}. \tag{3}$$

For other molecular shapes, the relaxation is more complex. In general, the anisotropy decay for an asymmetric rotor is given by a sum of five exponential terms:

$$\begin{aligned} r(t) = {} & \frac{6}{5}\sum_{i=1}^{3} c_i \exp\left(\frac{-t}{\tau_i}\right) + \frac{3(F+G)}{10}\exp(-(6\overline{D} - 2\Delta)t) \\ & + \frac{3(F-G)}{10}\exp(-(6\overline{D} + 2\Delta)t). \end{aligned} \tag{4}$$

In the preceding expression,

$$\overline{D} = \frac{D_1 + D_2 + D_3}{3}, \qquad \tau_i = \frac{1}{3\overline{D} + 3D_i},$$

and

$$\Delta = \sqrt{D_1^2 + D_2^2 + D_3^2 - D_1D_2 - D_1D_3 - D_2D_3},$$

where D_i is the diffusion coefficient for rotation about the principal axis i. The other parameters incorporate the direction of the transition dipole(s) in the molecular frame, such that

$$c_i = \alpha_j \alpha_k \beta_j \beta_k, \qquad F = \sum_{i=1}^{3} \alpha_i^2 \beta_i^2 - \frac{1}{3},$$

and

$$G\Delta = \sum_{i=1}^{3} D_i(\alpha_i^2 \beta_i^2 + \alpha_j^2 \beta_k^2 + \alpha_k^2 \beta_j^2) - \overline{D} \text{ for } i \neq j \neq k \neq i,$$

where the α_i's are the direction cosines defining the transition dipole orientation of the first electronic transition (pumped) and the β_i's are the direction cosines defining the transition dipole orientation of the electronic transition that is probed. Also, by probing different transition dipoles, the relative statistical weights of D_i in $r(t)$ can be changed, and one can extract them individually. Under appropriate conditions, this expression simplifies considerably.[11–12]

The rotational relaxation of a solute molecule in solution is determined by both repulsive and attractive intermolecular forces. Wolynes[13] has suggested dividing the friction coefficient into components that relate to their origin in the intermolecular potential—for example,

$$\zeta = \zeta_{HH} + \zeta_{HS} + \zeta_{SH} + \zeta_{SS}. \tag{5}$$

In this formulation, the friction ζ_{HH} arises from hard collisions on the repulsive wall of the interparticle potential, and the friction ζ_{SS} arises from soft collisions involving the attractive part of the interparticle potential. The other two contributions to the friction coefficient account for correlations between the repulsive and attractive parts of the potential. In this spirit, it has been common in experimental studies to assume that the total frictional coupling can be written approximately a sum of two parts: the electrostatic friction ζ_{df} and the "mechanical friction" ζ_{mech}. The ζ_{mech} term accounts for the friction mechanisms that are found in nonpolar solvents and corresponds primarily to the ζ_{HH} term. The ζ_{df} term accounts for friction resulting from electrostatic interactions and corresponds primarily to the ζ_{SS} term. In this approximation, one can write

$$D_i = \frac{kT}{\zeta_{df,i} + \zeta_{mech,i}}, \tag{6}$$

where D_i is the diffusion constant for rotation about axis i and the friction coefficients are evaluated for rotation about axis i. This formulation ignores coupling between the electrostatic forces and the forces contributing to ζ_{mech},—that is, the cross terms in Eq. 5. Recent molecular dynamics (MD) results[14–15] show that the resulting approximation does not always hold and that the coupling between these terms can significantly affect the relaxation. Nevertheless, the approximation simplifies the analysis and has been useful.

2.1 Continuum Models for the Friction

Continuum models treat the solvent as a continuous medium that can be characterized by a few parameters, such as the shear viscosity η or the dielectric constant ε. Although such models do not capture the full molecular character of the solute–solvent interactions, they greatly simplify the theoretical considerations and do capture important primary features of the solute–solvent frictional coupling. Three different continuum-based models are sketched here. First, the hydrodynamic model is outlined. This model has been the primary reference for most considerations of the solute–solvent frictional coupling and is used frequently. Second, dielectric friction models are outlined. These models aim to capture the friction that arises from the interaction between a solute molecules' electrostatic field and the polarization field it creates in the solvent. Last, the ion atmosphere model for frictional coupling, relevant to electrolyte solutions, is discussed.

Hydrodynamic Friction

The standard hydrodynamic [Debye–Stokes–Einstein (DSE)] model[11–12,16] treats the solvent surrounding the probe molecule as a continuous viscous medium. In this approximation, the diffusion coefficient of the solute about axis i is given by

$$D_{\mathrm{i}} = \frac{kT}{6V\eta f_{\mathrm{i}} C_{\mathrm{i}}}, \tag{7}$$

where V is the volume of the solute, η is the viscosity of the solvent, f_{i} is a shape factor that accounts for the nonsphericity of the solute, and C_{i} is a parameter that is determined by the hydrodynamic boundary conditions and the shape of the solute. The value of C_{i} can range from zero in the slip limit for a spherical solute to one in the stick limit. The stick boundary condition requires that the relative solute–solvent velocity at the surface of the solute be zero—that is, it corresponds to strong solute–solvent coupling (Fig. 3.1). In contrast, slip hydrodynamics is appropriate when the solvent exerts no tangential forces on the solute and $C_{\mathrm{i}} = 0$ would indicate no

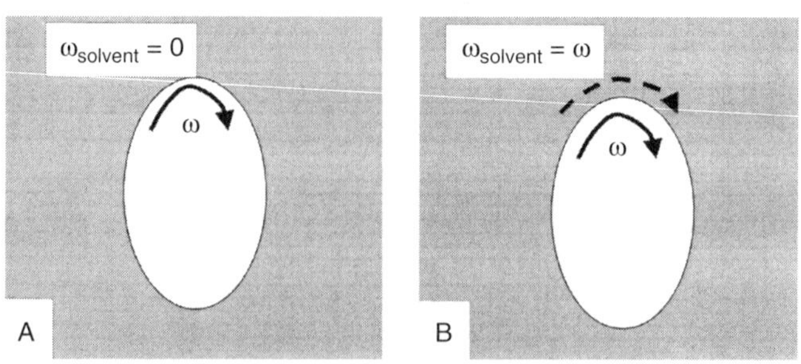

Figure 3.1. Two boundary conditions for the mechanical friction of a solute rotating in a solvent. Panel A illustrates the slip boundary condition, where the velocity of the first layer of the solvent is zero, but the molecule rotates at ω. Panel B illustrates the stick boundary condition, where the angular velocity of the solute and the solvent are the same.

interaction between solute and solvent. The rotational relaxation of nonpolar solutes in nonpolar solvents is reasonably described by a slip hydrodynamic model,[11,17–18] as long as the solvent molecules are not significantly larger than those of the solute.

The evaluation of the friction coefficients for the rotational relaxation is straightforward for smooth solute shapes. For the case of a spherical solute rotating under a stick boundary condition, both f_i and C_i are equal to unity. For symmetric ellipsoids, one requires two unique diffusion coefficients: $D_\parallel$, for rotation about the axis of symmetry, and $D_\perp$, for rotation about the other axes. In this case, the C_i's are still equal to unity, but the shape factors are given by

$$f_\perp = \frac{2(\rho^4 - 1)}{3\rho[(2\rho^2 - 1)S - \rho]} \tag{8}$$

and

$$f_\parallel = \frac{2(\rho^2 - 1)}{3\rho(\rho - S)}, \tag{9}$$

where ρ is the axial ratio of the ellipsoid and

$$S = \frac{\tan^{-1}\left(\dfrac{\sqrt{1-\rho^2}}{\rho}\right)}{\sqrt{1-\rho^2}} \tag{10}$$

for an oblate rotor and

$$S = \frac{\ln[\rho + \sqrt{\rho^2 - 1}]}{\sqrt{\rho^2 - 1}} \tag{11}$$

for a prolate rotor. For an asymmetric ellipsoid, C_i is again unity, but the f_i's are more complicated. For rotation about each of the three axes ($i = 1, 2, 3$) with radii a, b, and $c (a > b > c)$, one finds that

$$f_a = \frac{2}{3abc}\,\frac{b^2 + c^2}{b^2\displaystyle\int_0^\infty \frac{1}{(b^2+s)r(s)}ds + c^2\int_0^\infty \frac{1}{(c^2+s)r(s)}ds}, \tag{12}$$

$$f_b = \frac{2}{3abc}\,\frac{a^2 + c^2}{a^2\displaystyle\int_0^\infty \frac{1}{(a^2+s)r(s)}ds + c^2\int_0^\infty \frac{1}{(c^2+s)r(s)}ds}, \tag{13}$$

$$f_c = \frac{2}{3abc}\,\frac{b^2 + a^2}{b^2\displaystyle\int_0^\infty \frac{1}{(b^2+s)r(s)}ds + a^2\int_0^\infty \frac{1}{(a^2+s)r(s)}ds}, \tag{14}$$

and

$$r(s) = \sqrt{(a^2 + s)(b^2 + s)(c^2 + s)}. \tag{15}$$

These elliptic integrals must be evaluated numerically; this is a straightforward operation with modern mathematical software such as Mathematica or MathCAD. The evaluation of the friction for the slip boundary condition is more involved than that for the stick limit. Typically, workers have computed the friction coefficients in the stick boundary condition and converted it to the slip case with tabulated conversion factors. Earlier tables were found to be in error, but Hochstrasser and Sension recently published a corrected table of slip friction coefficients.[18]

The relationship of these idealized shapes (smooth ellipsoids) to more realistic molecular shapes has been addressed in a few studies.[19–22] A bumpy cylinder model and bumpy sphere models have been used to show that a slip boundary condition applied locally to a rough rotating body can give an overall boundary condition that appears to be stick.[19] Ben-Amotz and co-workers[17,23] investigated how the change in roughness features (corrugations expected to arise from the discrete atoms) of a solute changes the overall coupling. As the size of a solute molecule increases, the effect of roughness features of the molecule becomes more significant, and the observed hydrodynamic boundary condition evolves from the slip limit to the stick limit.

A number of workers have developed models for the hydrodynamic boundary condition that interpolate between the stick and slip limits.[16,22,24] For example, in Hynes, Kapral, and Weinberg's model,[22] the solute's linear and angular momenta are transferred to a layer of nearest neighbor solvent molecules by collisions between the molecules. In the rotational diffusion limit for rough spheres, the rotational relaxation time is expressed as

$$\frac{1}{\tau} = \frac{kT}{\eta V_{\text{solute}}}\left(1 + \frac{3\eta}{\beta}\right), \tag{16}$$

where β is a parameter that is determined by the solvent–solute volume ratio, solvent number density, and solute–solvent pair distribution function. When $V_{\text{solute}}/V_{\text{solvent}} \gg 1$ and, correspondingly, $\beta \gg 1$, this equation approaches the limit of a stick boundary condition. In the opposite limit, it approaches the slip boundary condition. Hence, the Hynes–Kapral–Weinberg model uses the molecular properties of the solute and the solvent (e.g. relative volumes) to replace the traditional stick and slip boundary conditions.

Dielectric Friction

A number of workers have used rotational diffusion measurements of organic solutes in organic solvents to probe the importance of dielectric friction[25–26] The early work demonstrated that the total friction did not correlate with a single solvent parameter, such as the shear viscosity. Later work investigated the rotational relaxation of similarly shaped, but electrostatically different, solutes[27–32] or the same solute in different electronic states.[28c,33] This work has demonstrated the importance of the solute molecule's charge distribution on the solute–solvent frictional coupling. The initial attempts to quantify the influence of the electrostatic field on the frictional coupling concentrated on the Nee–Zwanzig model or other point-source models of the frictional coupling.[34–36] More recent work shows that the extended nature of the solute molecule's charge distribution must be considered in evaluating the magnitude of the dielectric friction.[28,30–31,37] Although such models perform significantly better than point-source models, they are still limited by the treatment of the solvent as a continuum.

The idea of dielectric friction dates from the work of Born[38] on the translational diffusion of ions through a continuous dielectric medium. In 1970, Nee and Zwanzig[34] developed a model for the dielectric friction on a point dipole rotating in a spherical cavity. In the zero-frequency limit, they found that

$$\zeta_{\mathrm{NZ}} = \zeta = \frac{6\mu^2}{a^3}\frac{(\varepsilon_s - 1)}{(2\varepsilon_s + 1)^2}\tau_{\mathrm{D}}, \tag{17}$$

where ε_s is the static dielectric constant, τ_D is the Debye relaxation time, a is the radius of the spherical cavity, and μ is the dipole moment. In this model, a point dipole is placed in a spherical cavity that is immersed in a dielectric continuum with a frequency-dependent dielectric constant $\varepsilon(\omega)$. The dipole creates a polarization field in the solvent, which, in turn, couples to the dipole. The cartoon in Figure 3.2 illustrates the essential physics of this model. The diagram shows a point dipole μ that is rotating with an angular frequency ω inside a spherical cavity that is immersed in a frequency-dependent dielectric medium. The interaction between the induced solvent polarization and the solute is described by the dot product of the reaction field R and the solute dipole moment.[37] Because the dielectric has a frequency-dependent response, a phase lag $\phi(\omega)$ is generated between the rotating dipole and the reaction field. This lag causes a torque to act on the rotating dipole, coupling the solute to the solvent and leading to dissipation of energy.

Nee and Zwanzig[34] solved this model without any consideration of hydrodynamic effects (i.e., the hydrodynamic boundary condition of the spherical cavity was taken to be slip). The model was later extended to incorporate hydrodynamic effects by Felderhof and others,[35] who included the polarization of a point dipole into the hydrodynamic stress tensor and solved the linearized Navier–Stokes equation. Despite such considerations, these models cannot account for medium-sized organic

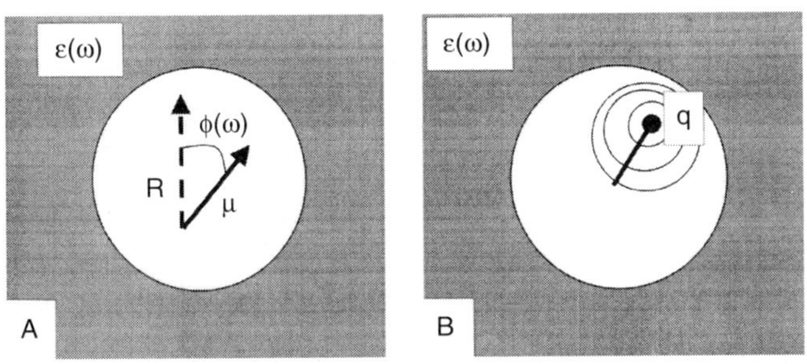

Figure 3.2. Two models for the dielectric friction of a body rotating in a dielectric continuum described by $\varepsilon(\omega)$. Panel A corresponds to the point dipole model and panel B corresponds to a charge that is displaced from the center of the cavity.

molecules, because the models do not provide a realistic description of the molecular charge distribution and the polarization field the charge creates in the solvent. Largely for this reason, such models underestimate the magnitude of the dielectric friction.

Alavi[37] generalized Nee and Zwanzig's treatment by considering a distribution of point charges inside a cavity immersed in a dielectric medium. Figure 3.2B illustrates this idea by showing the polarization created by a charge that is displaced from the center of the cavity. By the principle of superposition, one is then able to describe the polarization for an arbitrary charge distribution in the cavity and obtain an expression for the dielectric friction. By assigning partial charges q_i to the different atomic positions (r_i, θ_i, ϕ_i) of a solute molecule, one can mimic its electrostatic potential.[39] In the zero-frequency limit, Alavi found that the friction is given by

$$\zeta_{AW} \equiv \zeta = \frac{2\tau_D}{a}\sum_{j=1}^{N}\sum_{i=1}^{N}\sum_{L=1}^{\infty}\sum_{M=1}^{L}\left(\frac{2L+1}{L+1}\right)\frac{(L-M)!}{(L+M)!}M^2 q_i q_j\left(\frac{r_i}{a}\right)^L\left(\frac{r_j}{a}\right)^L \times\left[\frac{\varepsilon_s-\varepsilon_c}{\varepsilon_s+\varepsilon_c\dfrac{L}{L+1}}\right]P_L^M(\cos\theta_i)P_L^M(\cos\theta_j)\cos M\phi_{ji}, \tag{18}$$

where the $P_L^M(x)$ are the Legendre polynomials, ε_c is the dielectric constant of the cavity, and N is the number of atomic sites in the molecule. For highly polar solvents ($\varepsilon_s > 15$) and letting $\varepsilon_c = 1$, this result is well approximated by

$$\zeta_{AW} \equiv \zeta = \frac{8(\varepsilon_s-1)\tau_D}{(2\varepsilon_s+1)^2 a}\sum_{j=1}^{N}\sum_{i=1}^{N}\sum_{L=1}^{\infty}\sum_{M=1}^{L}\left(\frac{2L+1}{L+1}\right)\frac{(L-M)!}{(L+M)!} \times M^2 q_i q_j\left(\frac{r_i}{a}\right)^L\left(\frac{r_j}{a}\right)^L P_L^M(\cos\theta_i)P_L^M(\cos\theta_j)\cos M\phi_{ji} \tag{19}$$

The model predicts values of friction that are much larger than those predicted by point-source models. For example, a system with two point charges of $0.1e$ that are 4.5 Å apart in a 5.0-Å cavity has a value of friction that is 90-times larger than the point-dipole result.[37] Because of its more realistic treatment of the solute's charge distribution, this model provides a better agreement with experiment.

Van der Zwan and Hynes developed a "semiempirical" model that relates the dielectric friction ζ_{ZH} with the solvation time τ_S and the Stokes shift S:[36]

$$\zeta_{ZH} = \frac{\mu^2 S\tau_S}{\Delta\mu^2}, \tag{20}$$

Here, μ is the solute's dipole moment, $\Delta\mu$ is the change in the dipole moment between the ground state and the excited state of the solute, and S is the Stokes shift, which describes the energy difference between the 0–0 transitions for absorption and fluorescence. The Stokes shift reflects the difference in solvation of the ground electronic state and the excited electronic state that is caused by the change in dipole moment upon excitation of the solute. Initially, the excited solute is in the equilibrium polarization field of the solvent for the solute's ground electronic state. If the solute dipole moment changes upon excitation, the solvent polarization will readjust to the new value of the dipole moment. In the limit of linear response, the response function may be used to find the dielectric friction. In the language of the continuum model, the Stokes shift measures the strength of the reaction field and τ_S measures its relaxation time. Hence, this method uses empirical measures from the molecule's spectrum and solvation dynamics to predict the dielectric friction. The model represents an improvement over the Nee–Zwanzig approach, but it is still based on a point-dipole description of the electronic properties of the solute.

For the organic systems that are described here, the mechanical and dielectric contributions to the frictional coupling can be similar in magnitude. Hence, the way in which the mechanical contribution to the frictional coupling ζ_{mech} is modeled can have a significant impact on the conclusions one draws concerning the accuracy of different electrostatic models for the coupling. A number of studies[23,30,40–41] indicate that the mechanical part of the frictional coupling can be reasonably described with a slip boundary condition, and this approach has been used in research. A second method that is used to model the mechanical part of the friction is to measure the relaxation time of a nonpolar molecule that is of similar size and shape to the molecule of interest. As long as the polar and nonpolar molecule have similar solvation environments, this method should provide a reliable measure of the mechanical component of the frictional coupling. The third method, employed in a couple of studies,[28c,33] is to measure the rotational relaxation of a single solute in different electronic states. This approach allows one to assess how the difference in the electrostatic field of the solute's electronic states affects the dielectric friction.

Ion Atmosphere Friction

Electrostatic contributions to frictional coupling are expected to be important in electrolyte solutions; however, experimental studies of the rotational relaxation of organic solutes in electrolytes are quite limited.[25,27,42] A nonuniform distribution of charged species in the solution creates an additional mechanism for friction, termed ion atmosphere friction. This type of friction is caused by the finite response time of the ion atmosphere about the solute, which creates a torque that opposes the probe molecule's rotational motion and increases the friction (Fig. 3.3). The ion atmosphere has a particularly slow relaxation time compared with the solute rotation, because it requires the displacement of ions through translational diffusion. However, the ion atmosphere also screens the electrostatic potential of the solute, which decreases the solute–solvent coupling and reduces the friction. A quantitative description of ion atmosphere friction was proposed by van der Zwan and Hynes,[43] who treated a solute molecule as a point dipole μ that rotates inside a spherical cavity of radius a and is surrounded by an ion atmosphere. In their model, the ion atmosphere friction is given by

$$\zeta_{IA} = \frac{3\mu^2}{2a^2} \frac{\varepsilon_S y}{D(2\varepsilon_S + 1)(2\varepsilon_S + 1 + \varepsilon_S y)}, \tag{21}$$

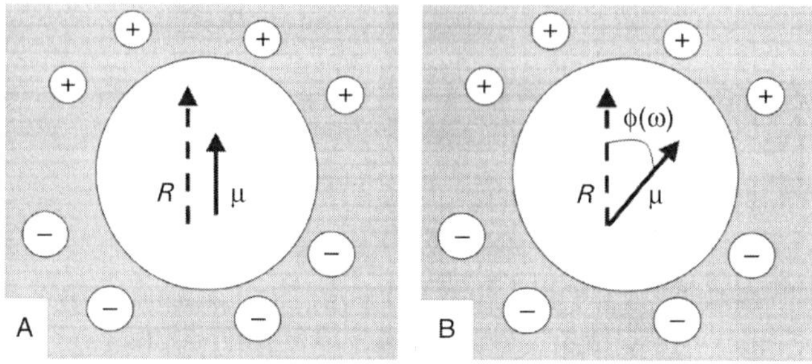

Figure 3.3. Mechanism of ion atmosphere friction, showing the direction of the solute's dipole moment μ and the reaction field *R* before (Panel A) and after (Panel B) rotation of the solute.

where $y = (\kappa a)^2/(1 + \kappa a)$, D is the ions' ambipolar diffusion constant, and κ is a reciprocal Debye length. The Debye length characterizes the spatial extent of screening for the electrostatic potential of the solute by the "ion cloud" and is given by

$$\text{Debye length} = \frac{1}{\kappa} = \sqrt{\frac{\varepsilon_S kT}{4\pi q^2 c}}, \tag{22}$$

where q is the charge and c is the concentration of ions in the solution. Figure 3.4 plots the ion atmosphere friction constant versus κa, scaled by μ^2/Da for a DMSO electrolyte solution.[44] The plot demonstrates that at small κa (and therefore low concentration of ions) the ion atmosphere friction increases as the ion concentration increases. The effect of changing the dielectric parameters of the solution (in particular, the decrease in the solution's dielectric constant) dominates at smaller concentrations of electrolyte ($\kappa a < 2.5$ or $c < 0.8$ M in DMSO). At higher concen-

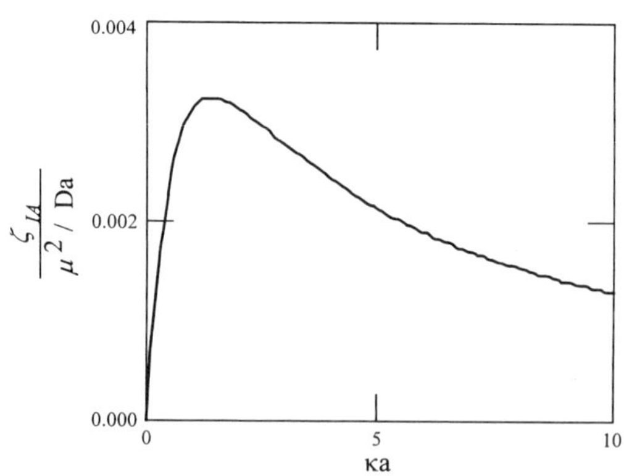

Figure 3.4. The ion atmosphere friction constant, scaled by μ^2/Da, is plotted against κa for DMSO electrolyte solution. Taken from Reference 44.

trations, the screening of the electrostatic potential by the ion atmosphere becomes more significant, which causes the ion atmosphere friction to decrease. In addition, it appears that the value of the ion atmosphere friction is low compared to the hydrodynamic and dielectric friction for the system studied. At an ion concentration of 0.8 M in DMSO, which corresponds to the maximum value of the ion atmosphere friction, this effect accounts for only 3% of the overall rotational relaxation time of solutes that have been studied (resorufin and thionine).

2.2 Molecular Models for the Friction

The description of solute–solvent interactions in polar associated systems and electrolytic solution requires appropriate consideration of the solvent structure. For strongly interacting systems, a detailed analysis of the specific solute–solvent interactions should be addressed in describing the relaxation processes. The work that attempts to include solvent structure has proceeded in two ways: via the development of analytically tractable models and through molecular dynamics simulations.

Models

An early development that sought to include the discontinuous nature of the solvent in the determination of the friction was proposed by Gierer and Wirtz (GW),[16,45] who represented the solvent around a spherical solute as a combination of several layers, each following the solute rotation less strongly as the distance from the solute molecule increases. This model provides a more general description of the boundary conditions, and the parameter C in Eq. 7 becomes

$$\frac{1}{C} = \frac{6}{\sigma} \sum_{m=1}^{\infty} \frac{\sqrt[3]{V_s/V_p}}{\left[1 + 2m\sqrt[3]{V_s/V_p}\right]^4}, \tag{23}$$

where

$$\sigma = \frac{\sum\limits_{m=1}^{\infty} \left[1 + 2m\sqrt[3]{V_S/V_p}\right]^{-4}}{\sum\limits_{m=0}^{\infty} \left[1 + 2m\sqrt[3]{V_S/V_p}\right]^{-4}}, \tag{24}$$

V_S and V_p are the volumes of the solvent and the solute molecules, respectively, and m is an index for the solvent shell. Comparison of this model with experimental rotational diffusion studies is becoming more common,[46] and recently the model was extended to describe the rotational relaxation of particles with ellipsoidal shapes.[47]

The concept of free volume and its impact on diffusion can be traced over many years.[48] Dote, Kivelson, and Schwartz (DKS) used this concept to construct a quasi-hydrodynamic model for rotational relaxation.[16] They assumed that the solvent-free space plays a dominant role in determining the hydrodynamic boundary condition when the solvent and solute size are similar. In their treatment, the boundary condition depends on both the free volume available to the solute and the relative size of solute and solvent in such a way that

$$C = \frac{\phi V_{\mathrm{p}}}{\phi V_{\mathrm{p}} + \gamma V_{\mathrm{p}}}, \tag{25}$$

where ϕV_{p} is the "effective rotational volume" of the solute and γV_{p} represents the free space surrounding the solute molecule. The DKS model successfully describes the rotational correlation time of small to intermediate-size solutes in alkane solvents, as well as the dependence of solute rotational correlation times on the solvent size in alkanes.[16,46,49] The model has been used to account for the observed change in boundary condition between alkane and alcohol solvents on a qualitative level, but is unable to predict the change quantitatively.

Dielectric Response Models

The simple continuum treatments of dielectric friction that were described earlier ignore important aspects of the solvent structure. In principle, these treatments can be generalized to include structural features of the solvent through a wave-vector-dependent dielectric response. The extended charge distribution model, Eq. 19, can be generalized to account for the different dielectric response of the solvent, which arises from spatial variations of the solute molecule's charge distribution on small length scales. The bulk dielectric constant of the solvent is unable to describe the solvent's response when the length scale of the solute's charge variation becomes similar to the characteristic length scale of solvent structures (e.g., the interatomic distances in the solvent molecules). In associated solvents (such as alcohols with their hydrogen-bonding network), one expects the dielectric response to depend on the longer-length-scale structural features that arise from the intermolecular association, which, for alcohols, would correspond to hydrogen-bonded clusters.[29,31,37]

It is possible to crudely capture the effect of the solvent's structural features by assuming that the static dielectric constant has the form

$$\varepsilon(L) = \varepsilon_{\mathrm{S}} \text{ for } L \leq L_{\max}; \qquad \varepsilon(L) = 1 \text{ for } L > L_{\max}, \tag{26}$$

where $L_{\max}$ is a cutoff value for L.[29,31,37] This approximation assumes that for values of L above $L_{\max}$, the solvent is unable to respond to spatial variations in the charge distribution of the solute.

With this approximation in mind, Eq. 19 becomes

$$\zeta_{\mathrm{AW}} \equiv \zeta = \frac{8(\varepsilon_{\mathrm{S}} - 1)\tau_{\mathrm{D}}}{(2\varepsilon_{\mathrm{S}} + 1)^2 a} \sum_{\mathrm{j}=1}^{N} \sum_{\mathrm{i}=1}^{N} \sum_{L=1}^{L_{\max}} \sum_{M=1}^{L} \left(\frac{2L+1}{L+1}\right) \frac{(L-M)!}{(L+M)!} \times M^2 q_{\mathrm{i}} q_{\mathrm{j}} \left(\frac{r_{\mathrm{i}}}{a}\right)^L \left(\frac{r_{\mathrm{j}}}{a}\right)^L P_{\mathrm{L}}^{\mathrm{M}}(\cos\theta_{\mathrm{i}}) P_{\mathrm{L}}^{\mathrm{M}}(\cos\theta_{\mathrm{j}}) \cos M\phi_{\mathrm{ji}} \tag{27}$$

In unassociated solvents (like DMSO or benzonitrile), one can estimate the expected magnitude of $L_{\max}$ by comparing it with the characteristic diameter of a solvent molecule. In other words, if the charge distribution varies on shorter distance scales than the molecular size, no polarization is induced, and if the charge distribution varies on longer distance scales than the molecular size, the polarization is described by the static dielectric constant (Eq. 26). Using this approach to choose the value for L_{max} in unassociated solvents one finds that, the value obtained is significantly higher than that needed for the sum over L to converge (e.g., L_{max} was found to be 25 in DMSO and 15 in benzonitrile). For alcohol solvents, the value of L_{max} should be chosen to match the length scale of the hydrogen-bonded clusters of the solvent molecules.

This approximation for the dielectric constant's dependence on solvent structure is quite crude. More realistic models[50–52] are now available for nonlocal, or wave-vector-dependent, dielectric constants of liquids. It will be worthwhile to incorporate these more molecular descriptions of a solvent's dielectric response into a description of the dielectric friction with an extended charge distribution. Bagchi[51] proposed a model for the wave-vector-dependent dielectric constant that expresses its longitudinal and transverse components in terms of polarization correlation functions, which include the effects of intermolecular interactions. He explored the different wave-vector regimes and showed their effect on the solvation dynamics and dielectric friction. The continuum model description of the dielectric constant was found to be appropriate for small values of the wave vector **k** ($\mathbf{k} < 0.4\,\text{Å}^{-1}$). The model breaks down completely at larger values of **k** when nearest-neighbor interactions and the microscopic structure of the liquid become important. Friedman has developed a molecular theory of solvation dynamics in which the solute–solvent coupling is formulated in terms of intermolecular interactions.[52] Both the solute and the solvent are represented by using an interaction site model (ISM), and the solute–solvent interactions are represented by site–site correlation functions between the partial charges on the particular sites of the molecules. The potential energy of solute–solvent interaction is found as a sum of site–site terms, including Coulomb interactions between partial charges located on the molecular sites. The implementation of such models to experimentally studied systems will be an important requirement for a realistic theory of solute–solvent interactions and relaxation rates.

Molecular Dynamics

In associated solvents and electrolytic solutions, specific association between the solute and the solvent may occur and play an important role in the solute's rotational relaxation. It is clear from both experimental studies and molecular dynamics simulations that association between solvent and solute will be essential in quantitatively accurate models.[15,27,31,53–55] The importance of association has been appreciated for some time,[11] although it is difficult to spectroscopically identify intermolecular complexes and probe their dynamics. The effect of solvent attachment on the rotational motion of tryptophan was discussed by Hu and Fleming.[53] In a molecular dynamics study of that molecule's rotational relaxation, these researchers observed that the solvation shell did not move with the solute, but that several solvent molecules did rotate with the charged groups of the probe. From the simulation, they were able to conclude that both polar interactions and deviations from a simple ellipsoidal shape contribute to the rotational relaxation time of tryptophan. Limitations of the continuum approach to describing the frictional coupling are also evident in molecular dynamics simulations of the rotational relaxation of anthracene in cyclohexane and isopropanol by Johnson et al., who analyzed the rotational diffusion coefficients about the individual molecular axes and showed that changes in temperature and in the type of solvent caused nonuniform variation of the diffusion coefficients. Such variations reflect specific solute–solvent interactions and the influence of solvent structure.

Simulation studies show that the coupling between electrostatic and mechanical contributions to the friction can be significant.[14,31] It is possible to perform MD simulations with a solute molecule that contains no partial charges and with one that contains partial charges. By comparing the orientational correlation function of the solute in both cases, one can identify how the electrostatic field of the solute changes the total friction. Kurnikova[14] showed that the friction kernel can be divided into two components. The first piece is associated with torques that arise from the van der Waals part of the interatomic potentials, and the second piece is associated with torques that arise from the Coulomb forces (partial charges on the atoms). These two components of the friction kernel were associated with the mechanical and electrostatic contributions to the friction and allowed the influence of the electrostatic field on the mechanical friction to be evaluated. Kurnikova found that in some cases no coupling between the electrostatic and mechanical contributions to the friction were evident, but that in other cases such coupling was highly significant. The origin of the coupling in the local (first solvation shell) structure about the solute, in local density changes, and in solvent association has been discussed.[14–15,31,55]

2.3. Experimental Studies of Solute–Solvent Frictional Coupling

The studies described here exploit the rotational motion of medium-sized (100–300 Å^3) solute molecules to probe solute–solvent interactions. Although some exceptions are known,[56] the rotational relaxation time τ_{or} of a solute in a single solvent has generally been found to be proportional to the solvent shear viscosity η divided by the temperature T; that is,

$$\tau_{or} = \frac{B\eta}{T} + \tau_0, \tag{28}$$

where B and τ_0 are adjustable parameters.[26,31,41,57] The emphasis of many studies is on understanding how the parameter B is related to the solute–solvent interactions and usually proceeds through direct comparison of the relaxation times with calculations from particular models. The parameter τ_0 has been interpreted by some as a free rotor time;[58] however, it is often obtained by extrapolating data at high densities to the limit of zero viscosity (or low density), and its interpretation is not clear. Some theoretical work has probed how the relaxation time behaves at low viscosities.[59] Although many studies have found a good correlation with Eq. 28 for a single solute in a range of solvents,[31,57,60] this correlation is often found to fail when the relaxation time is studied in homologous series of solvents, in solvent mixtures, in electrolytic solutions, and in other situations (vide infra).

The Role of Electrostatic Interactions

Recent work by our group at the University of Pittsburgh has investigated a range of solute molecules in different solvents. The theme of this work is to keep the size and shape of the solute molecules relatively constant, but to vary their electrostatic properties through chemical modification. Over the past decade, a range of solute molecule functionalities (NH_2, NH_3^+, $=O$, O^-, F, Cl, CH_3, OH, CN, OCH_3, CO_2^-, and SO_3^-) and their impact on the solute–solvent frictional coupling have been explored. The structures of some of the solute molecules that have been investigated are shown in Chart 1.

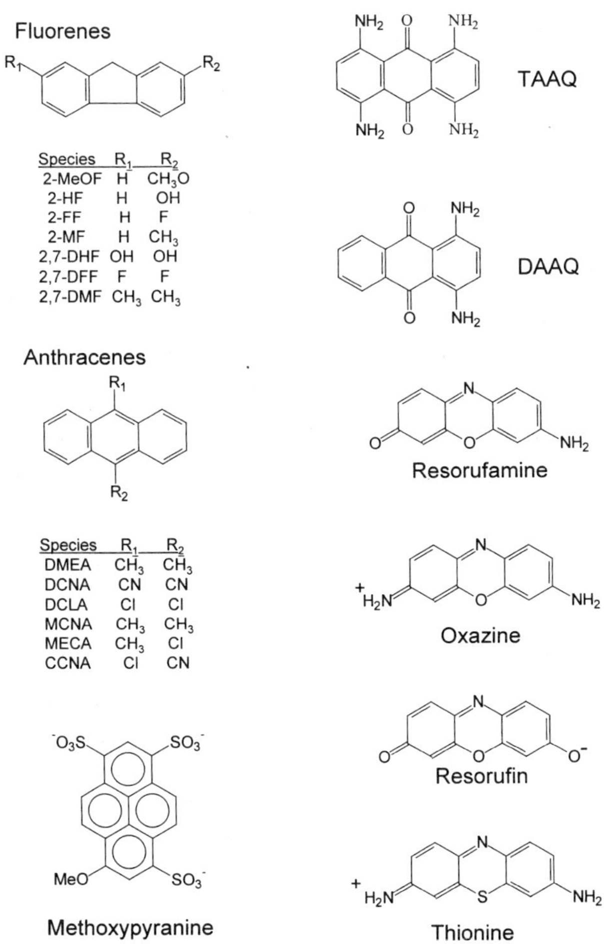

Table 1 compiles relaxation times for a subset of different solute molecules in dimethylsulfoxide (DMSO) at room temperature. Despite the solute molecules' similarity in size and shape, the rotational relaxation times range over more than an order of magnitude in DMSO. Hydrodynamic models for the friction with a single boundary condition predict that these relaxation times should be similar; however, it is not even possible to rationalize the results by changing the boundary condition from slip to stick. It is also clear that recourse to the molecule's dipole moment is unable to explain the data. For example, the solute molecule CCNA (9-chloro,10-cyanoanthracene) has the largest dipole moment of the anthracenes, but rotates faster than the 9-methyl, 10-cyanoanthracene solute, which has a smaller dipole moment. Also, the dipole moment of CCNA is similar to that of resorufamine, which rotates four times slower. Other inconsistencies of this qualitative sort can be drawn.[28,30–31] These observations stand in direct contradiction to the predictions of point-source models for the dielectric friction.

The extended charge distribution model (Eq. 19 or Eq. 27) is able to rationalize the trends in the data, however. Because this model is quite sensitive to the radius of the cavity, it is used as a fitting parameter. The best fit radii for these solutes (see Table 1) fall within the range of 5.3 Å$< a <$ 7.3 Å and do not differ very much from the expected van der Waals radius for the long axis of the molecule. Although the average radius for the different solutes in DMSO is 6.05 Å, with a standard deviation of 10%, anomalies in the best fit radii are evident. For example, dichloroanthracene

Table 1 Relaxation times and best-fit radii for solute molecules in dimethylsulfoxide (DMSO)[a]

Solute/DMSO	a (Å)	τ_{or}(ps)	Volume (Å^3)	Axial ratios	μ(D)[b]
Resorufin	6.7	83	190	6.5:3.5:2.0	2.6
Resorufamine	6.3	168	190	6.5:3.5:2.0	5.6
Oxazine	6.3	290	190	6.5:3.5:2.0	2.0
Thionine	6.6	302			
DMEA	5.6	26	204	6.0:5.2:1.8	0.0
DCLA	6.0	26	200	6.0:4.9:1.8	0.0
MCNA	5.4	44	207	6.0:5.7:1.8	2.6 (3.4)
MECA	5.6	27	202	6.0:5.0:1.8	2.8 (3.8)
CCNA	5.6	35	205	6.0:5.6:1.8	5.4 (7.3)
2,7-DMF	7.3	64	203	6.9:3.7:1.9	0.66
2,7-DHF	5.7	206	188	6.4:3.7:1.9	2.9
2,7-DFF	6.6	44	182	6.2:3.7:1.9	0.65
2-MF	6.1	52	185	6.3:3.7:1.9	1.3
2-HF	5.3	88	177	6.0:3.7:1.9	2.3
2-FF	5.7	46	174	5.9:3.7:1.9	0.83
2-MeOF		58			

See Chart 1 for a legend of solute names.

[a] The data in this table are taken from Refs. 28,30,31,55. The error in the reported time constants is 10% or less.

[b] The dipole moments were computed at the Hartree–Fock level. The values in parentheses were found using self-consistent reaction field calculations in a spherical cavity of 5.3 A and a dielectric constant of 46.45 for DMSO.

has an anomalously larger effective radius than the other anthracenes, and hydroxyfluorene has an anomalously smaller radius than the other fluorenes. These anomalies in the best fit radius are an artifact of the data fitting and likely arise from limitations of the continuum description for the solvent. Despite these limitations, the inclusion of the extended nature of the molecular charge distribution represents a significant improvement over point-source models for the dielectric friction and provides some insight into the different boundary conditions that are generally observed.

In high dielectric solvents like DMSO, the electrostatic contribution to the friction simplifies to a product of a solvent term $[(\varepsilon_S - 1)\tau_D/(2\varepsilon_S + 1)^2]$ and a solute term that describes the solute's electrostatic potential. In turn, one often finds that the solvent parameter is proportional to the solvent viscosity. To motivate the relationship with the observed boundary condition, consider the case of a spherical solute, for which one may write the rotational relaxation time as the sum

$$\tau_{or} = \tau_{diel} + \tau_{hyd}, \tag{29}$$

where τ_{hyd} is given by Eqs. 3 and 7 and τ_{diel} is the electrostatic friction's contribution to the relaxation time. In high dielectric solvents, τ_{diel} may be written as

$$\tau_{diel} = \frac{(\varepsilon - 1)\tau_D}{(2\varepsilon + 1)^2 T} Q, \tag{30}$$

where Q depends on the solute's properties. When the solvent parameter $(\varepsilon - 1)\tau_D/(2\varepsilon + 1)^2 T$ is proportional to η/T with slope b, it is possible to write

$$\tau_{or} \approx A + \frac{\eta}{T}(H + bQ), \tag{31}$$

where A is a constant, H is the hydrodynamic contribution to τ_{or}, and bQ is the electrostatic contribution to τ_{or}.[61] A comparison of Eqs. 31 and 28 shows that the quantity $(H + bQ)$ corresponds to the slope in the simple DSE model. Although these considerations are not rigorous for the systems studied, they provide some insight into how the observed hydrodynamic boundary condition depends on the electrostatic properties of the solute. On a more general level, it is clear that this approach will fail when the strength of the solute–solvent interactions becomes strong or when the solvent possesses long-range structure.

Although the extended charge distribution model is more realistic than point-source models, it remains limited in a number of important ways. The friction, and hence relaxation time, one obtains can depend sensitively on the cavity radius that is chosen[30–31] The friction also depends on the accuracy of the charge distributions used to obtain the best-fit radii. For example, the analysis resulting in the best-fit radius of 6.7 Å for resorufin (see Table 1) was found for a charge distribution that was calculated using the Hartree–Fock method with an STO-3G basis. By contrast, a more recent calculation of the resorufin charge distribution using the Hartree–Fock method with 3-21G basis in a dielectric medium and the assignment of the charge using the Tomasi algorithm gives a different charge distribution and a best-fit radius of 6.9 Å. The extended charge distribution model maintains simplicity by the use of a dielectric continuum treatment for the solvent, but this feature costs considerably in realism when solute–solvent association or solvent structure is important.

Size Effects on the Frictional Coupling

Molecular size plays an important role in the strength of the solute–solvent friction and the measured rotational relaxation times.[23,46,62–63] The importance of solute size is evident in the standard hydrodynamic model for which the friction (and relaxation time) is directly proportional to the solute volume. The solute size—in particular, the relative solute-to-solvent size—also influences the hydrodynamic boundary condition. A number of theoretical studies[18–21] on bumpy spheres and cylinders have shown how the application of a "locally" slip hydrodynamic boundary condition can result in a relaxation time for a bumpy-shaped object that appears to be stick. The transition from slip to stick relaxation times depends on the shape of the solute and the number and size of its roughness features. Ben-Amotz and co-workers[23] have explored this effect by analyzing a range of data for differently sized nonpolar solute molecules in alkane solvents. Although they do not find quantitative agreement between simple models and the experimental results, the general evolution of a slip boundary condition at small solute-to-solvent-size ratios to a stick boundary condition at large solute-to-solvent size ratios is evident.

Figure 3.5 illustrates the change in the observed solute–solvent frictional coupling with the solute molecule size. This figure shows a plot of the observed rotational relaxation time divided by the relaxation time expected for the stick hydrodynamic boundary condition as a function of the volume of the solute. The data points are for observed relaxation times of nonpolar or weakly polar solutes in alkane solvents. The data are taken from a range of measurements made by different experimenters. A clear evolution of the ratio from a value of 0.1 to a value near unity can be observed. This sort of behavior is exactly that predicted by a consideration of

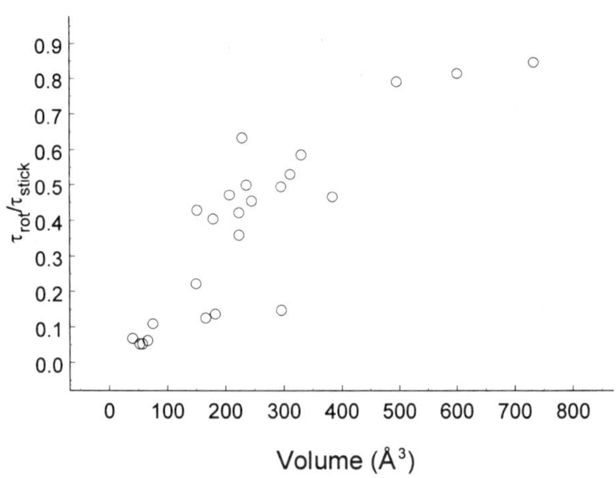

Figure 3.5. The ratio of the measured relaxation time of a solute to its expected value for a stick hydrodynamic boundary condition is plotted as a function of the solute molecule's van der Waals volume.

solute roughness; that is, the application of a "locally" slip boundary condition gives rise to an overall relaxation time that is stick when the size of the solute molecule becomes large. Although some of the scatter in the plot arises from experimental error, other contributions arise from the differences in the relative solute-to-solvent size ratio through the series of *n*-alkanes. Despite this less-than-rigorous analysis, the trend remains evident.

The influence of relative solute–solvent size is apparent in a number of different systems. A few workers have studied the rotational relaxation of solute molecules through homologous series of solvents, such as *n*-alkanes[64] and *n*-alcohols.[49,57a,65–66] Many of these studies reveal a good correlation between the measured relaxation time over limited viscosity ranges, but an overall downward curvature in the plot of relaxation time versus viscosity. One possible interpretation of the data is that the hydrodynamic boundary condition has changed (to be more "sliplike") as the solvent size increases. This behavior is manifest in "subslip" rotation, which has been reported in a number of cases.[67] Several other studies[25,29a,68] have investigated the rotational relaxation in mixed solvents that display long-range structural features. In these cases, there also appears to be a change in the "effective" hydrodynamic boundary condition that is correlated with the solvent's structural features. Because the contribution of dielectric (or electrostatic) friction is also present in these systems, it has not yet been possible to identify whether the change in the total frictional coupling results from hydrodynamic effects, dielectric effects, or a combination of the two.

Molecular Aspects of Solute–Solvent Friction

Since the very early studies[11,69] of rotational relaxation, the importance of specific association between solute molecules and the solvent has been appreciated. However, it has been difficult to assess that importance precisely. Many of the experimental studies provide information on only the overall rotational relaxation of the solute in the solvent, and the specific association is manifest only as an overall increase in the rotational drag of the solute molecule in the fluid. A few studies have looked at the rotational relaxation of the specific complexes (e.g., Coumarin 102 with trifluorethanol molecule[70] in nonpolar solvents). Although useful in a general sense, these results do not apply in a direct way to relaxation in interacting solvents in which there is competition for hydrogen bonding. From the experimental perspective, the need exists for direct studies of solute–solvent complexes that provide information on the structure and dynamics of those complexes.[71]

Molecular dynamics simulations have been used to provide insight into the role of solute–solvent complexes in rotational relaxation.[15,53] An early study by Fleming[53] on the rotational relaxation of tryptophan in water showed that water molecules associated with the tryptophan were present for times that were long compared to the rotational relaxation time of the tryptophan. This observation supports the assignment of an effective hydrodynamic volume to the solute molecule in the analysis of experimental data.[72] More recently, Kurnikova and collaborators have used molecular dynamics simulations to identify important molecular aspects of solute–solvent frictional coupling.[15,31] Their work investigated a set of solutes (resorufin, resorufamine, and thionine; see Chart 1) in DMSO and found good qualitative agreement with the experimentally observed rotational relaxation rates. Kurnikova's simulations identified specific solute–solvent association (in particular, between the DMSO and the solute's amino functionality) that could account for the changes in frictional coupling. They found a residence time for a DMSO molecule at an amino site to be about 20 ps, which is similar to the solute's relaxation time in the model DMSO solvent. These studies suggest a relationship between the "residence time" of solute–solvent complexes and the increase in the friction; however, this important area is largely unexplored.

Friction in Electrolytic Solutions

The nature of solute–solvent frictional coupling is important in the dynamics of solutes in electrolytic solutions. Recent studies in this area have identified the important considerations. In particular, the studies have shown that the electrolyte concentration needs to be on the order of 0.1 M before significant effects on the rotational relaxation time of organic ions become apparent, and which suggests that the ion atmosphere friction represents a small contribution to the total frictional coupling. However, these investigations show that ionic association (ion pairing) plays a large role in the rotational dynamics when the ion pair lifetime is long compared to the rotational timescale (less than a few hundred picoseconds). For the organic anion resorufin in DMSO, Hartman[27b] was able to identify the formation of a resorufin–Li ion pair spectroscopically and to evaluate its association constant. By considering the ion pair and its percentage contribution to the total chromophore population, Hartman was able to rationalize the changes in the observed rotational relaxation time. Later, Balabai demonstrated the dependence of the relaxation characteristics of resorufin on the charge-to-radius ratio of the cation.[27a]

More recent work explores the qualitatively different relaxation behavior of organic cations from that of organic anions in LiCl electrolytic solutions. Figure 3.6 shows the normalized relaxation times for the organic cation thionine and the

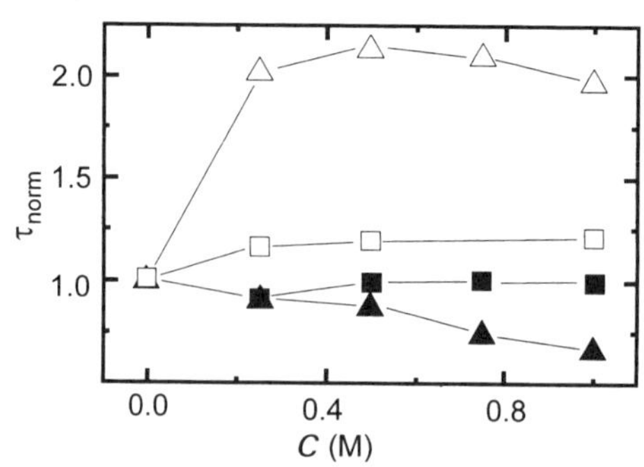

Figure 3.6. The normalized rotational relaxation times of thionine cation and resorufin anion are plotted against the concentration of LiCl in DMSO solution [thionine (filled triangles) and resorufin (open triangles) in DMSO; thionine (filled squares) and resorufin (open squares) in methanol].

organic anion resorufin in LiCl/DMSO solutions. In order to remove intrinsic differences in the rotational relaxation times of the solute molecules, the figure uses a normalized relaxation time. First, the observed relaxation time was normalized by the solution's viscosity, so that the data should fall along a horizontal line if hydrodynamics provides an adequate description of the molecule's relaxation. Second, the viscosity-normalized relaxation time was further reduced by dividing it by the viscosity-normalized relaxation time of the solute in the pure solvent. This procedure rescales the horizontal lines to begin at a value of unity. It is clear from the plot that the cation shows a decrease in the normalized relaxation time with increasing electrolyte concentration, whereas the anion shows an *increase* in the normalized relaxation time with increasing electrolyte concentration.

A combination of experimental studies and molecular dynamics simulations reveals that these trends occur because of the interplay between the formation of ion pairs and the decrease in dielectric friction with the increase in electrolyte concentration that is expected from a continuum model. For the case of the anion resorufin, the solute in the pure solvent has weak interactions with the solvent and rotates with a relaxation time that is well described by a slip boundary condition. When LiCl is added to the solution, ion pairs can form, which causes a significant increase in the relaxation time.[27] In contrast, the cation thionine interacts strongly with the DMSO solvent. When electrolyte is added to the solution, a competition exists between ion pair formation and solvent association with the amino groups on the thionine. Molecular dynamics simulations[55] suggest that the relaxation of the solvent–solute complex and the ion-paired form of thionine are similar. The decrease in the normalized relaxation time of the thionine with electrolyte concentration was interpreted to result from the enhanced shielding of the solute's electrostatic potential in the electrolyte solution, which causes a decrease in the electrostatic friction.

Although substantial progress has been made in understanding the nature of solute–solvent frictional coupling and its impact on the rotational relaxation of solute molecules, considerable work remains to be done before a comprehensive and predictive understanding is achieved. The most important need lies with the understanding of molecular characteristics of the solute–solvent frictional coupling—in particular, those interactions between the molecules in the first solvation shell and the solute. Molecular dynamics simulations of solute–solvent systems are beginning to address these issues with the degree of realism that is required. Experimental studies need to move beyond the simple measurement of an average diffusion coefficient and begin to look at the different components of the diffusion tensor and the different types of specific solute–solvent interactions and associated complexes. The next section discusses some recent studies of the dynamics of weakly bound bimolecular complexes, which provides one perspective on specific solute–solvent interactions.

3 Orientational Relaxation of Bimolecular Complexes

Understanding the interactions that control the relative motion of two molecules that are in close proximity is an important component in understanding bimolecular chemical reactions and catalysis. Such motions are important in the evolution of a reactive encounter pair (i.e., its lifetime and the volume of configuration space that it samples). For example, consider the function of a biological enzyme. The lock-and-key view of enzyme activity captures crucial features of enzyme action.[73] But what is the lock designed to fit—the reactant, the transition state(s), or the product? Or maybe all three? Clearly, stabilization of the transition state is critical, but the enzyme must also bind the substrate. However, the substrate must not bind so rigidly that neither it nor the enzyme can rearrange itself during the course of the reaction. The interaction between the host (enzyme) and guest (substrate) as the latter evolves from reactant to product is likely to have important dynamical features.

Before proceeding, it is useful to introduce the common approach for characterizing bimolecular complexes. One may consider the formation of a complex to be a reversible bimolecular reaction, viz.,

$$\mathrm{H} + \mathrm{G} \Leftrightarrow \mathrm{HG}, \tag{32}$$

where H (host) and G (guest) represent the two species that combine to form the complex. Considerable progress has been made over the past two decades in designing the properties of H and G so that well-defined bimolecular complexes are formed with controlled average geometry.[74] The experimental characterization of such complexes usually encompasses the measurement of the association constant K,

$$\mathrm{K} = \frac{[\mathrm{HG}]}{[\mathrm{H}][\mathrm{G}]}, \tag{33}$$

where [HG], [H], and [G] are the species' concentrations, or, more precisely, their activities. The measurement is usually carried out by means of a titration curve (or binding curve), with a spectroscopic probe of the association (e.g., IR spectroscopy, UV–VIS absorption, fluorescence, ^{1}H-NMR, and other methods).[76] A very useful measurement of the association is accomplished by titration calorimetry, which provides the enthalpy of association directly. Otherwise, one must determine the binding constant as a function of temperature to characterize the thermodynamics of complexation. In addition, other spectroscopies (NMR) and X-ray diffraction are used to determine structural aspects of the complexes. Although useful, such data provide only a static picture of the binding event and do not yield any insight into the complex's dynamics.

Some effort has focused on characterizing the kinetics of bimolecular association and dissociation. A variety of different experimental probes (e.g., stopped-flow measurements,[77] NMR methods,[78] ultrasonic relaxation,[79] quenching studies,[80] and transient spectroscopies[81–82] have been brought to bear on this problem. The measured rates vary widely; for instance, the complex between a 1-naphthyl-1-ethanol complex and β-CD has "on" and "off" rate constants of about $5 \times 10^{8}\,\mathrm{M^{-1}\,s^{-1}}$ and $5 \times 10^{5}\,\mathrm{s^{-1}}$, but a "tennis ball" hydrogen-bonded dimer in chloroform[78] has formation and dissociation rate constants of $2 \times 10^{9}\,\mathrm{M^{-1}s^{-1}}$ and $0.62\,\mathrm{s^{-1}}$, respectively. Such kinetic studies are critical to understanding the behavior and reactivity of bimolecular complexes, but do not reveal the internal dynamics of the complexes (e.g., vibrational and orientational motions).

Much of our understanding of the dynamics of bimolecular complexes in solution has evolved from studies performed by the biophysical community. A number of analytical models for the relative motion of a guest molecule with respect to a host molecule (usually considered to be a polymer molecule or protein) have been developed.[83–84] However, current efforts have centered on developing realistic numerical simulations of such systems.[73,85–86] Experimental work on the internal motions of intermolecular complexes has focused on proteins and DNA, usually with intercalated or covalently linked fluorophores.[87–89] The measurement of the orientational correlation function for electronically excited states of chromophores (i.e., the fluorescence depolarization method) has been a powerful probe of the local dynamics in such biological systems.

Previous work on small systems has lain mainly with complexes in which dispersion interactions dominate the host–guest interaction (e.g., fluorescence depolarization or NMR relaxation studies of solutes included in molecular cavities).[90–92] The common implementation of NMR relaxation experiments probes the correlation times for the relaxation processes. Hence, one can determine the characteristic time that a bimolecular complex is bound or the characteristic time of rotational diffusion, but the method does not monitor the dynamics of the complex in any direct way. Time-resolved studies that directly measure the orientational correlation function of one of the species (either host or guest) with respect to the other can probe important dynamical issues (e.g., they can determine the range of motions and their time scales and whether the motion is diffusive or inertial in nature). The two studies described later[5,93] investigate small bimolecular systems in which the properties of the complex can be tuned synthetically. The motivation for studies of this sort is to develop an understanding of which structural and energetic features of the complex control the orientational relaxation. The quantitative description of such systems can be incorporated into a description of more complex biological systems and applied to the development of design principles for supramolecular systems.

3.1 Theoretical Models

A number of workers[83–84] have developed models to describe the relaxation of a chromophore that is confined (or weakly bound) in an intermolecular potential. Most of the work has focused on a description that ignores the influence of the potential, except as a means to restrict the range of angles over which the chromophore can diffuse. Szabo[83] developed a more comprehensive description that is able to include the presence of a potential and that incorporates inertial effects on the motion. The latter treatment is similar to those found in the study of polymer dynamics.[94] All of these models express the orientational correlation function of the chromophore in terms of the anisotropy $r(t)$. For a macroscopically isotropic sample with a single excited state and orientationally independent population kinetics, the anisotropy is given by

$$r(t) = \frac{2}{5}\langle P_2(\boldsymbol{\mu}_{\mathrm{LE}}(t)\cdot\boldsymbol{\mu}_{\mathrm{LA}}(0))\rangle, \tag{34}$$

where $\boldsymbol{\mu}_{LA}(0)$ is the orientation of the absorption dipole at time zero, $\boldsymbol{\mu}_{LE}(t)$ is the orientation of the emission dipole at time t and $P_2(x) = (3x^2 - 1)/2$ is the second Legendre polynomial. Such models have been used extensively in understanding the motion of molecules in microheterogeneous and nanostructured media, such as liquid crystals,[95] micelles,[96] membranes,[97] proteins,[98] polymers,[99] and other media.[100]

Szabo[83] provides a unified formalism for describing the decay of the fluorescence emission anisotropy in systems where both excited-state population relaxation and orientational relaxation contribute to the depolarization. However, the analysis can be greatly simplified by making several assumptions. First, one assumes that the population and orientational dynamics are uncoupled, with the former being much slower than the latter. Second, if one assumes that the overall motion of the complex is isotropic, then its contribution to the anisotropy cleanly separates from the contribution of the internal motion of the chromophore. This assumption can be removed if necessary and Szabo has analyzed the more complex case. Third, one assumes that the transition moment direction in the guest molecule has a well-defined direction with respect to its molecular coordinate system, given by (X_G, Y_G, Z_G). Last, one must define a coordinate system for the host molecule, (X_H, Y_H, Z_H). The laboratory coordinate system is defined by the polarization of the applied field (e.g., the electric field direction in an optical experiment or the magnetic field direction in an NMR experiment).

Szabo[83] has analyzed orientational guest dynamics for a number of different geometries and degrees of freedom. For example, his treatment allows for motion of the substrate around two different axes with respect to the host molecule and incorporates a range of possible dipole orientations with respect to the axes of internal motion. Figure 3.7 illustrates how the dipole orientation selects different motions of the guest chromophore with respect to the host. If the direction of the transition dipole moment of the chromophore coincides with the "effective" intermolecular bond axis, motions about the axis transverse to this direction will lead to the reorientation of the transition dipole. If the opposite case holds and the dipole moment is perpendicular to the bond axis, one can probe the torsional motion about that axis, as well as the "in-plane" motions of planar guest molecules. The

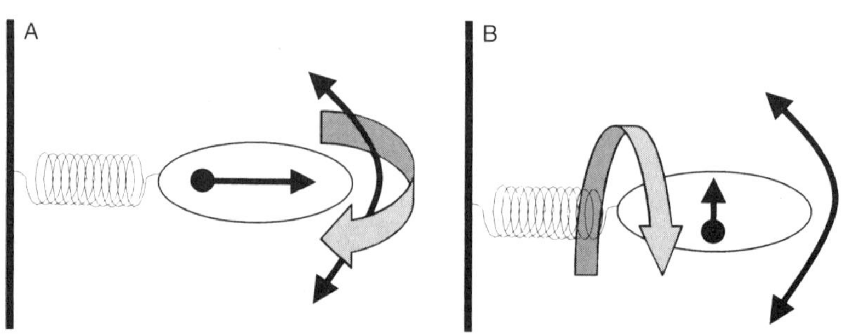

Figure 3.7. Possible observable torsional motions of the guest molecule bound to the host. In panel A; the transition dipole moment (shown by the straight arrow) is directed along the bond axis. In panel B; the transition dipole moment is directed perpendicular to the bond.

timescales for these low-frequency vibrations can be very different, depending on the particular host–guest system and the solvent in which it is studied.

Now, consider a system like that illustrated by Figure 3.8, in which the guest molecule (or substrate) is allowed to rotate about the axis labeled Z_G. This model geometry will be used to treat an experimental system later. The figure shows a view of the complex in which the Z_G and Z_H axes are parallel and oriented out of the plane of the paper. The X_G and X_H axes are perpendicular to the plane of the host and guest molecules, respectively, and lie in the plane of the paper. The Y_G and Y_H axes are oriented perpendicular to the X and Z axes. The angle ϕ specifies the orientation of the guest's transition dipole with respect to its Z_G axis. With this geometry and these assumptions, the time-dependent anisotropy can be written as

$$r(t) = \frac{2}{5}\exp(-6Dt)\left[\Gamma_{00}(t)P_2^2(\cos\phi) + \frac{3}{4}\Gamma_{11}(t)\sin^2\phi + \frac{3}{4}\Gamma_{22}(t)\sin^4\phi\right], \tag{35}$$

where D is the diffusion constant for the whole complex.[33] The functions $\Gamma_{mn}(t)$ are angular correlation functions for motion by the angle θ and are given by

$$\Gamma_{mn}(t) = \langle\exp[-im\theta(t)]\exp[-in\theta(0)]\rangle, \tag{36}$$

where m and n are integers and the angular bracket indicates an ensemble average. If $\phi = 90^\circ$, $r(t)$ simplifies, further and one finds that

$$r(t) = \frac{2}{5}\exp(-6Dt)\left[\frac{1}{4} + \frac{3}{4}\Gamma_{22}(t)\right]. \tag{37}$$

These considerations are determined by the nature of the spectroscopic experiment, its particular implementation (i.e., the geometry), and the structural features of the complex. The function of these considerations is to relate the experimental

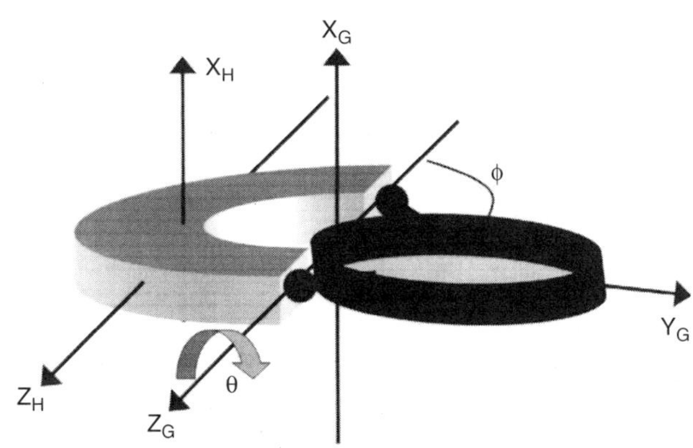

Figure 3.8. Coordinate systems for the host (H) and guest (G) molecules. The angle ϕ specifies the orientation of the transition dipole moment of the chromophore with respect to the Z_G molecular axis. The orientational motion of the guest frame of coordinates (X_G, Y_G, Z_G) with respect to the host frame (X_H, Y_H, Z_H) is specified by the angle θ.

observable $r(t)$ to the time correlation functions that are used to describe the orientational dynamics. What remains is to develop a model for the time evolution of the torsional angle θ and use it to predict the time dependence of the anisotropy via Eq. 35.

One approach to the torsional dynamics is to treat it as diffusion over a restricted angular range, usually approximated as a cone defined by the angle θ of its apex.[101–103] In this case, the anisotropy decays monotonically and is given by a sum of exponentials. If the internal relaxation decays more rapidly than the overall tumbling of the complex, then the loss in amplitude that results from the fast decay can be used to estimate the angle of the cone. Wahl[101] considered a one-dimensional cone model that corresponds to rotational diffusion of the chromophore about an axis and has an angular range of $-\theta_{max}$ to θ_{max} (See Fig. 3.9). The conical surfaces act as reflecting boundaries in the solution of the diffusion equation. More elaborate (necessarily two-dimensional) models that allow the axis to also wobble could be exploited. By defining an intermediate timescale τ_{int} that is much longer than the time of internal motion, but much shorter than the time for overall tumbling of the complex, one can write the anisotropy as

$$\frac{r(\tau_{int})}{r(0)} = \left[P_2(\cos^2\phi)\right]^2 + \frac{3}{4}\sin^4\phi\,\frac{\sin^2(2\theta_{max})}{4\theta_{max}^2} + \frac{3}{4}\sin^2 2\phi\,\frac{\sin^2\theta_{max}}{\theta_{max}^2}, \tag{38}$$

where ϕ is the angle between the transition dipole and the axis of internal rotation. If one knows the orientation of the transition dipole ϕ from an independent measurement, then the measurement of the anisotropy can be used to determine the angle of the cone, θ_{max}. Figure 3.10 shows the dependence of the anisotropy at τ_{int} on the cone angle θ_{max} for different values of ϕ. It is straightforward to obtain an estimate of the root-mean-square angular deviation from this model as well. Since all angles from $-\theta_{max}$ to θ_{max} have equal statistical weight, one has an average angular deviation of zero and a root-mean-square deviation given by $\theta_{rms} = \theta_{max}/\sqrt{3}$.

An alternative approach is to model the guest molecule's dynamics by using the Langevin equation. This approach allows one to proceed from an inertial, or underdamped, limit to a diffusive, or overdamped, limit in a seamless manner. By

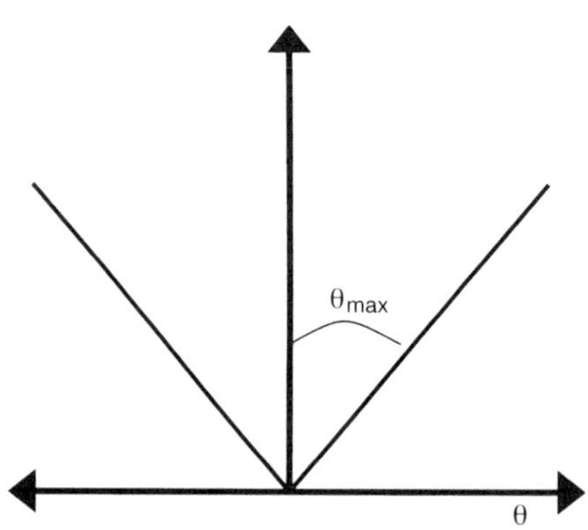

Figure 3.9. A one-dimensional cone. Inside the cone the particle diffuses freely.

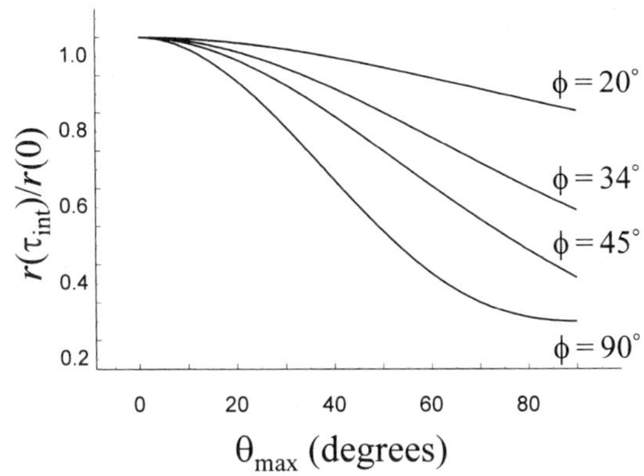

Figure 3.10. Plots of the anisotropy ratio predicted by the Wahl model for diffusion about a fixed axis as a function of the maximum angular deviation (θ_{max}), for four values of the transition dipole moment orientation (ϕ).

approximating the intermolecular potential as harmonic with a frequency ω, one can simplify the analysis. The adequacy of this approximation for such a low-frequency motion is open to question; nevertheless, it provides a starting point for a discussion of the dynamics and offers an avenue for assessing the reasonableness of the potential parameters. In this approximation, the Langevin equation becomes

$$I\ddot{\theta} + \zeta\dot{\theta} + I\omega^2\theta = F(t), \tag{39}$$

where ζ is the friction coefficient, I is the reduced moment of inertia for the motion, and $F(t)$ is a random force. It is straightforward to solve Eq. 39 for the time evolution of the angle and obtain expressions for the angular correlation functions,[83] such that

$$\Gamma_{mn}(t) = \exp\left[-\frac{kT}{2I\omega^2}\left(m^2 + n^2 - 2mn\exp(-\lambda t)\right.\right.$$
$$\left.\left.\times\left(\cosh\delta t + \frac{\lambda}{\delta}\sinh\delta t\right)\right)\right], \tag{40}$$

where $\lambda = \zeta/2I$ and $\delta = (\lambda^2 - \omega^2)^{1/2}$. In the inertial limit when the friction $\zeta = 0$, this expression reduces to

$$\Gamma_{mn}(t) = \exp\left[-\frac{kT}{2I\omega^2}(m^2 + n^2 - 2mn\cos(\omega t))\right]. \tag{41}$$

In this case, Γ_{mn} oscillates with a fundamental frequency ω. In the overdamped limit, in which the friction dominates, one finds that

$$\Gamma_{mn}(t) = \exp\left[-\frac{kT}{2I\omega^2}\left(m^2 + n^2 - 2mn\exp\left(\frac{-I\omega^2 Dt}{kT}\right)\right)\right], \tag{42}$$

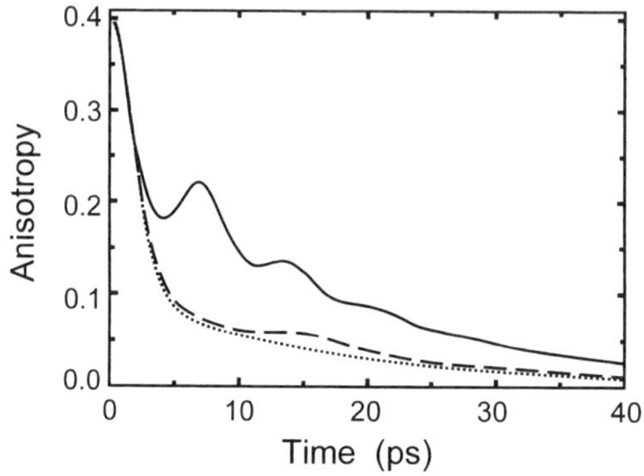

Figure 3.11. The dependence of the simulated anisotropy decay on the frequency ω of the potential well [ω is $30\,cm^{-1}$ (solid), $13\,cm^{-1}$ (dashed), and $6.6\,cm^{-1}$ (dotted line)]. The friction coefficient ζ is 8×10^{-33} Js, the diffusion coefficient is $D = 10^{10}\,s^{-1}$, and the reduced moment of inertia is $I = 4 \times 10^{-44}\,kg\,m^2$. Taken from Reference 61.

where $D = kT/\zeta$ is the guest diffusion constant. In this case, Γ_{mn} displays a monotonic decay in time, and when $I\omega^2 Dt/kT$ is small enough, the correlation function decays exponentially. Equations 41 and 42 represent the two limiting behaviors of the Langevin treatment of the dynamics.

Figure 3.11 shows anisotropy decay curves, using Eqs. 41 and 42, as a function of the frequency of the intermolecular potential.[61] To generate these curves, the friction coefficient, inertial moment, and diffusion constant of the complex were held constant. The expected anisotropy decay is shown for three different frequencies that are chosen in order to display the range of behaviors expected. In the case where the frequency is the highest, one can clearly see oscillations in the anisotropy that slowly dampen with time. This limit corresponds to the underdamped regime. In the case where the frequency is the lowest, no oscillations are displayed, and the anisotropy decays exponentially at long times. This curve corresponds to the overdamped limit. It is evident from these simulations that the shape of the anisotropy decay provides important qualitative insight into the nature of the guest's motion. In particular, one can determine whether the motion is diffusive or inertial in character.

3.2 Experimental Examples

In this section, we discuss two recent experimental studies carried out by our group that compare time-resolved measurements of the anisotropy with the models we have discussed so far. In the first example, the relaxation of a strongly bound complex between an anthracene dicarboxylate and a bisguanidinium receptor is described. In the second example, the relaxation of dye molecules that are intercalated into the cavity of cyclodextrin is examined.

Carboxylate Complex: The torsional relaxation of an anthracene-1,3-dicarboxylate (ADC) complexed to a *bis*-guanidinium receptor was reported by Sukharevsky[5] (see Figure 3.12). The binding constant of the ADC with the *bis*-guanidinium is very high, greater than $10^5\,M^{-1}$, which makes it difficult to quantify precisely. However

1 2

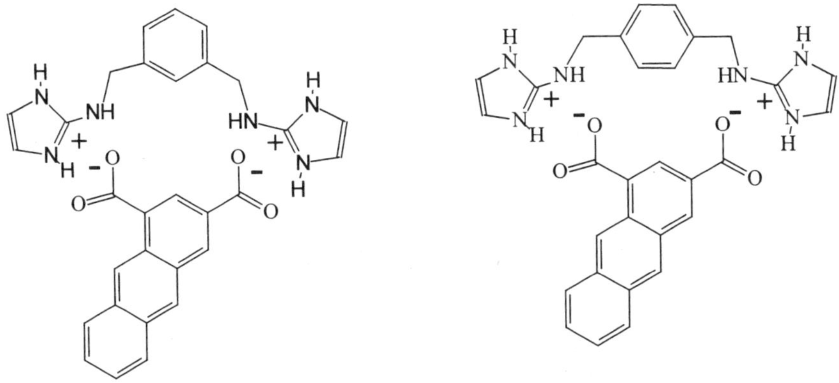

Figure 3.12. Molecular structures for the two different complexes of ADC and the *bis*-guanidinium receptors.

fluorescence, NMR, and calorimetric titration curves indicate that it is of the order of $10^6\ M^{-1}$. The calorimetric measurements also show that the enthalpy of binding is on the order of 4 kcal/mol for these complexes. The two complexes shown display some differences from one another that are presumably associated with the different geometries of the guanidinium units about the benzene moiety. However, the qualitative features of the binding and the anisotropy relaxation are similar in each complex.

An anisotropy decay for complex **2** dissolved in dry DMSO is shown in Figure 3.13 for the case in which the *bis*-guanidinium is in hundred-fold excess concentra-

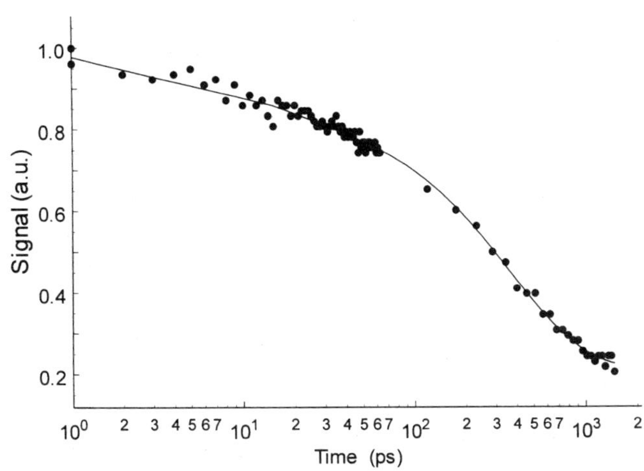

Figure 3.13. An experimental anisotropy decay for the complex **2** (see Figure 12) in DMSO at 303 K.[5]

tion. The decay is plotted against a logarithmic time axis in order to elucidate the wide dynamic range required by these experiments. It is clear from the form of the decay law that the relaxation dynamics is diffusive, rather than inertial. It was shown that the relaxation is not well described by an exponential decay law, but that a double exponential decay law agrees with the data. The double exponential fit provides a short time constant of 14 ps and a long time constant of 356 ps, which should be contrasted with the 30-ps relaxation time of ADC in DMSO.[5] A study of how the relaxation times depend on the relative concentrations of the host and guest molecules demonstrates that the short time component in the observed decay results from the torsional motion of the complex, rather than the uncomplexed guest. A comparison of the relative amplitudes of the short- and long-lifetime components via Eq. 36 indicates that θ_{max} is 40° for the complex with **2** and 60° for the complex with **1**.[5] These values for θ_{max} give root-mean-square angular displacements of 23° for the complex with **2** and 35° for the complex with **1**.

It is also possible to fit the decay law to the analytical expression obtained from the Szabo treatment (see Refs. 5 and 61 for details concerning the analysis). The values obtained for the parameters of the complex formed with **2** are 14 ns^{-1} for the diffusion constant (18 ns^{-1} for that formed with **1**), a diffusion constant of 0.46 ns^{-1} for the internal motion (0.45 ns^{-1} for the complex formed with **1**), and a characteristic frequency of 32 cm^{-1}(27 cm^{-1} for the complex formed with **1**). Three features of these parameters are noteworthy.[5] First, the diffusion constant D for the internal motion in the two complexes is similar and is in good agreement with the diffusion constants that are measured for nonpolar anthracenes in DMSO. This result is reasonable, since the diffusion constant should depend primarily on the interaction between the guest chromophore and the solvent and is consistent with the carboxylate moiety being localized in the binding site and having the nonpolar end of the molecule exposed to the solvent. Second, the overall diffusion is similar for the two complexes, which is expected because the complexes have a similar size and shape. Sukharevsky[5] discusses how the overall rotation of the complex can be rationalized by using a hydrodynamic model with a boundary condition that lies between the stick and slip limits. Third, the frequencies that are extracted from the fit are reasonable for a torsional motion. The root-mean-square angular deviations that are found from these parameters are in good agreement with those found by treating the motion as diffusion over a restricted angular range. Last, the parameters extracted from the data are reasonable on a physical level.

A primary aim of studying small systems of this sort is the ability to compare the observations with molecular dynamics simulations. To this end, a molecular dynamics trajectory was calculated for the complex (in vacuo) in order to assess the frequency of the motion and the range of the angular motion.[5,61] A sample trajectory is shown in Figure 3.14A for the host–guest complex **1** in vacuo. The corresponding angular correlation function is shown in Figure 3.14B. The torsional angle θ represents the motion of the guest about an effective intermolecular axis that is defined as running between the two "point" contacts where the carboxylate moieties and the guanidiniums meet. Such a trajectory gives information about the amplitude and frequency of the angular displacement, and the angular correlation

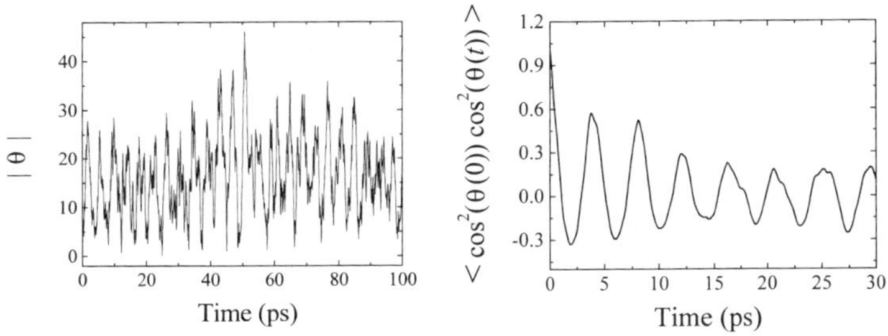

Figure 3.14. (A, left) Simulated angular trajectory for the torsional motion of the guest about the "effective" intermolecular axis of the complex **1** (see Figure 12). This plot corresponds to the $\cos^2(\theta)$ trajectory in Ref. 5. (B, right) Normalized correlation function derived from the trajectory in A. Taken from Reference 61.

function can be directly compared with the guest rotational relaxation measured in the experiment. The underdamped oscillation apparent in Figure 14B occurs because no solvent is present in the simulation, and the host and guest molecules were treated as if they were rigid. One would expect the oscillation amplitude to greatly diminish or even disappear completely as a result of diffusive guest motion in a solvent and coupling to other low-frequency modes of the complex. A root-mean-square angular deviation obtained from this trajectory yields a value of ± 28 degrees. This angular fluctuation is consistent with those found by modeling the anisotropy and is also consistent with the open architecture of this intermolecular complex.

These three approaches to modeling the internal dynamics of the chromophore provide a consistent view of the complex's flexibility. The analysis suggests a large-amplitude motion for the chromophore, on the order of 30 to 40° (in the case of ADC: **2**), with a frequency that is consistent with a librational or torsional mode ($30\,\text{cm}^{-1}$). In general, the orientational relaxation of the complex is double exponential or even more complicated, because the hindered rotation of the guest molecule in the complex is faster than the overall motion of the complex. It will be important to identify how the relaxation parameters depend on properties of the complex—in particular, structural and energetic factors involved in binding. A distinguishing feature of these studies is that they directly probe the torsional dynamics of the complex and are able to identify the motion as diffusive. It will be quite interesting to study complexes in which the guest is not as exposed to the solvent and see whether inertial motion is evident. For example, the study of the rotational relaxation of molecules trapped in calixarenes may be useful in this regard.[4,104]

Cyclodextrin: The dynamics of guest molecules in cyclodextrin cavities has been demonstrated to be strongly affected by the relative size and geometry of the host and guest molecules.[91] β-cyclodextrin (β-CD) is a cyclic oligomer of seven 1,4-linked-glucose monomers that form a truncated cone. The interior of the cyclodextrin is hydrophobic, and the exterior, laced with hydroxyl groups, is hydrophilic. Both the internal motion and the overall tumbling of β-cyclodextrin complexes have been studied using NMR relaxation methods.[90–91] Our group has examined[93] the dynamics of three organic dye molecules [resorufin (an anion), oxazine 118 (a cation) and oxazine 725 (a cation)] that are included in β-cyclodextrin (see Fig. 3.15).

The ability of the β-cyclodextrin to form inclusion complexes with resorufin and oxazine-118 was demonstrated with a variety of methods. Both absorption spectral measurements and calorimetric measurements show that 1:1 binding is dominant in all three cases, with binding constants greater than or equal to $10^3\,M^{-1}$. A combination of steady-state absorption and fluorescence spectroscopy with fluorescence lifetime measurements was able to show that the resorufin molecules position themselves in the cyclodextrin cavities and that the complex is long lived on the electronic-state timescale (a few nanoseconds). In addition, the NMR spectrum of the resorufin/CD complex indicates that the exchange with the solvent is rapid on the NMR timescale. These observations are consistent with the temperature jump measurements of the dissociation rate constant of cyclodextrin complexes, which ranges from $10\,s^{-1}$ to $10^5\,s^{-1}$.[105] The change in the spectra and photophysics of the organic cations is more subtle; however, all three chromophores bind to cyclodextrin under the experimental conditions.[93]

The rotational relaxation data look very similar for both the resorufin and the oxazine-118/β-cyclodextrin complexes, and both are well described by a double exponential decay law. Figure 3.16 shows decay curves for oxazine-118 in free solution and complexed with β-CD. The relaxation in water is well described as single exponential with a lifetime of 67 ps. When oxazine-118 is complexed to β-CD, the anisotropy decay requires a double exponential decay law (the dashed curve) and is not well fit by a single exponential (the solid line). The fast time constant of the decay law is similar in size to that for the free molecule in solution (a bit shorter), and the long time component is significantly longer. To demonstrate that the double exponential character does not originate from two species, free and bound chromophore, the anisotropy decays were studied as a function of the CD-to-chromophore

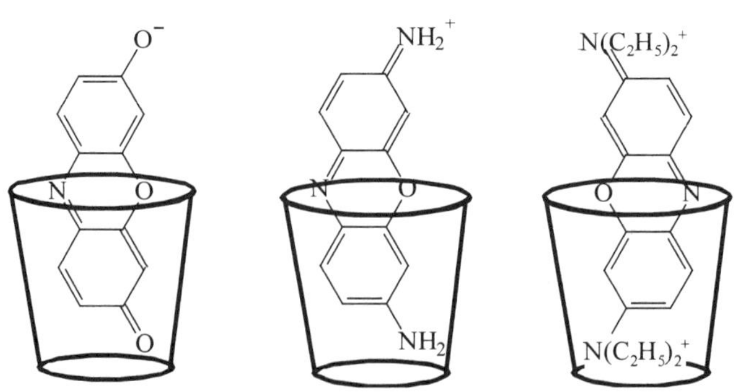

Figure 3.15. The type of inclusion complexes envisioned for the three dye molecules with β-cyclodextrin. (Resorufin is on the left, oxazine-118 is in the middle, and oxazine-725 is on the right.)

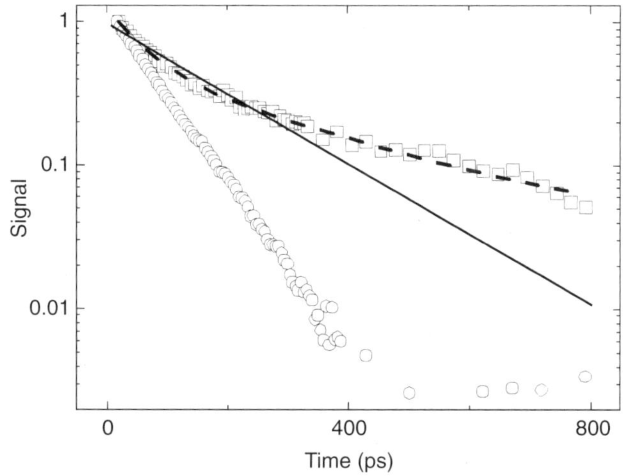

Figure 3.16. Anisotropy decays for oxazine-118 in DMSO (○) and complexed with β-CD in DMSO (□). The curves represent fits to the relaxation data for the complex (solid line is a single exponential fit and the dashed line is a double exponential fit).

ratio.[93] The anisotropy parameters that are reported in Table 2 were taken at a 1:400 (guest : host) concentration ratio, so that more than 99% of the guest is bound.

The relaxation characteristics of the resorufin and oxazine-118 display similar time constants, but different relative contributions of the short and long components to the decay law (i.e., the ratio of A_2 to A_1). The short time component is interpreted as the wobbling of the chromophore inside of the cyclodextrin cavity. The similarity of the time constants reflects the similarity of the relaxation time found for the free chromophores. The long time component of the decay is interpreted as the overall tumbling of the complex in solution, and the similarity of this component is consistent with the two complexes having similar size and shape. The large amplitude of the long time component suggests that the chromophores are "caged" inside the cavity. These data show that neither the internal nor the overall motion of the host–guest complex is sensitive to the electrostatic character of the

Table 2 Fitting parameters for some dye–β-CD complexes

	Resorufin	Oxazine-118
	no CD	
τ_{or} (ps)	63 ± 5	67 ± 6
	β-CD	
τ_{or1} (ps)	59 ± 5	56 ± 9
τ_{or2} (ps)	301 ± 25	281 ± 29
A_2/A_1	1.6 ± 0.4	1.2 ± 0.4

guest or its hydrogen-bonding ability. Rather, the complex's geometry and rotational behavior are determined mainly by the relative sizes of the guest and the cyclodextrin cavity.

Balabai[93] used diffusion in a cone model to describe the fast anisotropy decay. She allowed the chromophore orientation, defined by the solid angle $\Omega = (\theta, \phi)$, to diffuse freely with a diffusion coefficient D_θ over a restricted set of angles, such that $0^\circ < \theta < \theta_{max}$ and $0 < \phi < 2\pi$. Assuming a separation of timescales, as with the one-dimensional model previously described, Balabai was able to use the amplitude ratio of the fast and long time constants to determine the range of angles for the chromophore relaxation. She found a $\theta_{max} = 32^\circ \pm 10^\circ$ for the resorufin and $\theta_{max} = 35^\circ \pm 11^\circ$ for the oxazine-118. The root-mean-square angle θ_{rms} was then found to be 18° for the resorufin and 20° for the oxazine-118. The similarity of this parameter in the two complexes also suggests that their average structures are similar.

4 Isomerization and the Importance of Frictional Coupling

The influence of the medium on geometric transformations has important implications, both fundamentally and practically. On the fundamental side, the ability to tune the friction properties of the medium provides an avenue for exploring and describing the dynamics of condensed-phase reactions. A major aim of modern condensed-phase reaction dynamics is to understand the important atomic motions, energy flow, and intermolecular couplings of reacting systems and to control them, either with applied fields or by means of other methods.[106] Understanding complex biological processes also provides an incentive for understanding large-amplitude motions. Conformational transitions are quite important in natural processes, such as peptides and proteins,[107] the initial steps in vision,[108] and other photochemically initiated processes.[109] On the practical side, it is possible to exploit the friction dependence of these transformations to modify the outcome or yield of reactions. Work of this sort has proceeded through the study of systems in zeolites,[110] cyclodextrins,[111] monolayer films,[112] and other microheterogeneous media.[113]

For a number of decades, the isomerization of polyenes and other conjugated systems has provided an avenue for investigating the role that solute–solvent frictional coupling plays in unimolecular chemical reactions, and these studies will be discussed here. However, solute–solvent friction plays an important role in a wide range of reactions. For example, friction effects are important in reactions that involve ring opening[114] and ring closure.[115] Friction plays a role in the conformational isomerization of aliphatic rings[116] and chains.[117] Both dielectric and mechanical friction can be important in the formation of twisted intramolecular charge transfer (TICT) states[118] and conformational changes that occur with electron transfer reactions.[119] Solute–solvent friction is also important in bimolecular reactions, not only in determining the rate of diffusional encounter,[120] but also in the dynamics of the encounter complex.[121–122]

Figure 3.17 illustrates the common view of the reaction coordinate for photoisomerization reactions of polyenes. The barrier crossing involves a large amplitude motion (for stilbene it is commonly taken to be twisting about the ethylenic moiety) but it could involve other modes of the system as well.[123–124] For excited-state photoisomerization, the initial isomer of the molecule is excited to the S_1 state. If the barrier (E_0) is large enough, the molecule will vibrationally equilibrate with the environment. For *t*-stilbene, a number of time-resolved spectroscopic studies of the excited state show that the vibrational equilibration is largely complete within a few tens of picoseconds in liquids[125] and in clusters.[126] Hence, the system proceeds across the barrier to a twisted geometry by thermal activation. The crossing from the twisted state to the ground electronic surface is usually taken to be rapid, and the population decay of the initially excited state can be used to monitor the reaction kinetics. An alternative process is illustrated in the diagram for the excitation of the *cis* form. In this case no barrier is present, and the reaction proceeds before it is equilibrated with the environment (e.g., as with the isomerization of *cis*-stilbene). The adequacy of this one-dimensional picture is not wholly clear. Even for a "simple" system such as stilbene, a number of different studies have suggested that more than one internal degree of freedom may be important to the isomerization dynamics.[127–130]

The dependence of the reaction rate on the frictional coupling between the reacting system and its surroundings has been understood qualitatively for some time. At low values of the solvent friction, the reaction rate increases with increasing friction because collisions provide the energy needed to attain the threshold energy for reaction (i.e., the Lindemann mechanism).[131] In the opposite limit, when the friction or the collision frequency becomes very high, the collisions (or friction)

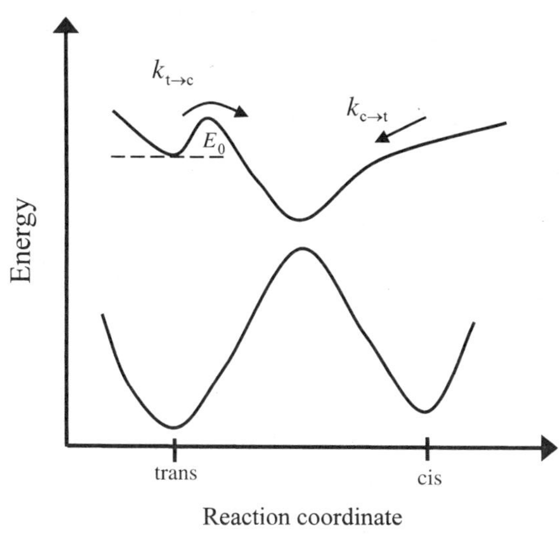

Figure 3.17. One-dimensional projection of a multidimensional potential surface that is commonly used to illustrate photoisomerization reactions (see text for details).

impede the reaction, and the rate decreases with increasing friction. Clearly, the dependence of the rate on the frictional coupling must change in character between these two limits. This turnover in the behavior of the rates has been observed in a number of systems,[123,132–134] and models that account for the friction effects over the entire regime are now available.[135]

Traditionally, the reaction rate constant k_{rxn} is written as the product of a dynamical factor (κ) and the transition-state theory rate constant (k_{TST}):

$$k_{\text{rxn}} = \kappa k_{\text{TST}}. \tag{43}$$

The k_{TST} rate constant for unimolecular reactions is given by

$$k_{\text{TST}} = \nu \frac{Q^{\neq}}{Q_{\text{r}}} \exp\left(\frac{-E_0}{k_B T}\right), \tag{44}$$

where k_{B} is Boltzmann's constant, T is the temperature, E_0 is the energy required for the reactant to reach the transition state, ν is the frequency of reactant molecules moving through the transition-state region, Q_{r} is the canonical partition function for the reactants, and $Q^{\neq}$ is the partition function for the transition state. The transition-state theory is primarily an equilibrium theory of reaction rates. It assumes that the reactant and the activated complex are in equilibrium and that the population of the other molecular and bath degrees of freedom may be characterized by a temperature T.[136] Equation 43 places the nonequilibrium aspects of the rate constant in the dynamical factor, and it is this term that is the focus of modern interest. An alternative formulation of the rate constant in terms of the reactive flux through the transition state can also be used to find Eq. 44 and, more importantly, to include nonequilibrium effects ("recrossings").[137]

4.1 Models for the Rate Constant

The earliest model that relates the frictional coupling of the reacting system with the bath to the transmission factor κ was proposed by Kramers, and that treatment has provided a point of reference for reactions involving activation barriers.[138–140] Kramers[138] treated the reactive motion as the escape of a particle over a potential barrier in one dimension (see Figure 3.18.) The particle is initially bound in the potential well on the left at point A, and it must pass over the barrier at point B to reach the product at point C on the right. Kramers described the dynamics of the particle as diffusion along the potential and was able to calculate the flux of population across the barrier—the rate of reaction—through the following assumptions:

1. A delta-correlated random force drives the motion.
2. $E_0 \gg kT$, so that a quasi-stationary state exists for the particles in the well at A.
3. A Boltzmann distribution of velocities exists near A.
4. The potential is piecewise harmonic with a frequency ω_0 near A and ω' near B.
5. At some point past the top of the barrier, passage to the product is certain (i.e., there is no appreciable back reaction).

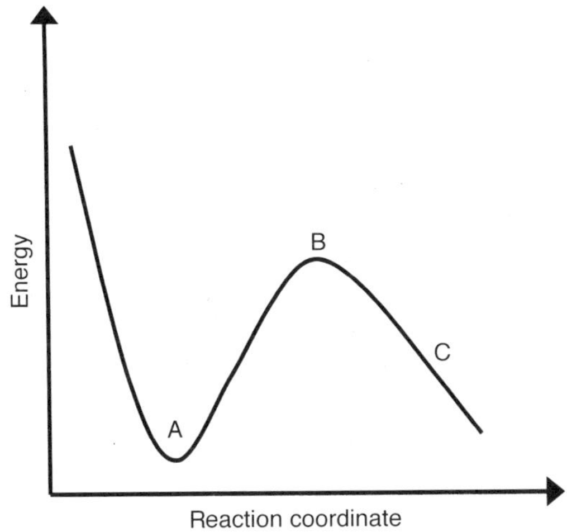

Figure 3.18. A one-dimensional potential surface for an activated barrier crossing.

Kramers found that the rate constant is given by

$$k_{\mathrm{rxn}} = k_{\mathrm{TST}}\left[\frac{\sqrt{1+(2\omega'\tau_{\mathrm{v}})^2}-1}{2\omega'\tau_{\mathrm{v}}}\right], \tag{45}$$

where τ_{v} is the characteristic time for the velocity distribution to thermalize. The term in square brackets is the transmission coefficient κ. The velocity relaxation time is given by $\tau_{\mathrm{v}} = 1/\zeta$, where ζ is the friction. Equation 45 applies to the "intermediate" and high-friction regimes, where the rate constant decreases as the friction increases. It has been common to write the rate constant as

$$k_{\mathrm{rxn}} = F(\zeta)\exp\left(-\frac{E_0}{RT}\right), \tag{46}$$

in which the dynamical factor $F(\zeta)$ contains the transmission factor and the pre-exponential terms in the k_{TST} expression. The use of Eq. 46 reflects the observation that many unimolecular isomerization rates are described well by the Arrhenius form.

One expects that a significant fraction of the conformational transitions in condensed phases will occur in the limit of very high friction. This limit is attained when the velocity relaxation time τ_{v} is short compared to the characteristic timescale of the particle's free motion on the potential surface. In this limit, Eq. 45 can be simplified by an expansion about $\omega'\tau_{\mathrm{v}}$. One finds the Smoluchowski expression

$$k_{\mathrm{rxn}} = \frac{\omega_0\omega'\tau_{\mathrm{v}}}{2\pi}\exp\left(-\frac{E_0}{RT}\right) = \frac{\omega_0\omega'}{2\pi\zeta}\exp\left(\frac{-E_0}{RT}\right) \tag{47}$$

for the rate constant, in which it is inversely proportional to the friction. An important subtlety of the criterion for this limit is that the reaction rate constant k_{rxn} does not define the timescale for comparison with τ_{v}. Rather, the inverse of the barrier frequency ω' is the relevant timescale, and it is determined by the curvature of the potential surface (i.e., the force constant). Hence, reactions with slow rate constants that are caused by high and steep reaction barriers may occur in the intermediate friction regime, rather than the high-friction regime.

Kramers's view can be generalized by relaxing the first assumption, namely, allowing the stochastic force that drives the motion to have a finite response time (i.e., a memory). This description was developed by Grote and Hynes[139,141] for the barrier-crossing problem. They found that

$$k_{\mathrm{rxn}} = k_{\mathrm{TST}} \frac{\lambda}{\omega'}, \tag{48}$$

where λ, the reactive frequency, is related to the transmission coefficient κ by $\kappa = \lambda/\omega'$. λ is determined from the self-consistency relation

$$\lambda = \frac{\omega'^2}{\lambda + \zeta(\lambda)}. \tag{49}$$

The frequency-dependent friction $\zeta(\lambda)$ is determined at the reactive frequency and defined by

$$\zeta(\lambda) = \int_0^\infty dt \zeta(t) \exp(-\lambda t), \tag{50}$$

where the time-dependent friction $\zeta(t)$ is given by the force–force autocorrelation function

$$\zeta(t) = \frac{1}{\mu k_B T} \langle F(0)F(t) \rangle. \tag{51}$$

This equation allows for correlations in the random force term, or memory in the friction. Combining Eqs. 48 and 49 and rearranging to eliminate λ gives

$$k_{\mathrm{rxn}} = k_{\mathrm{TST}} \left[\sqrt{1 + \left(\frac{\zeta(\lambda)}{2\omega'} \right)^2} - \frac{\zeta(\lambda)}{2\omega'} \right], \tag{52}$$

which corresponds to the Kramers expression with a frequency-dependent friction. In the limit that $\zeta(\lambda) = \zeta$, the Kramers result (Eq. 45) is recovered. An impediment to the use of this relationship has been the absence of an adequate model for $\zeta(\lambda)$.

The generalization of this one-dimensional description of the reaction rate constant to many dimensions has been undertaken.[142–146] It becomes necessary to view the reaction as occurring on a multidimensional potential-energy surface when degrees of freedom other than the reaction coordinate are able to evolve slowly. When the other degrees of freedom respond rapidly compared to the reactive mode, they constitute the bath that keeps the nonreactive modes in thermal equilibrium and provides the friction for the reaction. This corresponds to the one-dimensional model of Kramers, with or without frequency-dependent friction. If one or more nonreactive modes relax slowly with respect to the reactive mode, then it will not stay in equilibrium during the course of the reaction. With the reaction occurring on a multidimensional energy surface, motion along one coordinate could have a stronger coupling to solvent friction than motion along some other coordinate. In that case it is possible for the reaction coordinate to change as the friction changes.[144,146] The process may also be strongly influenced by anisotropy in the potential-energy surface, and the interplay between these effects can be complicated.

One circumstance in which multidimensional effects can be observed is the case of highly anisotropic friction, as discussed by Berezhkovskii[124,146] for a two-dimensional system. Consider the potential-energy surface illustrated in Figure 3.19, where q is the reactive coordinate and x is the nonreactive coordinate. For the situation in which motion along x is rapid compared to motion along q, the reactive trajectory will proceed through the saddle point region. This situation corresponds to the one-dimensional description of the reaction. If the particle's motion along x is slow compared to its motion along q, reactions may occur at different values of x and saddle points may be avoided. This behavior is distinct from the one-dimensional view of the reaction. For the situation in which x is extremely slow and the system–bath interaction along x is strong, one would observe an inhomogeneous distribution of reaction rates $k_{rxn}(x)$. Although considerable theoretical efforts have been expended in this area, no quantitative comparisons with experiment have yet been performed. Another important feature of chemical reactions in solution is the possible dependence of the frictional coupling on the reaction coordinate q. For example, the friction for the *trans* form of 4,4′-dihydroxystilbene, which is nonpolar, may be different than the friction at the transition state, which is dipolar. The inclusion of the spatial dependence of the friction has been appreciated for some time; however, recent work in this area has extended the considerations to include both the time and frequency dependence of

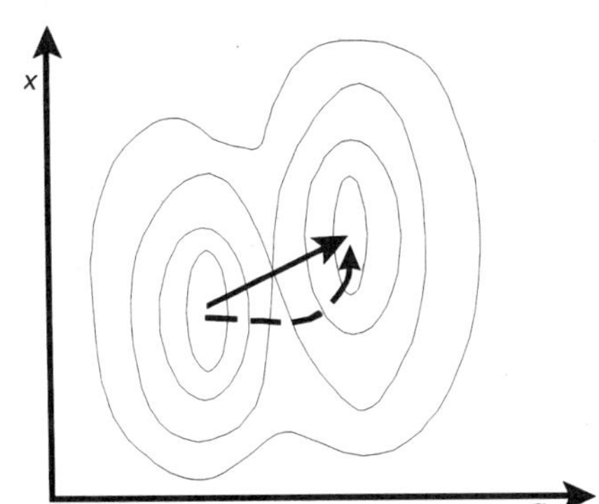

Figure 3.19. A two-dimensional 'effective' potential energy surface. The solid line represents a reactive trajectory that passes through the saddle point, whereas the dashed curve represents a reactive trajectory that avoids the saddle point.

the frictional coupling.[147] It remains to be seen how important these considerations are in experimental systems.

4.2 Identifying an Intrinsic Barrier

Before addressing the friction dependence of the reaction, one must characterize the energy barrier to reaction. A number of methods have been developed for trying to disentangle the energy effects from the friction effect on the reaction. The primary idea is to change either the solvent or solvent conditions so that one can obtain a temperature dependence for the rate at constant values of the frictional coupling and solvation. In many cases this separation appears to be possible; however, it should be made clear that this may not always be so. A number of workers have found that the degree of solvation for the system can be difficult to constrain. Also, one may find that the reaction barrier appears to be friction dependent, and this could be a signature to the effect that a multidimensional view of the reaction is required. In any case, the purpose of identifying the reaction barrier is to extract the dynamical factor $F(\zeta)$ from the expression for the rate. This factor will be referred to as the *reduced isomerization rate*.

The best method for determining the reaction barrier in solution is to study the rate in a single solvent as a function of temperature and pressure.[132,148–149] In this way, one can obtain the temperature dependence of the reaction rate, but keep the collision frequency, or solvent viscosity, constant through control of the pressure. Such an approach requires that the activation volume not be large[123,150] (i.e., the energy barrier must not be a strong function of the pressure). The dependence of the activation energy on pressure can be evaluated by performing activation plots over different pressure and temperature ranges. If aforesaid conditions hold, then it is plausible to associate the activation energy obtained from such measurements to the energy barrier to reaction.

A more common approach measures the isomerization rate in a homologous solvent series (e.g., *n*-alkanes or *n*-alkylnitriles) so that differences arising from changing the solvent molecules are minimized. This method is easier to carry out experimentally than are temperature and pressure studies. In most cases, workers adjust the temperature in each member of the series so that the frictional coupling (as measured by viscosity, collision frequency, or some other parameter) is constant. From these data, an Arrhenius plot at constant friction, or an "isofriction" plot, can be constructed. This method assumes that the solvation energy and the form of the solute–solvent friction do not change significantly through the homologous series; however, that assumption does not always hold. For example, the assumption that the shear viscosity is a measure of the solute–solvent friction could fail for a relatively small solute (such as stilbene) through the series of alkane solvents.[167] Certainly, the effective hydrodynamic boundary condition is expected to change for such a system, (see Section 2). Fortunately, it is possible to evaluate the accuracy of this assumption by obtaining isofriction plots at different viscosities, with different measures of the friction (eg., viscosity and collision frequency) and with different sets of solvent molecules in the series. If the assumed measure of the friction is robust, then the slopes of the curves should give similar activation barriers. More commonly, it has been found that the activation barrier through a homologous series of solvents changes because of differences in the solvation for different solvent molecules. The effect of solvation on the barrier can be significant[123,151] and requires careful treatment, since it has such a strong impact on the rate constant. Early studies used empirical measures of the solvent polarity[124,151] to quantify how solvation modifies the barrier height. Recently, Kauffman[152] has proposed the use of "isodielectric" plots to account for changes in the solvation through a homologous series. In this approach, he assumes that the solvation will scale with the solvent's dielectric constant. By fitting the data to the Kramers–Hubbard relation or by obtaining Arrhenius-type plots at a constant dielectric constant, he was able to determine the dependence of the activation barrier on the solvent's dielectric constant in a quantitative manner.

The photoisomerization of *trans*-stilbene in different solvents displays the range of "nonspecific" behavior that has been observed in photoisomerization reactions.[123–134] The threshold to isomerization of the isolated *trans*-stilbene has been measured to be 3.2–3.8 kcal/mol.[123,153] Isoviscosity plots for the isomerization in the shorter chain alkane solvents (i.e., propane through hexane) give an activation energy of 3.5 kcal/mol, in good agreement with the value for the isolated molecule. However, isoviscosity plots in the longer chain alkane solvents (decane to hexadecane) display an activation energy of 4.0 kcal/mol (see Table 7 in Ref. 123). Using a somewhat different method, Sun and Saltiel[154] found an intrinsic barrier of 2.9 kcal/mol, close to the value for the isolated molecule. Troe and co-workers, using temperature and pressure[132,148] to study the isomerization of stilbene in ethane, propane, and butane, found activation energies of about 1.8 kcal/mol. The disagreement between this result and those obtained from the other methods might suggest a pressure dependence of the barrier; however, similar activation energies were found over different pressure and temperature ranges. Troe and co-workers have attributed this drop in the energy barrier to a change in the solvation of stilbene and its barrier to isomerization, although other explanations for these observations have also been offered.[123,155–156]

The separation of the barrier from the friction effects appears to be possible for *trans*-stilbene in the homologous series of *n*-alkylnitrile solvents, from propanenitrile to decanenitrile. Sivakumar, Hoburg, and Waldeck[157] found that the activation energy obtained from isoviscosity plots did not change significantly over the range of 5 cP. The activation energy they observed was 2.6 kcal/mol, which is significantly lower than that reported in the alkanes. This reduction presumably results from differential solvation of the transition state and the reactant state. Apparently, the variation of the solvation through the series of alkyl nitriles was not significant enough to affect the barrier significantly. However, the analysis did not include isomerization rates in acetonitrile, because doing so led to a variation in the activation energy of isoviscosity plots.[124] This effect is caused by differential solvation and has also been observed in other systems.[151] The method recently outlined by Kauffman[152] may be useful for systems of this sort. The comparison of these results[124,151,152,157] with those found in the alkane solvents clearly shows that solvation can have an important impact on the reaction barrier for *t*-stilbene isomerization.

The isomerization of *t*-stilbene in *n*-alcohol solvents displays yet another behavior. Fleming[158] studied the isomerization of *t*-stilbene in the alcohol solvents from methanol to decanol. The isoviscosity plots he obtained displayed an apparent decrease in the activation energy with increasing viscosity, which also correlates with increasing size of the alcohol molecules and increasing dielectric relaxation time of the solvent. Later, Troe[159] performed pressure and temperature studies of *t*-stilbene isomerization in the alcohols and observed similar behavior. Recently, Kauffman used isodielectric plots in an attempt to extract a solvated reaction barrier for the stilbene system.[160] These observations[158,159] have been explained as resulting from incomplete solvation of the reactant in the alcohol solvents. Because the alcohols have dielectric relaxation properties that proceed over multiple timescales, the molecules are not completely solvated during the course of the reaction when the reaction and solvation timescales overlap. Related studies on dimethoxystilbene and dihydroxystilbene in alcohols[161] came to similar conclusions, and in those cases the spectral evolution resulting from the solvation was evident. This dependence of the observed activation energy on the solvation dynamics complicates the description of the reaction in terms of a one-dimensional, solvated reaction coordinate and necessitates a multidimensional description.[124,161]

4.3 Modeling the Friction

An important feature of the comparison between measured reaction rates and rates calculated from theoretical models is the modeling of the friction between the isomerizing moiety and the solvent. It is clear from the discussion in Section 2 that the proper description of solute–solvent friction is not yet in hand. Beyond this evident limitation, other complexities enter into the determination of how the friction should be modeled. For example, the presence of many degrees of freedom in the isomerizing moiety can complicate the issue. If other nonreactive modes in the molecule act as dissipation channels for energy from the reactive mode, they can effect the reaction rate and contribute to the friction. Whether they increase or decrease the reaction rate depends on whether one is in the high- or low-friction limits.[98,116] This issue is related to a second complexity that relates to the proper modeling of the frequency-dependent friction. A proper modeling of the friction kernel would incorporate both the solvent and internal solute bath modes, but such models are not available for realistic systems. The different approaches used in modeling the friction between the isomerizing moiety and the solvent are described next.

The simplest modeling of the frictional coupling ζ is to assume that it is proportional to the solvent shear viscosity η, the DSE relation of Eq. 7. The applicability of the DSE relation to a wide variety of rotational diffusion and translational diffusion data provides a motivation for this approach, which is commonly discussed and on whose adequacy many experimental studies have explicitly focused. If one makes the assumption embodied therein for the friction coefficient, then the Kramers expression becomes

$$F(\zeta) = A_1\left[\sqrt{1+\left(\frac{\eta}{A_2}\right)^2} - \frac{\eta}{A_2}\right], \tag{53}$$

where $A_1 = \omega_0/2\pi$ and $A_2/\eta = 2\omega'\tau_v$. With some exceptions (e.g., the isomerization of diphenylbutadiene in alcohols,[159,162] the isomerization of binaphthyl in alcohols,[163] and stiff stilbene in alkanes[164]), the comparison of isomerization rate data with Eq. 53 fails. The failure could arise from a variety of different sources. It may be that the friction is not proportional to the solvent shear viscosity and that a more adequate modeling of the frictional coupling needs to be developed. In particular, it may be that the friction needs to be evaluated on microscopic time and length scales. Considerable experimental effort has tried to address this issue. An alternative explanation of the model's failure could be the need for more than one dimension in describing the reaction.

One approach to correcting the hydrodynamic version of the Kramers expression (Eq. 53) is to use a microscopic measure of the solvent response or a microscopic treatment of the solute–solvent friction or both. To this end, Zeglinski[161] used the light-scattering relaxation time τ_{ls} of alkanes as a measure of the friction, rather than the solvent viscosity. Although τ_{ls} scales linearly with η/T for a given alkane, it is not proportional to η through the homologous series of alkanes.[165] Note that whereas this approach partly accounts for molecular effects, comparison with the Kramers model still displays significant disagreement. Other workers[166–167,176a] have applied the Gierer–Wirtz model and the Dote–Kivelson–Schwartz quasi-hydrodynamic models to a description of the frictional coupling between the isomerizing moiety and the solvent. Once again, the correspondence between the predictions of the Kramers formula and the measured reaction rates is improved, but significant differences between the predicted and experimental rates are still evident.

A number of studies have compared measured isomerization rates in a single solvent with the predictions of the hydrodynamic Kramers expression and found disagreement between the two. For example, the isomerization of *t*-stilbene in decanenitrile is not well described by Eq. 53,[157] even though one can reasonably expect the rotational diffusion constant to scale linearly with viscosity in this system. From the earlier discussion (Section 2), it is clear that the frictional coupling can vary with solvent size, for both mechanical and dielectric components, and the failure of the model in the homologous series can have a contribution from this size dependence of the solute–solvent friction. This conclusion is corroborated by the improved agreement found by more microscopic measures of the solute–solvent frictional coupling. Nevertheless, the failure of the model in a single solvent for which the solute–solvent friction is well characterized by η/T is strong evidence that the simple Kramers form is not adequate.

A more successful approach to modeling the friction term in the Kramers expression (Eq. 45) has been to measure the diffusion coefficient of the isomerizing moiety, or of the whole molecule, in the solvent of interest. This approach has been used with both translational (D_T)[154,163,168] and rotational (D_{or})[163,169–170] diffusion coefficients and is moderately successful, especially in describing the reaction rate behavior for a number of different systems in which the hydrodynamic approach fails, namely, binaphthyl in alcohols,[163] DPB in alkanes,[169] and DiIC2 in alcohols.[171] However, the aforesaid approach clearly fails in other cases (vide infra). This procedure is sometimes referred to as the Hubbard–Kramers treatment[152,169,171] because one can use the rotational diffusion time to provide a measure of the angular velocity relaxation time through the Hubbard relation

$$\tau_\omega = \frac{I}{6k_B T \tau_{or}}. \tag{54}$$

This measure of the friction is somewhat closer to a collision frequency, and hence the velocity relaxation time, in Kramer's treatment; however, for strong solute–solvent coupling (a few kT), the relation may not hold.[172]

Despite the success of the preceding approach in some systems, a large number of examples exist in which such approaches fail. Most notable is the case of DODCI[170,173,174] and other cyanine dyes in alcohol[171,175] and nitrile solvents,[176] where the isomerization deviates significantly from the hydrodynamic Kramers equation. The deviation between the hydrodynamic Kramers expression and the measured viscosity-dependent rate is correlated with the barrier height for reaction in these systems. As the barrier height increases, the deviations become more severe. A number of workers have tried to use microviscosity and other measures of the friction that incorporate molecular features or the free volume in solution. To some extent, these considerations improve the agreement between the prediction from the Kramers model and the measured rates, but not wholly so. For example, in the dye DiIC2,[171] such considerations bring the observed rates in line with the observations. In many other systems, however, those considerations fail[170,173–176] (e.g., deviations are found for DODCI isomerization rates, even though the rotational diffusion time is linear in viscosity),[170,173] and comparisons are made for a single solvent.[174] Perhaps a more detailed description of the isomerizing motion (one that includes, for example, volume displacement and the relative proportions of translation and rotation) could further improve the correspondence, but it seems clear that it will not be able to fully account for the change in timescale between the overall rotational motion and the internal motion along the potential surface. The interpretation from these works is that the timescale of the isomerization near the top of the barrier determines the frequency at which the friction response of the solvent should be evaluated.

The use of a power-law dependence of the reaction rate on the solvent shear viscosity has been quite popular as an alternative to the Kramers model. Although the origin of this dependence in molecular considerations is clouded, the procedure is useful because it provides a way to characterize the data and draw comparisons between data sets. In addition, the approach can be used to draw qualitative comparisons with theoretical models. The power-law form for the friction dependence is

$$k_{\mathrm{rxn}} = \frac{B}{\eta^{a}} \exp\left(- \frac{E_0}{RT} \right), \tag{55}$$

where B and a are parameters, and E_0 is the activation barrier, typically obtained from isofriction plots. A number of models have been proposed for the origin of the power-law form.[177–181] The earliest model relates to the consideration of the available free volume in the solution compared with the critical free volume needed for isomerization.[177] Closely related to this model is the medium-enhanced barrier model developed by Saltiel and successfully applied to a number of different systems, including alkylstilbenes.[178] More recently, Robinson[179] exploited the form of the Grote–Hynes relationship for the rate law and critical scaling ideas to develop a model for the power-law form, which he then applied to the isomerization of merocyanine.[176b] A recent calculation for the frequency-dependent friction by Bagchi[180] using mode-coupling theory is very appealing, and it appears to give a behavior similar to the power-law form. The origin of the power law form for the rate constant's viscosity dependence is still under investigation.

Equation 55 has been found to fit the isomerization rate data quite well in a wide variety of systems.[123,182] If the power-law model is used to fit the isomerization data for the cyanine dyes (see Table 3), one observes a correlation between the a parameter and the barrier height that is extracted from isofriction plots. The decrease in the a parameter with increasing barrier height reflects a weaker dependence of the reaction rate on the zero-frequency shear viscosity. The correlation has motivated the consideration of models that incorporate the frequency-dependent response of the solvent to describe the frictional coupling.

The necessity of properly considering both the spatial and frequency scales of the frictional coupling between the solute and solvent is evident in a number of other systems as well. The isomerization of alkenylanthracenes has been used as a probe of these effects, because it involves the motion of a small moiety (a vinyl or propenyl group) relative to the anthracene. Although using a realistic measure of the friction for the isomerizing moiety helps the correspondence with the prediction of friction dependence from the Kramers model, it does not give full agreement therewith.[183] It appears that the proper incorporation of frequency-dependent friction is able to mimic the data with reasonable potential parameters.[184] However, one will clearly need to be able to realistically model both the frequency dependence and the spatial dependence of the friction if one intends to extract realistic potential parameters from the data.

The Grote–Hynes treatment of the barrier-crossing dynamics suggests the importance of using a frequency-dependent friction in modeling the reactive event. This result arises because the motion of the isomerizing moiety in the region of the barrier crossing occurs on timescales significantly shorter (ps and shorter) than the typical timescale of rotational diffusion for these species (ps and longer). The data in Table 3 reflect these considerations. The viscosity dependencies for rotational diffusion are linear, but the viscosity dependencies of the isomerization

Table 3 Barrier heights, isomerization rates, and the power law parameter a for various cyanine dyes

System	Barrier Height (kJ/mol)	Rate at 1 cP (Hz)	a	Reference
DODCI				
GS (n of 1–10)	57.3	4×10^{12}	0.26	170
GS (methanol)	54.5	1×10^{13}	0.48	174
GS (ethanol)	54.5	8×10^{12}	0.38	174
GS (propanol)	54.5	9×10^{12}	0.26	174
GS (butanol)	54.5	8×10^{12}	0.25	174
ES (n of 1–10)	11.3	8×10^{11}	0.43	170
Rotation (n of 1–10)	0	10^{10}	0.99	170
Rotation ($n = 1, 2, 5, 8, 10$)	0	4×10^{9}	1	175d
DTCI				
GS ($n = 1, 2, 5, 8, 10$)	60 ± 1 (51^c)	6×10^{12}	0.17 (0.31^c)	175(g)
GS ($n = 1 - 8$)	54 ± 3	8×10^{11}	0.20 ± 0.02 (0.25^b)	176(a)
GS (nitriles $n = 1 - 10$)	54 ± 3	7×10^{11}	0.38 ± 0.02 (0.51^b)	176(a)
ES ($n = 1, 2, 5, 8, 10$)	10 ± 1	2×10^{11}	0.82	175(g)
ES (nitriles $n = 1 - 10$)	12.6 ± 0.8	3×10^{11}	0.65 ± 0.02 (0.85^b)	176(a)
Rotation	0	4×10^{9}	1	175(d)
DOCI				
GS ($n = 1, 2, 5, 8, 10$)	61 ± 2 (52^c)	2×10^{12}	0.30 (0.52^c)	175(g)
GS ($n = 1, 2, 3, 4, 6$)	62		0.37	175(e)
ES ($n = 1, 2, 3, 4, 6, 8$)	6.9	10^{11}	0.66	175(e)
ES ($n = 1, 2, 5, 8, 10$)	10 ± 1	2×10^{11}	0.92	175(g)
Rotation ($n = 1, 2, 5, 8, 10$)	0	3×10^{9}	1	175(d)
DTDCI				
GS ($n = 1, 2, 5, 8, 10$)	47 ± 1	5×10^{11}	0.44	175(g)
ES ($n = 1, 2, 5, 8, 10$)	24 ± 1	6×10^{12}	0.51	175(g)
Rotation	0	4×10^{9}	1	175(d)
DOTCI				
GS ($n = 1, 2, 5, 8$)	43 ± 2	3×10^{11}	0.36	175(g)
ES ($n = 1, 2, 5, 8$)	20 ± 1	5×10^{11}	0.66	175(g)
HIDCI				
GS ($n = 1, 2, 5, 8, 10$)	47 ± 3	2×10^{11}	0.32	175(g)
ES ($n = 1, 2, 5, 8, 10$)	20 ± 1	4×10^{12}	0.72	175(g)
Rotation	0	4×10^{9}	1	175(d)
DiIC2				
ES (alcohols n = 1–12)	7.5^a	1×10^{11}	0.66 (0.73^b)	175(a)
Rotation (alcohols n = 1–12)	0	7×10^{9}	1	175(b)
Merocyanine 540				
ES (n = 1 to 8)	5.9 ± 0.9	2.5×10^{10}	0.62	176(b)
ES (nitriles)	15.2 ± 0.3	2.5×10^{11}	1.0	176(b)

Unless stated otherwise, the solvent is an n-alkyl alcohol.

[a] The barrier height was found to be solvent dependent.

[b] Found when a 'microviscosity' was used.

[c] These values include a solvation correction for the barrier height that was found to be important for the ground-state isomerization of these shorter chain cyanine dyes.

are sublinear. These findings are also evident in polyene systems (see Table 4, which compiles isomerization rate data from a number of different polyene systems). For the species in which the barrier heights and frequencies are large, the Kramers treatment fails, whereas for those of moderate size, it seems to work better.

Although the frequency-dependent friction in the Grote–Hynes formulation includes other degrees of freedom of the solute, most comparisons with experiment have been made by accounting for the frequency response of the solvent degrees of freedom. Modeling of the frequency-dependent friction in nonpolar solvents has followed the work of Bagchi and Oxtoby.[185] These treatments represent the isomerization as the motion of a sphere that is attached to the axis of rotation and that moves through the fluid with a slip hydrodynamic boundary condition. The model for the frequency-dependent shear viscosity $\eta(\lambda)$ of the solvent is the Maxwell form

$$\eta(\lambda) = \frac{\eta_0}{1 + \lambda\tau}, \tag{56}$$

where η_0 is the zero-frequency shear viscosity and the relaxation time τ is given by η_0/G_∞. (G_∞ is the infinite-frequency elastic modulus of the solvent.) This model for the frequency-dependent friction response and its incorporation into the Grote–Hynes relation has been able to mimic the observed rates for a number of different systems,[157,169,176a,184] but the potential parameters one obtains from the analysis are not always physically reasonable. A variety of other models for the frequency-dependent friction could be constructed. Two such models are described next. The

Table 4 Summary of some isomerization data for polyenes

System	Barrier height (kJ/mol)	a	Hydrodynamic Kramers	Hubbard–Kramers	Reference
Alkanes					
Binaphthyl	4.8	—	NO	YES	163
Stiff Stilbene	6.3	1.0	YES	YES	207
Stilbene	14.6 to 16.6	0.32	NO	NO	65, 154, 169, 167
4,4′-DMS	18.0	0.26	NO	—	129, 178
DPB	19.7	0.59	NO	—	208
4,4′-*t*-BuS	19.7	0.33	NO	—	178
4,4′-DMeOS	23.9	0.22	NO	—	161
Nitriles					
Stilbene	10.9	0.51	NO	—	161
4,4′-DHS	16.8	0.48	NO	—	161
4,4′-DMeOS	17.6	0.44	NO	—	161
Alcohols					
DPB	4 to 8	1.0	YES	YES	152, 162
Binaphthyl	5.2	—	YES	YES	163
4,4′-DHS	8.8	0.61	NO	—	161
4,4′-DMeOS	11.7	0.67	NO	—	

Stiff Stilbene–1,1′-biinylidene; 4,4′-DMS is 4,4′-dimethylstilbene; 4,4′-DMeOS is 4,4′-dimethoxystilbene. 4,4′-t-BuS is 4,4′-t-butylstilbene; 4,4′-DHS is 4,4′-dihydroxystilbene; DPB is diphenylbutadiene first has found some success in comparison with data in polar solvents, and the second was recently proposed by Bagchi.[180]

In polar solvents, dielectric friction (see Section 2) can be a significant damping mechanism for isomerizations that involve the movement of polar or ionic moieties. For solvents, like the alkylnitriles, that have simple Debye responses, it is possible to include the frequency dependence of the dielectric friction in a direct way.[34,37] For such systems, the frequency-dependent friction will have a Maxwell form, with the time constant given by the Debye relaxation time τ_D. Van der Zwan and Hynes[186] developed a treatment of this sort for the isomerization of two dipolar moieties in a dipolar solvent. Sivakumar, Hoburg, and Waldeck[157] incorporated the Maxwell form of the friction into the Grote–Hynes formula to obtain an expression for the rate constant. It is somewhat unclear what to use for the reactive frequency in that equation; however, they replaced the reactive frequency with $F(\zeta)$ and found that

$$\omega_b^2 - F^2(\zeta) = \frac{\omega_b^2 \beta F(\zeta)\tau_D}{1 + F(\zeta)\tau_D}, \tag{57}$$

where ω_b is the barrier frequency for the solvated potential and β is a measure of the solute–solvent coupling strength. In a comparison with experiment that was performed for *t*-stilbene in the *n*-alkylnitrile solvents, Sivakumar et al. were able to fit the data very well and obtained reasonable potential parameters; however, systematic deviations between different solvents were apparent. Others[179a] have used a similar approach in analyzing their isomerization results in polar solvents.

Another model for the frequency-dependent viscosity of the medium recently has been proposed and compared with some of the earlier experimental data.[180] In this model, the frequency-dependent friction coefficient is composed of three terms: a "bare" collisional part that is modeled with a friction that decays as a Gaussian in time [i.e., $\exp(-at^2)$], a part that arises from density fluctuations in the liquid, and a part that corresponds to the hydrodynamic shear viscosity. For dense, highly viscous liquids, the first two terms can be very important in processes that occur at high frequencies, (or rates) and these terms were not included in earlier models. In addition, the viscosity term has a frequency dependence that is clearly different from that of the Maxwell form (Eq. 56). Using this formulation of the frequency dependent friction, Bagchi was able to mimic the viscosity behavior of the reaction rate and demonstrated that the correlation with the zero-frequency shear viscosity could be well described by a power law. When the results were compared with experiment, he was able to obtain reasonable potential parameters for the isomerization coordinate. It should be interesting to see how this model performs in comparisons with other experimental data.

4.4 The Importance of Other Degrees of Freedom

The influence of more than one mode on the reaction has been considered by a number of workers and is usually implemented for two explicit degrees of freedom. From these considerations, two physically appealing models have arisen: situations in which the reaction coordinate changes with the degree of frictional coupling and situations in which a solvent coordinate serves as one of the reaction coordinates.

One manifestation of multidimensional effects is that proposed by Agmon and Kosloff[144] with regard to the photoisomerization of *t*-stilbene. These researchers considered both ethylenic twisting and phenyl torsion. If the anisotropy in the frictional coupling for these two motions is large enough, then one could imagine the mechanism of the reaction changing as the friction of the medium changes. For example, at low frictions the motion by ethylenic twisting dominates, and at high frictions that of phenyl torsion dominates. The presence of such an effect requires the proper balance between the characteristics of the potential surface and the frictional coupling. One can imagine that such considerations may be relevant to a wide number of conformational transitions (e.g., the isomerization of methylstilbenes[129] and azobenzenes[187]).

The second effect concerns reactions in which the solvent polarization can act as a reaction coordinate. Polar solvents have a dramatic effect on the isomerization of molecules that have dipolar and ionic moieties, have a highly polarizable transition state, or involve significant charge displacement. Differential solvation of the molecule at the transition state and in the reactant state will cause changes in the activation energy of the reaction and can significantly affect the reaction rate. This issue is appreciated, and efforts are being made to model these properties (vide supra). A second effect is the dielectric friction between the solute and solvent, which causes a decrease in the reaction rate. Theoretical models capable of treating such systems are available, but comparisons with experiment have been rare.[124]

If the frictional coupling between the solute and the solvent is strong and anisotropic, then the friction and solvation effects will be coupled in a nontrivial way. The isomerization of dihydroxystilbene and dimethoxystilbene in the alkylnitriles and alcohol solvents display some of these features.[124,161] In the nitrile solvents it seems possible to separate the effects of the friction and the solvation, but in alcohol solvents it is not possible to separate them cleanly. The fluorescence kinetics of these systems clearly show that the dynamics of solvation and the isomerization are occurring on similar timescales. For the nitrile solvents, one can consider the reaction as occurring along an "equilibrium-solvated" reaction coordinate and incorporate nonequilibrium solvation effects into a frequency-dependent friction.[139] For alcohol solvents, it appears that one must use a multidimensional model that includes a solvent coordinate (or polarization coordinate) explicitly.[124,145] These multidimensional effects could be probed more fully by studying the photoisomerization of asymmetrically substituted ("push–pull") stilbenes.[188–190] It is found that two competing reaction channels are present in such systems: one associated with charge displacement from the donor end to the acceptor end and the other with the *trans* to *cis* isomerization. Although considerable debate exists over the details of the mechanisms, it is clear that the relative importance of the two channels is sensitive to the solvent viscosity and polarity.

4.5 Barrierless Reactions

The isomerization process without a barrier is considerably more complex. In this situation, the optically prepared state does not necessarily thermalize with the bath before the reaction proceeds. Hence, the reaction occurs from a nonequilibrium state that is determined by the preparation and the solute-bath dynamics. Considerable work in this area has focused on triphenylmethanes[191] and other dyes.[192] It is generally found that the BFO model[193] is able to reasonably describe the relaxation properties. The photoisomerization of *cis* stilbene is also barrierless (barrier $< kT$) and has received considerable experimental[194–196] and theoretical[197–199] attention. The initial excited state proceeds to relax to the ground-state surface by a conformational change. The process occurs very rapidly, at about 1 ps or less—faster than vibrational relaxation. The kinetics is somewhat correlated with the solvent viscosity; however, it seems clear that solvent polarity plays an important role in modifying the excited-state potential. This system is complex, because there are multiple photoproducts and the excited-state potential must be treated as multidimensional. The key to understanding the system is to understand the details of the potential surface, the system preparation by the optical field, and the solute–solvent friction. Although this represents a daunting degree of complexity, the dynamics of *cis*-stilbene isomerization may display interesting features found in reaction dynamics, such as conical intersections and coherent vibrational motion.

5 Conclusion

Solute–solvent frictional coupling clearly plays a key role in determining the dynamics of the large-amplitude motion of organic molecules in liquid solution. The rotational relaxation of medium-sized solutes acts as a useful probe of these solute–solvent interactions, and comparisons of such data with numerical and analytical theory should lead to the development of a realistic description of solute–solvent friction on molecular time and length scales. On a more practical level, the rotational relaxation of dye chromophores in microheterogeneous media can be used to empirically measure the "micro- and nano viscosity" of the chromophores. Such approaches have been used for polymers and glasses[200–201] and in liquid crystals.[202] The location of dye probes in micelles, vesicles, and membranes is used to investigate local motions and restricted relaxation.[203] The intercalation and covalent attachment of chromophores to biological molecules is also used to investigate their motion.[204] Recently, some workers have utilized rotational relaxation to investigate molecular features (e.g., density augmentation) of supercritical fluids[205] and at critical points.[206]

The investigation of the solute–solvent friction and its importance for the dynamics of bimolecular complexes is much less well developed. Dynamical studies of such systems are needed to understand the timescales of the motion and how they are influenced by the structural and energetic details of the complex. Studies of this sort not only should provide important information concerning the dynamics of bimolecular complexes, but also should act as a useful probe of the solute–solvent friction and its dependence on the frequency of the motion. By tuning the strength of the intermolecular binding in such complexes, one can vary the characteristic frequency of the torsional potential in which the chromophore moves (either by libration or diffusion) and probe the friction at different frequencies.

A wide number of investigations into the importance of friction on photoisomerization reactions have now been carried out. It appears that a major obstacle to developing a quantitative model for such reactions lies in the proper modeling of the solute–solvent friction on the relevant time and space scales. We still do not have a clear view of the form of the frequency-dependent friction, either from experiment or from theory. However, a recent model of the frequency-dependent friction proposed by Bagchi[180] may represent substantial progress on this front. A further complication to understanding the friction in these reactive systems is the involvement of other degrees of freedom (multidimensional effects) and the important changes in the reaction barrier that can arise from solvation.

Acknowledgment

I would like to thank my students D. S. Alavi, N. Balabai, G. B. Dutt, R. S. Hartman, E. A. Hoburg, W. Konitsky, M. Kurnikova, Z. Lin, N. S. Park, I. Read, N. Sivakumar, B. Strazisar, A. Sukharevsky, and D. Zeglinski, as well as R. D. Coalson and the NSF, which supported a significant amount of work on this topic in my laboratory.

References

1. a) Reichardt, C. *Solvents and Solvent Effects in Organic Chemistry*, VCH:New York, **1988**; b) Tapia, O.; Bertrán, J. eds. *Solvent Effects and Chemical Reactivity*, Dordrecht: Boston, **1996**; c) Gauduel, Y.; Rossky, P.J. *Ultrafast Reaction Dynamics and Solvent Effects*, AIP:New York, **1994**; d) Dogonadze, R.R. *The Chemical Physics of Solvation* Elsevier:New York, **1985**; e) Marcus, Y. *Ion Solvation*, Wiley: New York, **1985**.

2. a) Huntress, W.T., *Adv. Mag. Res.*, **1970**, *4*, 1; b) Debye, P. *Polar Molecules*, Dover: New York, **1929**.

3. Baskin, J.S.; Chachisvilis, M.; Gupta, M.; Zewail, A.H. *J. Phys. Chem: A*, **1998**, *102*, 4158.

4. Parola, A.J.; Pina, F.; Maestri, M.; Armaroli, N.; Balzani, V. *New J. Chem.*, **1994**, *18*, 659.

5. Sukharevsky, A.P.; Read, I.; Linton, B.; Hamilton, A.D.; Waldeck, D.H. *J. Phys. Chem.*, **1998**, *102*, 5394.

6. a) Birge, R. *Scientific American*, **1995**, *222*, 90; b) Birge, R.R.; Parsons, B.; Song, Q.W.; Tallent, J.R. In *Molecular Electronics* Jortner, J.; Ratner, M., eds., (Blackwell, Malden MA), **1997**; c) Evans, S.D.; Johnson, S.R.; Ringsdorf, H.; Williams, L.M.; Wolf, H. *Langmuir*, **1998**, *14*, 6436.

7. a) Sauvage, J.P. *Acc. Chem. Res.*, **1998**, *31*, 611; b) Rappoport, Z.; Biali, S.E. *Acc. Chem. Res.*, **1997**, *30*, 307.

8. a) Cong, P.; Chang, Y.J.; Simon, J.D. *J. Phys. Chem.*, **1996**, *100*, 8613; b) Chang, Y.J.; Castner, Jr., E.W. *J. Phys. Chem.*, **1996**, *100*, 3330; c) Cho, M.; Rosenthal, S.J.; Scherer, N.F.; Ziegler, L.D.; Fleming, G.R. *J. Chem. Phys.*, **1992**, *96*, 5033.

9. a) Einstein, A. *Investigations on the Theory of the Brownian Movement*, Dover: New York, **1956**; b) Hynes, J.T. *Ann. Rev. Phys. Chem.*, **1977**, *28*, 301.

10. a) Burshtein, A.I.; Temkin, S.I. *Spectroscopy of Molecular Rotation in Gases and Liquids*, Cambridge: New York, **1994**; b) Berne, B.J.; Pecora, R. *Dynamic Light Scattering*, Wiley: New York, **1976**; c) Rothschild, W.G. *Dynamics of Molecular Liquids*, Wiley: New York, **1984**; d) Dorfmüller, T.; Pecora, R., eds. *Rotational Dynamics of Small and Macromolecules: Proceedings of a Workshop Held at the Zentrum für interdisziplinäre Forschung*, Springer-Verlag: New York, **1987**.

11. Fleming, G.R. *Chemical Applications of Ultrafast Spectroscopy*, Oxford, New York, **1986**.

12. a) Small, E.W.; Libertini, L.J.D.; Brown, W.; Small, J.R. *Optical Engineering*, **1991**, *30*, 345; b) Small, E.W.; Isenberg, I. *Biopolymers*, **1977**, *16*, 1907.

13. Wolynes, P.G. *Ann. Rev. Phys. Chem.*, **1980**, *31*, 345.

14. Kurnikova, M.G.; Waldeck, D.H.; Coalson, R.D. *J. Chem. Phys.*, **1996**, *105*, 628.

15. Kurnikova, M.G.; Balabai, N.; Waldeck, D.H.; Coalson, R.D. *J. Am. Chem. Soc.*, **1998**, *120*, 6121.

16. a) Dote, J.L.; Kivelson, D.; Schwartz, R.N. *J. Phys. Chem.*, **1981**, *80*, 2171; b) Dote, J.L.; Kivelson, D. *J. Phys. Chem.*, **1983**, *87*, 3889.

17. Williams, A.M.; Jiang, Y.; Ben-Amotz, D. *Chem. Phys.*, **1994**, *180*, 119.

18. a) Hu, C.M.; Zwanzig, R. *J. Chem. Phys.*, **1974**, *60*, 4354; b) Youngren, G.K.; Acrivos, A. *J. Chem. Phys.*, **1975**, *63*, 3846; c) Sension, R.J.; Hochstrasser, R.M. *J. Chem. Phys.*, **1993**, *98*, 2940.

19. a) Zwanzig, R. *J. Chem. Phys.*, **1978**, *68*, 4325; b) O'Dell, J.; Berne, B.H. *J. Chem. Phys.*, **1975**, *63*, 2376.

20. a) Pastor, R.W.; Karplus, M. *J. Phys. Chem.*, **1988**, *92*, 2636; b) Pastor, R.W.; Zwanzig, R. *J. Chem. Phys.*, **1989**, *90*, 5729.

21. a) Mikosch, W.; Dorfmüller, T.; Eimer, W. *J. Chem. Phys.*, **1994**, *101*, 11044; b) Eimer, W.; Niermann, M.; Eppe, M.A.; Jockusch, B.M. *J. Mol. Biol.*, **1993**, *229*, 146.

22. a) Hynes, J.T.; Kapral, R.; Weinberg, M.; *Chem. Phys. Lett.*, **1977**, *47*, 575; b) Hynes, J.T.; Kapral, R.; Weinberg, M. *J. Chem. Phys.*, **1977**, *67*, 3256; c) Hynes, J.T.; Kapral, R.; Weinberg, M. *J. Chem. Phys.*, **1978**, *68*, 2725.

23. a) Ben-Amotz, D.; Drake, J.M. *J. Chem. Phys.*, **1988**, *89*, 1019; b) Ben-Amotz, D.; Scott, T.W. *J. Chem. Phys.*, **1987**, *87*, 3739.

24. a) Ahn, M. *Chem. Phys. Lett.*, **1977**, *52*, 135; b) Basset, A.B. *A Treatise on Hydrodynamics*, Vol. 2, Dover: New York, **1961.**

25. a) Gudgin-Templeton, E.F.; Kenney-Wallace, G.A. *J. Phys. Chem.*, **1986**, *90*, 5441; b) Gudgin-Templeton, E.F.; Quitevis, E.L.; Kenney-Wallace, G.A., **1985**, *89*, 3238; c) Kenney-Wallace, G.A. *Faraday Discuss. Chem. Soc.*, **1988**, *85*, 185; d) Philips, L.A.; Webb, S.P.; Clark, J.H. *J. Chem. Phys.*, **1985**, *83*, 5810.

26. a) Kivelson, D.; Spears, K.G. *J. Phys. Chem.*, **1985**, *89*, 1999; b) Kovacs, H.; Kowalski, J. Maliniak, A. *Chem. Phys. Lett.*, **1988**, *152*, 427; c) Maroncelli, M. *J. Chem. Phys.*, **1996**, *106*, 1545; d) Horng, M.L. Gardecki,; J.A.; Maroncelli, M. *J. Phys. Chem.: A*, **1997**, *101*, 1030; e) Blanchard, G.J. *J. Phys. Chem.*, **1988**, *92*, 6303; f) ibid. *J. Chem. Phys.*, **1987**, *87*, 6802; g) Blanchard, G.J.; Cihal, C. *J. Phys. Chem.*, **1988**, *92*, 5950; h) Khudiakov, D.V.; Rubtsov, I.V.; Nadtochenko, V.A.; Moravskii, A.P. *Russ. Chem. Bull.*, **1996**, *45*, 560; i) Rubtsov, I.V.; Khudiakov, D.V.; A.P.; Nadtochenko, V.A. *Chem. Phys. Lett.*, **1996**, *249*, 101; j) Imeshev, G.; Khundkar, L.R. *J. Chem. Phys.*, **1995**, *103*, 8322.

27. a) Balabai, N.; Waldeck, D.H. *J. Phys. Chem.*, **1997**, *101*, 2339; b) Hartman, R.S.; Konitsky, W.M.; Waldeck, D.H. *J. Am. Chem. Soc.*, **1993**, *115*, 9662.

28. a) Hartman, R.S.; Alavi, D.S.; Waldeck, D.H. *J. Phys. Chem.*, **1991**, *95*, 7872; b) Hartman, R.S.; Waldeck, D.H. *J. Phys. Chem.*, **1994**, *98*, 1386; c) Alavi, D.S.; Hartman, R.S.; Waldeck, D.H. *J. Chem. Phys.*, **1991**, *94*, 4509.

29. a) Alavi, D.S.; Waldeck, D.H. *J. Phys. Chem.*, **1991**, *95*, 4848; b) Alavi, D.S.; Hartman, R.S.; Waldeck, D.H. *J. Chem. Phys.*, **1991**, *95*, 6770.

30. Hartman, R.S.; Konitsky, W.M.; Waldeck, D.H.; Chang, Y.J.; Castner, Jr., E.W. *J. Chem. Phys.*, **1997**, *106*, 7920.

31. a) Balabai, N.; Sukharevsky, A.; Read, I.; Strazisar, B.; Kurnikova, M.; Hartman, R.S.; Coalson, R.D.; Waldeck, D.H. *J. Mol. Liq.*, **1998**, *77*, 37; b) Hartman, R.S.; Alavi, D.S.; Waldeck, D.H. *Israel J. Chem.*, **1993**, *33*, 157.

32. a) Dutt, G.B.; Doraiswamy, S. *J. Chem. Phys.*, **1992**, *96*, 2475; b) Dutt, G.B.; Doraiswamy, S.; Periasamy, N. *J. Chem. Phys.*, **1991**, *94*, 5360; c) Srivastava, A.; Doraiswamy, S. *J. Chem. Phys.*, **1995**, *103*, 6197.

33. Simon, J.D.; Thompson, P.A. *J. Chem. Phys.*, **1990**, *92*, 2891.

34. Nee, T.W.; Zwanzig, R. *J. Chem. Phys.*, **1970**, *52*, 6353.

35. a) Hubbard, J.B. *J. Chem. Phys.*, **1978**, *69*, 1007; b) Hubbard, J.B.; Wolynes, P.G. *J. Chem. Phys.*, **1978**, *69*, 998; c) Felderhof, B.U. *Mol. Phys.*, **1983**, *48*, 1269; d) Felderhof, B.U. *Mol. Phys.*, **1983**, *48*, 1283; e) Nowak, E. *J. Chem. Phys.*, **1983**, *79*, 976; f) Madden, P.; Kivelson, D. *J. Chem. Phys.*, **1982**, *86*, 4244; g) Brito, P.; Bordewijk, P. *Mol. Phys.*, **1980**, *39*, 217.

36. van der Zwan, G.; Hynes, J.T. *J. Phys. Chem.*, **1985**, *89*, 4181.

37. a) Alavi, D.S.; Waldeck, D.H. *Understanding Chemical Reactivity*, Kluwer: Amsterdam, **1994**, 294; b) Alavi, D.S.; Waldeck, D.H. *J. Chem. Phys.*, **1991**, *94*, 6196; erratum, **1993**, *98*, 3580.

38. Born, M. *Z. Phys.*, **1920**, *1*, 221.

39. a) Besler, B.H.; Merz, K.M.; Kollman, P.A. *J. Comput. Chem.*, **1990**, *11*, 431; b) Singh, U.C.; Kollman, P.A. *J. Comput. Chem.*, **1984**, *5*, 129.

40. a) Lettenberger, M; Emmerling, F.; Gottfried, N.H.; Laubereau, A. *Chem. Phys. Lett.*, **1995**, *240*, 324; b) Viovy, J.L. *J. Phys. Chem.*, **1985**, *89*, 5465; c) Uphoff, R.; Potthast, R.; Dorfmüller, T. *J. Chem. Phys.*, **1989**, *90*, 6008; d) Fleming, G.R.; Knight, A.E.W.; Morris, J.M.; Robbins, R.J.; Robinson, G.W. *Chem. Phys. Lett.*, **1977**, *51*, 399; e) Dou, X.; Shang, Q.; Hudson, B.S. *Chem. Phys. Lett.*, **1992**, *189*, 48; f) Barkley, M.D.; Kowalczyk, A.A.; Brand L. *J. Chem. Phys.*, **1981**, *75*, 3581.

41. Pauls, S.W.; Hedstrom, J.F.; Johnson, C.K. *Chem. Phys.*, **1998**, *237*, 205.

42. a) Rasimas, J.P.; Blanchard, G.J. *J. Phys. Chem.*, **1996**, *100*, 11526; b) Blanchard, G. *J. Phys. Chem.*, **1991**, *95*, 5293.

43. van der Zwan, G.; Hynes, J.T. *Chem. Phys.*, **1991**, *152*, 169.

44. Balabai, N.; PhD. dissertation, University of Pittsburgh, **1999**.

45. Gierer, A.; Wirtz, K. *Z. Naturforsch. Teil A*, **1953**, *8*, 532.

46. Roy, M.; Doraiswamy, S. *J. Chem. Phys.*, **1993**, *98*, 3213.

47. Klüner, R.P.; Dölle, A. *J. Phys. Chem. A.*, **1997**, *101*, 1657.

48. a) Doolittle, A.K. *J. Appl. Phys.*, **1951**, *22*, 1471; b) Gegiou, D; Muszkat, K. A.; Fischer, E. *J. Am. Chem. Soc.*, **1968**, *90*, 12; c) Pollack, G.L.; Enyeart, J.J. *Phys. Rev.: A.*, **1985**, *31*, 980.

49. Anderton, R.M.; Kauffman, J.F. *J. Phys. Chem.*, **1994**, *98*, 12117.

50. Madden, P.; Kivelson, D. *J. Chem. Phys.*, **1982**, *86*, 4244.

51. a) Ravichandran, S.; Bagchi, B. *Int. Rev. Phys. Chem.*, **1995**, *14*, 271; b) Chandra, A.; Bagchi, B. *J. Chem. Phys.*, **1989**, *90*, 1832; c) Bagchi, B. *J. Chem. Phys.*, **1991**, *95*, 467.

52. a) Raineri, F.O.; Resat, H.; Perng, B.-C.; Hirata, F.; Friedman, H.L. *J. Chem. Phys.*, **1994**, *100*, 1477; b) Friedman, H.L.; Raineri, F.O.; Perng, B.-C.; Newton, M.D. *J. Mol. Liq.* **1995**, *65/66*, 7.

53. Hu, Y.; Fleming, G.R. *J. Chem. Phys.*, **1991**, *94*, 3857.

54. Jas, G.S.; Wang, Y.; Pauls, S.W.; Johnson, C.K.; Kuczera, K. *J. Chem. Phys.*, **1997**, *107*, 8800.

55. Balabai, N.; Kurnikova, M.G.; Coalson, R.D.; Waldeck, D.H. *J. Am. Chem. Soc.*, **1998**, *120*, 7944.

56. Ancian, B.; Tiffon, B.; Dubois, J.E. *J. Chem. Phys.*, **1981**, *74*, 5857.

57. a) Waldeck, D.H.; Fleming, G.R. *J. Phys. Chem.*, **1981**, *85*, 2614; b) Greenfield, S.R.; Sengupta, A.; Stankus, J.J.; Terazima, M.; Fayer, M.D. *J. Phys. Chem.*, **1994**, *98*, 313; c) Bessire, D.R.; Quitevis, E.L. *J. Phys. Chem.*, **1994**, *98*, 13083; d) Dutt, G.B.; Doraiswamy; Periasamy, N.; Venkataraman, B. *J. Chem. Phys.*, **1990**, *93*, 8498; e) Biermann, U.M.; Mikosch, W.; Dorfmüller, T.; Eimer, W. *J. Phys. Chem.*, **1996**, *100*, 1705; f) Patron, M.; Kivelson, D.; Schwartz, R.N. *J. Phys. Chem.*, **1982**, *86*, 516; g) Megens, M.; Sprik, R.; Wegdam, G.H.; Lagendijk, A. *J. Chem. Phys.*, **1997**, *107*, 493.

58. a) Alms, G. R.; Bauer, D. R.; Brauman, J. I.; Pecora, R. *J. Chem. Phys.*, **1973**, *58*, 5310; b) ibid, 5321.

59. Evans, G. T.; Kivelson, D. *J. Chem. Phys.*, **1986**, *84*, 385.

60. a) De Backer, S.; Dutt, G. B.; Ameloot, M.; De Schryver, F.C.; Müllen, K.; Holtrup, F. *J. Phys. Chem.*, **1996**, *100*, 512; b) Megens, M.; Sprik, R.; Wegdam, G.H.; Lagendijk, A. *J. Chem. Phys.*, **1997**, *107*, 493; c) Srivastava, A.; Doraiswamy, S. *J. Chem. Phys.*, **1995**, *103*, 6197; d) Laitinen, E.; Korppi-Tommola, J.; Linnanto, J. *J. Chem. Phys.*, **1997**, *107*, 7601.

61. Sukharevsky, A. PhD. dissertation, University of Pittsburgh, **1999.**

62. Bhattacharyya, S.; Bagchi, B. *J. Chem. Phys.*, **1997**, *106*, 1757.

63. Dorfmüller, T.; Daum, B.; Hanschmidt, A. *J. Chem. Phys.*, **1991**, *95*, 813.

64. a) Benzler, J.; Luther, K. *Chem. Phys. Lett.*, **1997**, *279*, 333; b) Jiang, Y.; Blanchard, G.J. *J. Phys. Chem.*, **1994**, *98*, 6436; c) Canonica, S.; Schmid, A. A.; Wild, U.P. *Chem. Phys. Lett.*, **1985**, *122*, 529; d) Goldie, S. N.; Blanchard, G.J. *J. Phys. Chem.: A.*, **1999**, *103*, 999.

65. Courtney, S.H.; Kim, S.K.; Canonica, S.; Fleming, G.R. *J. Chem. Soc., Faraday Trans. 2*, **1986**, *82*, 2065.

66. Goldies, S.N.; Blanchard, G.J. *J. Phys. Chem.*, **1999**, *103*, 999.

67. a) Brocklehurst, B.; Young, R.N. *J. Phys. Chem.*, **1995**, *99*, 40; b) Kim, Y.R.; Hochstrasser, R.M. *J. Phys. Chem.*, **1992**, *96*, 9595.

68. a) Dutt, G.B.; Doraiswamy, S. *J. Chem. Phys.*, **1992**, *96*, 2475; b) Mikosch, W.; Dorfmüller, T.; Eimer, W. *J. Chem. Phys.*, **1991**, *101*, 11044; c) Rice, S. A.; Kenney-Wallace, G.A. *Chem. Phys.*, **1980**, *47*, 161; d) Ko, D.-S. *J. Chem. Phys.*, **1994**, *100*, 746.

69. Fleming, G.R.; Morris, J. M.; Robinson, G.W. *Chem. Phys.*, **1970**, *17*, 91.

70. Moog, R.S.; Bankert, D.L.; Maroncelli, M. *J. Phys. Chem.*, **1993**, *97*, 1496.

71. a) Yu, J.; Berg, M. *Chem. Phys. Lett.*, **1993**, *208*, 315; b) Benigno, A. J.; Ahmed, E.; Berg, M. *J. Chem. Phys.*, **1996**, *104*, 7382.

72. Harrison, A. K.; Zwanzig, R.J. *Chem. Phys.*, **1985**, *83*, 5861.

73. a) Kohen, A.; Klinman, J.P. *Acc. Chem. Res.*, **1998**, *31*, 397; b) Stryer, L. *Biochemistry*, 3rd ed., Freeman: New York, **1988**; c) Fersht, A. *Enzyme Structure and Mechanism*, 2nd ed., Freeman: New York, **1985**; d) Welch, G.R., ed., *The Fluctuating Enzyme*, Wiley, New York, **1986**; e) McCammon,

J.A.; Harvey, S.C. *Dynamics of Proteins and Nucleic Acids*, Cambridge: New York, **1987**; f) Brooks, III, C. L.; Karplus, M.; Pettit, B. M. *Adv. Chem. Phys.*, **1988**, *71;* g) *Chem. Phys.*, **1991**, *158*, 191–531 special issue with guest editors Eaton, W.A. and Szabo, A.

74. a) Beer, P.D. *Acc. Chem. Res.*, **1998**, *31*, 71; b) Balzani, V.; Gómez-López, M.; Stoddart, J.F. *Acc. Chem. Res.*, **1998**, *31*, 405; c) Ikeda, A.; Shinkai, S. *Chem. Rev.*, **1997**, *97*, 1713; d) Zimmerman, S.C. *Topics in Curr. Chem.*, **1993**, *165*, 71; e) Cram, D.J.; Cram, J.M., *Container Molecules and Their Guests*, Royal Society: Cambridge **1994**; f) Vögtle, F. *Supramolecular Chemistry*, Wiley: New York, **1991**); g) Atwood, J.L.; Davies, J.E.D.; MacNicol, D.D.; Vögtle, F.; Lehn, J.-M. eds, *Comprehensive Supramolecular Chemistry*, Pergamon: Amsterdam, **1996**; h) Gellman, S.H., *Chem. Rev.*, **1997**, *97*, #5; i) Schneider, H.-J. *Angew. Chem. Int. Ed. Engl.*, **1991**, *31*, 1417; j) Chapman, R.G.; Sherman, J.C. *Tetrahedron*, **1997**, *53*, 15911.

75. Hamilton, A. D. *Advances in Supramolecular Chemistry*, Vol. 1, p. 1–64, **1990.**

76. O'Connor, K. *Binding Constants*, Wiley: New York, **1987.**

77. a) Yoshida, N.; Yamaguchi, H.; Higashi, M. *J. Phys. Chem.; A*, **1998**, 102, 1523; b) Yoshida, N.; Seiyama, A.; Fujimoto, M. *J. Phys. Chem.*, **1990**, *94*, 4246.

78. Szabo, T.; Hilmersson, G.; Rebek, J., Jr. *J. Am. Chem. Soc.*, **1998**, 120, 6193.

79. Rohrbach, R.P.; Rodrigues, L.J.; Eyring, E.M.; Wojcik, J.F. *J. Phys. Chem.*, **1977**, *81*, 944.

80. a) Liao, Y.; Bohne, C. *J. Phys. Chem.*, **1996**, 100, 734; b) Barros, T.C.; Stefaniak, K.; Holzwarth, J.F.; Bohne, C. *J. Phys. Chem. A.*, **1998**, *102*, 5639;) Yoshida, N.; Fujita, Y. *J. Phys. Chem.*, **1996**, *100*, 3073.

81. Yoshida, N.; Seiyama, A.; Fujimoto, M. *J. Phys. Chem.*, **1990**, *94*, 4254.

82. a) Rathore, R.; Hubig, S. M.; Kochi, J.K. *J. Am. Chem. Soc.*, **1997**, 119, 11468; b) Walker II, L.A.; Jarrett, J.T.; Anderson, N.A.; Pullen, S.H.; Matthews, R.G.; Sension, R.J. *J. Am. Chem. Soc.*, **1998**, *120*, 3597.

83. Szabo, A. *J. Chem. Phys.*, **1984**, *81*, 150.

84. a) Schurr, J.M. *Chem. Phys.*, **1982**, *65*, 417; b) Toptygin, D.; Svobodova, J.; Konopasek, I.; Brand; L. *J. Chem. Phys.*, **1992**, *96*, 7919.

85. a) Kirchoff, P. D.; Dutasta, J.-P.; Collet, A.; McCammon, J. A. *J. Am. Chem. Soc.*, **1997**, *119*, 8015; b) Sheu, C.; Houk, K. N. *J. Am. Chem. Soc.*, **1996**, *118*, 8056; c) Kirchoff, P.D.; Bass, M.B.; Hanks, B.A.; Briggs, J.M.; Collet, A.; McCammon, J.A. *J. Am. Chem. Soc.*, **1996**, *118*, 3237.

86. a) Cartling, B. *J. Chem. Phys.*, **1991**, *94*, 6203; b) Brönger; Karplus, M. *Acc. Chem. Res.*, **1991**, *24*, 54; c) Brooks, III; C.L.; Karplus, M. *J. Mol. Biol.*, **1989**, *208*, 159; d) Jardetzky, O. *Prog. Biophys. Molec. Biol.*, **1996**, *65*, 171.

87. a) Lakowicz, J.R. ed. *Time-Resolved Laser Spectroscopy in Biochemistry*, Proceedings of the SPIE, Volume 1204, SPIE: Bellingham, WA, **1990**; b) Hoppe, W.; Lohmann, W.; Markl, H.; Ziegler, H. eds., *Biophysics*, Springer; New York, **1983**.

88. a) van der Meulen, D.L.; Nealon, D.G.; Gratton, E.; Jameson, D.M. *Biophysical Chemistry*, **1990**, *36*, 177; b) Palmer, III, A.G.; Hochstrasser, R.A.; Miller, D.P.; Rance, M. *J. Am. Chem. Soc.*, **1993**, *115*, 6333; c) Rich, R.L.; Gai, F.; Lane, J.W.; Petrich, J.W.; Schwabacher, A.W. *J. Am. Chem. Soc.*, **1995**, *117*, 733; d) Lim, K.; Jameson, D.M.; Gentry, C.A; Herron, J.N. *Biochemistry*, **1995**, *34*, 6975.

89. a) Millar, D.P.; Robbins, R.J.; Zewail, A.H. *J. Chem. Phys.*, **1982**, *76*, 2080; b) Millar, D.P.; Ho, K.M.; Aroney, M.J. *Biochemistry*, **1988**, *27*, 8599; c) Nuutero, S.; Fujimoto, B.S.; Flynn, P.F.; Reid, B.R.; Ribeiro, N.S.; Schurr, J.M. *Biopolymers* **1994**, *34*, 463; d) Cuniberti, C.; Guenza, M. *Biophysical Chemistry*, **1990**, *38*, 11; e) Barcellona, M.L.; Gratton, E. *Biochemistry*, **1996**, *35*, 321; f) Fujimoto, B.S.; Clendenning, J.B.; Delrow, J. J.; Heath, P.J.; Schurr, M. *J. Phys. Chem.*, **1994**, *98*, 6633.

90. a) Catoire, L.; Michon, V.; Monville, L.; Hocquet, A.; Jullien, L.; Canceill, J.; Lehn, J.-M. Piotto, M.; Du Penhoat, C.H. *Carb. Res.*, **1997**, *303*, 379; b) Ivanov, P.M.; Salvatierra, D.; Jaime, C. *J. Org. Chem.*, **1996**, *61*, 7012; c) Kowalewski, J.; Widmalm, G. *J. Phys. Chem.*, **1994**, *98*, 28; d) Ceccato, M.; Nostro, P.L.; Rossi, C.; Bonechi, C.; Donati, A.; Baglioni, P. *J. Phys. Chem. B.*, **1997**, *101*, 5094; Azaroual-Bellanger, N.; Perly, B. *Mag. Res. Chem.*, **1994**, *32*, 8.

91. Connors, K. A. *Chem. Rev.*, **1997**, *97*, 1325, and references therein.

92. a) Paci, B.; Deleuze, M.S.; Caciuffo, R.; Tomkinson, J.; Ugozzoli, F.; Zerbetto, F. *J. Phys. Chem.: A.*, **1998**, *102*, 6910; b) Brouwer, E. B.; Ripmeester, J. A.; Enright, G. D. *J. Inc. Phen. and Mol. Recog. in Chem.*, **1996**, *24*, 1.

93. Balabai, N.; Linton, B.; Napper, A.; Priyadarshy, S.; Sukharevsky, A.P.; Waldeck, D.H. *J. Phys. Chem. B*, **1998**, *101*, 9617.

94. Doi, M.; Edwards; S.F. *The Theory of Polymer Dynamics*, Oxford University Press: New York, **1989.**

95. a) Choi, M.; Jin, D.; Kim, H.; Kang, T.J.; Jeoung, S.C.; Kim, D. *J. Phys. Chem. B*, **1997**, *101*, 8092; b) Deeg, F.W.; Greenfield, S.R.; Stankus, J.J.; Newell, V.J.; Fayer, M.D. *J. Chem. Phys.*, **1990**, *93*, 3503; c) ibid, *J. Chem. Phys.*, **1989**, *90*, 6893.

96. a) Maiti, N.C.; Krishna, M.M.G.; Britto, P.J.; Periasamy, N. *J. Phys. Chem. B*, **1997**, *101*, 11051; b) Wittouck, N.; Negri, R.M.; Ameloot, M.; de Schryver, F.C. *J. Am. Chem. Soc.*, **1994**, 116, 10601; c) Tamura, K.; Nii, N. *J. Phys. Chem.*, **1989**, *93*, 4825; d) Quitevis, E.L.; Marcus, A.H.; Fayer, M.D. *J. Phys. Chem.*, **1993**, *97*, 5762; e) Matzinger, S.; Hussey, D.M.; Fayer, M.D. *J. Phys. Chem. B*, **1998**, *102*, 7216; f) Riter, R. E.; Undiks, E. P.; Kimmel, J. R.; Levinger, N.E. *J. Phys. Chem.: B*, **1998**, *102*, 7931.

97. a) Smith, L. M.; Weis, R. M.; McConnell, H. M. *Biophys. J.*, **1981**, *36*, 73; b) Muller, J.M.; van Faassen, E.E.; van Ginkel, G. *Chem. Phys.*, **1994**, 185, 393; c) Heyn, M. P. *FEBS Lett.*, **1979**, *108*, 359; d) Binder, H.; Lantzsch, G.; Dittes, K. *Wiss. Z. Karl-Marx-Univ. Leipz, Math-Nat.Wiss.*, **1989**, *38*, 6.

98. Visser, N.V.; van Hoek, A.; Visser, A.J.W.G.; Frank, J.; Apell, H.-J.; Clarke, R.J. *Biochemistry*, **1995**, *34*, 11777.

99. Sato, N.; Sugiura, K.; Ito, S.; Yamamoto, M. *Langmuir.*, **1997**, *13*, 5685.

100. Tulock, J.J.; Blanchard, G.J. *J. Phys. Chem. B*, **1998**, 102, 7148.

101. Wahl, P. *Chem. Phys.*, **1975**, *7*, 210.

102. Fisz, J.J.; *Chem. Phys.*, **1994**, *181*, 425.

103. Lipari, G.; Szabo, A. *J. Chem. Phys.*, **1981**, *75*, 2971.

104. a) Mason, S.; Clifford, T.; Seib, L.; Kuczera, K.; Bowman-James, K. *J. Am. Chem. Soc.*, **1998**, *120*, 8899; b) Paci, B.; Deleuze, M.S.; Caciuffo, R.; Tomkinson, J.; Ugozzoli, F.; Zerbetto, F. *J. Phys. Chem.: A*, **1998**, *102*, 6910.

105. Cramer, F.; Saenger, W.; Spatz, H.-Ch. *J. Am. Chem. Soc.*, **1967**, *89*, 14.

106. Gaspard, P.; Burghardt, I. *Adv. Chem. Phys.*, **1997**, *101.*

107. a) Lednev, I.K.; Karnoup, A.S.; Sparrow, M.C.; Asher, S.A. *J. Am. Chem. Soc.*, **1999**, *121*, 4076; b) Li, P.; Chen X.G.; Shulin, E.; Asher, S.A. *J. Am. Chem. Soc.*, **1997**, *119*, 1116; c) Ansari, A. *J. Chem. Phys.*, **1999**, *110*, 1774; d) Chirico, G.; Beretta, S.; Baldini, G. *J. Chem. Phys.*, **1999**, *110*, 2297; e) Ghosh, I.; McCammon, J.A. *J. Phys. Chem.*, **1987**, *91*, 4878; f) Klimov, D.K.; Thirumalai, D. *Phys. Rev. Lett.*, **1997**, *79*, 317.

108. a) Yamaguchi, S.; Hamaguchi, H. *J. Chem. Phys.*, **1998**, *109*, 1397; b) Liu, R.S.H.; Browne, D.T. *Acc. Chem. Res.*, **1986**, *19*, 42; c) Bonaić-Koutecký, V.; Koutecký, J.; Michl, J. *Angew. Chem. Int. Ed. Engl.*, **1987**, *26*, 170; d) Deckert, V.; Iwata, K.; Hamaguchi, H. *J. Photochem. Photobiol. A: Chem.*, **1996**, *102*, 35.

109. a) Macpherson, A.N.; Gillbro, T. *J. Phys. Chem.: A*, **1998**, *102*, 5049; b) Müller, A.M.; Lochbrunner, S.; Schmid, W.E.; Fuss, W. *Angew. Chem. Int. Ed. Engl.*, **1998**, *37*, 505; c) Mathies, R.A.; Cruz, C.H.B.; Pollard, W.T.; Shank, C.V. *Science*, **1988**, *240*, 777; d) Pollard, W.T.; Cruz, C.H.B.; Shank, C.V.; Mathies, R.A. *J. Chem. Phys.*, **1989**, *90*, 199; e) Haran, G.; Wynne, K.; Xie, A.H.; He, Q.; Chance, M.; Hochstrasser, R.M. *Chem. Phys. Lett.*, **1996**, *261*, 389.

110. a) Kuriyama, Y.; Takeya, H.; Oishi, S.; Kojima, M. *Chem. Lett.*, **1998**, 843; b) Kojima, M.; Takeya, H.; Kuriyama, Y.; Oishi, S. *Chem. Lett.*, **1997**, 997; c) Takeya, H.; Kuriyama, Y.; Kojima, M. *Tet. Lett.*, **1998**, *39*, 5967.

111. a) Douhal, A.; Fiebig, T.; Chachisvilis, M.; Zewail, A.H. *J. Phys. Chem.: A*, **1998**, *102*, 1657; b) Bortolus, P.; Monti, S. *J. Phys. Chem.*, **1987**, *91*, 5046; c) Herrman, W.; Wehrle, S.; Wenz, G. *Chem. Commun.*, **1997**, 1709; d) Davidson, R.S.; Liu, J. *J. Photochem. Photobiol. A.: Chem.*, **1993**, *76*, 121; e) Costanzo, L.L.; Giuffrida, S.; Sortino, S.; Chiacchio, U.; De Guidi, G. *J. Photochem. Photobiol. A: Chem.*, **1993**, *76*, 127.

112. a) Blinov, L.M. *J. Nonlinear Opt. Phys. and Materials*, **1996**, *5*, 165; b) Yamazaki, I.; Ohta, N. *Pure Appl. Chem.*, **1995**, *67*, 209.

113. a) Strashnikova, N.; Papper, V.; Parkhomyuk, P.; Likhtenshtein, G.I.; Ratner, V.; Marks, R. *J. Photochem. Photobiol. A:Chem.*, **1999**, *122*, 133; b) Datta, A.; Mandal, D.; Pal, S.K.; Bhattacharyya, K. *Chem. Phys. Lett.*, **1997**, *278*, 77; c) Sasai, R.; Shinya, N.; Shichi, T.; Takagi, K.; Gekko, K. *Langmuir*, **1999**, *15*, 413; d) Grieser, F.; Lay, M.; Thistlethwaite, P.J. *J. Phys. Chem.*, **1985**, *89*, 2065.

114. a) Reid, P.J.; Lawless, M.K.; Wickham, S.D.; Mathies, R.A. *J. Phys. Chem.*, **1994**, *98*, 5597; b) Anderson, N.A.; Pullen, S.H.; Walker, L.A.; Shiang, J.J.; Sension, R.J. *J. Phys. Chem.:A*, **1998**, *102*, 10588.

115. Lipson, M.; Peters, K.S. *J. Phys. Chem:A*, **1998**, *102*, 1691.

116. a) Campbell, D.M.; Mackowiak, M.; Jonas, J. *J. Chem. Phys.*, **1992**, *96*, 2717; b) Jonas, J. *Acc. Chem. Res.*, **1984**, *17*, 74.

117. a) Depaepe, J.-M.; Ryckaert, J.-P. *Chem. Phys. Lett.*, **1995**, *245*, 653; b) Cates, D.A.; MacPhail, R.A. *J. Phys. Chem.*, **1991**, *95*, 2209; c) Koda, S.; Iwasawa, A.; Hirao, H.; Nomura, H. *J. Chem. Soc. Faraday Trans.*, **1995**, *91*, 445.

118. a) Rettig, W. *Topics in Curr. Chem.*, **1994**, *169*, 254; b) Maus, M.; Rettig, W. *J. Inf. Rec.*, **1998**, *24*, 461; c) Dobkowski, J.; Waluk, J.; Yang, W.; Rulliére, Rettig, W. *New. J. Chem.*, **1997**, *21*, 429; d) Maus, M.; Rettig, W.; Depaemelaere, S.; Onkelinx, A.; de Schryver, F.C.; Iwai, K. *Chem. Phys. Lett.*, **1998**, *292*, 115; e) Bulgarevich, D.S.; Kajimoto, O.; Hara, K. *J. Phys. Chem.*, **1995**, *99*, 13356; f) Müller, S.; Heinze, *J. Chem. Phys.*, **1991**, *157*, 231.

119. a) Scherer, T.; van Stokkum, I.H.M.; Brouwer, A.M.; Verhoeven, J.W. *J. Phys. Chem.*, **1994**, *98*, 10539; b) Jäger, W.; Schneider, S.; Verhoeven, J.W. *Chem. Phys. Lett.*, **1997**, *270*, 50; c) Changenet, P.; Plaza, P.; Martin, M.M.; Meyer, Y.H. *J. Phys. Chem., A*, **1997**, *101*, 8186.

120. a) Ovchinnikov, A.A.; Timashev, S.F.; Belyy, A.A. *Kinetics of Diffusion Controlled Chemical Processes*, Nova: New York; 1989; b) Andre, J.C.; Baros, F.; Bouchy, A.; Bouchy, M.; Dong, W.; Duhamel, J.; Reis, A.T.; Souosa, E.; Viriot, M.L. *Laser Chem.*, **1990**, *10*, 431; c) Keizer, J. *Chem. Rev.*, **1987**, *87*, 167.

121. a) Pilling, M.J.; Seakins, P.W. *Reaction Kinetics*, Oxford: New York, **1995:** b) Li, Y.S.; Wilson, K.R. *J. Chem. Phys.*, **1990**, *93*, 8821; c) Gertner, B.J.; Whitnell, R.M.; Wilson, K.R.; Hynes, J.T., *Am. Chem. Soc.* **1991**, *113*, 74; d) Wang, W.N.; Nelson, K.A.; Xiao; L.; Coker, D.F., *J. Chem. Phys.*, **1994**, *101*, 9663; e) Benjamin, I.; Barbara, P.F.; Gertner, B.J.; Hynes, J.T. *J. Phys. Chem.*, **1995**, *99*, 7557.

122. a) Harris, A.L.; Brown, J.K.; Harris, C.B., *Ann. Rev. Phys. Chem.*, **1988**, *39*, 341; b) Wan, C.Z.; Gupta, M.; Zewail, A.H. *Chem. Phys. Lett.*, **1996**, *256*, 279; b) Gnanakaran, S.; Lim, M.; Pugliano, N.; Volk, M.; Hochstrasser, R.M. *J. Phys. Cond. Matter*, **1996**, *8*, 9201; c) Chang, Y.J.; Simon, J.D. *J.*

Phys. Chem., **1996**, *100*, 6406; d) Walhout, P.K.; Alfano, J.C.; Thakur, K.A.M.; Barbara, P.F. *J. Phys. Chem.*, **1995**, *99*, 7568.

123. Waldeck, D.H. *Chem. Rev.*, **1991**, *91*, 415.

124. Waldeck, D.H. *J. Mol. Liq.*, **1993**, *57*, 127.

125. a) Iwata, K.; Hamaguchi, H. *J. Phys. Chem.:A*, **1997**, *101*, 632; b) Qian, J.; Schultz, S.L.; Jean, J.M. *Chem. Phys. Lett.*, **1995**, *233*, 9; c) Qian, J.; Schultz, S.L.; Bradburn, G.R.; Jean J.M. *J. Phys. Chem.*, **1993**, *97*, 10638; d) Weaver, W.L.; Huston, L.A.; Iwata, K.; Gustafson, T.L. *J. Phys. Chem.*, **1992**, *96*, 8956; e) Hester, R.E.; Matousek, P.; Moore, J.N.; Parker, A.W.; Toner, W.T.; Towrie, M. *Chem. Phys. Lett.*, **1992**, *196*, 496; f) Nikowa, L.; Schwarzer, D.; Troe, J. *Chem. Phys. Lett.*, **1995**, *233*, 303; g) Schultz, S.L.; Qian, J.; Jean, J.M. *J. Phys. Chem.:A*, **1997**, *101*, 1000; h) Nakabayashi, T.; Okamoto, H.; Tasumi, M. *J. Phys. Chem.:A*, **1998**, *102*, 9686.

126. a) Lienau, C.; Heikal, A.A.; Zewail, A.H. *Chem. Phys.*, **1993**, *175*, 171; b) Heikal, A.A.; Chong, S.H.; Baskin, J.S.; Zewail, A.H. *Chem. Phys. Lett.*, **1995**, *242*, 380.

127. a) Heikal, A.A.; Baskin, J.S.; Bañares, L.; Zewail, A.H. *J. Phys. Chem.:A* **1997**, *101*, 572; b) Bañares, L.; Heikal, A.A.; Zewail, A.H. *J. Phys. Chem.*, **1992**, *96*, 4127; c) Siewert, S.S.; Spangler, L.H. *J. Phys. Chem.*, **1995**, *99*, 9316; b) Chiang, W.-Y.; Laane, J. *J. Chem. Phys.*, **1994**, *100*, 8755; d) Takahashi, M.; Kimura, K. *J. Phys. Chem.*, **1995**, *99*, 1628.

128. Gano, J.E.; Garry, P.A.; Sekher, P.; Schliesser, J.; Kim, Y.W.; Lenoir, D. *J. Am. Chem. Soc.*, **1997**, *119*, 3826.

129. a) Park, N.S.; Waldeck, D.H. *Chem. Phys. Lett.*, **1990**, *168*, 379; b) Dutt, G.B.; Konitsky, W.; Waldeck, D.H. *Chem. Phys. Lett.*, **1995**, *245*, 437; c) Park, N.S.; Waldeck, D.H. *J. Chem. Phys.*, **1989**, *91*, 943.

130. a) Agmon, N.; Kosloff, R. *J. Phys. Chem.*, **1987**, *91*, 1988; b) Kwac, K.; Lee, S.; Shin, K. *J. Bull. Korean Chem. Soc.*, **1995**, *16*, 427.

131. Meyer, A.; Schroeder, J.; Troe, J.; Botsmeier, M. *J. Photochem. Photobiol. A:Chem.*, **1997**, *105*, 345.

132. a) Schroeder, J. *Ber. Bunsenges. Phys. Chem.*, **1997**, *101*, 643; b) Schroeder, J.; Troe, J.; Vöhringer, P. *Chem. Phys. Lett.*, **1993**, *203*, 255; c) Troe, J. *Pure & Appl. Chem.*, **1997**, *69*, 841; d) Schroeder, J.; Troe, J.; Vöhringer, P. *Z. Phys. Chem.*, **1995**, *287;* e) Courtney, S.H.; Fleming, G.R. *J. Chem. Phys.*, **1985**, *83*, 215; f) Lee, M.; Holtom, G.R.; Hochstrasser, R.M. *Chem. Phys. Lett.*, **1985**, *118*, 359.

133. a) Hasha, D.L.; Eguchi, T.; Jonas, J. *J. Chem. Phys.*, **1981**, *75*, 1571; b) ibid. *J. Am. Chem. Soc.*, **1982**, *104*, 2290; c) J. Jonas, *Acc. Chem. Res.* **1984**, *17*, 74.

134. a) Hara, K.; Ito, N.; Kajimoto, O. *J. Chem. Phys.*, **1999**, *110*, 1662; b) Hara, K.; Kiyotani, H.; Kajimoto, O. *J. Chem. Phys.*, **1995**, *103*, 5548.

135. a) Nitzan, A. *J. Chem. Phys.*, **1987**, *86*, 2734; b) Pollak, E.; Grabert, H.; Hänggi, P. *J. Chem. Phys.*, **1989**, *91*, 4073; c) Hershkowitz, E.; Pollak, E. *J. Chem. Phys.*, **1997**, *106*, 7678.

136. a) Hill, T.L. *Introduction to Statistical Thermodynamics* Dover:New York, **1986**; b) Johnston, H.S. *Gas Phase Reaction Rate Theory*, Ronald Press:New York, **1966**; c) Troe, J. *Physical Chemistry: An Advanced Treatise* (Academic:New York, **1975**; d) Wardlaw, D.M.; Marcus, R.A. *Adv. Chem. Phys.* Volume 70, Wiley:New York, **1988**, 231.

137. a) Chandler, D. *Introduction to Modern Statistical Mechanics*, Oxford:New York, **1987**; b) Montgomery, J.A.; Chandler, D.; Berne, B.J. *J. Chem. Phys.*, **1979**, *70*, 4056; c) Tannor, D. J.; Kohen, D. *J. Chem. Phys.*, **1994**, *100*, 4932.

138. a) Kramers, H.A. *Physica*, **1940**, *7*, 284; b) Chandrasekhar, S. *Rev. Mod. Phys.*, **1943**, *15*, 1.

139. a) Hynes, J.T. *J. Stat. Phys.*, **1986**, *42*, 149; b) Hynes, J.T. *Theory of Chemical Reactions*, Volume 4, CRC:New York **1985.**

140. a) Hänggi, P.; Talkner, T.; Borkovec, M. *Rev. Mod. Phys.*, **1990**, *62*, 251; b) Nitzan, A. Adv. Chem. Phys. Volume 70, Wiley:New York, **1988**, c) Berne, B.J.; Borkovec, M.; Straub, J.E.; *J. Phys. Chem.*, **1988**, *92*, 3711.

141. Grote, R.F.; Hynes, J.T. *J. Chem. Phys.*, **1980**, *73*, 2715; b) ibid, **1981**, *74*, 4465.

142. a) Langer, J.S. *Ann. Phys.*, **1969**, *54*, 258; b) Larson, R.S.; Kostin, M.D. *J. Chem. Phys.*, **1982**, *77*, 5017; c) Kaufman, A.D; Whaley, K.B. *J. Chem. Phys.*, **1989**, *90*, 2758; d) Matkowsky, B.J.; Nitzan, A.; Schuss, Z. *J. Chem. Phys.*, **1988**, *88*, 4765; e) Klosek-Dygas, M.M.; Hoffman, B.M.; Matkowsky, B.J.; Nitzan, A.; Ratner, M.A.; Schuss, Z. *J. Chem. Phys.*, **1989**, *90*, 1141; f) Pollak, E. *J. Chem. Phys.*, **1990**, *93*, 1116.

143. a) Moro, G.J.; Nordio, P.L.; Polimeno, A. *Mol Phys.*, **1989**, *68*, 1131; b) Ulitsky, A.; Shalloway, D. *J. Chem. Phys.*, **1998**, *109*, 1670; c) Basilevsky, M.V.; Davidovitch, G.V. *J. Chem. Phys.*, **1995**, *102*, 1607; d) Kim, H.J.; Hynes, J.T. *J. Photochem. Photobiol. A: Chem.*, **1997**, *105*, 337.

144. a) Agmon, N.; Hopfield, J.J. *J. Chem. Phys.*, **1983**, *78*, 6947; b) ibid, **1983**, *79*, 2042.

145. a) Sumi, H.; Marcus, R.A. *J. Chem. Phys.*, **1986**, *84*, 4272; b) ibid, 4894; c) Nadler, W.; Marcus, R.A. *J. Chem. Phys.*, **1987**, *86*, 3906; d) ibid. *Israel J. Chem.*, **1990**, *30*, 69.

146. a) Berezhkovskii, A.M.; Pollak, E.; Zitserman, V.Y. *J. Chem. Phys.*, **1992**, *97*, 2422; b) Berezhkovskii, A.M.; Zitserman, V.Y.; Sheu, S.-Y.; Yang, D.-Y.; Kuo, J.; Lin, S.H. *J. Chem. Phys.*, **1997**, *107*, 10539; c) Berezhkovkii, A.M.; Zitserman, V.Y. *Physica A*, **1990**, *166*, 585; d) Berezhkovskii, A.M.; Zitserman, V.Y. *Chem. Phys.*, **1992**, *164*, 341; e) Berezhkovskii, A.M.; Berezhkovskii, L.M.; Zitzerman, V.Y. *Chem. Phys.*, **1989**, *130*, 55.

147. a) Haynes, G.R.; Voth, G.A.; Pollak, E. *J. Chem. Phys.*, **1994**, *101*, 7811. This work provides a nice overview of earlier work; b) Pollak, E.; Berezhkovskii, A.M. *J. Chem. Phys.*, **1993**, *99*, 1344; c) Haynes, G.R.; Voth, G.A. *J. Chem. Phys.*, **1995**, *103*, 10176; d) Neria, E.; Karplus, M. *J. Chem. Phys.*, **1996**, *105*, 10812.

148. a) Schroeder, J.; Troe, J. *Ann. Rev. Phys. Chem.*, **1987**, *38*, 163; b) Troe, J. *J. Phys. Chem.*, **1986**, *90*, 357; c) Gehrke, C.; Schroeder, J.; Troe, J.; Voss, F. *J. Chem. Phys.*, **1990**, *92*, 4805; d) Maneke, G.; Schroeder, J.; Troe, J.; Voss, F. *Ber. Bunsenges. Phys. Chem.*, **1985**, *89*, 896.

149. Rettig, W.; Fritz, R.; Braun, D. *J. Phys. Chem.: A*, **1997**, *101*, 6830.

150. Drljaca, A.; Hubbard, C.D.; van Eldik, R.; Asano, T.; Basilevsky, M.V.; le Noble, W.J. *Chem. Rev.*, **1998**, *98*, 2167.

151. a) Åkesson, E; Sundström, V.; Gillbro, T. *Chem. Phys.*, **1986**, *106*, 269; b) Harju, T.O.; Korppi-Tommola, J.E.I; Huizer, A.H.; Varma, C.A.G.O. *J. Phys. Chem.*, **1996**, *100*, 3592; c) Hicks, J.; Vandersall, M.; Babarogic, Z.; Eisenthal, K.B. *Chem. Phys. Lett.*, **1985**, *116*, 18.

152. a) Anderton, R.M.; Kauffman, J.F. *J. Phys. Chem.*, **1995**, *99*, 14628; b) Anderton, R.M.; Kauffman, J.F. *J. Phys. Chem.*, **1994**, *98*, 12125.

153. a) Syage, J.A.; Felker, P.M; Zewail, A.H. *J. Chem. Phys.*, **1984**, *81*, 4706; b) Felker, P.M.; Zewail, A.H. *J. Phys. Chem.*, **1985**, *89*, 5402; c) Syage, J.A.; Lambert, W.R.; Felker, P.M.; Zewail, A.H.; Hochstrasser, R.M. *Chem. Phys. Lett.*, **1982**, *88*, 266; d) Troe, J.; Schroeder, J. *J. Phys. Chem.*, **1986**, *90*, 4215; e) Troe, J. *Chem. Phys. Lett.*, **1985**, *114*, 241; f) Majors, T.J.; Even, U.; Jortner, J. *J. Chem. Phys.*, **1984**, *81*, 2330; g) Amirav, A.; Jortner, J. *Chem. Phys. Lett.*, **1983**, *95*, 295.

154. Sun, Y.-P.; Saltiel, J. *J. Phys. Chem.*, **1989**, *93*, 8310.

155. a) Gershinsky, G.; Pollak, E. *J. Chem. Phys.*, **1998**, *108*, 9186; b) Gershinsky, G.; Pollak, E. *J. Chem. Phys.*, **1997**, *107*, 812; c) Talkner, P.; Pollak, E.; Berezhkovskii, A. M. *Chem. Phys.*, **1998**, *235*, 131.

156. Leitner, D.M.; Wolynes, P.G. *Chem. Phys. Lett.*, **1997**, *280*, 411.

157. Sivakumar, N.; Hoburg, E.A.; Waldeck. D.H. *J. Chem. Phys.*, **1989**, *90*, 2305.

158. Kim, S.K.; Courtney, S.H.; Fleming, G.R. *Chem. Phys. Lett.*, **1989**, *159*, 543.

159. a) Schroeder, J.; Schwarzer, D.; Troe, J.; Vöhringer, P. *Chem. Phys. Lett.*, **1994**, *218*, 43; b) Mohrschladt, R.; Schroeder, J.; Schwarzer, D.; Troe, J.; Vöhringer, P. *J. Chem. Phys.*, **1994**, *101*, 7566.

160. Anderton, R.M.; Kauffman, J.F. *Chem. Phys. Lett.*, **1995**, *237*, 145.

161. a) Zeglinski, D.M.; Waldeck, D.H. *J. Phys. Chem.*, **1988**, *92*, 692; b) Park, N. S.; Waldeck, D.H. *J. Phys. Chem.*, **1990**, *94*, 662.

162. Keery, K.M.; Fleming, G.R. *Chem. Phys. Lett.*, **1982**, *93*, 322.

163. a) Bowman, R.M.; Eisenthal, K.B.; Millar, D.P. *J. Chem. Phys.*, **1988**, *89*, 762; b) Bowman, R.M.; Eisenthal, K.B. *Chem. Phys. Lett.*, **1989**, *155*, 99; c) Millar, D.P.; Eisenthal, K.B. *J. Chem. Phys.*, **1985**, *83*, 5076.

164. Rothenberger, G.; Negus, D.K.; Hochstrasser, R.M. *J. Chem. Phys.*, **1983**, *79*, 5360.

165. Patterson, G.D.; Carroll, P.J. *J. Phys. Chem.*, **1985**, *89*, 1344 and references therein.

166. Harju, T.O.; Eorstyák, J.; Chow, Y.L.; Korppi-Tommola, J.E.I. *Chem. Phys.*, **1994**, *181*, 259.

167. Kim, S.K.; Fleming, G.R. *J. Phys. Chem.*, **1988**, *92*, 2168.

168. McCaskill, J.S.; Gilbert, R.G. *Chem. Phys.*, **1979**, *44*, 389.

169. a) Lee, M.; Bain, A.J.; McCarthy, P.J.; Han, C.H.; Haseltine, J.N.; Smith, III, A. B. Hochstrasser, R.M. *J. Chem. Phys.*, **1986**, *85*, 4341; b) Lee, M.; Haseltine, J.N.; Smith, III, A.B.; Hochstrasser, R.M. *J. Am. Chem. Soc.*, **1989**, *111*, 5044.

170. Velsko, S.P.; Waldeck; D.H.; Fleming, G.R. *J. Chem. Phys.*, **1983**, *78*, 249.

171. Åkesson, E.; Hakkarainen, A.; Laitinen, E.; Helenius, V.; Gillbro, T.; Korppi-Tommola, J.; Sundström, V. *J. Chem. Phys.*, **1991**, *95*, 6508.

172. Polimeno, A.; Freed, J.H. *Chem. Phys. Lett.*, **1990**, *174*, 338.

173. Velsko, S.P.; Fleming, G.R. *Chem. Phys.*, **1982**, *65*, 59.

174. Hara K.; Akimoto, S. *J. Phys. Chem.*, **1991**, *95*, 5811.

175. a) Korppi-Tommola, J.E.I.; Hakkarainen, A.; Hukka, T; Subbi, J. *J. Phys. Chem.*, **1991**, *95*, 8482; b) Laitinen, E.; Ruuskanen-Järvinen, Rempel, U.; Helenius, V.; Korppi-Tommola, J.E.I. *Chem. Phys. Lett.*, **1994**, *218*, 73; c) Murphy, S.; Sauerwein, B.; Drickamer, H.G.; Schuster, G.B. *J. Phys. Chem.*, **1994**, *98*, 13476; d) Levitus, M.; Negri, R.M.; Aramendía *J. Phys. Chem.*, **1995**, *99*, 14231; e) Ponterini, G.; Caselli, M. *Ber. Bunsenges. Phys. Chem.*, **1992**, *96*, 564; f) Rodrígues, J.; Scherlis, D.; Estrin, D.; Aramendía, P.F.; Negri, R.M. *J. Phys. Chem.: A*, **1997**, *101*, 6998; g) Aramendía, P.F.; Negri, R.M.; Román, E.S, *J. Phys. Chem.*, **1994**, *98*, 3165.

176. a) Vauthey, E.; *Chem. Phys.*, **1995**, *196*, 569; b) Onganer, Y.; Yin, M.; Bessire, D.R.; Quitevis, E.L. *J. Phys. Chem.*, **1993**, *97*, 2344.

177. a) Doolittle, A.K. *J. Appl. Phys.*, **1951**, *22*, 1471; b) Gegiou, D.; Muszkat, K.A.; Fischer, E. *J. Am. Chem. Soc.*, **1968**, *90*, 12.

178. a) Saltiel, J.; D'Agostino, J.T. *J. Am. Chem. Soc.*, **1972**, *94*, 6445; b) Saltiel, J.; Waller, A.S.; Sears, Jr. D.F.; Hoburg, E.A.; Zeglinski, D.M.; Waldeck, D. H. *J. Phys. Chem.*, **1994**, *98*, 10689.

179. a) Vedamuthu, M.; Singh, Surjit; Onganer, Y.; Bessire, D.R.; Yin, M.; Quitevis, E.L.; Robinson, G.W. *J. Phys. Chem.*, **1996**, *100*, 11907; b) Singh, S.; Robinson, G.W. *J. Chem. Phys.*, **1995**, *103*, 4920; c) Singh, S.; Robinson, G.W. *J. Phys. Chem.*, **1994**, *98*, 7300; d) Singh, S.; Robinson, G.W. *Chem. Phys.*, **1996**, *212*, 125.

180. a) Murarka, R.K.; Bhattacharyya, S.; Biswas, R.; Bagchi, B. *J. Chem. Phys.*, **1999**, *110*, 7365; b) Biswas, R.; Bagchi, B. *J. Chem. Phys.*, **1996**, *105*, 7543.

181. Sumi, H. *Chem. Phys.*, **1996**, *212*, 9.

182. Saltiel, J.; Sun, Y.P. Dürr, H.; Bouas-Laurent, H. eds. *Photochromism-Molecules and Systems*, Elsevier: Amsterdam, **1990.**

183. a) Kang, T.J.; Etheridge, T.; Jarzeba, W.; Barbara, P.F. *J. Phys. Chem.*, **1989**, *93*, 1876; b) Barbara, P.F. *Acc. Chem. Res.*, **1988**, *21*, 195.

184. a) Hara, K.; Kiyotani, H.; Bulgarevich, D.S. *Chem. Phys. Lett.*, **1995**; *242*, 455; b) Hara, K.; Ito, N.; Kajimoto, O. *J. Phys. Chem: A*, **1997**, *101*, 2240.

185. Bagchi, B. *Int. Rev. Phys. Chem.*, **1987**, *6*, 1; b) Bagchi, B.; Oxtoby, D. *J. Chem. Phys.*, **1983**, *78*, 2735.

186. van der Zwan, G.; Hynes, J.T. *Chem. Phys.*, **1984**, *90*, 21.

187. a) Dillow, A.K.; Brown, J.S.; Liotta, C.L.; Eckert, C.A. *J. Phys. Chem.: A*, **1998**, *102*, 7609; b) Lednev, I.K.; Ye, T.Q.; Hester, R.E.; Moore, J.N. *J. Phys. Chem.*, **1996**, *100*, 13338; c) Asano, T.; Okada, T. *J. Org. Chem.*, **1984**, *49*, 4387; d) Lednev, I.K.; Ye, T.Q.; Matousek, P.; Towrie, M.; Foggi, P.; Neuwahl, F.V.R.; Umapathy, S.; Hester, R.E.; Moore, J.N. *Chem. Phys. Lett.*, **1998**, *290*, 68; e) Sumi, H.; Asano, T. *Electrochim. Acta*, **1997**, *42*, 2763; f) ibid. *J. Chem. Phys.*, **1995**, *102*, 9565.

188. a) Il'ichev, Y.; Kühnle, W.; Zachariasse, K. A. *Chem. Phys.*, **1996**, *211*, 441; b) Il'ichev, Y.V.; Zachariasse, K.A. *Ber. Bunsenges. Phys. Chem.*, **1997**, *101*, 625.

189. a) Rettig, W.; Majenz, W. *Chem. Phys. Lett.*, **1989**, *154*, 335; b) Lapouyade, R.; Kuhn, A.; Letard, J.F.; Rettig, W. *Chem. Phys. Lett.*, **1993**, *208*, 48; c) Lapouyade, R.; Czeschka, K.; Majenz, W.; Rettig, W.; Gilabert, E.; Rullière, C. *J. Phys. Chem.*, **1992**, *96*, 9643; d) Rettig, W.; Majenz, W.; Lapouyade, R.; Haucke, G. *J. Photochem. Photobiol. A: Chem.*, **1992**, *62*, 415; e) Letard, J.F.; Lapouyade, R.; Rettig, W. *Chem. Phys.*, **1994**, *186*, 119; f) Strehmel, B.; Seifert, H.; Rettig, W. *J. Phys. Chem.: B*, **1997**, *101*, 2232; g) Rettig, W. *Topics Curr. Chem.*, **1994**, *169*, 254.

190. Gruen, H.; Görner, H. *J. Phys. Chem.*, **1989**, *93*, 7144.

191. a) Vogel, M.; Rettig, W. *Ber. Bunseng. Phys. Chem.*, **1987**, *91*, 1241; b) Martin, M.M.; Plaza, P.; Meyer, Y.H. *J. Phys. Chem.*, **1991**, *95*, 9310; c) Ben-Amotz, D.; Jeanloz, R.; Harris, C.B. *J. Chem. Phys.*, **1987**, *86*, 6119; d) Sundström, V.; Gillbro, T.J. *Chem. Phys.*, **1984**, *81*, 3463.

192. a) Alvarez, J.-L.; Yartsev, A.; Åberg, U.; Åkesson, E.; Sundström, V. *J. Phys. Chem.: B.*, **1998**, *102*, 7651; b) Changenet, P.; Zhang, H.; van der Meer, M.J.; Glasbeek, M.; Plaza, P.; Martin, M.M. *J. Phys. Chem.: A*, **1998**, *102*, 6716; c) Tredwell, C.J.; Osborne, A.D. *J.C.S. Faraday II*, **1980**, *76*, 1627; d) Pines, D.; Goldberg, S.Y.; Huppert, D. *Israel J. Chem.*, **1993**, *33*, 215.

193. Bagchi, B.; Fleming, G. R.; Oxtoby, D.W. *J. Chem. Phys.*, **1983**, *78*, 7375.

194. a) Todd, D.S.; Fleming, G.R. *J. Chem. Phys.*, **1993**, *98*, 269; b) Sension, R.J., Repinec, S.T.; Szarka, A.Z.; Hochstrasser, R.M. *J. Chem. Phys.*, **1993**, *98*, 6291; c) Rice, J.K.; Baronavski, A.P. *J. Phys. Chem.*, **1992**, *96*, 3359; d) Sension, R.J., Repinec, S.T.; Hochstrasser, R.M. *J. Phys. Chem.*, **1991**, *95*, 2946; e) Szarka, A.Z.; Pugliano, N.; Palit, D.K.; Hochstrasser, R.M. *Chem. Phys. Lett.*, **1995**, *240*, 25; f) Repinec, S.T.; Sension, R. J.; Hochstrasser, R.M. *Ber. Bunsenges. Phys. Chem.*, **1991**, *95*, 248.

195. a) Frederick, J.H.; Fujiwara, Y.; Penn, J.H.; Yoshihara, K.; Petek, H. *J. Phys. Chem.*, **1991**, *95*, 2845; b) Petek, H.; Yoshihara, K.; Fujiwara, Y.; Lin, Z.; Penn, J.H.; Fredérick, J.H. *J. Phys. Chem.*, **1990**, *94*, 7539; c) Petek, H.; Fujiwara, Y.; Kim, D.; Yoshihara, K. *J. Am. Chem. Soc.*, **1988**, *110*, 6269.

196. a) Myers, A.B.; Mathies, R.A. *J. Chem. Phys.*, **1984**, *81*, 1552; b) Rodier, J.M.; Myers, A.B. *J. Am. Chem. Soc.*, **1993**, *115*, 10791.

197. a) Berweger, C.D.; van Gunsteren, W.F.; Müller-Plathe, F. *J. Chem. Phys.*, **1998**, *108*, 8773; b) Vachev, V.D.; Frederick, J.H.; Grishanin, B.A.; Zadkoc, V.N.; Karoteev, N.I. *J. Phys. Chem.*, **1995**, *99*, 5247; c) Fuss, W.; Lochbrunner, S.; Müller, A.M.; Schikarski, T.; Schmid, W.E.; Trushin, S.A. *Chem. Phys.*, **1998**, *232*, 161.

198. a) Todd, D.C.; Fleming, G.R.; Jean, J.M. *J. Chem. Phys.*, **1992**, *97*, 8915; b) Molina, V.; Merchán, M.; Roos, B.O. *Spectrochim. Acta A*, **1999**, *55*, 33; c) Arenas, J.F. Tocón, I.L.; Otero, J.C.; Marcos, J.I. *J. Phys. Chem.*, **1995**, *99*, 11392.

199. a) McGill, R.A.; Rice, J.K.; Baronavski, A.P.; Owrutsky, J.C.; Lowrey, A.H.; Stavrev, K.K.; Tamm, T.; Zerner, M. C. *Int. J. Quant. Chem.*, **1996**, *30*, 383; b) Amatatsu, Y. *Theochem. J. Mol. Struc.*, **1999**, *462*, 311; c) Bearpark, M. J.; Bernardi, F.; Clifford, S.; Olivucci, M.; Robb, M.A.; Vreven, T. *J. Phys. Chem.: A.*, **1997**, *101*, 3841.

200. a) Stein, A.D.; Hoffman, D.A.; Frank, C.W.; Fayer, M.D. *J. Chem. Phys.*, **1992**, *96*, 3269; b) Hyde, P.D.; Ediger, M.D. *J. Chem. Phys.*, **1990**, *92*, 1036; c) Hyde, P.D.; Evert, T.E.; Ediger, M.D. *J. Chem. Phys.*, **1990**, *93*, 2274; d) Brocklehurst, B.; Young, R.N. *J. Phys. Chem.*, **1995**, *99*, 40.

201. a) Nikiel, L.; Zerda, T.W. *J. Chem. Phys.*, **1990**, *93*, 8464; b) Liu, G.; Mackowiak, M.; Li, Y.; Jonas, J. *J. Chem. Phys.*, **1991**, *94*, 239; c) Roduner, E.; Stolmár, M.; Dilger, H.; Reid, I.D. *J. Phys. Chem. A*, **1998**, *102*, 7591; d) Kowert, B.A.; Higgins, E.J.; Mariencheck, W.I.; Stemmler, T.L.; Kantorovich, V. *J. Phys. Chem.*, **1996**, *100*, 11211.

202. Dong, R.Y.; Shen, X. *J. Chem. Phys.*, **1996**, *105*, 2106.

203. a) Hink, M.A.; van Hoek, A.; Visser, A.J.W.G. *Langmuir*, **1999**, *15*, 992; b) Das, T.K. *J. Phys. Chem.*, **1996**, *100*, 20143; c) Wirth, M. J.; Burbage, J.D. *J. Phys. Chem.*, **1992**, *96*, 9022.

204. Small, E.W.; Libertini, L.J.; Brown, D.W.; Small, J.R. *Opt. Eng.*, **1991**, *30*, 345.

205. a) Heitz, M.P.; Bright, F.V. *J. Phys. Chem.*, **1996**, *100*, 6889; b) Anderton, R.M.; Kauffman, J.F. *J. Phys. Chem.*, **1995**, *99*, 13759; c) Rice, J.K.; Niemeyer, E.D.; Bright, F.V. *J. Phys. Chem.*, **1996**, *100*, 8499.

206. Tsunomori, F.; Ushiki, H.; Hamano, K. *Chem. Phys. Lett.*, **1996**, *252*, 121.

CHAPTER

4

Structural Aspects of Exciplex Formation

Fred Brouwer

Institute of Molecular Chemistry, Amsterdam, The Netherlands

1 Exciplexes and Other Excited-State Complexes

The word **exciplex** is an interesting example of a scientific word that is not very well defined and certainly does not mean what it seems to mean.[1] Words with multiple meanings have an important role in natural languages, because they allow us to communicate at different levels, to negotiate, to cheat, to have fun, etc. For the language of science, on the other hand, such ambiguity is not a desirable property. **Exciplex** is a contraction of the words "**exci**ted-state com**plex**," which allows a broad range of species to be called exciplexes. Most chemists would not use the word "exciplex" in this sense. Certain features are associated with exciplexes that do not follow directly from the apparent meaning of the word. First of all, in an exciplex, there is a considerable extent of *charge transfer* (CT) between a component that is an *electron donor* in its ground state and another component that is an *electron acceptor* in the ground state. Thus, exciplexes are formed from two (or more) chemically different molecules, in contrast to *excimers*, which are excited-state complexes composed of identical species.[2] Both can be formed by a bimolecular interaction of an excited molecule with a ground-state molecule, which is accompanied by quenching of the fluorescence of the former (Fig. 4.1).

In this chapter, we will describe exciplex formation in general as the result of the interaction of a photoexcited species A^* and a ground-state species D, leading to what is best described as a radical ion pair $D^{+\cdot}A^{-\cdot}$. In practice, the photoexcited species might as well be the electron donor, and the reactants are not necessarily neutral in the ground state.

Conformational Analysis of Molecules in Excited States, Edited by Jacek Waluk
ISBN 0-471-29707-0

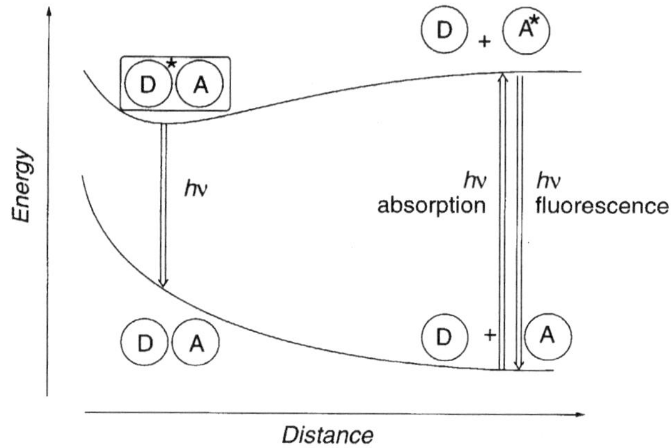

Figure 4.1. Formation of exciplexes (A = electron acceptor, D = electron donor; charge transfer) and excimers (A = D). Either A or D may be the initially excited reactant.

Although they are usually described in terms of the same basic electronic configurations or valence bond resonance contributions (Fig. 4.2), excimers and exciplexes ("heteroexcimers") have very different wave functions (Fig. 4.3). In the former, exciton resonance is usually the most important. For the exciplex, the charge transfer configuration $D^{+\cdot}A^{-\cdot}$ is dominant; that is, the exciplex resembles a radical ion pair (RIP). The mixing in of locally excited (LE) configurations ($D^{*}A$ and DA^{*}), however, can be very important for the properties of exciplexes.

Excimers of polycyclic aromatic compounds were experimentally discovered[3] due to their characteristic fluorescence in the visible range of the spectrum. A characteristic fluorescence can often be observed in exciplexes as well, but this is not always the case. Some authors consider fluorescence a characteristic property of an exciplex, and the term "nonfluorescent exciplex" can be found in the literature to describe what these authors consider to be the exception.[4]

It is commonly accepted that excimers and exciplexes of aromatic molecules have a sandwich-type geometry. Since the exciplex wave function is mostly radical ionic, the term "contact radical ion pair" is an appropriate description for most exciplexes. Another type of radical ion pair is the so-called solvent-separated radical ion pair (SSRIP), also known as a solvent-shared or loose ion pair (Fig. 4.4). This is perhaps not a species with a very well-defined geometric structure, but it is definitely a

excimer:
- nonpolar
- exciton interaction *(charge resonance)* $\mathbf{M^{*}M} \leftrightarrow \mathbf{MM^{*}} \leftrightarrow M^{+\cdot}M^{-\cdot} \leftrightarrow M^{-\cdot}M^{+\cdot}$

exciplex:
- polar
- charge *transfer*: $\mathbf{D^{+\cdot}A^{-\cdot}} \leftrightarrow \mathbf{DA^{*}} \leftrightarrow \mathbf{D^{*}A} \leftrightarrow DA$

Figure 4.2. Characteristics of electronic wave functions of excimers and exciplexes.

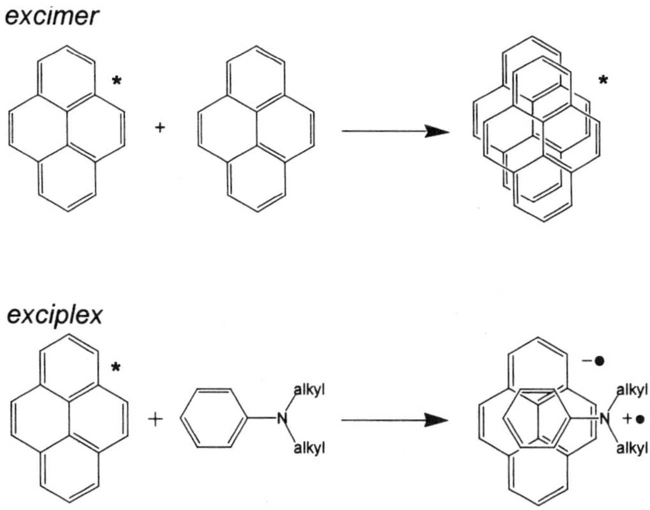

Figure 4.3. Classical examples of an excimer and an exciplex.

distinct kinetic intermediate, necessary in the mechanistic description of the dynamics of radical ion pairs. In contrast to the **CRIP**, the **SSRIP** is considered to be nonfluorescent.

When electron donor and acceptor form part of the same molecule, intramolecular charge-transfer excited states can be formed. When these happen to emit detectable fluorescence, they have sometimes been called "exciplex" states. In studies of intramolecular donor–acceptor systems with a flexible linkage, those conformations that have a face-to-face geometry (which, until the end of the 1980s, were considered by most chemists to be the only fluorescent species) have been termed "exciplex conformations." Other conformations were denoted as "loose exciplex" or "loose ion pair." Later, it was shown that a sandwich conformation was not at all a prerequisite for fluorescence in such systems.[5] Clearly, the use of the word "exciplex" in connection with intramolecular charge transfer can lead to confusion.

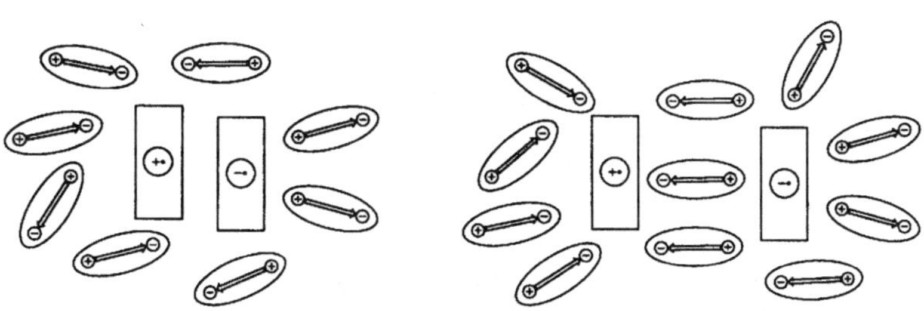

Figure 4.4. Schematic representation of a contact radical ion pair and a solvent-separated radical ion pair.

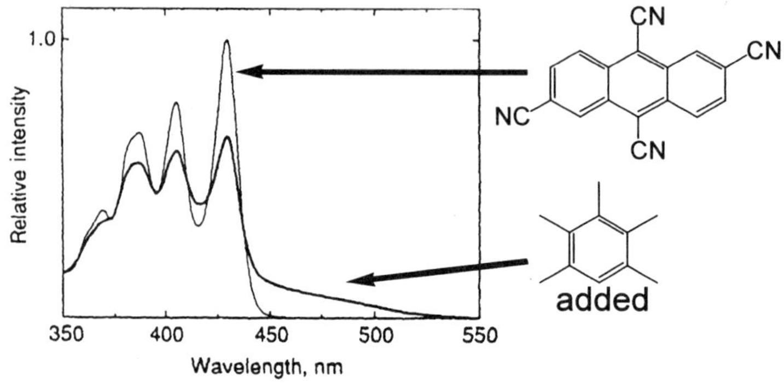

Figure 4.5. Example of a CT absorption. When pentamethylbenzene (0.56 M) is added to 2,6,9,10-tetracyanoanthracene in trichloroethylene, a new absorption band at a long wavelength arises. The band is due to the CT complex. (Reproduced with permission from Ref. 8. Copyright American Chemical Society.)

Another property commonly attributed to excited-state complexes is that they are not stable in the ground state. This is generally true for excimers in solution, but exciplexes have a ground-state counterpart in *charge-transfer complexes* [CT complexes, also known as electron donor–acceptor (EDA) complexes].[6,7] These are complexes formed in the ground state between strongly electron-donating and electron-accepting species, as exemplified in Fig. 4.5. They derive much of their stabilization energy from the mixing of a CT electronic configuration into the ground-state wave function. This mixing is effective only when the energy of the CT configuration $D^{+\cdot}A^{-\cdot}$ is low, which usually means that it is lower than the locally excited states of the donor and acceptor chromophores. CT complexes are therefore often characterized by low-energy optical transitions that are not present in the individual donor and acceptor species. These transitions can lead to visible color of mixtures of otherwise colorless compounds. It is therefore not surprising that CT complexes have been known much longer than exciplexes.[6]

Exciplexes can be formed by a bimolecular process involving a photoexcited electron acceptor A and an electron donor D (or vice versa), but also by excitation of the corresponding EDA complexes. This requires that D and A have sufficiently strong electron-donating and -accepting properties in their ground states to form such a complex and that the transition moment of the CT complex is large enough (see, for example, Fig. 4.5). There is no formal difference between a relaxed CT excited state of an EDA complex and an exciplex (CRIP). Although fluorescence from CT complexes had been observed well before exciplex emission, in solutions at room temperature fluorescence it is invariably weak and easily overlooked, for reasons that will become clear later. (see Section 2). For fast time-resolved spectroscopy, excitation of CT complexes is the method of choice to produce exciplexes (contact radical ion pairs) because the time lag in exciplex formation due to diffusion is avoided (Fig. 4.6).

In this chapter, we shall use the term "exciplex" to denote excited-state complexes that meet the following criteria:

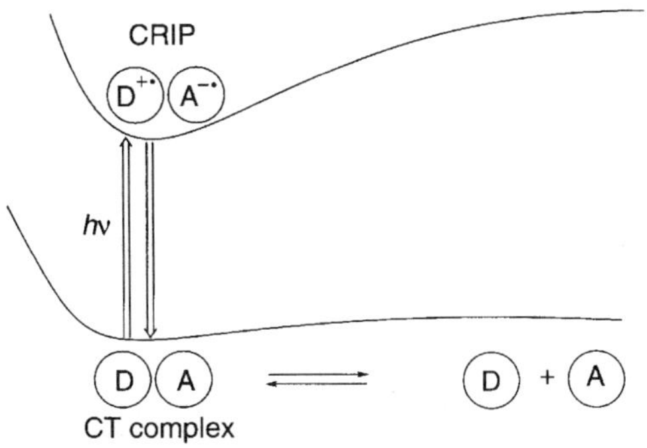

Figure 4.6. Formation of an exciplex (CRIP) by excitation of a charge-transfer complex.

- The ground state species involved are an electron donor and an electron acceptor.
- The excited state has considerable charge transfer character.
- The molecules that make up the complex are in close contact. (The orbitals directly overlap, with no solvent molecules in between.)

The other requirements mentioned, namely lack of stability of the complex in the ground state, exciplex fluorescence, and the geometric criterion, are too restrictive. The difference between an exciplex as previously defined and a contact radical ion pair is very small, being determined only by the amount of locally excited state character in the electronic wave function.

We will restrict ourselves here to exciplexes in which both partners are molecules. We will not discuss exciplexes involving atoms, even though these are of great practical importance in the so-called excimer lasers. In the next section, we shall discuss the formation and decay of exciplexes in the context of contemporary electron transfer theories. Section 3 is concerned with the structure of exciplexes, both in terms of geometry and electronic wave function. In the last section, we will discuss molecules in *intramolecular* CT states, which have sufficient flexibility that geometries can be attained which are considered typical of exciplexes (i.e., face-to-face sandwich-type orientations).

2 Formation and Decay of Exciplexes

2.1 Formation of Exciplexes

In an early application of flash photolysis transient absorption experiments, Leonhardt and Weller showed[9,10] that the quenching of perylene fluorescence by electron-donating amines in polar solvents is accompanied by the formation of the radical anion of perylene. This confirmed the occurrence of a fluorescence-quenching mechanism via electron transfer (Scheme 1), a possibility proposed already in the 1930s.[11,12] The study of the dependence of the extent of quenching on the concentration of the quencher revealed quenching rate constants of up to

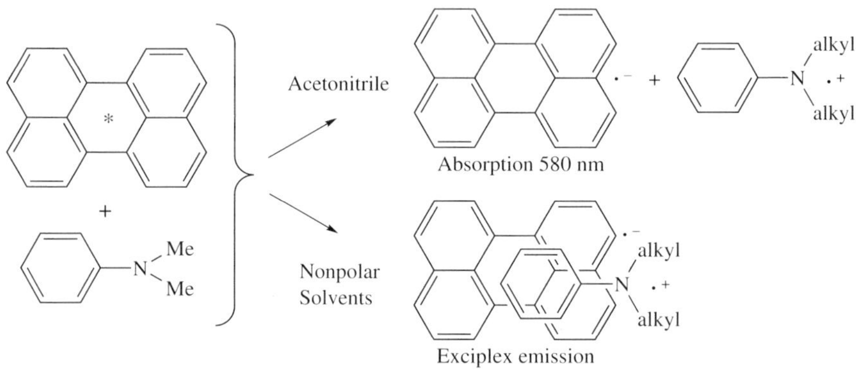

Scheme 1

$2 \times 10^{10}\,M^{-1}\,s^{-1}$ in polar solvents. Based on estimates of the diffusion constants, Leonhardt and Weller concluded that, in this case, the intermolecular encounter distance at which quenching takes place is about 7 Å.

In their 1963 paper, Leonhardt and Weller also published the first emission spectra showing *exciplex fluorescence*. While in polar solvents the fluorescence of perylene is merely quenched by electron-donating (or -accepting) molecules, in nonpolar solvents such as benzene and methylcyclohexane a new fluorescence appeared at long wavelengths (Fig. 4.7). Such a charge transfer fluorescence had previously been observed only upon excitation of CT complexes in low-temperature glasses.[6] Leonhardt and Weller concluded that in nonpolar solvents electron transfer is slightly endothermic and becomes apparent only at relatively high concentrations of electron donors. Moreover, the quenching distance was found to be much shorter, and radical ions were not observed in the transient absorption experiment. Instead,

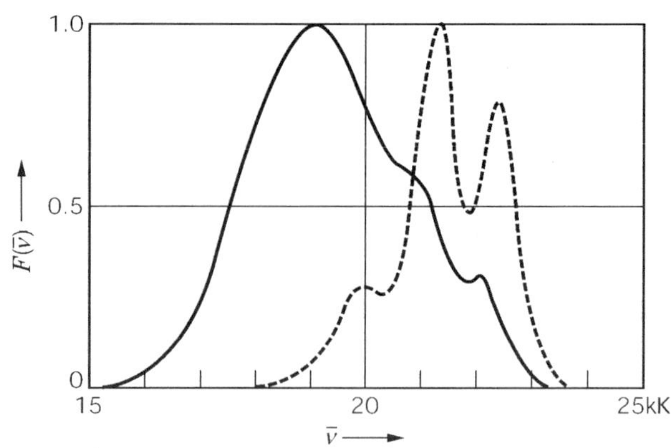

Figure 4.7. Fluorescence spectra of perylene (7×10^{-5} M) (dashed line) in methylcyclohexane and with dimethylaniline (2.0 M) added (solid line). The exciplex emission maximum is at about 19,000 cm^{-1}. At higher energy some of the structured perylene emission remains. (Reproduced with permission from Ref. 10. Copyright Wiley/VCH.)

flash photolysis showed the formation of triplet excited perylene in high yield. Since perylene itself does not have a strong tendency towards intersystem crossing (the fluorescence quantum yield is 89% in benzene),[13] and the singlet excited state is strongly quenched by electron transfer, it is evident that the formation of the triplet species is a result of the electron transfer quenching process. This led Leonhardt and Weller to propose the schemes shown in Fig. 4.8, describing the photophysical processes in polar and nonpolar solvents, respectively. Although considerable further insight into all the steps shown in these schemes has been gained since 1963, most of the basic ideas that were put forward then still hold today.

The conclusion of Leonhardt and Weller that electron transfer quenching in polar solvents occurs at a distance of about 7 Å, leading to formation of a geminate ion pair that is solvent separated, was shown (30 years later) not to be *generally* valid. The key countervailing experiments were based on the possibility of forming the CRIP or exciplex by direct excitation of CT complexes. Insight into the efficiency of formation of the CRIP, as opposed to formation of the SSRIP, can be obtained by comparing the intensity of the exciplex emission that results from direct excitation of the CT complex with that resulting from electron transfer quenching of the fluorescent donor or acceptor. If the bimolecular quenching process leads to an SSRIP, not to a fluorescent CRIP, the apparent fluorescence yield should be lower. These experiments are rather difficult to perform, because the CT fluorescence from EDA complexes in polar solution is invariably very weak. The weakness arises from the rapid nonradiative decay of the excited state, which tends to occur as a consequence of the necessarily small energy gap between DA and D^+A^- for a

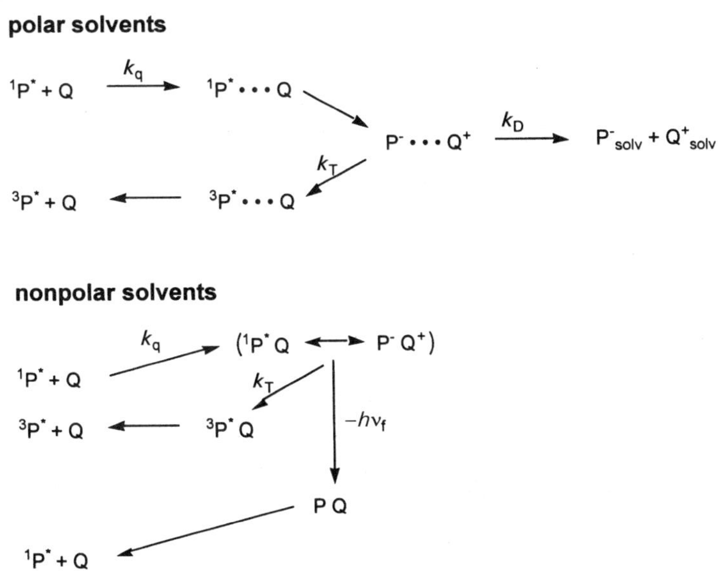

Figure 4.8. Reaction schemes for the quenching of singlet excited perylene (P) by a quencher Q according to Leonhardt and Weller.[10]

strong donor–acceptor couple. Although fluorescence of CT complexes had been observed even before exciplex emission, it was only in nonpolar media at low temperature.[6] As a part of their extensive studies of the chemistry of radical ion pairs,[14] Gould, Farid, and their co-workers could show that, in fact, direct formation of the SSRIP in electron transfer quenching is the exception, not the rule.[15] Like many of the phenomena in electron transfer quenching, this can be explained on the basis of a theoretical model of electron transfer, which we will discuss now in a simplified form.

2.2 Rates of Electron Transfer Quenching

The kinetics of fluorescence quenching and exciplex formation may in general be described as involving a diffusive process forming an encounter complex and subsequent electron transfer within the complex. In a few cases, the encounter complex, which is stabilized by charge-transfer interactions, has been observed by transient absorption spectroscopy.[16] In this section, we shall look at some of the characteristics of the electron transfer step. We give a description of that step for neutral initial reactants forming a radical ion pair (RIP) upon transfer of an electron, but the description is readily adapted to charged reactants. As a first approximation, the product state is assumed to be purely ionic, although we will see later that in many cases the electronic wave function is in fact more complex.

The relation between the kinetics and thermodynamics of molecular electron transfer processes can in many cases be rationalized by using a remarkably simple model that was first developed by Marcus.[17] To illustrate the model, parabolic free-energy curves are drawn that represent the initial and final states (Fig. 4.9). When an electron is transferred from an electron donor to an electron acceptor (in an encounter complex), changes occur in the geometry of the reactants and in the

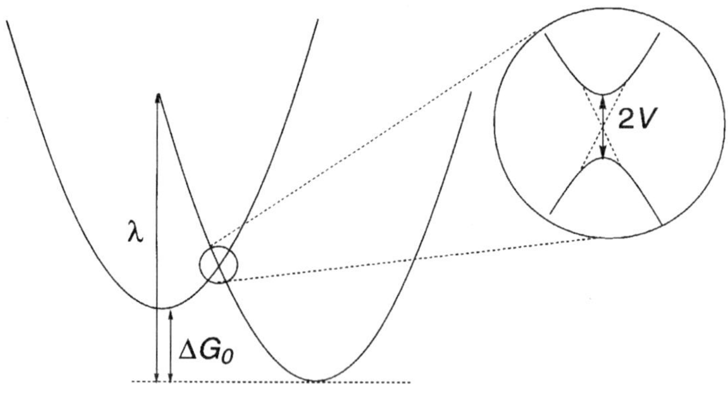

Figure 4.9. Free-energy curves along the reaction coordinate of electron transfer, and illustration of the free-energy change ΔG_0, the reorganization energy λ, and the electronic coupling V. Note that the reaction coordinate is a *nuclear* coordinate, involving the motion of solvent molecules and structural changes in the reactants.

solvation shell. The horizontal displacement of the curves (along the reaction coordinate) reflects the extent of the change in the geometry of the molecules and the solvent in the electron transfer process, while the vertical displacement is simply the free-energy difference of the two states. The system moves along the curve of the initial state. When it reaches the crossing point, there is a certain probability that an electronic transition occurs to the other state. Both the location (energy) of the crossing point and the electronic interaction energy between the two states at this point determine the rate of electron transfer. The theory was derived assuming that the electronic coupling, which lifts the degeneracy at the crossing point, is very small (nonadiabatic limit).

In this case, the rates of transitions between the free-energy wells are predicted to be

$$k_{\mathrm{et}} = \frac{2\pi^{3/2}}{h\sqrt{\lambda k_{\mathrm{B}} T}} V^2 e^{-(\Delta G_0 + \lambda)^2/4\lambda k_{\mathrm{B}} T}. \tag{2.1}$$

Apart from the temperature, three quantities determine the rate: the driving force ($-\Delta G_0$), the reorganization energy λ, and the electronic coupling V. The factor containing the exponential term is mostly responsible for the temperature dependence. It contains the energy of the crossing point of the two parabolas (easily found by simple geometry), which is much like an activation energy in the sense of Arrhenius theory or Eyring transition state theory. In the latter, the preexponential factor $\kappa k_{\mathrm{B}} T/h$ has a maximum value at room temperature of about $6 \times 10^{12}\,\mathrm{s}^{-1}$. For the preexponential factor in the Marcus rate expression to reach a similar value, an electronic coupling of about $100\,\mathrm{cm}^{-1}$ is required. Thus, as a rule of thumb, the velocity of nuclear motion may become rate limiting for $V > 100\,\mathrm{cm}^{-1}$. At an encounter distance of about 7 Å, V is less than $100\,\mathrm{cm}^{-1}$, but for two molecules in van der Waals contact, the electronic coupling is likely to be much larger than $100\,\mathrm{cm}^{-1}$. Thus, Eq. 2.1 is not universally valid. Nevertheless, the dependence of the exponential factor on the driving force and reorganization energies remains, although for large electronic coupling, the actual activation energy should be reduced to take the splitting $2V$ into account.

2.2.1 Role of the Electronic Coupling

It is illustrative to evaluate Eq. 2.1 with some values typically found for the rate-determining parameters. First of all, we note that when $-\Delta G_0 = \lambda$ the exponent vanishes, and a maximum rate is obtained, which is only weakly temperature dependent. In that case, for the electron transfer to compete favorably with the typical decay rate of a singlet excited organic molecule of $10^8\,\mathrm{s}^{-1}$, a very small electronic coupling of only about $1\,\mathrm{cm}^{-1}$ is already sufficient. In a more typical case (e.g., $\Delta G_0 = -0.1\,\mathrm{eV}$ and $\lambda = 0.8\,\mathrm{eV}$, an electron transfer rate of $3 \times 10^8\,\mathrm{s}^{-1}$ requires an electronic coupling $V \approx 20\,\mathrm{cm}^{-1}$. Much larger values of V are found in encounter complexes, which leads to the conclusion that *electronic coupling is not likely to be a rate-limiting factor in exciplex formation*. In practice, the thermodynamic feasibility, $\Delta G < 0$, is the most important factor, while the reorganization energy may have a significant effect on the energy barrier. This is most obvious when one considers the case of $\Delta G_0 = 0$, for which the activation energy is $\lambda/4$.

2.2.2 Reorganization Energies

The reorganization energy λ is composed of a solvation part λ_s and an intramolecular contribution λ_i, which is the change in energy associated with the intramolecular relaxation processes in donor and acceptor following electron transfer. The internal reorganization energy λ_i varies considerably with molecular structure. For example, the charging of polycyclic aromatic groups, such as anthracene, leads to negligible structural reorganization ($\lambda_i < 0.1$ eV),[18–20] while the oxidation of amines or hydrazines is accompanied by very large changes, with λ_i as large as 1.5 eV.[21] The internal reorganization energy λ_i of electron donors may be found by using photoelectron spectroscopy if the adiabatic ionization potential can be accurately determined. For complex molecules, this is often not the case, because the bands are broad and structureless. For radical anions, the spectroscopic determination of λ_i is even more difficult. Especially since the development of reliable density functional methods, computational chemistry may supply useful additional data here.[22] When charge transfer absorption and emission spectra of a donor–acceptor system can be measured in a nonpolar solvent, the internal reorganization energy can be estimated as half the Stokes shift. This is exemplified in Fig. 4.10.

The solvent reorganization energy λ_s reflects the energy associated with the change in the equilibrium configuration of solvent molecules upon transfer of

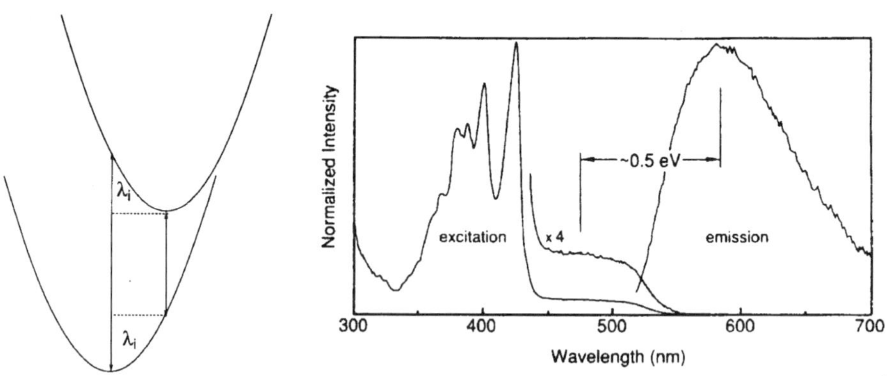

Figure 4.10. Left: The difference between the vertical absorption and emission energies is equal to twice the reorganization energy, provided that absorption and emission are respectively to and from the same electronic excited state. In nonpolar solvents, the solvent contribution to the reorganization may be negligible, so the internal reorganization energy is measured. This is illustrated on the right by the emission and excitation spectra of 2,6,9,10-tetracyanoanthracene (2×10^{-6} M) with hexamethylbenzene (0.026 M) in cyclohexane. (Reproduced with permission from Ref. 23. Copyright American Chemical Society.)

charge. This energy is often approximated with the following equation, derived from dielectric continuum theory with a model of the solute consisting of two spherical cavities with ion radii r^{ion} and a center-to-center distance of R_{DA}:

$$\lambda_s = 14.4\left(\frac{1}{2r_{\text{D}}^{\text{ion}}} + \frac{1}{2r_{\text{A}}^{\text{ion}}} - \frac{1}{R_{\text{DA}}}\right)\left(\frac{1}{\varepsilon_s} - \frac{1}{\varepsilon_\infty}\right) \text{eV}. \tag{2.2}$$

The solvent properties enter via ε_s, the static dielectric constant, and ε_∞, the high-frequency permittivity, experimentally measured as the square of the refractive index (n^2). When the distance between D and A is small relative to the ion radii, as is the case in exciplexes, Eq. 2.2 is not appropriate, as we shall see later. In practice, solvent reorganization energies for the formation of exciplexes vary from nearly zero (in alkane solvents) up to more than 1 eV for transfer over larger distances in polar organic media.

Earlier, it was mentioned that electron transfer quenching does *not* generally lead to an SSRIP. The electron transfer rate increases with decreasing distance, to some extent because the electronic coupling increases, but also because the barrier imposed by the solvent reorganization decreases. Thus, unless the driving force $-\Delta G_0$ is large enough to compensate for the reorganization energy, there is time enough for the reactants to diffuse more closely together.[14,15] When diffusion is slowed down—for example, by the use of a viscous solvent—complex fluorescence-quenching kinetics can be observed due to the electron transfer occurring with different rates at different distances.[24,25]

2.2.3 Thermodynamic Driving Force of Charge Separation

The energy level of the charge-separated state—the radical ion pair—in a polar solvent may be estimated from the redox potentials of D and A and the electrostatic interaction energy:

$$\Delta G_{\text{RIP,polar}} = e(E_{\text{D}}^{\text{ox}} - E_{\text{A}}^{\text{red}})_{\varepsilon_{\text{polar}}} - \frac{14.4}{\varepsilon_s R_{\text{DA}}}. \tag{2.3}$$

In acetonitrile, the electrostatic term is about -0.11 eV for a CRIP ($R_{\text{DA}} \approx 3.5$ Å) and -0.06 eV for an SSRIP (ca. 7 Å). Whether the RIP is more or less stable than the excited state, of course, depends on the excitation energy E_{00}:

$$\Delta G^*_{\text{RIP,polar}} = e(E_{\text{D}}^{\text{ox}} - E_{\text{A}}^{\text{red}})_{\varepsilon_{\text{polar}}} - \frac{14.4}{\varepsilon_s R_{\text{DA}}} - E_{00}(\text{eV}). \tag{2.4}$$

In a less polar solvent, the energy of the ions relative to their neutral precursors has to be corrected for the smaller solvation energy. This is usually done by using the Born equation:

$$\Delta G^*_{\mathrm{RIP},\varepsilon_s} = e(E_{\mathrm{D}}^{\mathrm{ox}} - E_{\mathrm{A}}^{\mathrm{red}})_{\varepsilon_{\mathrm{polar}}} - \frac{14.4}{\varepsilon_s R_{\mathrm{DA}}} + 14.4\left(\frac{1}{2r_{\mathrm{D}}^{\mathrm{ion}}} + \frac{1}{2r_{\mathrm{A}}^{\mathrm{ion}}}\right)\left(\frac{1}{\varepsilon_s} - \frac{1}{\varepsilon_{\mathrm{polar}}}\right)(\mathrm{eV}). \quad (2.5)$$

This equation, in principle, allows an estimate of the energy of a radical ion pair in any solvent. Unfortunately, it cannot easily be applied, because the Born ion radii can hardly be estimated with sufficient accuracy. For organic ions, values can be found in the literature ranging from about 3.3 to 4.5 Å. This implies a range of "reasonable" estimates of the solvent correction for nonpolar hydrocarbons ($\varepsilon_s \approx 2$) of 0.6 eV. In CRIP, the donor–acceptor distance R_{DA} is similar to the sum of the ionic radii, so independent solvation of the individual ions cannot occur. Under these circumstances, the additivity of the Born and Coulomb terms is not justified. For CRIP, it is more reasonable to use a model of dipole solvation, as explained in section 2.3. In a simple spherical cavity model (with radius ρ), the electrostatic energy of a dipole μ is given by

$$E_{\mathrm{electrostatic}} = -\frac{\mu^2}{\rho^3}\left(\frac{\varepsilon_s - 1}{2\varepsilon_s + 1}\right). \quad (2.6)$$

Finally, the electronic stabilization energy that results from mixing of the pure RIP configuration with the ground state and with locally excited states should be taken into account. Although this mixing is important for the transition probabilities, its energetic effect is usually negligible.[26,27]

Weller and co-workers used an elegant approach to experimentally investigate exciplex thermodynamics in alkane solvents. They made use of the reversibility of the process

$$\mathrm{A}^* + \mathrm{D} \rightleftharpoons \mathrm{D}^+\mathrm{A}^-.$$

By measuring the intensities of the emission bands of the locally excited species (in most cases an electron acceptor) and the exciplex as a function of temperature, they could determine the thermodynamic parameters. The enthalpy (relative to the excited-state energy) of exciplex formation was found to be related in a simple way to the redox properties (measured in polar solvents), as shown in Fig. 4.11. The entropy of exciplex formation was found to be essentially the same for all exciplexes.[28,29]

$$E_{00} - \Delta H^{\mathrm{alkane}}_{\mathrm{exciplex}} = e(E_{\mathrm{D}}^{\mathrm{ox}} - E_{\mathrm{A}}^{\mathrm{red}})_{\varepsilon_{\mathrm{polar}}} + 0.15\,\mathrm{eV}, \quad (2.7)$$

$$\Delta S^{n\text{-hexane}}_{\mathrm{exciplex}} = -18 \pm 1\ \mathrm{cal\ mol^{-1}\ K^{-1}}. \quad (2.8)$$

Interestingly, this temperature dependence of emission spectra has been exploited in spectroscopic "thermometers" for nonpolar media.[30,31]

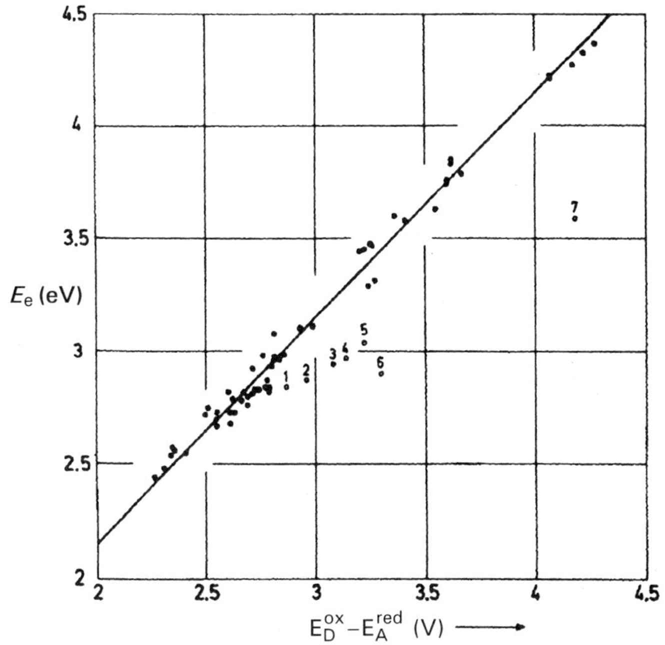

Figure 4.11. Singlet exciplex energies $E_e(= E_{00} - \Delta H)$ in alkane solvent plotted against the difference in donor oxidation potential and acceptor reduction potential. The symbols 1–7 represent (hetero)excimers with a much smaller CT character. (Reproduced with permission from Ref. 32. Copyright Academic Press.)

The entropy contribution to the free energy of exciplex formation $A^* + D \rightarrow (D^{+\cdot}A^{-\cdot})$ in alkanes is quite significant: 0.23 eV at room temperature. The energy of exciplexes in solvents of intermediate polarity may be estimated by combining the dipole solvation energy (Eq. 2.6) with the empirical free energy in *n*-hexane:[28]

$$\Delta G_{\text{exciplex}}^{\text{solvent}} = e(E_D^{\text{ox}} - E_A^{\text{red}})_{\varepsilon_{\text{polar}}} - 0.75\left(\frac{\varepsilon_s - 1}{2\varepsilon_s + 1} - 0.19\right) + 0.38 \text{ eV}. \tag{2.9}$$

The factor of 0.75 in Eq. 2.9 can be experimentally derived from the change in emission energy with solvent polarity, as discussed in the next section. With this average value for a number of donor–acceptor pairs, the difference in free energy between an exciplex in *n*-hexane and in acetonitrile amounts to about 0.22 eV. Although the energy in a polar solvent is lower, the coulombic and the solvation terms compensate for each other to a large extent. If this difference of 0.22 eV is compared with the energy difference that would follow from Eqs. 2.3 and 2.5, using $R_{DA} = 3.5$ Å, the average ionic radius is found to be 3.2 Å.

2.3 Fluorescence Solvatochromic Shift[33,34]

One of the most characteristic and fascinating features of exciplexes is their fluorescence. Because the excited state is highly polar, its energy is quite sensitive to the polarity of the environment. The ground state, on the other hand, is neutral, and its energy does not depend strongly on the solvent polarity. As a result, increasing the solvent polarity leads to a red shift of the exciplex emission as illustrated in Figure 4.12. If the ground-state dipole moment can be completely neglected, the solvatochromic shift of the emission maximum (wave number $\bar{\nu}_{CT}$) can be characterized as follows:[35]

$$\bar{\nu}_{CT} = \bar{\nu}^0_{CT} - \frac{2\mu^2}{\rho^3 hc} \cdot \Delta f; \tag{2.10}$$

$$\Delta f = \frac{(\varepsilon_s - 1)}{(2\varepsilon_s + 1)} - \frac{(n^2 - 1)}{(4n^2 + 2)}. \tag{2.11}$$

The interaction between the species and the medium is modeled by separating the space into a cavity containing the molecule and a dielectric continuum. The molecule itself is represented by some electrostatic potential, which induces a *reaction field* in the medium. In the simplest form, used in Eq. 2.10, the molecule is represented by a point dipole μ and the cavity is a sphere of radius ρ.

The solvent response can be separated according to its timescale into a fast and a slow component. The fast component corresponds to the electronic polarization of the solvent molecules. This polarization is able to keep up with the rate of the optical transition, which takes place on a femtosecond timescale. The slow component (lasting picoseconds in fluid solutions) is due to the reorientation of the solvent dipoles, which can occur only after the optical transition. The total solvent response is a function of the dielectric permittivity (static dielectric constant ε_s), and the electronic polarization of the solvent molecules is reflected in the high-frequency or "optical" dielectric constant (ε_∞ or ε_{op}) of the medium, which is experimentally measured as the square of the refractive index (n^2).

The solvatochromic shift reflects the entire solvent response in the excited state plus the reorientational part on the ground-state surface: The molecule reaches the ground state with a nonequilibrium (unfavorable) orientational distribution of the solvent molecules. In nonpolar solvents, which have negligible solvent reorganization energy, the Stokes shift is determined by the internal reorganization energy λ_i, or the "ground-state repulsion energy." For a large series of exciplexes involving aromatic donors and acceptors, Weller and co-workers found average ground-state repulsion energies of about 0.3 eV,[32] although there are clear variations with structure.[29] Together with Eq. 2.7 for the enthalpy of formation, these energies lead to Eq. 2.12 for the exciplex emission energy in nonpolar alkane solvents:

$$h\nu^{alkane}_{exciplex} = e(E^{ox}_D - E^{red}_A)_{\varepsilon_{polar}} - 0.15(\text{eV}). \tag{2.12}$$

Although the model given by Eqs. 2.10 and 2.11 is crude, it is useful for characterizing the experimentally observed shifts. Typically, when the emission maximum $\bar{\nu}_{CT}$ is plotted against Δf according to Eq. 2.10, exciplexes show a slope

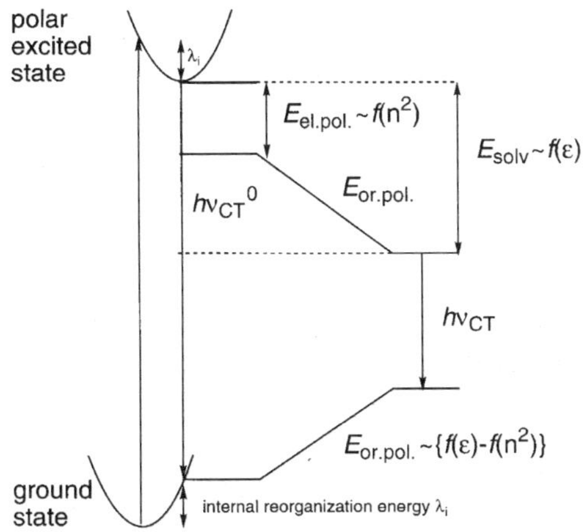

Figure 4.12. The contribution of electronic polarization and reorientation of solvent molecules to the solvatochromic shift in the emission of polar species such as exciplexes. In the dipole model, the solvent functions are $f(x) = (x - 1)/(2x + 1)$. The proportionality factor depends on the electric field generated by the solute and a geometric factor. For the dipole in a spherical cavity model, it is μ^2/ρ^3.

of about $10^4\ \text{cm}^{-1}$. An example is shown in Figure 4.13.[35] It is clear that a correlation exists, but at the same time, it is evident that the simple continuum model that is used has its limitations.

Some authors have extracted values of excited-state dipole moments from the solvatochromic shift plots, but this is dangerous, especially because the cavity model is too simple to be realistic. Other deficiencies of the model include the neglect of specific solute–solvent interactions and of solute polarizability. When the nature of the excited state changes with solvent polarity, nonlinear plots are observed (e.g., for the exciplex of cyanonaphthalene and naphthalene, which is only 30% charge

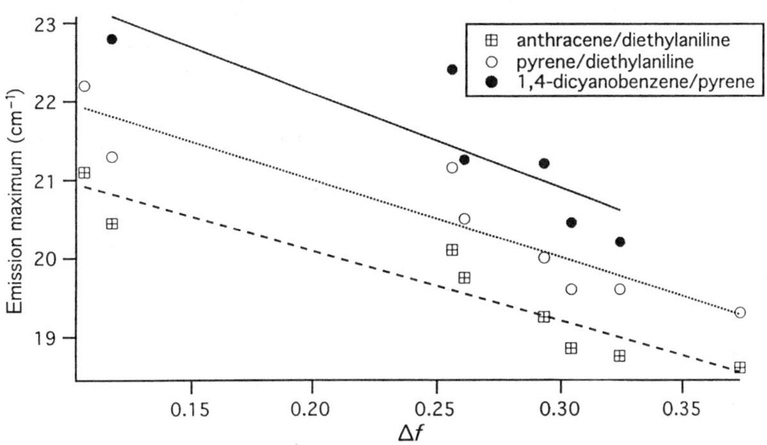

Figure 4.13. Exciplex emission maximum ($10^3\ \text{cm}^{-1}$) vs. polarity function Δf (Eq. 2.11). (Data from Ref. 35.)

separated in nonpolar solvents, but up to 70% in polar solvents).[36] Another important point is that some solvents are not well characterized by the macroscopic dielectric constant, which is measured using a homogeneous electric field. This field interacts only with the solvent dipole, but not with its higher electric moments. The latter do interact with the inhomogeneous electric field generated by a dipole at the molecular scale. Therefore, while solvents such as benzene and 1,4-dioxane have low dielectric constants (at 20°C, $\varepsilon_s = 2.3$ and 2.1, respectively), they appear as much more polar solvents in a solvatochromic shift analysis.

2.4 Decay of Exciplexes

Exciplexes may decay by several pathways:

- Nonradiative charge recombination to the ground state.
- Radiative charge recombination to the ground state (CT fluorescence).
- Charge recombination with spin inversion, forming a locally excited triplet state.
- Dissociation into separate ions.
- Formation of a reaction product.

2.4.1 Marcus Inverted Region

In many cases, the decay of exciplexes–radical ion pairs to the ground state is the most important pathway. The basic rules that govern the rate of the charge recombination process are described by Marcus theory. A crucial difference with the charge separation process previously discussed is that for charge recombination to the ground state, the driving force is usually much larger than the reorganization energy ($-\Delta G_0 > \lambda$). This leads to the so-called *inverted region* (Fig. 4.14), in which the predicted rate of electron transfer according to Eq. 2.1 decreases exponentially with increasing driving force.

For today's photochemists who are familiar with the energy gap law for radiationless transitions, the preceding result is no surprise, but for a long time there has been a real struggle among electron transfer chemists to *prove* the existence of the inverted region. It is practically impossible to do this using rates of

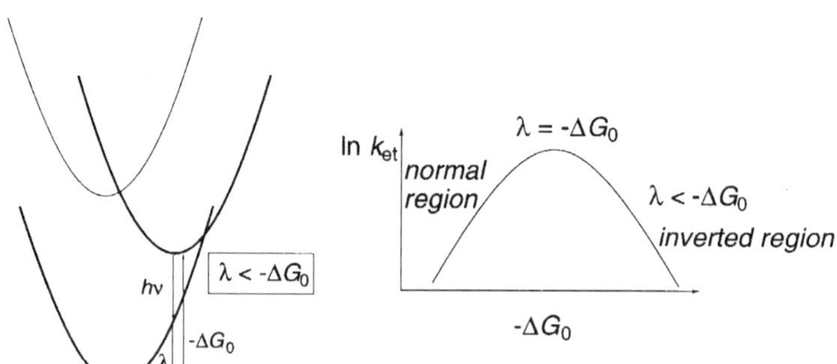

Figure 4.14. Photoinduced charge separation usually occurs in the Marcus "normal region"; charge recombination is a relative or nonradiative decay process in the "inverted region."

bimolecular reactions, because these are limited by the diffusion process,[37,38] but using intramolecular electron transfer[39] or charge recombination within radical ion pairs resulted in clear evidence for the existence of the inverted region, as will be discussed shortly.

Although the inverted-region effect was confirmed experimentally, the decrease in k_{et} with increasing energy gap was found to be smaller than that suggested by the classical Marcus formula (Fig. 4.15).

From a theoretical point of view, the classical description is not satisfactory in the inverted region, in which there is clearly not a normal activation barrier. The electron transfer rate theory can also be derived from time-dependent perturbation theory, which considers the transition between two (stationary) states to be induced by a (time-dependent) perturbation. This is a standard approach in spectroscopy, but it works also for other nonadiabatic processes, such as electron transfer with weak coupling.

In time-dependent perturbation theory, two electronic states A and B are considered.[40] Transitions occur from vibrational levels of A{Av} to those of B{Bw}, with conservation of total energy. The electronic coupling is V_{AB}. The transition probability per unit time is given by

$$W_{Av} = \frac{2\pi V_{AB}^2}{\hbar} \sum_{w} S_{Av,Bw}^2 \delta(\varepsilon_{Av} - \varepsilon_{Bw}), \tag{2.13}$$

where $S_{Av,Bw}$ is the vibrational overlap integral $< Av|Bw >$ and the delta function takes care of energy matching of the levels. Assuming a Boltzmann distribution for the levels {Av}, this leads to a rate

$$k = \frac{2\pi V_{AB}^2}{\hbar} (\overline{FC})$$

$$(\overline{FC}) = \frac{1}{Q_A} \sum_{v} \sum_{w} \exp\left(\frac{-\varepsilon_{Av}}{RT}\right) S_{Av,Bw}^2 \delta(\varepsilon_{Av} - \varepsilon_{Bw}). \tag{2.14}$$

$$Q_A = \sum_{v} \exp\left(\frac{-\varepsilon_{Av}}{RT}\right).$$

The symbol $(\overline{FC})$ is a thermally averaged Franck–Condon (vibrational overlap) factor.

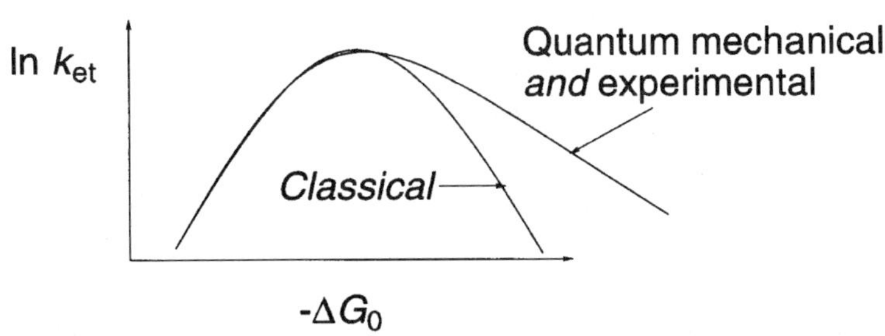

Figure 4.15. The decrease in k_{et} with increasing energy gap in the inverted region is not as dramatic as predicted by Marcus's classical Eq. 2.1.

Similar theoretical descriptions can be developed for spectroscopic transitions between vibronic states. The FC factor is based on the same vibrational wave functions for optical absorption,[41] spontaneous emission, or Raman scattering.

Obviously, a study of reaction rates will not provide insight into the vibrational wave functions that underlie the rate theory. Even a study of absorption or emission spectra cannot resolve the individual FC factors, because CT spectra are invariably broad and devoid of fine structure. It is possible, however, to make a useful connection between spectra and nonradiative recombination rates.[14]

In order to describe the experimentally found FC shape, it is sufficient to reduce the complete multimode picture to a model (Fig. 4.16) in which only one molecular vibrational quantum mode with an average frequency $\nu > k_B T/h$ needs to be considered, together with a classical low-frequency part that describes the reorganization energy λ_s of the solvent modes and low-frequency intramolecular modes. For the quantum mode, the harmonic oscillator model is applied. The vibrational overlap depends on the energy gap, the displacement of the curves (Δ), and the force constant, or equivalently, on the internal or vibrational reorganization energy λ_v and the dimensionless parameter $S = \lambda_v / h\nu_v$ (also known as the Huang–Ryss factor).

This results in the following expression:

$$k_{et} = \sqrt{\frac{\pi}{\hbar^2 \lambda_s k_B T}} V^2 \sum_{w=0}^{\infty} \frac{e^{-S} S^w}{w!} \exp\left[- \frac{(wh\nu_v + \Delta G_{et} + \lambda_s)^2}{4\lambda_s k_B T} \right] (S = \lambda_v / h\nu_v). \quad (2.15)$$

Equation 2.15 looks complicated at first sight and contains parameters that are no longer directly related to the actual vibrational potential-energy surfaces of the

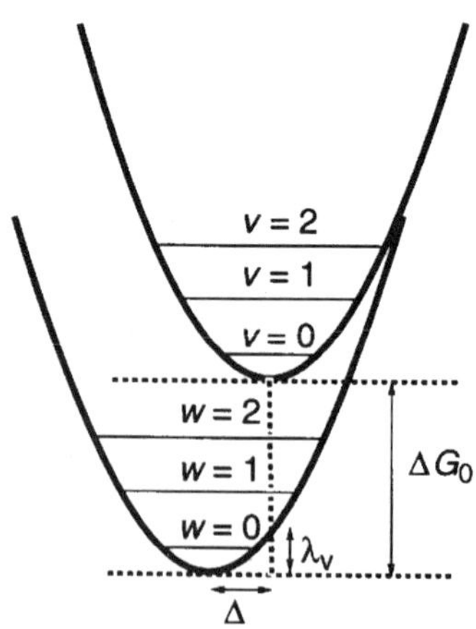

Figure 4.16. Electron transfer in the inverted region as described by a one-mode picture involves decay from $\nu = 0$ to different levels of the quantized vibrational mode in the ground state. This would result in large variations in k_{et} as a function of ΔG_0, showing peaks where energetic degeneracies occur with ground-state levels $w = n$. The "solvent" reorganization λ_s serves to broaden the distribution of occupied energy levels in such a way that the gaps between the peaks disappear and a smooth function results from Eq. 2.15 if λ_s is not too small.

multimode system. Nevertheless, it is a very useful expression, because it can be employed to rationalize many trends in electron transfer rates (e.g., the effects of changes in the energy gap, reorganization energies, and temperature).

Since the Franck–Condon factors that govern the kinetics and spectroscopy are the same, expressions similar to Eq. 2.15 can be derived for spectroscopic band shapes. For example, the corresponding expression for fluorescence intensity as a function of photon energy is shown in Eq. 2.16:[15]

$$I(\nu_{\rm f}) = \left[n \left(\frac{n^2+2}{3} \right)^2 \right] \left(\frac{16\pi^2}{3} \tilde{\nu}_{\rm f}^3 \right) \sqrt{\frac{\pi}{\hbar^2 \lambda_{\rm s} k_{\rm B} T}} M^2 \sum_{w=0}^{\infty} \frac{e^{-S} S^w}{w!} \times \exp\left[-\frac{(wh\nu_{\rm v} + \Delta G_{\rm et} + \lambda_{\rm s} + h\nu_{\rm f})^2}{4\lambda_{\rm s} k_{\rm B} T} \right] \quad (2.16)$$

Attempts have been made to obtain experimental information on the relevant energy surfaces from resonance Raman spectroscopy,[42,43] which provides data on the individual FC factors. This information can be used together with absorption and emission spectroscopic data to obtain a complete molecular picture of electron transfer.[44]

Closed-form expressions for the FC factor in Eqs. 2.15 and 2.16 have been derived in which the sum over vibrational levels is eliminated.[40] The relationship between spectral shapes and rates of electron transfer has been explored extensively in inorganic complexes.[45]

2.4.2 Experimental Observation of the "Inverted Region" in the Chemistry of Exciplexes and Radical Ion Pairs

The Marcus inverted-region effect is closely related to the energy gap law of radiationless transitions.[46,47] Studies of the rate of electron transfer versus the energy gap require a comparison of different systems in which the relevant parameters are systematically changed. In the case of charge transfer excited states, one can control the energy gap by altering the redox properties of the molecules involved or by changing the solvent polarity. Unfortunately, it is very hard to change only one parameter at a time. A unique aspect of electron transfer is the important role of reorganization energies. In comparing different molecules with different redox properties, one can try to minimize the differences in the internal reorganization energy and electronic coupling. In changing solvents, however, changes in the reorganization energy are inevitable. Note that the effect of changing the solvent polarity is different in the normal and inverted regions. In the former, changes in driving force and reorganization energy tend to compensate, but in the latter they are reinforcing: The decay rates of CT excited states increase strongly with an increase in the solvent polarity, because in more polar solvents the energy gap becomes smaller and the reorganization energy increases. The observation of the decay rates of the exciplex of pyrene and N,N-dimethylaniline in a range of solvents of different polarity[48] may have been one of the first published indications of the inverted-region effect, although it was not put into that context at the time.

Observation of the inverted-region effect in photoinduced charge separation processes is difficult. When the electronic coupling is sufficiently large, the maximum unimolecular rate of 10^{12}–10^{13} s^{-1} may be reached. The direct spectroscopic detection of such fast processes has been a technical problem until recently. In bimolecular processes, the maximum rate observed is limited by the diffusion process, typically on the order of 10^{10} M^{-1} s^{-1}. When the energy gap is such that the charge separation could be in the inverted region, several factors may still allow it to be as fast as a diffusion-limited process. For example, the reaction may occur at larger distances, where the driving force is smaller and the reorganization energy larger, or excited states of the ion pair may be formed.[49,50]

2.4.3 Charge Recombination in Geminate Pairs

In this chapter we focus on the exciplex, but one may also look at the fluorescence-quenching process in polar solvents with the production of (chemically reactive) free ions in mind. Early on,[51,52] it was realized that the yield of free ions was determined by the competition between separation of the geminate pair and unproductive charge recombination processes. The yield of free ions was found to be highest when the energy gap between the ion pair and the ground state was about 1.9 eV[52] and lower when the gap was smaller. This relationship was used as a measure of the rate of charge recombination processes in the Marcus inverted region.

The kinetic scheme (Fig. 4.17) for the formation and decay of exciplexes and the other ion-pair species is fairly complicated. Apart from the exciplex (contact RIP, CRIP) and free ions, the solvent-separated (or solvent-shared or loose) ion pair should also be taken into account as a kinetically distinguishable species. In polar solvents, the SSRIP may be the species formed primarily in fluorescence quenching. Furthermore, solvation of the exciplex and desolvation of the SSRIP has been shown to be a reversible process under certain conditions.[53,54]

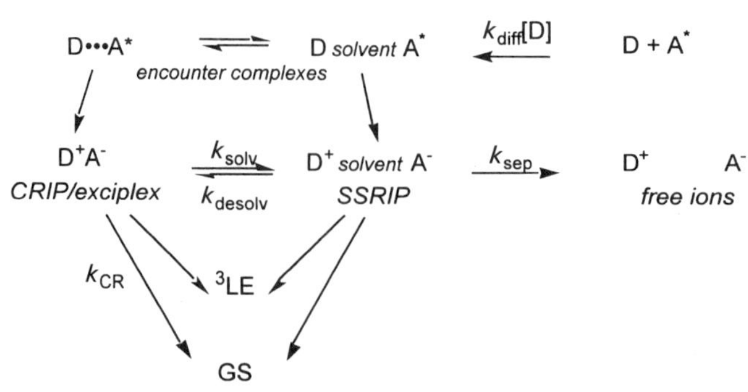

Figure 4.17. Transformations involved in the chemistry of photochemical formation and decay of radical ion pairs.

Clearly, the determination of all the rates in the scheme is a demanding task. Studies of the dependence of yields of fluorescence or of products on the chemical variables can be used to obtain information on rates, but assumptions are always required. For example, when the yield of free ions as a function of donor oxidation potential is taken to be a measure of the decay time of the corresponding geminate pairs, a constant separation rate has to be assumed. To obtain absolute rates, direct time-resolved experiments are required. The techniques that have been mostly applied are time-resolved fluorescence, absorption, and photoconductivity measurements. (Cf. Fig. 4.18.)

In the 1980s, Mataga and his co-workers applied picosecond time-resolved absorption spectroscopy to study the dynamics of geminate radical ion pairs formed from aromatic chromophores and a variety of electron acceptors and donors in acetonitrile. The decay of the radical ion absorption bands is biphasic, with a fast component due to the charge recombination and ion separation and a long-lived signal of the free ion pairs, which decay very slowly in a bimolecular process with second-order kinetics (Fig. 4.19). The relative amplitude of this contribution is a measure of the rate of ion separation k_{sep}. Separation rates of $5 \times 10^8\,s^{-1}$ to $2.5 \times 10^9\,s^{-1}$ in CH_3CN at room temperature were found, while Farid and co-workers always assumed a standard value[57,58] of $k_{sep} = 5 \times 10^8\,s^{-1}$ for reactions involving aromatic hydrocarbon donors and 9,10-dicyanoanthracene (DCA) or 2,6,9,10-tetracyanoanthracene (TCA) as acceptors.

Normally, charge recombination in exciplexes occurs in the inverted region. By using combinations of donors and acceptors with widely varying redox properties, Mataga and co-workers could also find some SSRIP cases in which the recombination process takes place in the normal region. This is possible due to the large reorganization energy associated with electron transfer over relatively long distances in a polar solvent. In that way, a large part of the bell-shaped Marcus curve, reproduced in Fig. 4.20, could actually be observed. The maximum rate of about $10^{12}\,s^{-1}$ was found to be near an energy gap $-\Delta G_0 \approx 1.5\,eV$.

At the end of the 1980s, two other groups addressed the problem of charge recombination in geminate pairs in acetonitrile and obtained quantitative data on Marcus's inverted region. In both cases, the energy gap was varied by using a series

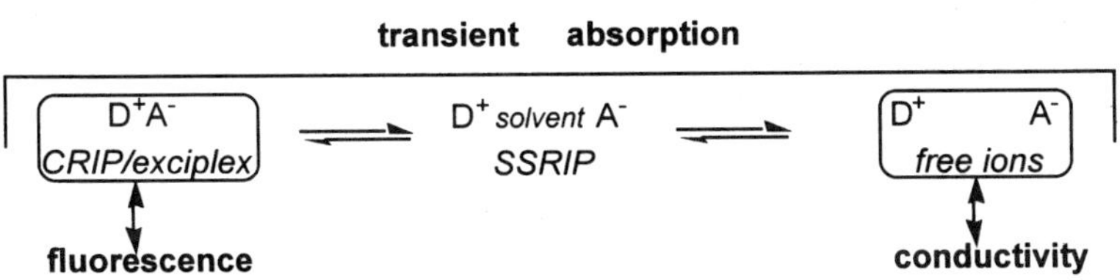

Figure 4.18. Commonly used methods for the detection of radical ion pair species in time-resolved experiments. All species have similar absorption spectra,[54,55] so transient absorption monitors the *total population*. Separate information on the exciplex population can be obtained from fluorescence, and the free-ion yield can be determined by using transient conductivity measurements or by monitoring the optical absorption on a relatively long timescale.

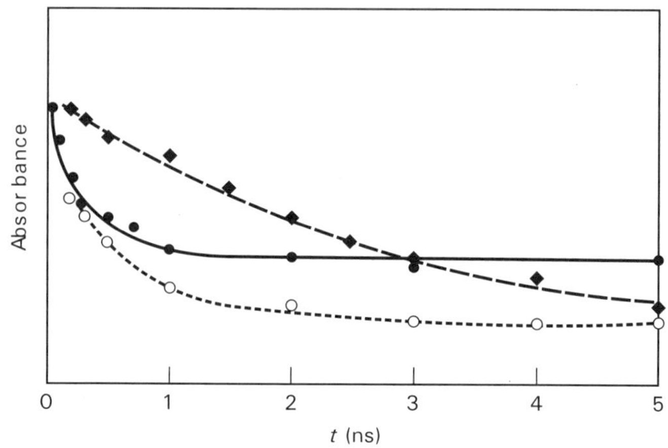

Figure 4.19. Decay of radical ion absorption bands following excitation of the tetracyanobenzene–toluene CT complex. Solvents are acetonitrile [•; rapid decay (0.3 ns), high yield of free ions (45%)], acetone [○; slower decay (0.45 ns), but lower ion yield (25%)] and 1,2-dichloroethane [◆; slow recombination of geminate pair (3.8 ns), no free ions formed] (Reproduced with permission from Ref. 56. Copyright International Union of Pure and Applied Chemistry.)

of (structurally related) electron donors interacting with only one or two different acceptors: 9,10-dicyanoanthracene and 2,6,9,10-tetracyanoanthracene. The recombination rates were determined indirectly by measuring the yield of free ions. Vauthey et al.[60a] used photoconductivity experiments for this purpose. Gould et al.[23,61] used a secondary donor to scavenge the free-radical cations that were formed and applied transient absorption spectroscopy to measure the yield of the radical cation of the scavenger. This technique avoids the need to know the absorption

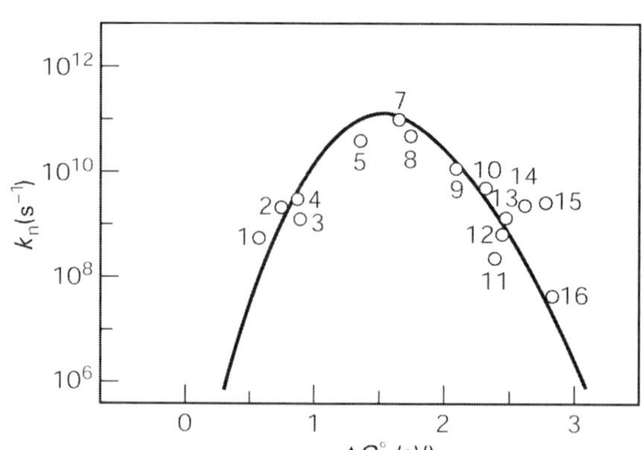

Figure 4.20. Rates of charge recombination versus driving force for geminate ion pairs formed by the quenching of excited aromatic hydrocarbons with aromatic donors and acceptors in acetonitrile. The points in the normal region ($-\Delta G_0 < 1$ eV) are for ion pairs involving the very strong acceptor tetracyanoethylene. (Reproduced with permission from Ref. 59. Copyright Elsevier Science Publishers)

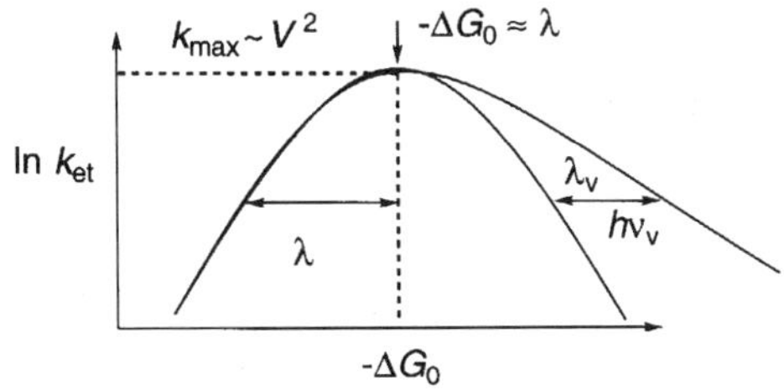

Figure 4.21. Although the curve of k_{et} versus ΔG_0 according to Eq. 2.15 depends on a complex interplay among all molecular parameters, some of the features are particularly sensitive to a subset of all parameters, as indicated.

coefficients of the radical cations of all the different donors used, but additional uncertainty may be introduced because the scavenging process has to be quantitative in all cases.

The analysis of the data by fitting to the mathematical form of Eq. 2.15 deserves some comment. Since the data are obtained by using a series of different D–A pairs, the molecular parameters are different for each data point. Therefore, a perfect fit to a single curve cannot be expected. In addition, there is some inevitable error in the experimental data. The shape of the theoretical curve contains four characteristic pieces of information, each of which is mostly related to one or two of the model parameters, as is schematically indicated in Fig. 4.21. The value of the maximum rate depends strongly on V^2, the energy at which the maximum occurs will be near $\lambda + \Delta G_0 = 0$. Considering that in the normal region the quantum effects are negligible, the width on that side of the curve depends on the total reorganization energy λ; the asymmetry (i.e., the weaker energy gap dependence in the inverted region) is sensitive to λ_v and $h\nu_v$.

Since almost all the experimental data points are in the inverted region, and rates greater than about $5\text{–}10 \times 10^{10}\ \text{s}^{-1}$ could not be determined, the model could not be fitted to the data unambiguously. Vauthey assumed that $\lambda_s = 0.97$ eV and fitted $V = 23\ \text{cm}^{-1}$, $h\nu_v = 2100\ \text{cm}^{-1}$ and $\lambda_v = 0.64$ eV. Gould and co-workers set $\lambda_v = 0.25$ eV (based on their measurement of the Stokes shift between absorption and emission maxima for some representative EDA complexes in cyclohexane) and $h\nu_v = 1500\ \text{cm}^{-1}$. They determined V and λ_s by fitting. The extensive data sets that were obtained permitted the separate determination of the parameters for different classes of electron donors. Gould et al. found that $V = 11.5\ \text{cm}^{-1}$ and $\lambda_s = 1.6$ eV for methyl benzenes, $V = 8.5\ \text{cm}^{-1}$ and $\lambda_s = 1.5$ eV for two-ring aromatic donors, and $V = 8.0\ \text{cm}^{-1}$ and $\lambda_s = 1.4$ eV for three-ring aromatic donors (Fig. 4.22). The small values of V and the large values of λ_s clearly indicate that the geminate pair is characterized by a relatively large distance between the ions, which implies that it is a solvent-separated ion pair, rather than a contact ion pair.[60b]

2.4.4 Charge Recombination in Contact Radical Ion Pairs

In the early time-resolved experiments, only the ion pairs that were formed upon quenching of a locally excited state were monitored. In most of these studies, a polar

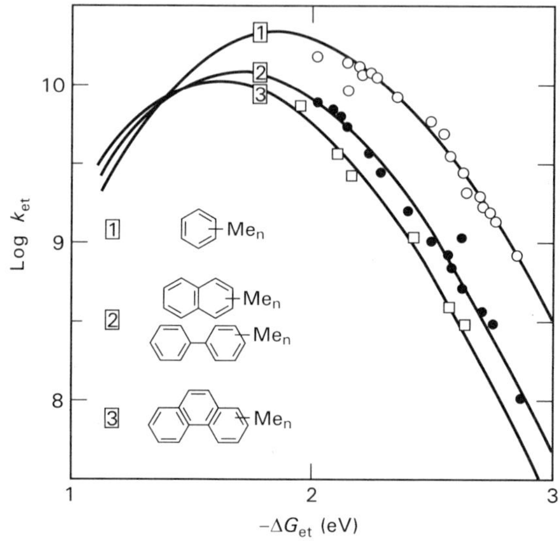

Figure 4.22. Rate constants for return electron transfer versus driving force in radical ion pairs of DCA and TCA with aromatic donors in acetonitrile. (Reproduced with permission from Ref. 14. Copyright American Chemical Society.)

solvent was used. It has become clear that the experimentally determined properties are those of *solvent-separated ion pairs*, as previously discussed. Study of the decay of *contact* ion pairs requires a different experimental approach. As mentioned earlier, the excitation of CT complexes offers a direct route to the preparation of contact radical ion pairs. The decay of some of these in polar solvents is very rapid, necessitating the use of subpicosecond techniques. Alternatively, exciplex fluorescence may be studied in solvents of intermediate polarity.

In 1989, Asahi and Mataga[62] studied the charge recombination process in CRIP by using subpicosecond transient absorption measurements. The energy gap dependence was found to be very different from that for geminate pairs in the same range of driving forces. For energy gaps $-\Delta G_0$ as small as 0.5 eV, decay times as short as 0.3 ps were observed, but no indication of a normal region could be found.[63]

In favorable cases, time-resolved fluorescence spectroscopy allows a straightforward measurement of the recombination rates of exciplexes on a picosecond-to-nanosecond timescale. A difficulty lies in determining the Franck–Condon parameters (i.e, the driving force) and reorganization energies. Gould, Farid, Young, and co-workers elegantly solved this problem by applying Eq. 2.16 for the band shape of the charge transfer emission spectrum.[14,64] From the fit to the experimental band shape, the Franck–Condon parameters were obtained, while the radiative rate constant contained information on the electronic coupling. In Fig. 4.23, we reproduce the result for cyanoanthracenes and alkyl benzenes. The CRIPs are characterized by a much larger electronic coupling $V = 750\,\text{cm}^{-1}$ and a much smaller reorganization energy $\lambda_s = 0.48$ eV. Note that, due to the small reorganization energy, even in acetonitrile, the normal region is predicted to show up at

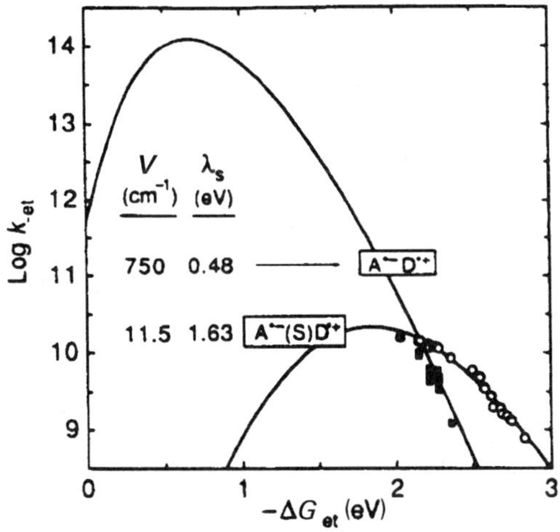

Figure 4.23. Rate constants for return electron transfer in contact radical ion pairs and solvent-separated radical ion pairs of DCA and TCA with alkylbenzenes. Curves are from Eq. 2.15 with $\lambda_v = 0.25$ eV and $h\nu_v = 1500\,\text{cm}^{-1}$. Reproduced with permission from Ref. 14. Copyright American Chemical Society.

$-\Delta G_0 < 0.6$ eV, with predicted rates exceeding $10^{13}\,\text{s}^{-1}$ because of the large electronic coupling.

So far, we have focused on radical ion pairs in the singlet state, but in fact, the Marcus inverted region for CRIP was first described by Levin et al. using exciplexes formed in the triplet state.[65] In this case, the electron transfer requires strong donors and acceptors, because the excitation energy is lower than in the singlet manifold. As a result, it is fairly easy to reach the normal region in charge recombination. The CR process is slow due to its spin forbiddenness and could be readily monitored by using transient absorption spectroscopy on a nanosecond timescale. The bell-shaped dependence of $\log(k_{et})$ on ΔG_0 in benzene and tetrachloroethane was fitted using λ_s of 0.4 and 0.5 eV values, and $V = 0.56\,\text{cm}^{-1}$ and $V = 0.34\,\text{cm}^{-1}$ in benzene and tetrachloroethane, respectively; in both solvents, $\lambda_v = 0.5$ eV and $h\nu_v = 2017\,\text{cm}^{-1}$ (Fig. 4.24). There is some scatter in the data, which is most likely due to variations in the electronic coupling V. In the slightly polar tetrachloroethane ($\varepsilon_s = 8.2$), the decay of the transient absorption signal is biexponential, which was attributed to the dissociation of the ion pair, occurring at a rate of $0.6 - 5.3 \times 10^5\,\text{s}^{-1}$. More recently, Arnold et al.[53,54] studied the decay of the singlet CRIP formed by excitation of the CT complex of tetracyanoanthracene with toluene in solvents of moderate polarity, using a combination of fluorescence and absorption spectroscopy. In this case, too, biexponential absorption decays were observed, but on a timescale of a few nanoseconds. After that, a constant absorption due to free ions remained. Also, the exciplex fluorescence decay was biexponential. All of the data could be rationalized, assuming that the CRIP dissociates into the SSRIP, but reversibly (see Fig. 4.17). This is one of the most direct experimental indications of the presence of two types of RIP present at the same time in a single system.

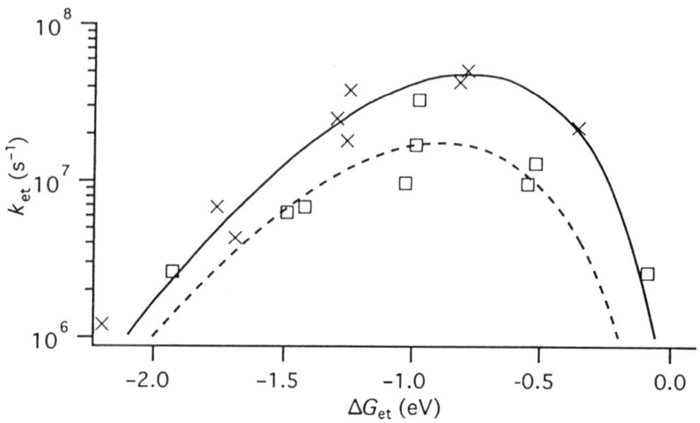

Figure 4.24. Rate versus driving force for triplet contact radical ion pairs in benzene (×, solid line) and tetrachloroethane (□, dashed line). Acceptors: 9,10-anthraquinone, 2-phenyl-1,4-benzoquinone, and *p*-fluoranil; donors: triphenylamine, 4-methoxy-*N*,*N*-dimethylaniline, and *N*,*N*,*N*′,*N*′-tetramethyl-*p*-phenylenediamine. (Data from Ref. 65.)

2.4.5 Physical Significance of the One-Mode Semiclassical Marcus Model

A word of caution concerning the physical significance of the parameters derived using Eqs. 2.15 and 2.16 is appropriate here. The model that is applied uses a single quantum mode, which accepts the excess energy in the radiationless decay process. This is illustrated in Fig. 4.25, in which we show how the band shape of a CT

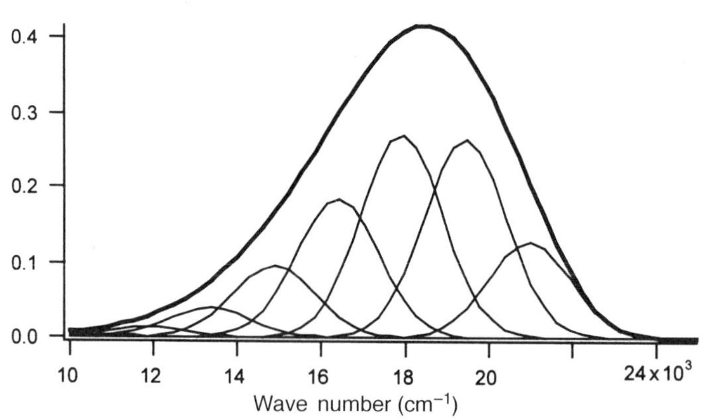

Figure 4.25. Reconstruction of a typical CT emission band according to Eq. 2.16 with

$$\Delta G = 23394\,\text{cm}^{-1}, \lambda_v = 3146\,\text{cm}^{-1}, h\nu_v = 1530\,\text{cm}^{-1} (S = 2.06),$$
$$\lambda_s = 2420\,\text{cm}^{-1}, \text{ and } k_B T = 200\,\text{cm}^{-1}.$$

The spectrum is a sum of Gaussians, each centered at $|wh\nu_v + \lambda_s + \Delta G_0|\,\text{cm}^{-1}$; the largest contribution occurs for $w = 2$; the structureless band shape is a result of the width of the individual Gaussian peaks, which is determined by λ_s.

fluorescence spectrum in the one-mode model actually arises. The FC pattern consists of a series of peaks, with a spacing $h\nu_v$, which are subject to a Gaussian broadening determined by λ_s. As long as λ_s is not chosen too small, this procedure models the overall band shape correctly, and the same procedure with the same parameters also describes the dependence of the electron transfer rate on the thermodynamic parameters. The physical reality, however, involves a number of active vibrations, which leads to progressions with different frequencies that require a smaller λ_s to fill in the gaps. The concepts of an "average vibrational energy" $h\nu_v$ and "solvent reorganization energy" λ_s are highly misleading. Both are artifical constructs. Because the experimental data barely permit the determination of the four parameters of the one-mode model, they certainly do not allow the determination of the unknown number of parameters of the full multimodal picture. An approach to obtain such data has been undertaken using resonance Raman spectra (Section 3).

2.4.6 An Application of the Effects of the Energy Gap: Generation of Reactive Radical Ions Using Cosensitization Methods

The insight gained from studies of the dependence of ion-pair recombination rates on the energy gap offers a useful rationale for optimizing the yields of reactive free-radical ions by photoinduced charge separation. It was already known in the early 1980s that the addition of a "cosensitizer" improved the efficiency of photochemical reactions initiated by a photoexcited electron acceptor.[66,67] This cosensitizer (C) should be an electron donor that can only just be oxidized by the photoexcited acceptor A*. In that way, the energy gap of the contact radical ion pair with the ground state is maximized and the rate of charge recombination minimized, allowing the efficient separation of the ion pair into free ions. The radical cation of the cosensitizer then has sufficient time to oxidize the substrate S, because the bimolecular charge recombination process of the free ions is very slow. The reactions are as follows:

Direct Electron Transfer

$$A^* + S \longrightarrow S^+A^- \xrightarrow{\text{Recombination}} A + S$$

Cosensitization

$$A^* + C \longrightarrow C^+A^- \xrightarrow{\text{Separation}} C^+ + A^-$$

$$C^+ + S \rightleftharpoons C + S^+$$

$$S^+ \longrightarrow \text{Products}$$

If the substrate would be oxidized directly, an ion pair would be formed that has a relatively small energy gap, which results in rapid recombination rather than the desired chemical reaction. A typical cosensitizer is biphenyl, used in combination with cyanoaromatic compounds. By using this scheme, it is even possible to perform endothermic electron transfer processes, such as the photooxygenation of epoxides. These substrates are unreactive if the cosensitizer is not present, because the endothermic electron transfer to singlet excited dicyanoanthracene does not occur to any significant extent within the nanoseconds lifetime of the excited electron acceptor. Once the biphenyl radical cation is formed, a much longer timescale becomes available for the reversible electron transfer to this cosensitizer, and the reaction goes to completion because the substrate radical cation reacts irreversibly with oxygen:[67]

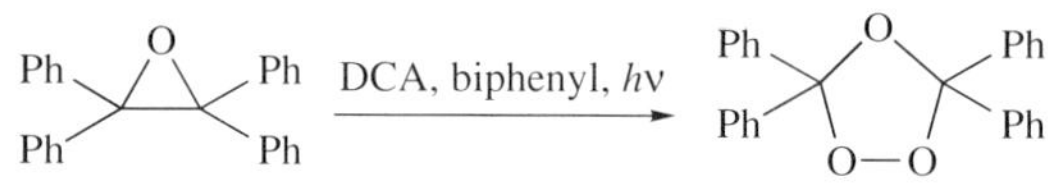

In one of the other possible cosensitization schemes, the cosensitizer is the initially excited chromophore. An example is the electron transfer reactions of furans and alkenes with dicyanobenzene, which are efficiently initiated by using phenanthrene, but not by using polycyclic aromatics with a lower oxidation potential.[66]

3 Electronic and Geometric Structure of Exciplexes

3.1 Basic Considerations

Many exciplexes are formed from two aromatic molecules, and it is commonly believed that such exciplexes adopt a sandwich structure in which the two aromatic units are oriented in parallel planes about 3.5 Å apart. Although this configuration is entirely reasonable, there is not much direct experimental evidence either for or against it. Moreover, even in the sandwich orientation, there are still rotational and translational degrees of freedom. In this section, we shall discuss some of the experimental and computational approaches to the problem of exciplex geometry.

The *electronic structure* of exciplexes is usually discussed as a simpler problem, in which only few degrees of freedom are taken into account. Basically, the exciplex is a species described by a dominant electronic configuration, viz., the charge-transfer configuration D^+A^-. The spin state may be either singlet or triplet. The ground-state configuration DA and the locally excited configurations D^*A and DA^* are mixed into the wave function with smaller coefficients. Other electronic configurations are ignored. The extent of mixing is determined by the electronic couplings between the configurations and their relative energies (Fig. 4.26).

Because the polar D^+A^- configuration has a much larger dipole moment than the other configurations, the relative energies of the configurations change with the polarity of the medium. As a result, the configuration mixing that determines the final exciplex wave function should also depend on the medium. This topic has been extensively investigated in the past several decades.

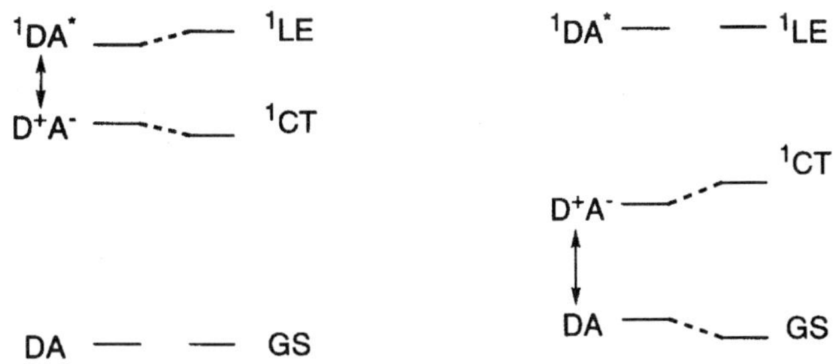

Figure 4.26. Dominant configuration mixing in exciplexes (left) and CT complexes (right) for a system in which D* is high in energy. An extension of the model involves the mixing of all three configurations.

3.2 Electronic Structure of Exciplexes

Weller and co-workers classified dipolar excited-state complexes as either exciplexes or heteroexcimers.[68–70] The former are characterized by essentially complete transfer of charge. Their energies (from thermodynamic studies)[29] as well as their emission energies[32] measured in nonpolar alkane solvents, show a direct correlation with the redox properties of the donor and acceptors involved.

$$E_{00} - \Delta H = e(E_{\mathrm{D}}^{\mathrm{ox}} - E_{\mathrm{A}}^{\mathrm{red}})_{\mathrm{polar}} + 0.15\ \mathrm{eV};$$

$$h\bar{\nu}_{\mathrm{exciplex}} = e(E_{\mathrm{D}}^{\mathrm{ox}} - E_{\mathrm{A}}^{\mathrm{red}})_{\mathrm{polar}} - 0.15\ \mathrm{eV}.$$

According to Weller, heteroexcimers, in which the wave function contains considerable contributions from DA* or D*A, can be recognized by the fact that their emission energy in *n*-hexane is lower than expected for a pure exciplex (Cf. Fig. 4.11).

It has long been known that the intensity of exciplex fluorescence decreases with increasing solvent polarity. Until recently, there was considerable controversy concerning the explanation of this phenomenon. In 1967, Mataga, Okada, and Yamamoto[48] described their study of the exciplex of pyrene and *N,N*-dimethylaniline by means of time-resolved and static fluorescence spectroscopy. In a series of solvents ranging from the nonpolar *n*-hexane to moderately polar solvents such as 1,2-dichloroethane and pyridine, the fluorescence quantum yield was shown to decrease strongly with increasing solvent polarity. The decay time also decreased, but much less than the fluorescence quantum yield. It was concluded that the electronic transition dipole moment diminishes substantially with increasing solvent polarity. Mataga et al. argued that this was due to the change in the electronic wave function of the exciplex with solvent polarity: In solvents of low polarity, the contribution of the locally excited pyrene configuration is larger than in solvents of higher polarity, where the wave function becomes more purely ionic. Because the transition dipole moment between the CT configuration and the ground state is small, intensity borrowing from the pyrene transition plays an important role.

In the same year, Knibbe et al. proposed an entirely different explanation. Based on the early finding[10] that in polar solvents electron transfer quenching takes place at a distance of about 7 Å, they argued that the quenching process leads directly to solvent-separated ion pairs, which reduces the yield of exciplex, so that there is obviously less exciplex fluorescence.[71]

In order to resolve the controversy, more insight was needed into both aspects: ion pair dynamics and exciplex electronic structure. As discussed in the preceding section, Gould, Farid, and co-workers showed that (at least for the class of compounds they studied), electron transfer quenching leads directly to exciplexes/CRIP, even in polar solvents.[8] This is because the charge separation reaction, which occurs in the Marcus normal region, is favored by large electronic coupling and small reorganization energy. Direct formation of an SSRIP by electron transfer over larger distances is disfavored by the smaller electronic coupling and larger reorganization energy; therefore, in most cases the reactants diffuse more closely together prior to electron transfer. Once it was established that the quantum yield of exciplex formation is essentially 100%, the radiative rate constants of exciplex emission could be determined from the fluorescence quantum yields and decay times. All the data from a large number of combinations of alkyl benzene donors with DCA (9,10-dicyanoanthracene) or TCA (2,6,9,10-tetracyanoanthracene) in a number of solvents follow the same trend, as shown in Fig. 4.27. The lower the average emission energy (determined by the redox energies and the solvent polarity), the lower is the radiative rate constant.

There appears to be a smooth transition from exciplex/CRIP species, which have a relatively small transition probability ($k'_f < 1 \times 10^7\,s^{-1}$), to the nearly purely locally excited acceptors in fluorobenzene and toluene, which have $k'_f \approx 2.6 \times 10^7\,s^{-1}$. From the radiative rate constants thus obtained, the electronic

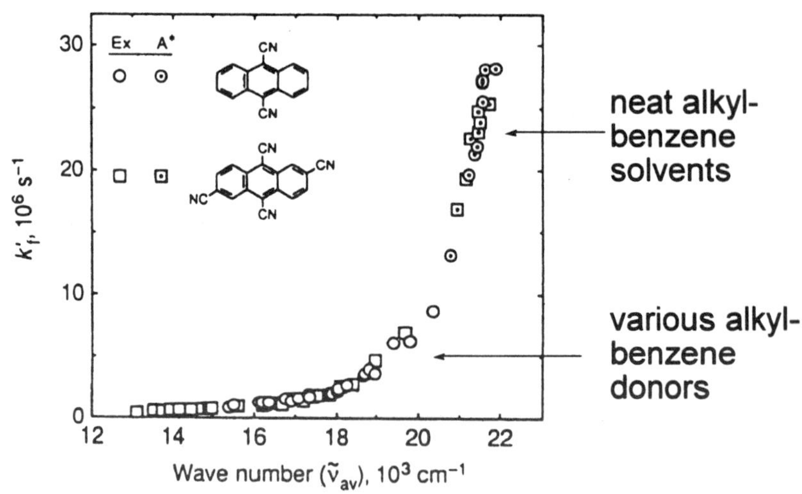

Figure 4.27. Plot of the corrected radiative rate constant $k'_f = 9k_f n^{-1}(n^2 + 2)^{-2}$ against the average emission energy of exciplexes of 9,10-dicyanoanthracene (DCA) and 2,6,9,10-tetracyanoanthracene (TCA), with various alkyl benzene donors in various solvents (open symbols) and DCA and TCA in alkylbenzene solvents. (Reproduced with permission from Ref. 8. Copyright American Chemical Society.)

transition dipole moments could be calculated. Just like the radiative rate constants, the transition dipole moments also increase smoothly with increasing average emission energy. This relationship is qualitatively easily understood as a result of the increased mixing of CT and LE configurations, which allows an increased intensity borrowing. A quantitative analysis based on a three-state mixing model allowed the authors to derive the percentage of charge transfer in each of the emissive CRIPs studied. Figure 4.28 shows the result for TCA exciplexes.

The figure clearly shows that the exciplex wave function in nonpolar solvents may have considerable contributions from nonpolar configurations. In such cases, the dipole moment of the exciplex depends significantly on the polarity of the medium. Estimates based on the Lippert–Mataga analysis (Section 2) for exciplexes with a high extent of charge separation typically give dipole moments of about 13 D. A clear drawback of the Lippert–Mataga analysis is that dipole moments can be derived only when assumptions are made about the shape and size of the cavity. Fortunately, other techniques are available that provide information on the dipole moments of excited states. Early measurements of exciplex dipole moments by electro-optical methods yielded a value of 9.8 D for DCA/HMB in *n*-hexane.[72] Since the amount of CT in DCA/HMB in alkanes was found to be only 83% on the basis of the analysis of the fluorescence data,[26] the electro-optical results are in good agreement with the typical value of 13 D corresponding to full charge transfer. The electro-optical experiments also provide information on the relative orientations of the dipole moment and the electronic transition moment. Groenen and van Velzen showed that, assuming a sandwich-type orientation, the transition dipole moment for

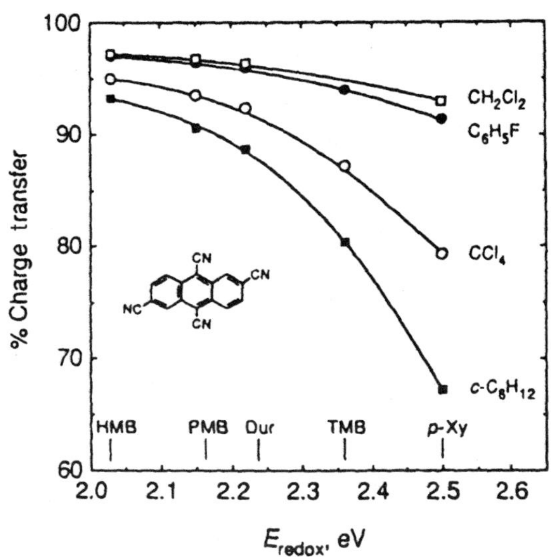

Figure 4.28. Percentage charge transfer in the electronic wave function of the CRIP of TCA with hexamethylbenzene (HMB), pentamethylbenzene (PMB), durene (Dur), 1,2,4-trimethylbenzene (TMB), and *p*-xylene (*p*-Xy). (Reproduced with permission from Ref. 26. Copyright American Chemical Society.)

exciplex emission is orthogonal to the dipole moment (i.e., in the plane of the rings, just like the LE emission). For the CT absorption in EDA complexes, in which mixing of the DA and D^+A^- configurations occurs, the transition dipole is along the z-axis (i.e., parallel to the dipole moment).

More recently, Smirnov and Braun applied their transient DC conductivity technique to exciplexes. The analysis of the data does not require assumptions about the shape of the cavity. For exciplexes of DCA and TCA with methylbenzenes in benzene, they found dipole moments of less than 11 D.[73] This maximum value was reached for relatively strong donors, for which the contribution of locally excited configurations to the exciplex wave function is negligible. The dipole moment expected for a complete separation of two units of charge across 3.5 Å, the typical sandwich exciplex distance, is 16.8 D. According to Mylon et al., the smaller dipole moment must be due to the mutual electrostatic polarization of the ions. Although the actual donor–acceptor distance in exciplexes is not known, it is highly unlikely that it could be as small as 2.3 Å, the distance of two unit charges forming a dipole of 11 D.

As a last example of a recent study of exciplex electronic structure, we mention the work of Chow and Johansson, who investigated the complexes of (dibenzoylmethanato)boron difluoride with benzene derivatives.[74,75] Under conditions where a full charge transfer is thermodynamically impossible, these researchers found exciplex-type emission, which they attributed to species with partial charge transfer. A smooth transition from such "heteroexcimers" to full charge transfer exciplexes occurs as the donor oxidation potential is reduced.

3.3 Singlet versus Triplet Exciplexes

Because the exciplex wave function is dominated by the biradicaloid D^+A^- contribution, one may, in general, expect that the energy difference between the singlet and triplet configurations is small: $E(^3D^+A^-) \approx E(^1D^+A^-)$. As we have seen, exciplexes formed by the quenching of singlet excited precursors have relatively short lifetimes—at most about 100 ns in solutions and often much shorter. In triplet RIP species, on the other hand, decay to the ground state is spin forbidden and may be expected to be considerably slower.

Interconversion between the singlet and triplet RIPs is usually not very fast. The most important mechanism that can induce the necessary spin inversion is hyperfine interaction of the unpaired (or weakly paired) electrons with magnetic nuclei, mostly ^1H. Singlet RIPs are usually too short-lived for intersystem crossing to the triplet RIP to be important. Note that intersystem crossing to a locally excited triplet state may be more rapid because it can be promoted by spin–orbit coupling, which is not the case in the interconversion of singlet and triplet ion pairs, because they have essentially the same spatial orbital occupation.

The formation of triplet RIPs in solution can be accomplished by starting from a triplet precursor excited state. To prevent interception of the singlet state, the primarily excited chromophore should undergo rapid intersystem crossing, and the concentration of the quencher should not be too high. Aromatic ketones such as benzophenone and quinones (e.g., the water-soluble anthraquinone sulfonate) are frequently used as electron acceptors in this approach, which proceeds as follows

$$^{1}A^* \rightarrow {}^{3}A^*;$$

$$^{3}A^* + D \rightarrow {}^{3}D^{+}A^{-}.$$

The triplet radical pairs thus formed are much longer lived than the corresponding singlet RIPs and can undergo diffusive separation and chemical reactions more easily. In addition to the cosensitization method (Section 2), this triplet route is one of the practically useful and general approaches to redox chemistry by photoinduced electron transfer.

Although the energy gap between singlet and triplet exciplexes is typically very small, significant energy differences may arise if the mixing with energetically nearby states of the same spin multiplicity is very different for singlet and triplet states.[69] This situation is schematically illustrated in Fig. 4.29. For example, when the energy of the RIP configuration is between the singlet and triplet locally excited configurations, the energy of the triplet RIP may be raised and that of the singlet exciplex lowered. When the exciplex state is of relatively low energy, stabilization of the triplet exciplex may result from the interaction of $^{3}D^{+}A^{-}$ with the ^{3}LE configuration, while the singlet exciplex may be raised in energy because the $^{1}D^{+}A^{-}$ configuration interacts more strongly with the ground-state configuration than with the ^{1}LE configuration.

Detailed information on the singlet–triplet splitting may be derived from the effects of magnetic fields.[58] The application of a magnetic field leads to splitting of the three levels of the triplet state, proportionally to the strength of the field. By varying the field strength, resonances may be created or destroyed between the triplet sublevels and the singlet state, which influences the decay kinetics.

Just as singlet exciplexes often emit fluorescence, triplet exciplexes can phosphoresce. Since phosporescence is typically visible only in low-temperature matrices, in which diffusive quenching processes are excluded, the triplet exciplex emission can be detected only by the excitation of a ground-state CT complex.[69,76]

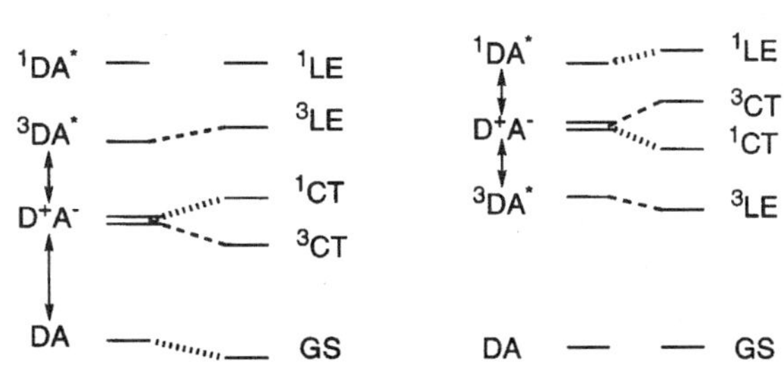

Figure 4.29. Conditions under which a significant singlet–triplet splitting may arise in CT states. Left: EDA complex (dominant mixing DA–$D^{+}A^{-}$); right: exciplex. The singlet–triplet splittings are strongly exaggerated compared with the splittings between ground and excited states.

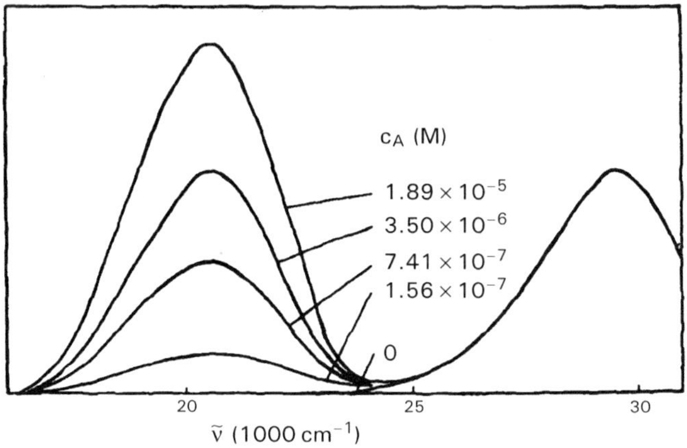

Figure 4.30. Emission observed from deaerated *N,N*-diethylaniline (1.26×10^{-4} M) in methylcyclohexane at room temperature, with low concentrations of dicyanobenzene as indicated. (Reproduced with permission from Ref. 70.)

Triplet exciplexes have also been studied using EPR spectroscopy.[77] The D parameter that is obtained, which is a measure of the average distance of the unpaired electrons, is about 0.05 cm^{-1} (i.e., roughly four times smaller than that of a triplet state of an aromatic molecule).[69] Accurate quantitative estimates of the distances, however, cannot be made from this result.

The slow process $^3D^+A^- \rightleftharpoons {}^1D^+A^-$ may give rise to a delayed CT fluorescence. The interaction of photoexcited *N,N*-diethylaniline (DEA) with 1,4-dicyanobenzene (DCB) in low concentrations provides a nice example of the importance of the timescale in photophysical processes in solution. Singlet excited DEA decays with a time constant less than 2 ns, producing the locally excited triplet state in a substantial yield. With concentrations of DCB on the order of 10^{-6} M, diffusion-controlled quenching of singlet excited DEA is not effective. Nevertheless, in addition to the normal DEA fluorescence, exciplex fluorescence is observed (Fig. 4.30)! The explanation is that triplet DEA, which decays only slowly, is intercepted by DCB, to yield the triplet CT species. Phosphorescence in liquid solution is not detectable, but slow, slightly endergonic intersystem crossing within the exciplex yields the singlet exciplex, which fluoresces!

3.4 Exciplex Geometry

As stated earlier, it is commonly accepted that exciplexes have a sandwichlike geometry, with an interplanar distance of about 3.5 Å. Because the two fragments are oppositely charged, they undergo a strong electrostatic attraction, which is counteracted by electron repulsion at small distances. The two forces should be equal roughly at a distance corresponding to the sum of the van der Waals radii. Structural information on exciplexes can be derived only from spectroscopic observations and computation. The ground-state counterparts, CT complexes, on the other hand, have been extensively investigated in the solid state, by means of diffraction methods. An

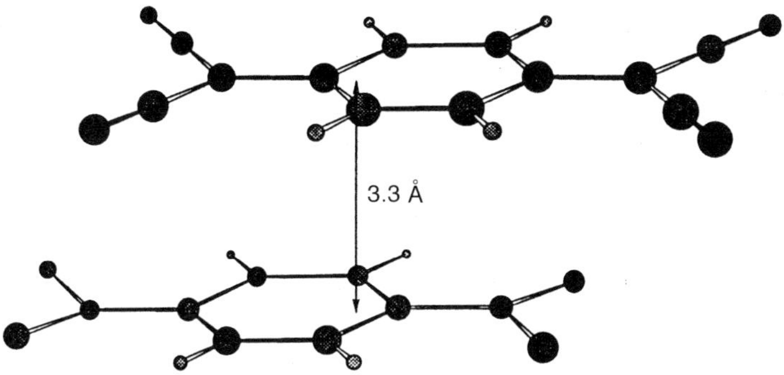

Figure 4.31. Crystal structure of the complex of *N,N,N′,N′*-tetramethyl-*p*-phenylenediamine and tetracyanoquinodimethane.[78]

example is shown in Fig. 4.31. It should be kept in mind that CT complexes are in essence ground-state species, with DA as the dominant electronic configuration. Thus, the strong electrostatic attraction that exists in the excited state is not reflected in the ground-state structure. Since D and A are only loosely bound, the structure of the CT complex may well change in going from the solid phase to solutions or to the gas phase.[6]

A link between the ground-state potential-energy surface and that of the excited states can be made by using resonance Raman spectroscopy. The intensities of the bands in a Raman spectrum measured in resonance with the CT transition are related to the changes in geometry on the excited-state surface along the ground-state normal modes.[42,43] Under the assumption that the potential-energy surfaces are harmonic, mode-specific internal reorganization energies can be derived. This has been done, for example, for the complex of hexamethylbenzene with tetracyanoethylene. In addition to the Raman data, the CT absorption and emission spectra were taken into account in an attempt to obtain detailed insight into the Franck–Condon factors. Unfortunately, the total internal reorganization energy that followed from the Raman intensity analysis was too small to account for the bandwidth of the absorption and emission spectra,[79–81] and a good agreement between all the data could be obtained only by assuming a large solvent reorganization energy. The Raman intensities contain information only on short-time dynamics. In some cases (e.g., a complex of 1,2,4,5-tetracyanobenzene and toluene), distinct structural rearrangements of the CT complex after excitation have been inferred from changes of the transient absorption and fluorescence spectra.[64,82]

Resonance Raman spectroscopy can also be used as a *transient* technique, to observe the relaxed exciplexes or RIPs in solution. Tahara and Hamaguchi studied complexes of triplet chloranil with alkylbenzenes.[83] When the oxidation potential of the benzene derivative is too high, electron transfer to form a RIP cannot occur, and the complexes are much like electron donor–acceptor complexes in the singlet ground state. Some of the Raman bands showed a change in frequency that correlates with the donor oxidation potential, indicating a variable amount of charge transfer character in the complex. When the benzene donor had a sufficiently low oxidation potential, a radical ion pair was formed, but all the pairs had the same spectra. Unfortunately, in this study only bands due to the electron acceptor could be observed. Vauthey and co-workers studied the geminate ion pair obtained by electron transfer from triplet anthraquinone and 1,2,4-trimethoxybenzene in tetrachloroethane.[84] Interestingly, they could resolve a shift in one of the Raman bands of the donor radical cation with time, which they interpreted as being due to separation of the geminate pair into free ions.

3.5 Exciplex Formation in Supersonic Jets

Vibrational spectroscopy provides more detailed information than electronic absorption or excitation spectroscopy in solution only because, in the latter, many more vibronic transitions are allowed that, together with inhomogeneous broadening, tend to lead to a broad overall spectrum. In supersonic jet spectroscopy, the inhomogeneous broadening is avoided. Moreover, the low temperature of the molecules in the jet favors the formation of ground-state complexes, which in turn allows high-resolution excitation spectroscopy in which van der Waals complexes are excited and the emission from their locally excited or charge transfer state can be monitored.[85,86]

The spectroscopy of molecules in supersonic jets has some characteristics that are very different from those in solution or low-temperature matrices. Because of the absence of collisions, the molecules cannot exchange energy. Relaxation processes following excitation of a molecule to a particular vibronic level consist primarily of a redistribution of vibrational energy over different modes. Return to the ground state occurs by fluorescence or by internal conversion from a vibronic level in the excited state to an isoenergetic level in the ground state. The fluorescence spectra are different for each vibronic level. Fluorescence excitation spectra can be very highly resolved, indeed, in many cases the line widths even permit the resolution of the rotational structure. As the excitation energy is progressively increased from the 0_0^0 level, the vibronic bands become congested, and the individual bands broaden because the lifetimes of the vibrational states decrease. Therefore, the spectral range in which highly resolved bands are observed is limited. In the case of van der Waals complexes, in which we are interested here, the spectra are complicated by the presence of different isomeric forms (to be discussed in more detail later) and the free molecules. It is possible to unravel some of the complexity in the experimental data by a hole-burning depletion technique[87] that allows one to identify bands that belong to a single species. The principle is simple: When one specific transition, which belongs to one species, is saturated using a pump laser, the bands that belong to the same species are no longer observed in the excitation spectrum. Saturation is very easy to achieve, because there are few molecules in the sample and the excited molecules do not return to the relaxed ground state anyway.

The feasibility of exciplex formation under solvent-free conditions depends on the ionization potential of the electron donor, the electron affinity of the acceptor, and the electrostatic attraction energy according to the equation

$$E_{\mathrm{CT}} = IP_{\mathrm{D}} - EA_{\mathrm{A}} + \frac{q_1 q_2}{R_{\mathrm{DA}}}. \tag{3.1}$$

The Coulomb term can be very important: For two unit charges at 3.5 Å in vacuum, it amounts to −4.1 eV. The spreading of the charges on the fragments, however, tends to reduce the Coulomb attraction. Moreover, since the CT energy is bound to be close to that of low-lying LE states, a mixed character of the exciplex wave function, with reduced charge transfer, is quite likely. Experimental data for the adiabatic ionization potentials and electron affinities of molecules are not always available. Vertical ionization potentials have been determined for many molecules, but they may be misleading because they overestimate the energy of the relaxed radical cation. Since we have seen that the energies of exciplexes in nonpolar solvents are not much higher than those in a polar solvent, we may expect that donor–acceptor pairs, which easily form exciplexes in nonpolar solvents, can also undergo charge separation under supersonic jet conditions.

When a very strong donor–acceptor pair is used (e.g., tetracyanoethylene/*p*-xylene),[88] a charge transfer transition may be the lowest energy transition. In such a case, the excitation profile of exciplex emission shows only a broad band, much like that in solution. This is because there are many active vibronic transitions and the individual bands are broadened due to the short lifetime of the vibronic states.

In most exciplex-forming systems, of which anthracene-dialkylanilines may once again serve as a prototype example, direct optical transitions from the ground state to the charge transfer state are not observed.[89,90] Excitation populates vibronic levels of the anthracene chromophore, and at low energies the spectral characteristics (vibrational progression in absorption and emission, and decay times) are essentially the same as those of anthracene. The 1:1 van der Waals complex is recognized by the fact that it has a different origin (0_0^0 excitation energy), and it is characterized in a separate experiment by its mass. Only when a certain excess energy is present can electron transfer occur, and exciplex emission is then observed. Because the transition is from a substantially excited vibrational level in the excited state to a correspondingly high density of ground-state levels, the spectrum is redshifted and broad. The width of the exciplex fluorescence band (ca. 3000 cm^{-1}) is almost the same as in nonpolar solutions.

The species excited in these experiments, the van der Waals complexes, may exist in several isomeric ground-state forms. A modeling approach commonly used in this field will be discussed shortly. In the case of anthracene/dimethylaniline, one isomer shows a behavior that is intermediate between that of CT complexes and the case discussed previously. Resolved excitation bands can be detected, but they have a width of 100–200 cm^{-1}. The potential-energy curves are as shown in Fig. 4.32 (right). The minimum corresponding to the LE state has (almost) disappeared, because the zero-order curves cross close to the LE minimum. Also, the electronic coupling must be strong for these species that are in close contact. Along coordinates other than the electron transfer coordinate, there is still a well-defined potential-energy well, so that distinct vibrational levels corresponding to the normal modes exist. The lifetimes of the vibronic states, however, become very short. The observed bandwidth of about 150 cm^{-1} corresponds to a decay time of approximately 40 fs. However, if the bands are also inhomogeneously broadened, the actual vibronic lifetimes may be longer.

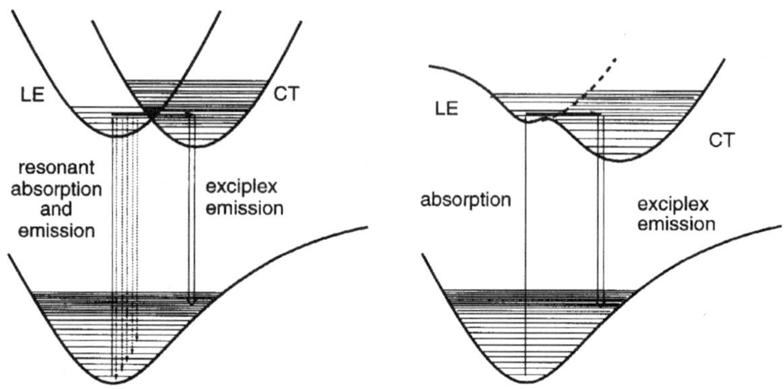

Figure 4.32. Exciplex formation upon excitation of a donor–acceptor van der Waals complex in a supersonic jet. Left: a barrier exists, and resonance fluorescence is observed at low E_{exc}. Right: barrierless electron transfer, with broadened absorption features; only exciplex emission is observed.

Interpretating the experimental data requires a reliable model to describe the three-dimensional structure of the ground and excited states of the complexes. The former could be approached using high-level quantum chemical methodology, but the potential-energy surface is shallow and complicated, with numerous minima separated by low barriers. A thorough exploration of the energy surface is currently not possible with the use of computationally intensive methods. Instead, a semi-empirical approach has been employed in which the interaction between the two parts of the complex is reduced to a set of pairwise interactions.[91–93] For the dispersion and short-range repulsion energy, analytical formulas are used. Standard atom–atom parameters are used, but the van der Waals radii are adjusted according to the charge on the atom. The electrostatic and polarization terms are based on multipole expansions. The quantum chemically computed wave function is used to derive monopole, dipole, and quadrupole moments at each atom and for each bond. When charge transfer interactions are important, they can be included in the model.[94] The geometries of the molecular subunits are held fixed during the exploration of the potential-energy surface defined by the total intermolecular interaction function. Because of the relatively small energy differences involved and the inevitable importance of steric interactions, the inability of the subunits to adjust their geometies is probably the most serious shortcoming of the method. For the van der Waals complex of anthracene and dimethylaniline, Tramer et al. found two conformations (Fig. 4.33), one of which shows resonance fluorescence ("*R*-conformation"), while the other directly decays upon excitation to the exciplex ("*E*-conformation")[90]

For the complexes of anthracene with other dialkylanilines, several *R*-conformations were found. The proper description of the excited-state energy surface is a difficult task. For the locally excited state, a configuration interaction treatment of anthracene allows the derivation of the multipole part of the interaction potential. Changes in polarizability were neglected. The resulting optimized complexes had structures that were similar to those of the neutral ground state. The exciplex state was

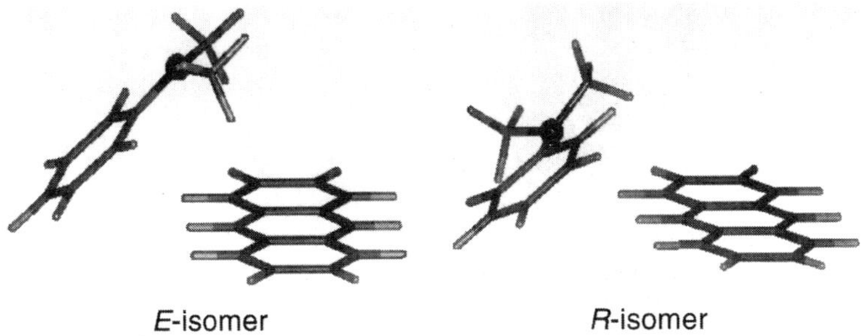

Figure 4.33. Ground-state isomeric forms of the van der Waals complex of anthracene with *N,N*-dimethylaniline. The model building was based on a thorough search of the potential-energy surface spanned by the intermolecular coordinates only. The energy function was based on pairwise interaction potentials, including dispersion, short-range repulsion, and electrostatic and polarization contributions ("Claverie method"[92]). (Reproduced with permission from Ref. 89. Copyright American Chemical Society.)

described assuming full transfer of charge (i.e., the interaction potentials were those of anthracene radical anion and dimethylaniline radical cation). In this case, a planar radical cation geometry was used.[95] On the potential-energy surface of the CT state, a single minimum was found: a sandwich configuration with an interplanar distance of 3.25 Å (Fig. 4.34). A comparison of the relative energies of the two states with the experimental emission energies revealed that the exciplex was only 1600 cm^{-1} too high in energy. The computed energy surfaces are represented in Fig. 4.34. In view of the large value of the electronic coupling for the close-contact complex, estimated at

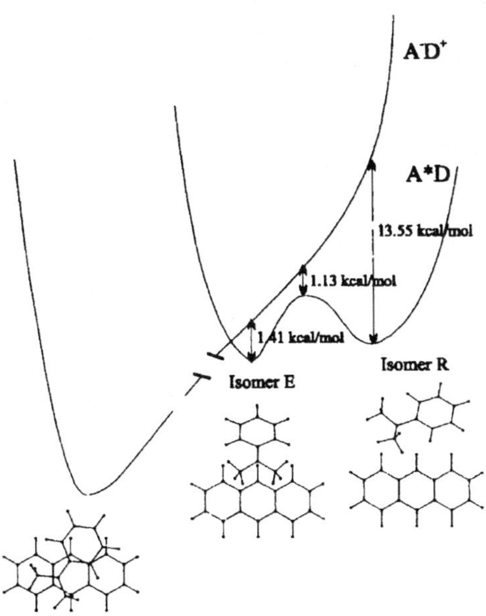

Figure 4.34. Energy curves derived for the complex of anthracene with *N,N*-dimethylaniline in the LE and CT states. (Reproduced with permission from Ref. 90. Copyright American Chemical Society.)

about $800\,cm^{-1}$, in good agreement with results from exciplex fluorescence in solution, the barrier between the *E*-isomer and the exciplex is meaningless.

The rather simple model used in the example discussed here is able to account for many of the experimental observations in different exciplexes. The combined use of modeling and supersonic jet spectroscopy thus gives a unique insight into the nature of exciplexes in their purest form (i.e., in isolation). It is to be expected that with the enormous advances in practical computational chemistry that are on the horizon more sophisticated electronic structure models will be applied to this subject in the near future.

3.6 The Solvent-Separated Ion Pair: What Is It?

Although the experimental evidence concerning the geometric structure of exciplexes or contact radical ion pairs is indirect, the hypothesis that sandwich-type geometries are usually formed is widely accepted, strongly supported by computational results, and not disputed by available experimental data. The situation is much less clear for the *solvent-separated radical ion pair*. There is no doubt that it exists as a crucial intermediate in the chemistry of radical ion pairs in polar solvents: There solvent-separated radical ion pairs show up as a distinct kinetic entity (see Section 2). Within the framework of electron transfer theory, their properties (small V and large λ_s) are consistent with a distance between D and A that is larger than in the CRIP. The idea that a favorable distance corresponds with the insertion of a solvent molecule between D^+ and A^- appears reasonable.

Computational studies of the free energy of ion pairs as a function of the distance have revealed the existence of minima corresponding to contact ion pairs and solvent-separated ion pairs. Clear results were obtained in molecular dynamics simulations of simple ions such as Na^+ and Cl^- in water[96] and in methanol,[97] but in a recent study of a model of an organic radical ion pair (dimethylaniline$^+$/anthracene$^-$) in acetonitrile and cyclohexane,[98] only the CRIP showed up as a minimum. Unfortunately, there are several difficulties associated with classical simulations. Since a flexible system is to be studied, with many degrees of freedom, the use of Monte Carlo or molecular dynamics sampling is mandatory. The large number of atoms involved rules out high-level quantum chemical methods. Classical simulation methods typically use fixed atomic charges and van der Waals parameters. For the free-radical ions D^+ and A^-, these charges and parameters can be derived from quantum chemical calculations of the separate radical ions. Owing to the recent progress in molecular density functional calculations, these calculations can nowadays be done with acceptable accuracy. However, in view of the very strong electric fields near the ionic particles, polarization of the solvent molecules, as well as the mutual polarization of the ions, must be taken into account. So far, this has not been done. Furthermore, the electronic description is limited to simple radical ion pairs, with complete charge transfer. Interactions with other electronic states cannot be taken into account in a classical simulation. Worse, such interactions are already very difficult to compute in a static quantum chemical calculation, because an accurately balanced representation of the different configurations is necessary, which requires the interactions with the medium to be taken into consideration.

In conclusion, solvent-separated ion pairs are a distinct kinetic intermediate in radical ion pair chemistry in polar solvents, but their structure remains elusive.

4 Intramolecular Exciplexes

4.1 Photoinduced Charge Separation in Linked-Electron Donor–Acceptor Systems

If one is interested in studying the role of structural factors such as orientation and distance on electron transfer processes, it is necessary to have some control over these factors. This requirement motivated early studies of intramolecular charge transfer starting in the 1970s. In these early studies, mostly alkane chains were used to spatially constrain the donor and acceptor groups; later on, numerous more or less rigidly bridged donor–acceptor systems of increasing complexity were described.[99–107] We will focus here on the "traditional" systems that have the necessary flexibility to attain a sandwich geometry like that of intermolecular exciplexes.

Among the most thoroughly investigated early examples are the systems composed of *N,N*-dimethylaniline as a donor and pyrene or anthracene as the electron-accepting chromophore, separated by oligomethylene chains.[36,108–111] The results obtained from fluorescence and transient absorption experiments on 1-(4-dimethylaminophenyl)-3-(1-pyrenyl)propane (**P**$_3$) can be summarized as follows:[56]

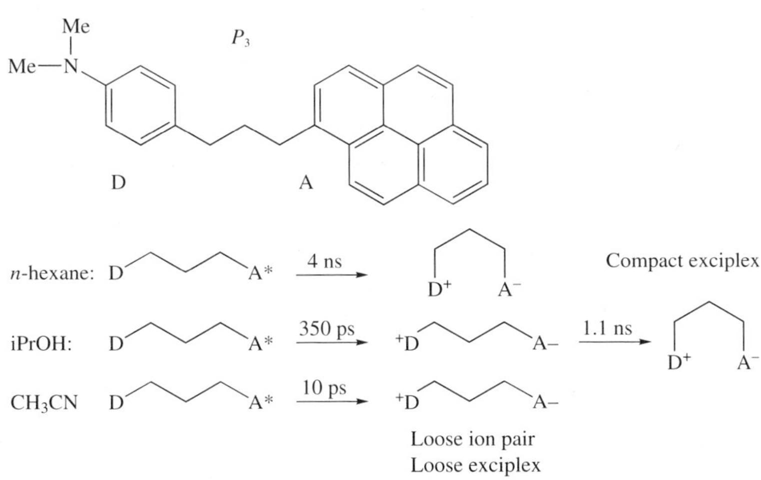

In *n*-hexane, the fluorescence spectrum[110] shows exciplex emission almost exclusively, in a broad band around 470 nm. The emission of the initially excited pyrene chromophore (LE emission) is largely quenched, but some of it is still visible around

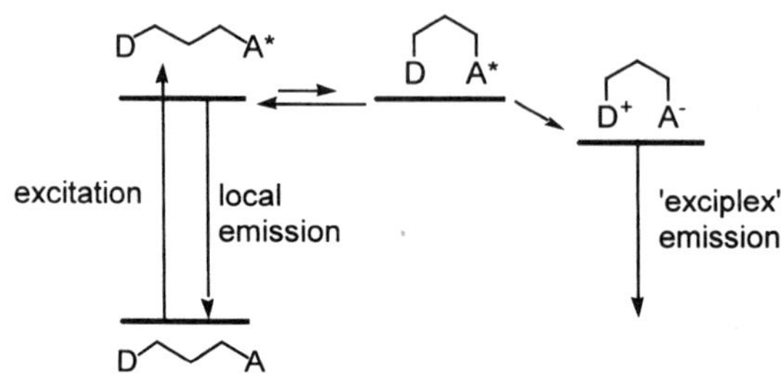

Figure 4.35. In nonpolar alkane solvents, a system such as $\mathbf{P}_3$ cannot easily undergo long-range intramolecular electron transfer because, in the predominantly populated all-*trans* conformation, it is endothermic. First, rotations of the C–C bonds have to bring A^* and D close enough together. Then, the coulomb attraction of D^+ and A^- must be large enough to bring the CT state below the LE state, so that electron transfer is feasible.

400 nm. Time-resolved measurements showed that the decay of the pyrene fluorescence and the rise of exciplex emission had similar time constants of about 4 ns, and it was concluded that a conformational change, bringing donor and acceptor closer together, had to occur before electron transfer would be thermodynamically feasible (Fig. 4.35). This is consistent with the finding that the related donor–acceptor systems with fewer than three CH_2 groups, in which a close-contact conformation is unattainable, cannot undergo electron transfer in alkane solvents at all.

In polar solvents, such as 2-propanol, photoinduced charge separation in $\mathbf{P}_3$ occurs in 0.35 ns, much more rapidly than in alkane solvents. This implies that the direct formation of a charge-separated state is possible, *without* prior extensive rotations around the C–C bonds. Subsequent relaxation of the initial CS species, termed a "loose heteroexcimer" by Mataga,[110] occurs with a time constant of around 1.1 ns, leading to a relaxed exciplex that has its emission maximum at a shorter wavelength and decays slowly (ca. 100 ns): The charge separation process is reversible, as indicated by the biexponential decay of the LE emission (0.35 and 1.1 ns). Interestingly, it was concluded that the loose "heteroexcimer" is fluorescent. Likewise, the intramolecular CS species with 0, 1, or 2 CH_2 groups showed CT fluorescence, so it was clearly established that a sandwich structure is not a prerequisite for exciplex-type fluorescence.

The transient optical absorption spectra of $\mathbf{P}_3$ (Fig. 4.36) are dominated by a band corresponding to the pyrene radical anion. The absorption rises with a time constant of about 0.35 ns in 2-propanol. At short times, the absorption spectrum is narrower. After a few nanoseconds it becomes broader, and then it does not change any more. This evolution can be described with the same time constants as those measured using fluorescence spectroscopy and is attributed to the formation of the compact exciplex. The behavior of $\mathbf{P}_3$ and $\mathbf{A}_3$ in acetone is qualitatively similar to that in 2-propanol, but the initial charge separation is faster (0.06 ns). In the very polar solvent acetonitrile, extremely rapid charge separation (10 ps) leads to a loose exciplex, characterized by a sharp absorption band, that does not undergo changes with time. Apparently, there is no formation of the sandwich exciplex.

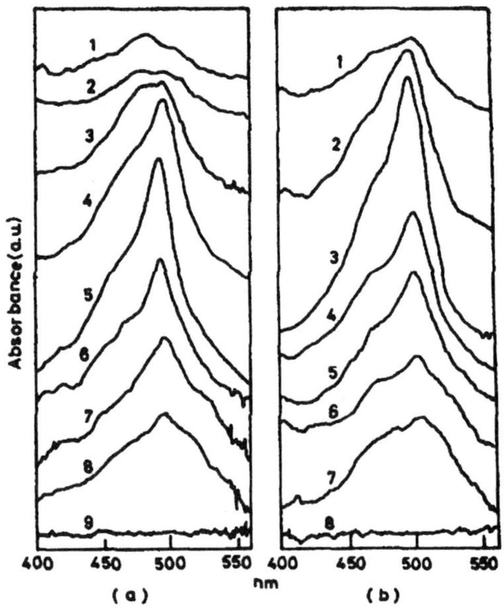

Figure 4.36. Transient absorption spectra of $\mathbf{P_3}$ in (a) 2-propanol and (b) acetone. Delay times with respect to the laser pulse in (a) (1)–(8) were −0.03, 0, 0.14, 0.4, 0.9, 1.8, 2.3, and 3.3 ns, respectively; in (b) (1)–(7), delay times were 0, 0.03, 0.16, 0.5, 1.5, 2.5, and 3.5 ns, respectively. The sharp feature at 490 nm at early times is attributed to the loose exciplex (absorption spectrum of the A^- fragment with little perturbation); at later times, a broader band remains due to the compact exciplex. (Reproduced with permission from Ref. 111. Copyright American Chemical Society.)

The formation of intramolecular excimers is favored for chain lengths of three CH_2 groups ("Hirayama's rule"),[112] and the same is true for exciplexes. With a shorter chain, fluorescent charge transfer states can be formed with a different geometry. With longer chains, the typical emission is the same as that of the sandwich exciplex, but the intensity is usually smaller.[113,114]

Since the 1980s, the importance of interactions through bonds has become widely recognized in the electron transfer field.[107] When the number of bonds between donor and acceptor is not too large and the orientation of the bonds is favorable, these interactions can easily be strong enough to allow significant transition probabilities in CT absorption and emission. Some examples are shown in Fig. 4.37, and many others are known.[115–117]

The electronic coupling V^* between the CT and LE zero-order states in systems such as those described by formulas **1** and **2**, in which donor and acceptor are separated by three bonds, is greater than $1000\,\text{cm}^{-1}$,[119] and even in the "six-bond" compound **3**, $V^* = 660\,\text{cm}^{-1}$,[121] (i.e., the same order of magnitude as in a face-to-face exciplex)[26] These results demonstrate that, for intramolecular electron transfer in donor–acceptor systems with a small number of bonds between D and A, the electronic coupling can easily be large enough that it does not limit the rate of electron transfer.

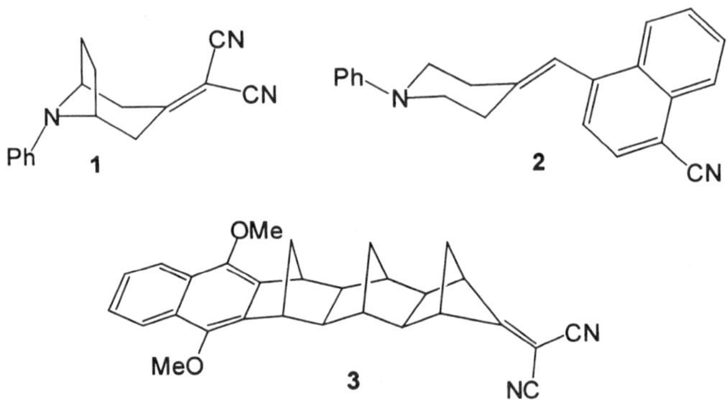

1 2 3

Figure 4.37. Some molecules showing strong CT absorption or emission. In **1**, the molar absorption coefficient at the maximum of the CT absorption band (360 nm) is 6100 L $mol^{-1} cm^{-1}$.[118] In **2**, the radiative rate constant of the CT emission is $10^8 s^{-1}$ in alkanes,[119] which is close to the typical value for an allowed transition in an organic molecule.[120]

4.2 Intramolecular Exciplexes in Isolated Molecules

The effect of the medium on the fluorescence behavior of $\mathbf{P}_3$ and the related $\mathbf{A}_3$ is very pronounced. In nonpolar solvents, exciplex formation is relatively slow (e.g., in $\mathbf{A}_3$ it takes about 2.5 ns in *n*-hexane at room temperature) In more polar solvents, direct charge separation is thermodynamically feasible and occurs rapidly. The anthracene compound $\mathbf{A}_3$ has also been studied in a supersonic jet expansion.[122] By means of this technique, the molecules are fully isolated and strongly cooled, as

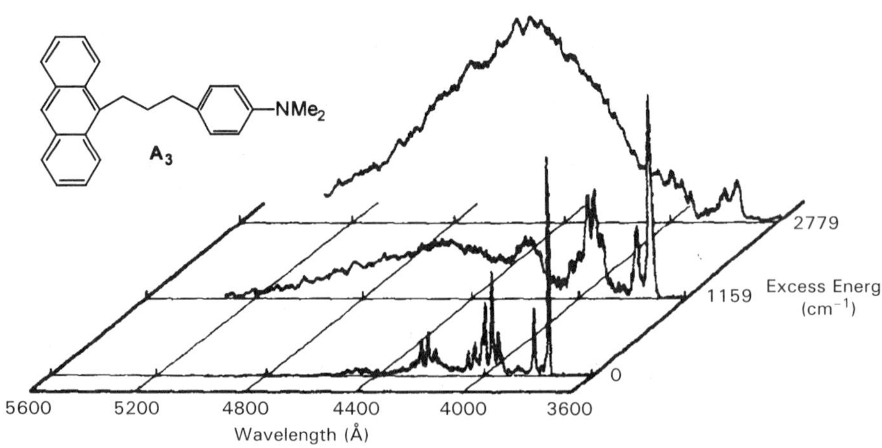

Figure 4.38. Emission spectra of A_3 in a supersonic jet, measured at different excitation wavelengths. (Reproduced with permission from Ref. 122. Copyright American Institute of Physics.)

discussed in Section 3. Fluorescence excitation spectra of organic compounds measured under such conditions are usually very well resolved. In the case of $\mathbf{A}_3$, LE emission was observed when excitation was at energies near the 0–0 transition. A sharply resolved excitation spectrum was detected. However, when the excess excitation energy was larger than 900 cm^{-1}, exciplex emission could be detected, as illustrated in Figs. 4.38 and 4.39.

The interpretation offered by Syage et al. is that the excess energy is needed to overcome the conformational barrier.[122] The energy of the CT state is higher than that of the LE state, except when the $(CH_2)_3$ chain adopts a gauche$^+$gauche$^-$ conformation. In more extended conformations, the electrostatic stabilization is not sufficient. Although the molecular physics of excited molecules in a supersonic jet are different from those in nonpolar solutions, the observations in terms of LE versus exciplex emission and the interpretations in terms of molecular structural changes are very similar.

When Wegewijs carried out similar experiments on compound **1r** in the mid-1980s, an onset of exciplex emission could be observed at an excess energy of 4.8 kcal/mol:[123]

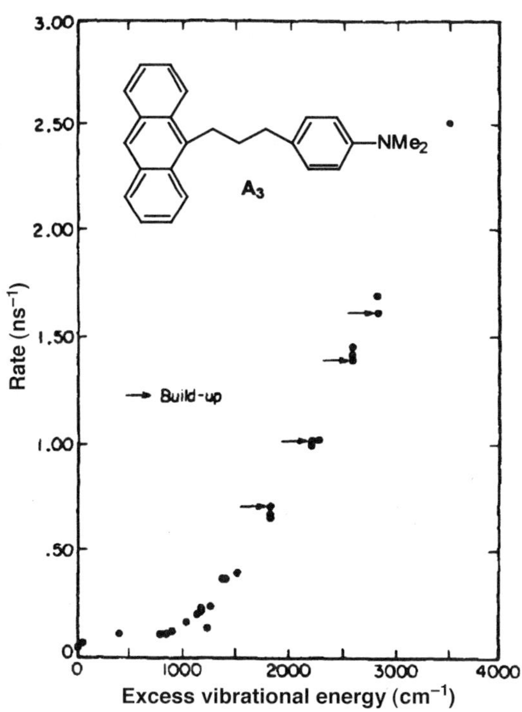

Figure 4.39. The change in decay time of the LE emission of $\mathbf{A}_3$ (which is equal to the buildup time of the exciplex emission) as a function of wavelength shows a threshold at around 900 cm^{-1}. At higher energies, the exciplex is formed progressively more rapidly. (Reproduced with permission from Ref. 122. Copyright American Institute of Physics.)

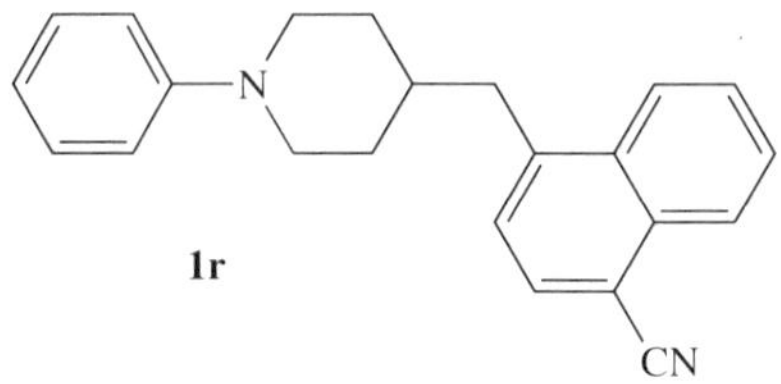

1r

This was a very surprising result, because the excess energy is much smaller than the conformational barrier that has to be overcome in order to achieve a sandwichlike conformation in this semirigid system. The preferred conformation in the ground state, which is the initial conformation in the LE state, has a chair form for the piperidine ring. Since a chair–boat conversion is necessary to bring donor and acceptor close together, a purely steric barrier of around 10 kcal/mol should be expected. In order to explain the observation, it was assumed that the onset which was observed was related to the excess energy required for long-range electron transfer. A subsequent conformational change leading to the species giving the exciplex emission then has a much lowered barrier, because a large amount of electrostatic energy is gained when the molecule is "folded" to bring the charged groups closer together. Thus, in this mechanism, called the "harpooning" mechanism, the order of events is reversed: Electron transfer preceeds and facilitates the conformational change (Fig. 4.40).

Persuasive evidence for the preceeding view was obtained by comparing three compounds that differ only in the ionization potential of the electron donor.[126] In **1r**, **2r**, and **3r** (Fig. 4.41), the LE state is the same, as demonstrated by the identical

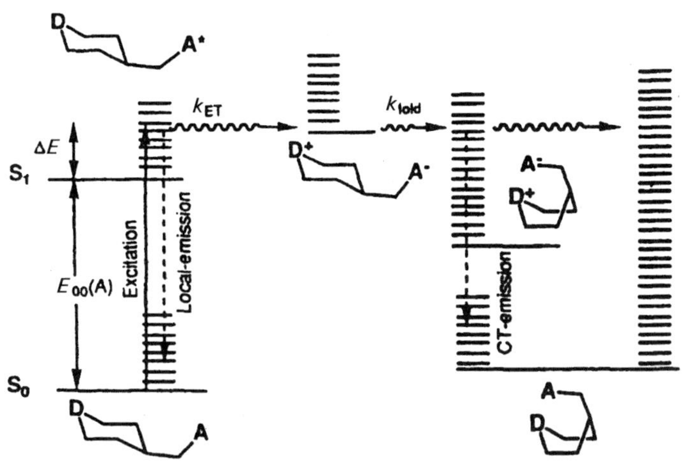

Figure 4.40. The harpooning mechanism in isolated **1r**. Electron transfer can occur (k_{ET}) when sufficient excess energy is available to reach the extended CT species. This sparse manifold cannot be significantly populated, but subsequent folding leads to the much denser manifold of the low-energy exciplexlike compact charge transfer species.[124] (Reproduced with permission from Ref. 125. Copyright John Wiley & Sons, Inc.)

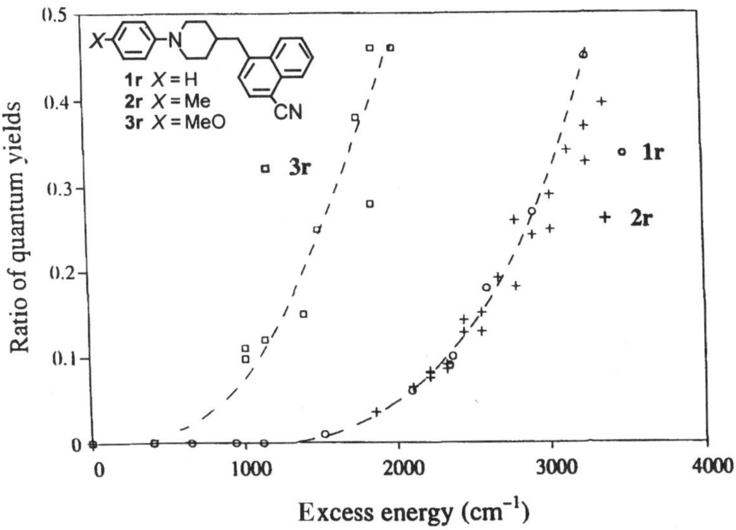

Figure 4.41. When the electron donor has a lower IP, a smaller excess energy is needed to induce the transition from the LE to the CT manifold in jet-cooled compounds. Subsequent conformational changes rapidly lead to the observed exciplex emission. (Reproduced with permission from Ref. 126. Copyright Elsevier Science Publishers.)

excitation spectra. The conformational barriers in these molecules also are the same. The onset of exciplex formation, however, shifts to much lower energy with decreasing donor ionization potential, as shown in Fig. 4.41. The maximum of the exciplex emission band also shifts to lower energy, by 0.05 eV and 0.19 eV for **2r** and **3r**, respectively. This reflects the differences in adiabatic ionization potentials.

4.3 Conformational Changes Following Intramolecular Charge Separation in Solution

In the jet experiments, the transition from the LE to the CT manifold occurs while the molecules are still in similar, more or less extended, conformations. Under these experimental conditions, when donor and acceptor are brought closer together, the potential energy in the CT state is lowered so dramatically that the intrinsic conformational barrier is easily overcome. If folding would not occur, the population of the extended CT species would be small because it has a much smaller density of vibrational states than the LE state, which is lower in potential energy. In contrast, in solution, even if the solvent is nonpolar, the extended CT species may be stabilized by the polarizable medium, and the excess energy is dissipated. Indeed, in alkanes an additional CT emission was observed, most clearly at temperatures below room temperature (Fig. 4.42). Note that also at low temperatures the LE emission is

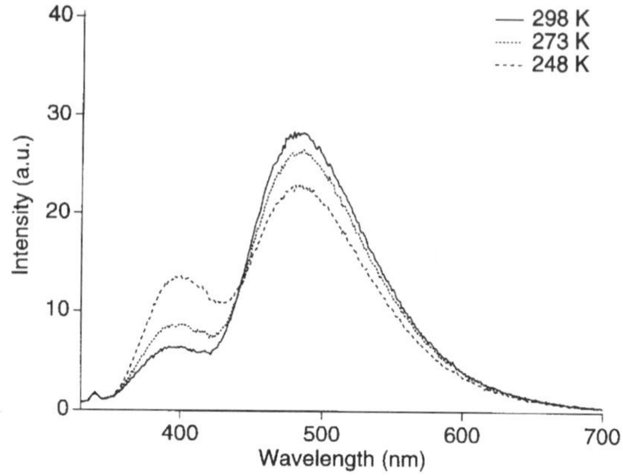

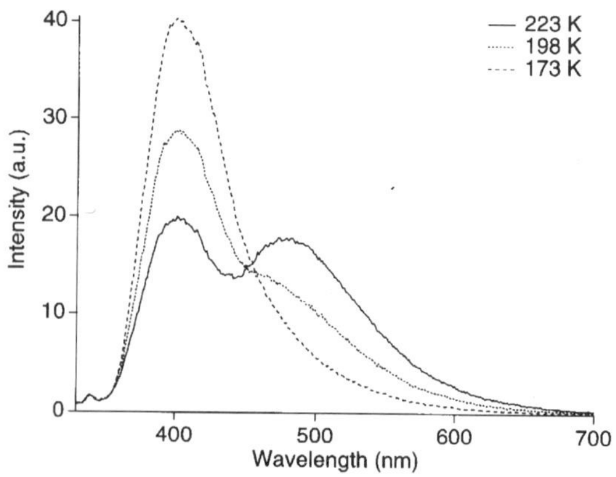

Figure 4.42. Dual emission of an alkyl derivative of **1r** in *trans*-decalin at different temperatures. The emission band at 400 nm is assigned to an extended conformation of the molecule in the CT state (ECT), and the band at around 490 nm is from a compact CT species. Local emission of the cyanonaphthalene chromophore (340 nm) is completely absent. (Reproduced with permission from Ref. 127. Copyright Wiley/VCH.)

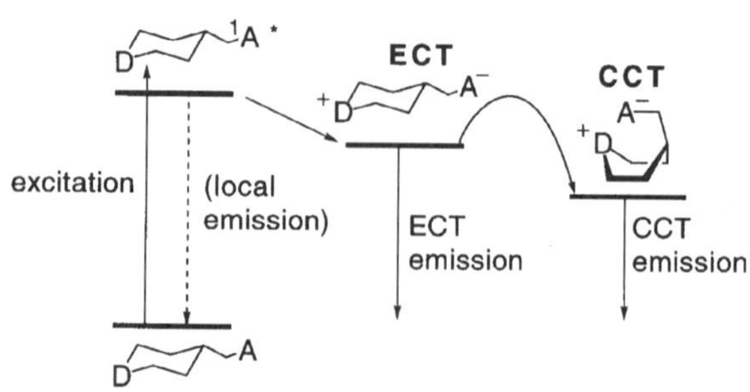

Figure 4.43. Reaction scheme describing the "harpooning" mechanism in solution. The decay time of the ECT emission (at short wavelength) is the same as the rise time of the longwavelength CCT (exciplexlike) emission.

completely quenched due to rapid electron transfer. The schematic description of the harpooning process in solution is depicted in Fig. 4.43.

Time-resolved measurements of the decay of the ECT and rise of the CCT band enabled Wegewijs to make an estimate of the thermal activation energy for the folding process in **1r** in MCH, which resulted in a value of 3.6 kcal/mol. It is not possible to derive this value without assumptions, because the total decay rate of ECT involves other processes, such as intersystem crossing to the triplet LE state and charge recombination to the ground state. The temperature dependence of the latter processes had to be neglected.

It is interesting to note that the radiative decay rate of the extended species, which is around $4 \times 10^6\ s^{-1}$ in alkanes,[128] is similar to radiative rates of typical exciplexes. This attests, once again, to the efficiency of electronic coupling through bonds vs. through space. In this case, there are five sigma bonds between donor and acceptor. For the systems studied by Wegewijs and Lauteslager, the radiative rate of the CCT conformers could not be determined precisely, because the efficiency of the folding process could not be derived from the available fluorescence data. The quantum yield of folding in alkanes, however, must be close to unity, because it occurs with a time constant of about $10^9\ s^{-1}$, which is at least 10-times faster than the decay of the ECT species under conditions that prevent harpooning. Thus, the radiative rates could be determined for the CCT species in alkane solvents. They are found to be about $2 \times 10^6\ s^{-1}$, even somewhat smaller than those of the ECT species!

The emission of an extended CT species, a loose exciplex, of $\mathbf{P_3}$ in 2-propanol had already been described by Migita et al. in 1981, but the actual spectra were not published. In the semiflexible systems studied by the Verhoeven group, ECT emission could be demonstrated much more clearly, because the folding process is much slower in these systems, and the difference in emission energies of ECT and CCT is much bigger. However, also in flexible systems, in which the barrier for folding is very low, ECT emission could be convincingly shown by Scherer et al.[5] The very simple propane derivatives **4** and **5** shown in Fig. 4.44 differ from $\mathbf{P_3}$ and $\mathbf{A_3}$ in the important respect that they can undergo long-range electron transfer in nonpolar solvents because the LE energy is higher, and the CT state lower, in energy. At room temperature, this difference may go unnoticed, because the only emission seen is the exciplex band. Upon lowering the temperature, however, it became very clear that the precursor of this emissive species is not the folded LE, but the extended CT species!

4.3.1 Conditions Required for Harpooning in Solution: Effects of Solvent Polarity

The harpooning phenomenon can occur if long-range charge separation is possible, if the compact CT species is lower in energy than the extended CT species, and if the rate of folding is sufficiently high in comparison with the rates of other decay processes of the ECT species. The first requirement is easily fulfilled by choosing electron donor and acceptor units with appropriate redox potentials and excitation

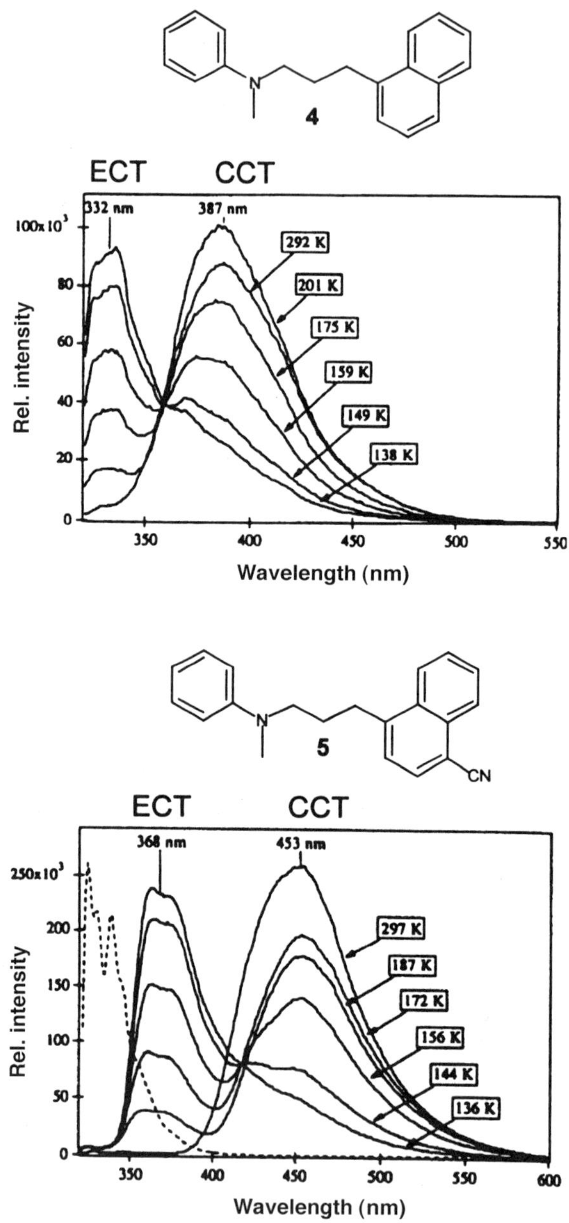

Figure 4.44. Emission spectra of the compounds **4** and **5** in methylcyclohexane at different temperatures. For comparison, the spectrum of cyanonaphthalene is included (dashed line) (Reproduced with permission from Ref. 5. Copyright American Chemical Society.)

energies. The energy difference between ECT and CCT will depend strongly on the polarity of the medium, as will the barrier height. Dynamical aspects of barrier crossing may become important in viscous media, as is discussed shortly.

We shall first consider the effects of solvent polarity. The dipole moments of ECT species are considerably greater than those of the corresponding CCT species. Upon increase of the solvent polarity, the former will be stabilized more than the latter, which will reduce the driving force for harpooning. If we follow Weller's reasoning regarding the relative energies of exciplexes versus ion pairs,[28] the ion-pair-like ECT is expected to become more stable than the exciplex-like CCT in solvents with dielectric constants greater than about 7. Because of the additional steric energy contained in the CCT species, the switch-over in the intramolecular case should occur in even less polar solvents. Moreover, the formation of the CCT species requires the crossing of a barrier that is lowered by the Coulomb attraction between D^+ and A^-. This barrier will increase with increasing solvent polarity. Therefore, a kinetic limitation may come into play and prevent the harpooning process from occurring within the ECT lifetime already in solvents of rather low polarity.

Lippert–Mataga plots (Fig. 4.45) may be used to estimate the relative energies of ECT and CCT species. The ECT species is much more solvatochromic than the CCT species. Separate bands can be seen in alkanes and in low polarity dialkyl ethers ($\Delta f < 0.2$).

Crossing of the lines in Fig. 4.45 occurs near $\Delta f = 0.25$ (i.e., for solvents such as diethyl ether or butylacetate). One might expect that a CCT emission maximum always occurs at lower energy than the ECT emission, because it produces the ground state in a (twist)boat conformation that has a higher energy (ca. 5.5 kcal/mol) than that of the extended ground state, while the CCT excited state

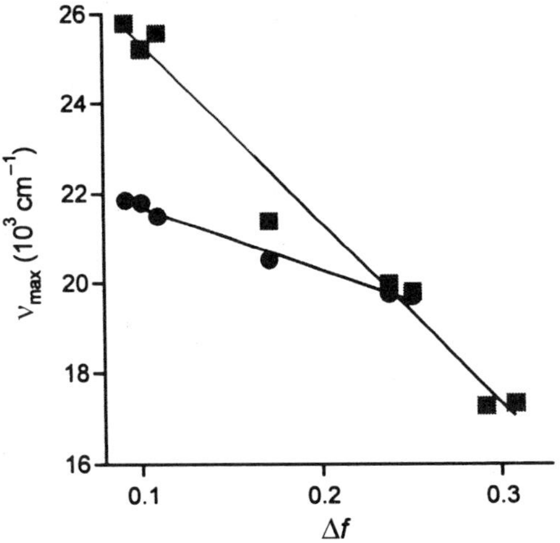

Figure 4.45. Plots of the emission energies against the solvent polarity parameter Δf for the ECT (squares) and CCT species (circles) of **1r**. (Reproduced with permission from Ref. 128. Copyright American Chemical Society.)

must be at a lower energy than (or at least must be of nearly equal energy to) the ECT species. However, the emission energies are also determined by the ground-state repulsion energies. There is a significant difference in the solvent reorganization energies λ_s for the two species, which contributes more to the redshift of the emission for the ECT than for the CCT species. According to the same model that forms the rationale behind the Lippert–Mataga plots, the solvent reorganization energy is

$$\lambda_s = \frac{\mu^2}{\rho^3}\left(\frac{\varepsilon - 1}{2\varepsilon + 1} - \frac{n^2 - 1}{2n^2 + 1}\right).$$

The values of μ^2/ρ^3 can be experimentally determined from the Lippert–Mataga plots. For the ECT species this amounts to around 19,000 cm^{-1}, and for the CCT species it is about 5,000 cm^{-1}. The estimated difference in λ_s for diethyl ether is then approximately 5.7 kcal/mol, roughly equal to the ground-state energy difference between the ECT and CCT product wells. Time-resolved emission spectroscopy of **1r** (and related compounds) in diethyl ether clearly showed that after excitation the emission band distinctly shifts to the red with a time constant of about 6 ns. The emission maxima are very close, but owing to the high-quality experimental data, the decisive conclusion could be drawn that there is a spectral change, attributed to harpooning.[128] In solvents of still higher polarity, however, simple exponential decay of the CT fluorescence is observed, indicating that harpooning does not occur.

4.3.2 Solvent Viscosity Effects on Harpooning

The solvent polarity has an influence on the thermodynamics of the harpooning process, in that it determines the driving force and the height of the barrier. An additional solvent effect is the dynamical one on the rate of barrier crossings that results from the viscosity of the medium. This may be expected to show up most prominently when the barriers on the potential-energy surface are low. Mataga and co-workers noted already that for $\mathbf{A_3}$ and $\mathbf{P_3}$, the rate of formation of the heteroexcimer was smaller in decalin than in the less viscous hexane.[108,109] In their experiment it was assumed that the rate-limiting step is a conformational change in the LE state, bringing D and A close enough together to permit electron

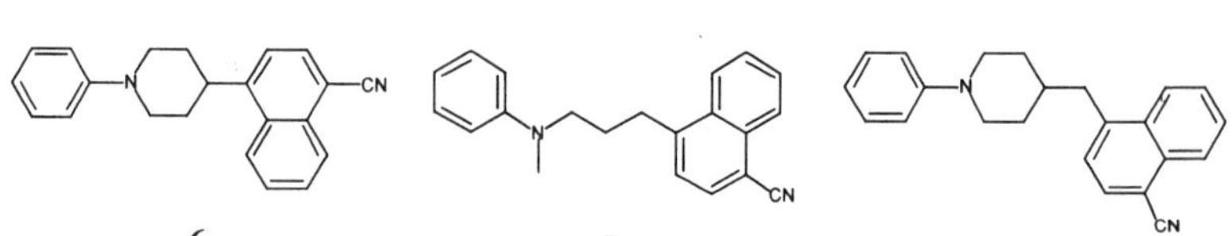

	6	5	1r
viscosity effect	*negligible*	*very large*	*intermediate*
steric barrier	*high*	*low*	*modest*

Figure 4.46. Comparison of the viscosity effects on the harpooning process in three different systems.

transfer to occur. In retrospect, it is not easy to exclude a mechanism of endergonic electron transfer leading to very low populations of the ECT species, which rearrange rapidly to the detected "heteroexcimer." Such a mechanism, which resembles the behavior of **1r** in the supersonic jet, was recently proposed for harpooning systems involving *p*-phenylenediamine chromophores and semiflexible bridges, in which the LE state is relatively low in energy.[129,130]

In most of the harpooning systems studied by Verhoeven and co-workers, the conversion of ECT to CCT can be monitored in a straightforward manner by using time-resolved fluorescence. Scherer et al.[5] made an attempt to separate effects of solvent viscosity and polarity by comparing different alkanes and varying the temperatures. For piperidine system **6** (see Fig. 4.46), no retarding effect of viscosity could be demonstrated, but the flexible systems such as **5** with the alkane spacers

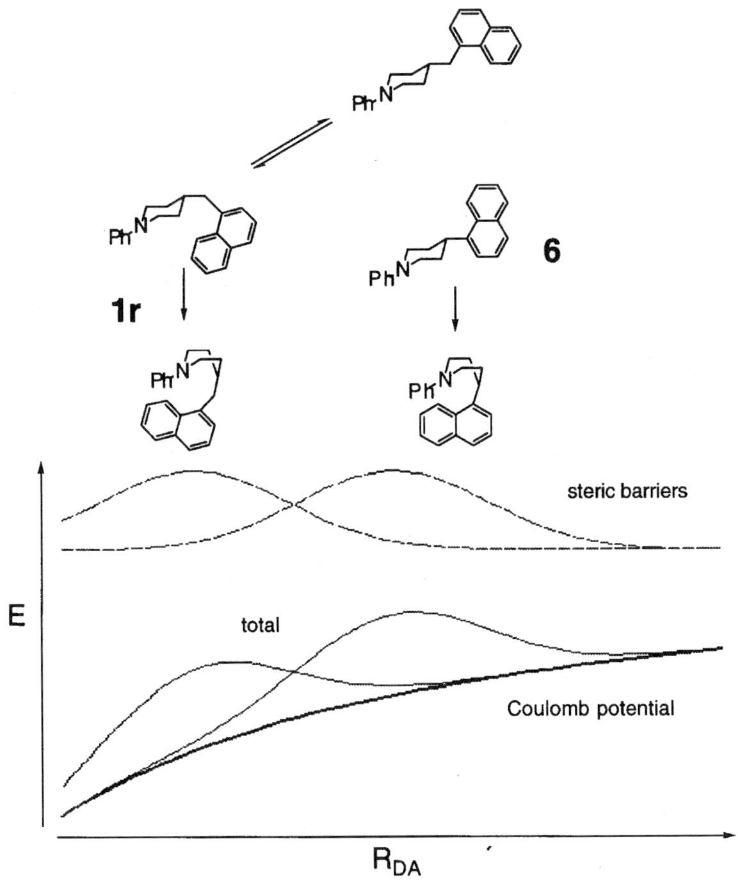

Figure 4.47. The chair-to-boat transition state for **1r** is located at a smaller donor–acceptor distance than that for **6**. The steric barrier (dashed lines) in the latter case is much less depressed by the electrostatic interaction (solid line) than in the former case.

showed a very clear effect, as expected. A problem in the approach used is that, upon changing solvents and temperatures, one changes not only the viscosity, but also the dielectric constant.

More detailed insight could be obtained by employing an additional experimental dimension, viz., pressure. The studies of Jäger et al. confirmed the absence of a viscosity effect in the case of **6**, but showed a clear effect in the case of **1r**.[131,132] The results are summarized in Fig. 4.46. At first sight, it may seem strange that there is such a pronounced difference in the behavior of **6** and **1r**. The clue to understanding this is in the added flexibility that is present in **1r**, but not in **6**. This flexibility has a pronounced effect on the location of the steric barrier along the coordinate defined by the donor–acceptor distance. In the extended conformation of the six-membered ring, a rotation around the exocyclic C–C bond can bring the acceptor group in **1r** effectively much closer to the donor than it is in compound **6**.[133] Because of the $1/r$ dependence of the electrostatic interaction energy, the depression of the steric barrier is more effective when the barrier is located at a smaller donor–acceptor distance (see Fig. 4.47).

Although an enormous amount of experimental data has been gathered over the past three decades on intramolecular charge transfer states and intermolecular exciplexes, the interpretation of the results is still mostly based on the concepts developed in the pioneering studies of the 1960s and 1970s. Despite the progress and increased insight into the factors that govern electron transfer processes, we are still far from an a priori prediction of the behavior of new chemical systems. If it is the goal of science to solve all problems in a field and to make itself superfluous by providing reliable and quantitatively accurate predictive models, then there is still a lot to do in the field of electron transfer.

References

1. Birks, J.B. *Nature*, **1967**, *214*, 1187–1190.

2. Förster, T. In Gordon, M. and Ware, W.R., eds. *The Exciplex*, Academic Press: New York, **1975**, 1–21.

3. Förster, T.; Kasper, K. *Z. Phys. Chem. Neue Folge*, **1954**, *1*, 275.

4. Caldwell, RA.; Creed, D. *Acc. Chem. Res.*, **1980**, *13*, 45–50.

5. Scherer, T.; van Stokkum, I.H.M.; Brouwer, A.M.; Verhoeven, J.W. *J. Phys. Chem.*, **1994**, *98*, 10539–10549.

6. Briegleb, G. *Elektron–Donator–Akzeptor–Komplexe*, Springer Verlag: Berlin–Göttingen–Heidelberg, **1961**.

7. Forster, R. *Organic Charge-Transfer Complexes*, Academic Press: London, New York, **1969**.

8. Gould, I.R.; Young, R.H.; Mueller, L.J.; Farid, S. *J. Am. Chem. Soc.*, **1994**, *116*, 8176–8187.

9. Leonhardt, H.; Weller, A. *Z. Physik. Chem. Neue Folge*, **1961**, *29*, 277–280.

10. Leonhardt, H.; Weller, A. *Ber. Bunsenges. Phys. Chem.* **1963**, *67*, 791–795.
11. Bauer, E. *Z. Physik. Chem., Abt. B*, **1932**, *16*, 465.
12. Weiss, J.; Fischgold, H. *Z. Physik. Chem., Abt. B*, **1936**, *32*, 135.
13. Melhuish, W.H. *J. Phys. Chem.*, **1961**, *65*, 229–235.
14. Gould, I.R.; Farid, S. *Acc. Chem. Res.*, **1996**, *29*, 522–528.
15. Gould, I.R.; Young, R.H.; Moody, R.E.; Farid, S. *J. Phys. Chem.*, **1991**, *95*, 2068–2080.
16. Rathore, R.; Hubig, S.M.; Kochi, J.K. *J. Am. Chem. Soc.*, **1997**, *119*, 11468–11480.
17. Marcus, R.A. *J. Chem. Phys.*, **1956**, *24*, 966–978.
18. Klimkans, A.; Larsson, S. *Chem. Phys.*, **1994**, *189*, 25–31.
19. Jakobsen, S.; Mikkelsen, K.V.; Pedersen, S.U. *J. Phys. Chem.*, **1996**, *100*, 7411–7417.
20. Cosa, G.; Chesta, C.A. *J. Phys. Chem.: A.*, **1997**, *101*, 4922–4928.
21. Nelsen, S.F. *Adv. Electr. Transfer Chem.*, **1993**, *3*, 168–189.
22. Brouwer, A.M. *J. Phys. Chem.: A.*, **1997**, *101*, 3626–3633.
23. Gould, I.R.; Ege, D.; Moser, J.E.; Farid, S. *J. Am. Chem. Soc.*, **1990**, *112*, 4290–4301.
24. Zelent, B.; Kusba, J.; Gryczynski, I.; Lakowicz, J.R. *Appl. Spectrosc.*, **1995**, *1*, 43–50.
25. Murata, S.; Tachiya, M. *J. Phys. Chem.*, **1996**, *100*, 4064–4070.
26. Gould, I.R.; Young, R.H.; Mueller, L.J.; Albrecht, A.C.; Farid, S. *J. Am. Chem. Soc.*, **1994**, *116*, 8188–8199.
27. Michl, J.; Bonačić-Koutecký, V. *Electronic Aspects of Organic Photochemistry*, Wiley: New York, **1990**.
28. Weller, A. *Z. Phys. Chem. Neue Folge*, **1982**, *133*, 93–98.
29. Knibbe, H.; Rehm, D.; Weller, A. *Ber. Bunsenges. Phys. Chem.*, **1969**, *73*, 839–845.
30. Parigger, C.; Plemmons, D.H.; Litchford, R.J.; Jeng, S.M. *Optics Lett.*, **1998**, *23*, 76–78.
31. Spivey, A.C.; Bissell, A.J.; Stammen, B. *Synth. Commun.*, **1998**, *28*, 623–632.
32. Weller, A. In *The Exciplex*, Gordon, M.; Ware, W.R., eds., Academic Press: New York, **1975**, 23–38.
33. Suppan, P. *J. Photochem. Photobiol.: A*, **1990**, *50*, 293–330.
34. Suppan, P.; Ghoneim, N. *Solvatochromism*, Royal Society of Chemistry: Cambridge, **1997**.
35. Beens, H.; Knibbe, H.; Weller, A. *J. Chem. Phys.*, **1967**, *47*, 1183–1184.
36. Mataga, N. In *The Exciplex*, Gordon, M.; Ware, W.R., eds., Academic Press: New York, **1975**, 113–144.
37. Rehm, D.; Weller, A. *Ber. Bunsenges. Phys. Chem.*, **1969**, *73*, 834–839.
38. Rehm, D.; Weller, A. *Isr. J. Chem.*, **1970**, *8*, 259–271.
39. Closs, G.L.; Miller, J.R. *Science*, **1988**, *240*, 440–447.
40. Brunschwig, B.S.; Sutin, N. *Comments Inorg. Chem.*, **1987**, *6*, 209–235.
41. Marcus, R.A. *J. Phys. Chem.*, **1989**, *93*, 3078.
42. Myers, A.B. *Chem. Rev.*, **1996**, *96*, 911–926.

43. Myers, A.B. *Acc. Chem. Res*, **1997**, *30*, 519–527.

44. Lilichenko, M.; Tittelbach-Helmrich, D.; Verhoeven, J.W.; Gould, I.R.; Myers, A.B. *J. Chem. Phys.* **1998**, *109*, 10958–10969.

45. Claude, J.P.; Williams, D.S.; Meyer, T.J. *J. Am. Chem. Soc.*, **1996**, *118*, 9782–9783.

46. Siebrandt, W. *J. Chem. Phys.*, **1967**, *47*, 2411–2422.

47. Siebrandt, W. *J. Chem. Phys.*, **1967**, *46*, 440–447.

48. Mataga, N.; Okada, T.; Yamamoto, N. *Chem. Phys. Lett.*, **1967**, *1*, 119–121.

49. Kikuchi, K.; Takahashi, Y.; Katagiri, T.; Niwa, T.; Hoshi, M.; Miyashi, T. *Chem. Phys. Lett.*, **1991**, *180*, 403–408.

50. Kikuchi, K.; Katagiri, T.; Niwa, T.; Takahashi, Y.; Suzuki, T.; Ikeda, H.; Miyashi, T. *Chem. Phys. Lett.*, **1992**, *193*, 155–160.

51. Mattes, S.L.; Farid, S. *J. Am. Chem. Soc.*, **1983**, *105*, 1386–1387.

52. Mattes, S.L.; Farid, S. *Science*, **1984**, *226*, 917–921.

53. Arnold, B.R.; Farid, S.; Goodman, J.L.; Gould, I.R. *J. Am. Chem. Soc.*, **1996**, *118*, 5482–5483.

54. Arnold, B.R.; Atherton, S.J.; Farid, S.; Goodman, J.L.; Gould, I.R. *Photochem. Photobiol.*, **1997**, *65*, 15–22.

55. Potashnik, R.; Goldschmidt, C.R.; Ottolenghi, M.; Weller, A. *J. Chem. Phys.*, **1971**, *55*, 5344–5348.

56. Mataga, N. *Pure Appl. Chem.*, **1984**, *56*, 1255–1268.

57. Mattes, S.L.; Farid, S. *J. Chem. Soc., Chem. Commun.*, **1980**, *3*, 126–128.

58. Weller, A. *Z. Phys. Chem. Neue Folge*, **1982**, *130*, 129–138.

59. Mataga, N.; Asahi, T.; Kanda, Y.; Okada, T.; Kakitani, T. *Chem. Phys.*, **1988**, *127*, 249–261.

60. (a) Vauthey, E.; Suppan, P.; Haselbach, E. *Helv. Chim. Acta*, **1988**, *71*, 93–99; (b) Recently, on the basis of direct measurements of ion-pair kinetics, Vauthey et al. argued that the charge recombination involves contact ion pairs rather than solvent separated ion pairs: Vauthey, E.; Högemann, C.; Allonas, X. *J. Phys. Chem.*, **1998**, *102*, 7362–7369.

61. Gould, I.R.; Ege, D.; Mattes, S.L.; Farid, S. *J. Am. Chem. Soc.*, **1987**, *109*, 3794–3796.

62. Asahi, T.; Mataga, N. *J. Phys. Chem.*, **1989**, *93*, 6575–6578.

63. Asahi, T.; Ohkohchi, M.; Mataga, N. *J. Phys. Chem.*, **1993**, *50*, 13132–13137.

64. Gould, I.R.; Noukakis, D.; Goméz-Jahn, L.; Young, R.H.; Goodman, J.L.; Farid, S. *Chem. Phys.*, **1993**, *176*, 439–456.

65. Levin, P.P.; Pluzhnikov, P.F.; Kuzmin, V.A. *Chem. Phys. Lett.*, **1988**, *147*, 283–287.

66. Majima, T.; Pac, C.; Nakasone, A.; Sakurai, H. *J. Am. Chem. Soc.*, **1981**, *103*, 4499–4508.

67. Schaap, A.P.; Siddiqui, S.; Prasad, G.; Palomino, E.; Lopez, L. *J. Photochem.* **1984**, *25*, 167–181.

68. Beens, H.; Weller, A. *Acta Phys. Polon.*, **1968**, *34*, 593–602.

69. Beens, H.; Weller, A. In *Molecular Luminescence*, Lim, E.C. ed., W.A. Benjamin: New York, **1969**, 203–207.

70. Beens, H., PhD. thesis, Free University, Amsterdam, **1970**.

71. Knibbe, H.; Röllig, K.; Schäfer, F.P.; Weller, A. *J. Chem. Phys.*, **1967**, *47*, 1184–1185.

72. Groenen, E.J.J.; van Velzen, P.N.T. *Mol. Phys.*, **1978**, *35*, 19.

73. Mylon, S.E.; Smirnov, S.N.; Braun, C.L. *J. Phys. Chem. A*, **1998**, *102*, 6558–6564.

74. Chow, Y.L.; Johansson, C.I. *J. Phys. Chem.*, **1995**, *99*, 17558–17565.

75. Chow, Y.L.; Johansson, C.J. *J. Phys. Chem.*, **1995**, *99*, 17566–17572.

76. Iwata, S.; Tanaka, J.; Nagakura, S. *J. Chem. Phys.*, **1967**, *47*, 2203–2209.

77. Hayashi, H.; Nagakura, S.; Iwata, S. *Mol. Phys.*, **1967**, *13*, 489–490.

78. Hanson, A.W. *Acta Crystallogr.*, **1965**, *19*, 610–613.

79. Markel, F.; Ferris, N.S.; Gould, I.R.; Myers, A.B. *J. Am. Chem., Soc.*, **1992**, *15*, 6208–6219.

80. Kulinowski, K.; Gould, I.R.; Ferris, N.S.; Myers, A.B. *J. Phys. Chem.*, **1995**, *99*, 17715–17723.

81. Kulinowski, K.; Gould, I.R.; Myers, A.B. *J. Phys. Chem.*, **1995**, *99*, 9017–9026.

82. Miyasaka, H.; Ojima, S.; Mataga, N. *J. Phys. Chem.*, **1989**, *93*, 3380–3382.

83. Tahara, T.; Hamaguchi, H. *J. Phys. Chem.* **1992**, *96*, 8252–8259.

84. Vauthey, E.; Parker, A.W.; Nohova, B.; Phillips, D. *J. Am. Chem. Soc.* **1994**, *116*, 9182–9186.

85. Haas, Y.; Anner, O. In *Photoinduced Electron Transfer, Part A*, Fox, M.A.; Chanon, M., eds., Elsevier: Amsterdam, **1988**, 305–361.

86. Brenner, V.; Millie, P.; Piuzzi, F.; Tramer, A. *J. Chem. Soc. Faraday Trans.*, **1997**, *93*, 3277–3287.

87. Kurono, M.; Takasu, R.; Itoh, M. *J. Phys. Chem.*, **1995**, *99*, 9668–9674.

88. Russell, T.D.; Levy, D.H. *J. Phys. Chem.*, **1982**, *86*, 2718.

89. Tramer, A.; Brenner, V.; Millié, P.; Piuzzi, F. *J. Phys. Chem.: A*, **1998**, *102*, 2798–2807.

90. Tramer, A.; Brenner, V.; Millié, P.; Piuzzi, F. *J. Phys. Chem.: A*, **1998**, *102*, 2808–2816.

91. Anner, O.; Haas, Y. *J. Am. Chem. Soc.*, **1988**, *110*, 1416–1423.

92. Langlet, J.; Claverie, P.; Caron, F.; Boeuve, J. C. *Int. J. Quantum Chem.*, **1981**, *19*, 299–338.

93. Deperasinska, I.; Prochorow, J. *J. Mol. Struct. (Theochem)*, **1995**, *343*, 11–23.

94. Prochorow, J. *J. Mol. Struct.*, **1997**, *404*, 199–211.

95. Brouwer, A.M.; Wilbrandt, R. *J. Phys. Chem.*, **1996**, *100*, 9678–9688.

96. Dang, L.X. *J. Chem. Phys.*, **1992**, *97*, 1919–1921.

97. Sesé, G.; Guàrdia, E.; Padró, J.A. *J. Phys. Chem.*, **1995**, *99*, 12647–12654.

98. Hilczer, M.; Tachiya, M. *Chem. Phys. Lett.*, **1998**, *295*, 337–346.

99. Joran, A.D.; Leland, B.A.; Geller, G.G.; Hopfield, J.J.; Dervan, PB. *J. Am. Chem. Soc.*, **1984**, *106*, 6090–6092.

100. Pasman, P.; Mes, G.F.; Koper, N.W.; Verhoeven, J.W. *J. Am. Chem. Soc.*, **1985**, *107*, 5839.

101. Oevering, H.; Paddon-Row, M.N.; Heppener, M.; Oliver, A.M.; Cotsaris, E.; Verhoeven, J.W.; Hush, N.S. *J. Am. Chem. Soc.*, **1987**, *109*, 3258–3269.

102. Wasielewski, M.R. *Chem. Rev.*, **1992**, *92*, 435–461.

103. Gust, D.; Moore, T.A.; Moore, A.L. *Acc. Chem. Res.*, **1993**, *4*, 198–205.

104. Sauvage, J.P.; Collin, J.P.; Chambron, J.C.; Guillerez, S.; Coudret, C.; Balzani, V.; Barigelletti, F.; De Cola, L.; Flamigni, L. *Chem. Rev.*, **1994**, *4*, 993–1019.

105. Van Dijk, S.I.; Wiering, P.G.; Groen, C.P.; Brouwer, A.M.; Verhoeven, J.W.; Schuddeboom, W.; Warman, J.M. *J. Chem. Soc. Faraday Trans.*, **1995**, *91*, 2107–2114.

106. Osuka, A.; Marumo, S.; Mataga, N.; Taniguchi, S.; Okada, T.; Yamazaki, I.; Nishimura, Y.; Ohno, T.; Nozaki, K. *J. Am. Chem. Soc.*, **1996**, *118*, 155–168.

107. Verhoeven, J.W. In *Electron Transfer—From Isolated Molecules to Biomolecules, Part One*, Jortner, J.; Bixon, M., eds., Wiley: New York, **1999**; 603–644.

108. Okada, T.; Fujita, T.; Kubota, M.; Masaki, S.; Mataga, N.; Ide, R.; Sakata, Y.; Misumi, S. *Chem. Phys. Lett.*, **1972**, *14*, 563–568.

109. Okada, T.; Saito, T.; Mataga, N.; Sakata, Y.; Misumi, S. *Bull. Chem. Soc. Jpn.*, **1977**, *50*, 331–336.

110. Migita, M.; Okada, T.; Mataga, N.; Sakata, Y.; Misumi, S.; Nakashima, N.; Yoshihara, K. *Bull. Chem. Soc. Jpn.*, **1981**, *54*, 3304–3311.

111. Okada, T.; Migita, M.; Mataga, N.; Sakata, Y.; Misumi, S. *J. Am. Chem. Soc.*, **1981**, *103*, 4715–4720.

112. Hirayama, F. *J. Chem. Phys.*, **1965**, *42*, 3162.

113. Borkent, J.H.; Verhoeven, J.W.; de Boer, T.J. *Tetrahedron Lett.*, **1972**, *32*, 3363–3366.

114. Vanderauwera, P.; De Schryver, F.C.; Weller, A.; Winnik, M.A.; Zachariasse, K.A. *J. Phys. Chem.*, **1984**, *88*, 2964–2970.

115. Pasman, P.; Rob, F.; Verhoeven, J.W. *J. Am. Chem. Soc.*, **1982**, *104*, 5127–5133.

116. Hermant, R.M.; Bakker, NAC.; Scherer, T.; Krijnen, B.; Verhoeven, J.W. *J. Am. Chem. Soc.*, **1990**, *112*, 1214–1221.

117. Brouwer, A.M.; Bakker, N.A.C.; Wiering, P.G.; Verhoeven, J.W. *J. Chem. Soc., Chem. Commun.*, **1991**, *16*, 1094–1096.

118. Krijnen, B.; Beverloo, H.B.; Verhoeven, J.W.; Reiss, C.A.; Goubitz, K.; Heijdenrijk, D. *J. Am. Chem. Soc.*, **1989**, *111*, 4433–4440.

119. Verhoeven, J.W.; Scherer, T.; Wegewijs, B.; Hermant, R.M.; Jortner, J.; Bixon, M.; Depaemelaere, S.; De Schryver, F.C. *Rec. Trav. Chim. Pays-Bas*, **1995**, *114*, 443–448.

120. Turro, N.J. *Modern Molecular Photochemistry*, Benjamin/Cummings: Menlo Park, **1978**.

121. Bixon, M.; Jortner, J.; Verhoeven, J.W. *J. Am. Chem. Soc.*, **1994**, *116*, 7349–7355.

122. Syage, J.A.; Felker, P.M.; Zewail, A.H. *J. Chem. Phys.*, **1984**, *81*, 2233–2256.

123. Wegewijs, B.; Hermant, R.M.; Verhoeven, J.W.; Kunst, A.G.M.; Rettschnick, R.P.H. *Chem. Phys. Lett.*, **1987**, *140*, 587.

124. Jortner, J.; Bixon, M.; Wegewijs, B.; Verhoeven, J.W.; Rettschnick, R.P.H. *Chem. Phys. Lett.*, **1993**, *205*, 451.

125. Wegewijs, B.; Verhoeven, J.W. In *Electron Transfer—From Isolated Molecules to Biomolecules, Part One*, Jortner, J.; Bixon, M., eds., Wiley: New York, **1999**; 221–264.

126. Wegewijs, B.; Ng, A.K.F.; Rettschnick, R.P.H.; Verhoeven, J.W. *Chem. Phys. Lett.*, **1992**, *200*, 357–363.

127. Lauteslager, X.Y.; Bartels, M.J.; Piet, J.J.; Warman, J.M.; Verhoeven, J.W.; Brouwer, A.M. *Eur. J. Org. Chem.*, **1998**, *11*, 2467–2481.

128. Lauteslager, X.Y.; Stokkum, I.H.M. van; Ramesdonk, H.J. van; Brouwer, A.M.; Verhoeven, J.W. *J. Phys Chem. A*, **1999**, *103*, 653–659.

129. Lauteslager, X.Y., PhD. thesis, University of Amsterdam, **1998**.

130. Lauteslager, X.Y.; Stokkum, I.H.M. van.; Ramesdonk, H.J. van; Bebelaar, D.; Fraanje, J.; Goubitz, K.; Brouwer, A.M.; Verhoeven, J.W. *Eur. J. Org. Chem.*, **2000**, *submitted.*

131. Schneider, S.; Jäger, W.; Lauteslager, X.Y.; Verhoeven, J.W. *J. Phys. Chem.*, **1996**, *199*, 8118–8124.

132. Jäger, W.; Schneider, S.; Verhoeven, J.W. *Chem. Phys. Lett.*, **1997**, *270*, 50–58.

133. Wegewijs, B.; Scherer, T.; Rettschnick, R.P.H.; Verhoeven, J.W. *Chem. Phys.*, **1993**, *176*, 349–357.

CHAPTER

5

Conformational Aspects of Organic Photochemistry

Giampiero Bartocci, Anna Spalletti, and Ugo Mazzucato

Dipartimento di Chimica, Università di Perugia, Perugia, Italy

1 Introduction

1.1 Internal Rotation and Conformational Equilibria

A conformational equilibrium occurs when a molecule can exist in different preferred conformations (stereoisomers) characterized by distinct minima of potential energy. The latter correspond to different spatial arrangements resulting from restricted internal rotation of a specific group around a single bond.[1]

Several physical methods can be useful in studying this phenomenon. X-ray diffraction and electron diffraction, as well as various kinds of spectroscopic techniques (microwave, infrared and Raman, electronic, and magnetic resonance), dielectric measurements, and optical rotatory dispersion,[2] have been used to determine the structure of conformational isomers (conformers or rotamers).

Since the conformers are structurally different species, they may differ in the properties of their electronically excited states. Therefore, the photophysical and photochemical behavior of the mixture and the abundances of the possible photoproducts depend on the initial composition of their ground-state mixtures.

1.2 Restricted Rotation About sp^2–sp^2 Carbon–Carbon Single Bonds

Torsional isomerism about hybridized sp^2–sp^2 carbon–carbon single bonds in unsaturated conjugated molecules, such as 1,2-diarylethenes (DAEs),[3–5] has been,

Conformational Analysis of Molecules in Excited States, Edited by Jacek Waluk
ISBN 0-471-29707-0

and is being, investigated in detail in the authors' laboratory.[3c,4] These compounds offer nice examples of the effect of rotamerism on the excited-state behavior and will be treated mainly in this chapter. The interplay between conjugative and steric effects leads to a type of diastereoisomerism (two isomers that are not mirror-image stereoisomers) and determines the stability of these species. The most stable conformation is generally the antiperiplanar (anti or *s-trans*, where *s* refers to conformations originated by rotation about single bonds)[6] conformer, which has a maximum of conjugation and a minimum of steric interaction. By rotation of 180°, one obtains a synperiplanar (syn or *s-cis*)[6] higher energy conformer characterized by greater steric repulsion (see, as an example, the conformers of 1-styrylnaphthalene (1-StN) in Scheme **1**).

Planarity is markedly favored when steric interference is not particularly strong. The existence of conformational isomers or rotamers, differing by about 180° rotation of one or more of their rotatable groups about one or more of their quasi-single C–C bonds, is a well-recognized phenomenon in solutions of organic compounds. These conformers can differ, more or less evidently, in their spectral absorption and emission properties and can, in addition, have a different photoreactivity. If their permanence in the excited state is long enough, they can also interconvert during its lifetime, thus leading to conformer mixtures of excited-state composition different from that in the ground state. However, since flexible molecules with conjugated double bonds undergo reversal of single–double-bond order in the excited states, as predicted by MO calculations,[7] the equilibration of conformational isomers, which has generally small torsional barriers in the ground state, is often inhibited by higher barriers in the lowest excited states of singlet and triplet multiplicity, S_1 and T_1. This consideration led Havinga and co-workers[8] to establish the "nonequilibration of excited rotamers" (NEER) principle, which assumes the existence of noninterconverting excited rotamers during their short lifetime. The consequence of this assumption is that the experimentally observed dependence of the fluorimetric behavior of such compounds on the excitation wavelength (λ_{exc}) reflects differences in the absorption spectra of distinct rotamers and their relative abundance in the ground state.[3–5]

During the last two-to-three decades, various research groups concerned with the excited-state properties of flexible molecules such as arylethenes, diarylethenes (DAEs), and diarylpolyenes (DAPs), paid a great interest in the study of the conformational equilibria existing in solutions of these compounds with the help of stationary and pulsed fluorescence techniques.[3–5]

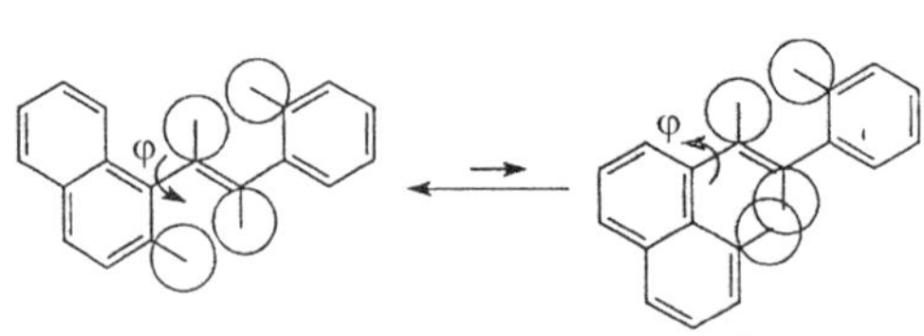

s-trans *s-cis*

trans-1-styrylnaphthalene

Scheme 1

Internal (restricted) rotation about the quasi-single Ar–C bonds connecting the aryl groups with the ethenic bridge, or the quasi-single C–C bonds of the conjugated chain of polyenes, leads to the existence of at least two discrete conformational isomers (quasi-coplanar and quasi-isoenergetic) separated by a relatively small energy barrier, which permits their interconversion and equilibration in the ground state. After pioneering work on molecules containing rotatable chromophoric groups such as vinylanthracenes,[9] biphenols, and biaryls,[10] an in-depth investigation into the rotamerism of ethene derivatives started at the end of the 1970s and focused on DAEs of the general formula Ar–CH=CH–Ar′ whenever Ar or Ar′ do not have C_{2v} symmetry. In a study of the *trans* isomer of 2-styrylnaphthalene (2-StN),[11] the existence of a dynamic equilibrium between distinct, yet quasi-isoenergetic, rotamers was hypothesized. Along these lines, other groups started to study rotamerism in DAEs, mainly by using the effect of λ_{exc} on the emission spectra of solutions of these compounds and assuming the validity of the NEER principle.[3–5,12] Most of the work done relies on the application of this principle. Much less investigated are systems in which changes in the relative abundancies of the rotamers occur in the lowest excited states of singlet or triplet multiplicities. However, when the interconversion torsional barrier for internal rotation is small or the excited-state lifetime is long, the behavior of the irradiated system is no longer related to the composition of the mixture in the ground state (see Section 3).

An exhaustive survey of the many series of ethene derivatives, for which examples of the rotameric behavior had been described, can be found in a review article on DAEs.[4] In this chapter, we intend to focus our attention on a few selected series of compounds, investigated in depth by various research groups (e.g., the naphthyl and anthryl derivatives), for which information available in the literature is so extensive as to allow an accurate description of their ground- and excited-state properties. We will show that these series represent two peculiar classes of compounds whose properties can be useful for the interpretation of the behavior of many other classes of more or less strictly related compounds.

Since the rotamers have different photophysical properties, they can have, in principle, different photoreactivities. Several systems were met whose radiative and reactive decays of the distinct rotamers have quite different quantum yields. However, it may happen that not only the yield, but also the nature of the photo-products, is different for the components of the rotamer mixture (see Section 4).

2 Methods of Investigation

2.1 Experimental Evidence of Ground-State Rotamers

The experimental study of the ground-state rotamers can be performed by direct spectroscopic techniques.[2] However, most of the studies reported in the literature were done by the indirect method of emission spectrometry (stationary and time-resolved), which can give the relative abundance of the rotamers only if the interconversion rate in the excited state is slow with respect to the other relaxation processes and does not change the ground-state composition of their mixtures. A brief review of the methods used in the studies follows.

2.1.1 Vibrational Spectrometry

If the IR spectrum is due to the superposition of two (or more) individual spectra, one could, in principle, separate the components absorbing at slightly different frequencies. Moreover, the relative intensities are expected to change with temperature due to a change in composition of the rotamer mixture. This technique allows the determination of the enthalpy difference between rotamers and the energy barriers for their interconversion.[2] The temperature effect on a pair of close Raman bands around $1630\,cm^{-1}$, assigned to the C=C stretching mode of the two rotamers, was studied for the well-known case of *trans*-2-styrylnaphthalene (2-StN).[13] The change in the relative intensity of the two bands over a large temperature range allowed a ΔH value of $0.71\,kcal\,mol^{-1}$ to be derived, in excellent agreement with the value obtained by fluorimetry ($0.68\,kcal\,mol^{-1}$ in *n*-hexane).[14] No sign of such doubling of the C=C stretching band was found for *trans*-1-styrylnaphthalene (1-StN), thus confirming the hypothesis that, in this case, because of the steric strain of one rotamer, the equilibrium is almost completely shifted towards the more stable form (see Scheme **1**).[3b,3c]

In addition to Raman spectroscopy of the ground states, the time-resolved resonance Raman spectroscopy of the lowest excited states can also be used to get information on the conformational equilibria and to test the validity of the NEER principle.[15]

2.1.2 Magnetic Resonance Spectrometry

Application of the NMR technique can be useful in studying the various nuclei of distinct rotamers, which are in slightly different magnetic environments only at low temperatures when the rate of their interconversion is slow with respect to the inherent timescale of the nuclear transitions. However, at room temperature, some information can also be obtained from the statistically averaged values of the spin–spin coupling constants J and chemical shifts δ (which, due to changes in the composition of the mixture, are expected to be dependent upon temperature). The $J(\delta)$ values of the individual rotamers and their thermodynamical parameters can be obtained. Several examples can be found in the literature wherein NMR spectroscopy gave important information in the conformational study of organic molecules, particularly of biological interest, such as tryptophan and its derivatives.[16] In the latter case, the technique was also used to get information on rotamer interconversion in the lowest excited singlet state (see Section 3.1.2).[16b]

On the other hand, few NMR studies of rotamerism of DAEs are described in the literature. One might recall that the preferential conformation of the geometrical isomers of substituted DAEs bearing pyrrole, furane, and thiophene pentatomic groups was argued on the basis of the chemical shifts of the ethenic and aromatic protons.[17] More examples can be found in the case of polyenes (Section 3.3).

The shorter timescale of the electron–nuclear double-resonance (ENDOR) spectroscopy, whereby the nuclear resonances are indirectly detected by their influence on the EPR lines, allows one to study the conformational equilibria even in the presence of low rotational barriers, of the order of 5 kcal mol^{-1}, as is generally found for DAEs.[4] However, it is applicable only to DAE radical ions. An ENDOR study of the 3,3′-dimethylstilbene radical anion at 183 K gave information on the relative stability of the three possible rotamers of this compound.[18]

2.1.3 Dipole Moments

The total electric dipole moment μ of a molecule can be useful in providing conformational information. Using the vectorial contributions of individual groups or measuring the μ-value of suitable model compounds, one can relate the experimental dipole moment to the contribution of the distinct conformers and then obtain information on the composition of the mixture, the equilibrium constant, and the enthalpy difference. Preferential conformations of DAE molecules in solution were first discussed in the literature for some *trans*-styrylpyridines and *trans*-dipyridylethenes on the basis of dielectric measurements.[19]

2.1.4 Flash Photolysis

Both laser (picosecond and nanosecond) and conventional (microsecond) flash photolysis can be useful in the direct study of rotamerism. Fast absorption techniques can give information similar to that obtained by pulsed fluorimetry on the lowest excited singlet state, while the conventional version can provide the properties (spectra, quantum yields, lifetimes) of the longer lived triplets. Conventional flash photolysis was used mainly by Fischer and co-workers to study the $T_1 \rightarrow T_n$ absorption of some DAEs, particularly those containing an anthryl group, in the microsecond time domain.[20] Nanosecond laser flash photolysis was used by Das et al.[21] Important conclusions about the exceptions to the NEER principle were thus obtained (see Section 3).[22] Faster techniques, which use mode-locked lasers to excite the compounds in the picosecond domain, allow one to observe the $S_1 \rightarrow S_n$ transitions and then to obtain complementary information on rotamerism a very short time after excitation.[23]

2.2 Computational Techniques

Theoretical calculations are of fundamental importance in the study of rotamerism. On their basis, one could predict the energy difference and then the composition of the mixture of the ground-state species and estimate the potential-energy barriers for internal rotation in both the ground and lowest excited states. Given the difficulty of using ab initio methods for such large molecules, the literature studies are concerned mainly with semiempirical MO calculations with different levels of sophistication. A survey of the methods (PPP, CNDO, INDO, MINDO, MNDO, QCFF/PI) used for the study of DAE rotamers is reported in a review article,[4] in which the successful applications of a particular modified method, the so-called CS-INDO, are described to show how they help in the study of the thermodynamic and kinetic aspects of rotamerism in the lowest excited states of singlet and triplet multiplicity. Chapter 6 of this text describes how theoretical calculations may help in the study and interpretation of the rotamerism phenomenon.

2.3 Methods Based on Fluorimetry

The fast fluorescence process has generally a response time short enough to avoid rotamer interconversion during the lifetime of the lowest excited singlet state. Indeed, fluorimetry is the most powerful and widely used technique for studying the rotamerism of compounds characterized by nonnegligible emission yield. The success of fluorimetry in the rotamer study relies on the validity of the NEER principle,[8] because only in such a case does the fluorescence behavior reflect the composition of the ground-state rotamer mixture. Since, in principle, different rotamers may have more or less different kinetic parameters of the competitive relaxation processes from the lowest excited state, including the reactive process, their relative abundance also may affect the overall quantum yield of photoreaction. Various approaches can be used, some of them based on stationary techniques, others on pulsed techniques. These methods, which lead to results that are more or less informative, are of different degrees of difficulty and sophistication.

2.3.1 Stationary Techniques: Excitation Wavelength Effect

Since the absorption and emission spectra of rotamers that differ in their electronic distribution and steric strain are generally slightly different, as are their enthalpies, the amount of light absorbed and emitted by each species can be modulated by changing λ_{exc} and the temperature (Figs. 5.1 and 5.2). The excitation spectrum also changes with the monitoring wavelength (λ_{em}) and differs from the absorption spectrum. These λ_{exc} and λ_{em} effects are the preliminary step of an investigation into conformational equilibria. A lot of examples can be found in the literature, particularly in the case of DAEs.[4] Since the absorption and fluorescence spectra of one rotamer are red-shifted compared with the others', selective excitation at the tail of the absorption spectrum can give, in favorable cases, its quasi-pure emission spectrum, while excitation at shorter wavelengths gives a spectrum that is the superposition of at least two components. Similarly, the quasi-pure excitation

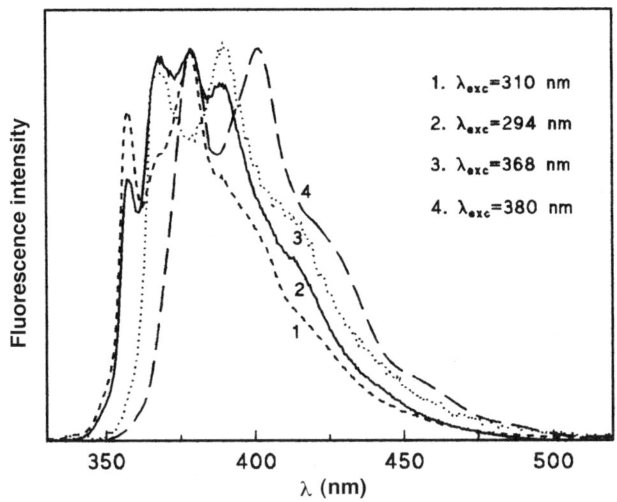

Figure 5.1. Normalized emission spectra of *trans*-1,2-di-(2-naphthyl)ethene in MCH/3MP as a function of λ_{exc} at room temperature.

spectrum of the hypsochromic rotamer can be obtained by using a monitoring wavelength, λ_{em}, near the blue edge of the emission spectrum. Recourse to temperature effects on the spectral properties or more complex analyses are needed to resolve the mixture spectrum when the absorption edges are practically coincident. The λ_{exc} effect on the fluorescence quantum yield can also lead to informative results on the radiative and nonradiative relaxations of the rotamers.

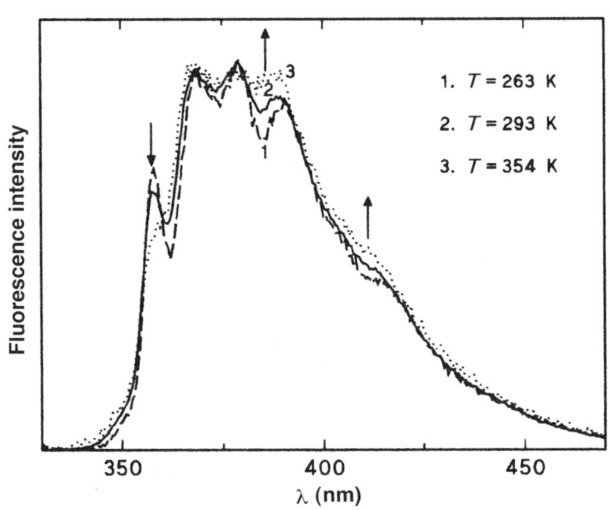

Figure 5.2. Normalized emission spectra of *trans*-1,2-di-(2-naphthyl)ethene in MCH/3MP at different temperatures ($\lambda_{exc} = 290\,nm$).

2.3.2 Time-Resolved Techniques: Lifetime Measurements

If the rotamers have different lifetimes, the bi(multi)exponential decay response function is highly informative, giving the separate lifetimes, information about which is necessary to estimate their relative abundancies. In the case of a two-component system, provided that the NEER hypothesis is valid, changes in the emission intensity with time can be expressed by the sum of two exponentials:

$$\bar{i}(t) = A_A \exp\left(-\frac{t}{\tau_{F,A}}\right) + A_B \exp\left(-\frac{t}{\tau_{F,B}}\right). \tag{1}$$

From the deconvolution treatment, one obtains the fluorescence lifetimes of the short-lived ($\tau_{F,A}$) and longer lived ($\tau_{F,B}$) rotamer and the preexponential factors, whose ratio, dependent on λ_{exc} and λ_{em}, is proportional to the relative contributions of the two species to the emission intensity. When the rotamers have different lifetimes, spectral resolution can be obtained by time-resolved fluorimetry. The late-gated spectrum is expected to reflect the shape of the longer lived rotamer.[3d,24]

2.3.3 Quantitative Methods: The Kinetic Fluorescence Analysis (KFA)

The kinetic fluorescence analysis (KFA), a method that was formulated by J. Birks in collaboration with our laboratory,[24,25] is particularly useful for the treatment of systems constituted by two fluorescent components having different lifetimes. Indicating the mixture properties by barred symbols, one can write the total molar concentration as $\bar{c} = c_A + c_B$ and the observed absorption spectrum of the mixture as

$$\bar{\varepsilon}(\lambda_{exc}) = \Sigma\varepsilon_i(\lambda_{exc})\frac{c_i}{\bar{c}}, \tag{2}$$

where $i = A, B$. The fractions of excited molecules of each species at different λ_{exc} are

$$f_i(\lambda_{exc}) = \frac{\varepsilon_i(\lambda_{exc})c_i}{\bar{\varepsilon}(\lambda_{exc})\bar{c}}. \tag{3}$$

The fluorescence quantum yield can be expressed as

$$\bar{\phi}_F(\lambda_{exc}) = \Sigma f_i(\lambda_{exc})\phi_{F,i}, \tag{4}$$

and the fluorescence intensity is

$$\bar{F}(\lambda_{exc}, \lambda_{em}) = \Sigma f_i(\lambda_{exc})F_i(\lambda_{em}). \tag{5}$$

The fluorescence response function (Eq. (1)), measured at λ_{em} following excitation at λ_{exc}, is

$$\bar{i}(t) = \Sigma f_i(\lambda_{exc})F_i(\lambda_{em})k_i \exp(-k_i t), \tag{6}$$

where $k_i = 1/\tau_{F,i}$. The equation of the fluorescence excitation spectrum of a dilute solution, describing the dependence of the emission intensity at a fixed λ_{em} on λ_{exc}, is given by

$$\bar{E}(\lambda_{exc}, \lambda_{em}) = \bar{\varepsilon}(\lambda_{exc})\bar{c}\Sigma f_i(\lambda_{exc})F_i(\lambda_{em}). \tag{7}$$

Deconvolution analysis gives the ratio of the preexponential factors of Eq. 1, viz.,

$$\alpha(\lambda_{exc}, \lambda_{em}) = \frac{A_A}{A_B} = \frac{f_A(\lambda_{exc})F_A(\lambda_{em})k_A}{f_B(\lambda_{exc})F_B(\lambda_{em})k_B} \tag{8}$$

and the derived parameter

$$\beta(\lambda_{exc}, \lambda_{em}) = \alpha(\lambda_{exc}, \lambda_{em})\frac{k_B}{k_A} = \frac{f_A}{f_B}(\lambda_{exc})\frac{F_A}{F_B}(\lambda_{em}). \tag{9}$$

The analysis is simplified if the overlapping emission spectra of similar quantum efficiency have one or more isoemissive wavelengths, λ'_{em}, at which $\bar{F}(\lambda'_{em})$ is independent of λ_{exc}, namely, when $\bar{F} = F_A = F_B$. These isoemissive points are derived by a set of fluorescence spectra, observed at different λ_{exc} and normalized to the same absorbed quanta in order to make the ratio of spectral area to the fluorescence quantum yield constant, independently of λ_{exc}. At any λ'_{em} the parameters

$$f_A(\lambda_{exc}) = \frac{\beta(\lambda_{exc})}{1 + \beta(\lambda_{exc})} \tag{10a}$$

and

$$f_B(\lambda_{exc}) = \frac{1}{1 + \beta(\lambda_{exc})} \tag{10b}$$

can be evaluated from the foregoing treatment.

In this way, the distinct (relative) absorption spectra (Eq. (3)) and fluorescence parameters can be derived.

At the isoemissive frequency, the contribution of each component to the overall fluorescence decay % (*i*) is the fraction of excited molecules at $\lambda_{exc}, f_i(\lambda_{exc})$. The observation of the fluorescence at λ'_{em} as a function of λ_{exc} allows the relative absorption spectrum, $\varepsilon_i(\lambda_{exc})c_i/\bar{c}$, of the two rotamers to be obtained. The emission spectra of the rotamers can also be directly obtained by measurements of the decay as a function of λ_{em}, using Eqs (5) and (9) at a fixed λ_{exc}.[25b]

Once the 0–0 absorption band of the first electronic transition is known from the relative absorption spectra of the two rotamers, if it is assumed that the Franck–Condon envelopes of the two species are similar in shape, one can derive, from the radiative lifetime relation, the equation

$$\frac{\varepsilon_{00,A}}{\varepsilon_{00,B}} \approx \frac{k_{F,A}}{k_{F,B}} = \frac{(\phi_{F,A}\tau_{F,B})}{(\phi_{F,B}\tau_{F,A})}, \tag{11}$$

where $k_{F,i}$ represents the radiative rate constant. The relative abundance of the two species is thus obtained from

$$\frac{c_A}{c_B} = \frac{(\varepsilon_A)_{00} c_A k_{F,B}}{(\varepsilon_B)_{00} c_B k_{F,A}} \tag{12}$$

From the abundance ratio, assuming that the ground-state equilibrium is established in solution and that the entropy term can be neglected, the enthalpy difference between the two species, ΔH, can be derived. This value can be compared with that obtained from the plot of $\log \beta(\lambda_{exc})$ measured at λ'_{em} against $1/T$, or the plot $\log \alpha(\lambda_{exc})$ versus $1/T$ in case the fluorescence lifetime remains constant in the T range explored, since α is then proportional to the abundance ratio.[25] In cases where the emission quantum yields of the two rotamers are dissimilar and an isoemissive wavelength cannot be found, a more complex procedure has to be adopted.[24]

2.3.4 Quantitative Methods: The Principal-Component Analysis (PCA)

The principal-component analysis (PCA) is a mathematical method based on the statistical treatment of data obtained from mixtures of two or more components.[26–29] The method offers a way to obtain the number and the excitation and emission spectra of the distinct components. Sets of fluorescence spectra obtained by varying an external parameter R affecting the composition of the mixture (e.g., λ_{exc}, temperature, or quencher concentration) are used to build an $m \times n$ data matrix such that each of the m row vectors corresponds to an experimental emission spectrum $\mathbf{S}(R)$ read at n different λ_{em}, namely, $\mathbf{S} = \Sigma \bar{F}(\lambda_{em})$, and n different column vectors correspond to changes in emission intensity, at each λ_{em}, produced by variations of the external parameter. If the variable R is λ_{exc}, the matrix column represents the excitation spectrum of the mixture observed at different λ_{em}. An example related to *trans*-2-styrylanthracene (2-StAn) in toluene at 293 K is shown in Figure 5.3.

Of course, the mixture spectra are linear combinations of the unknown spectra of the pure components. The number of the components corresponds to the number of "significant" eigenvectors necessary and sufficient to reconstruct the data matrix within the errors of the model, plus the instrumental noise in the data (see the example of *trans*-2-StAn in Figure 5.4). The term "significant" indicates the minimum number of the components necessary to detect how much of the variation in the data set is systematic (the explained percentage of the total variance of the measured spectra) and how much is "random noise."

The number of components can be found by several methods.[26b] The best results[26c] are obtained by using the NIPALS algorithm with cross-validatory estimation (a double cross procedure).[26d]

In the case of a two-component system, the experimental spectrum can be represented as a linear combination of two eigenvectors ($\mathbf{V}$):

$$\mathbf{S}(R) = \alpha(R)\mathbf{V}_\alpha + \beta(R)\mathbf{V}_\beta. \tag{13}$$

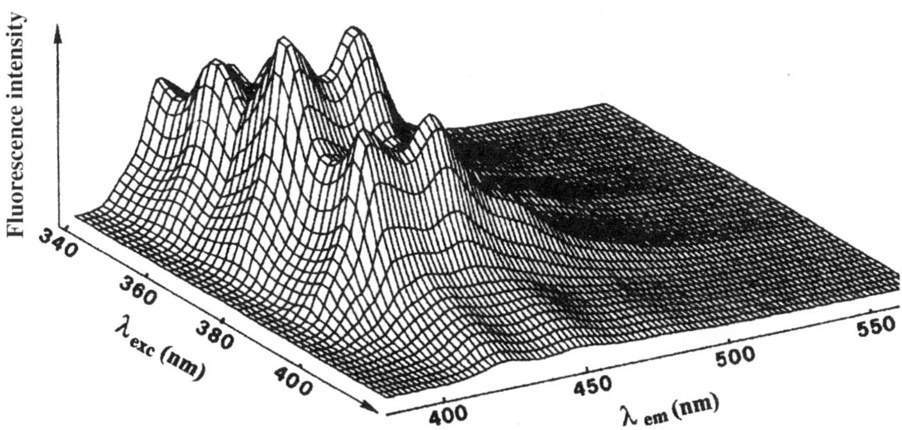

Figure 5.3. Fluorescence data matrix for *trans*-2-styrylanthracene in toluene at 293 K. (Reprinted from Ref. 29; copyright 1992 Elsevier Science.)

Here, the Greek letters represent the *R*th set of combination coefficients. Consequently, it is impossible to unambiguously establish the emission spectrum of the pure components by the PCA method alone. However, Lawton and Sylvestre (LS) combined the PCA method with the self-modeling technique (PCA-SM), imposing certain constraints to limit the ambiguity.[26a]

The λ_{exc}-dependent α and β coefficients, which generate the $\mathbf{S}(\lambda_{exc})$ emission spectra, must fall on a normalization line in the α, β plane defined by the two

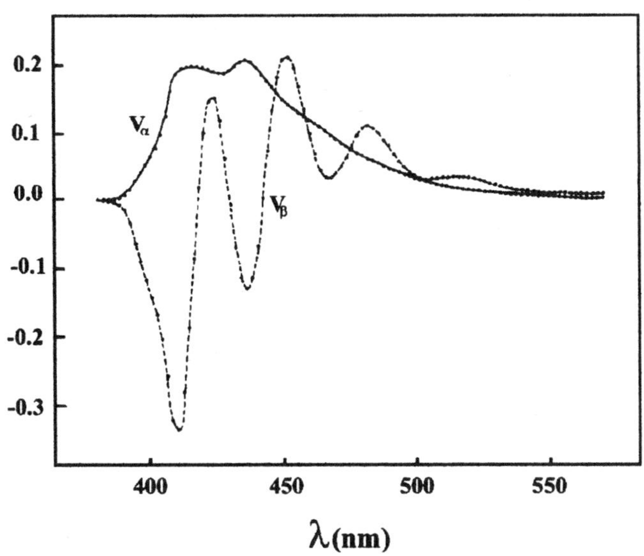

Figure 5.4. $\mathbf{V}_\alpha$ and $\mathbf{V}_\beta$ eigenvectors, as obtained from the fluorescence matrix of *trans*-2-styrylanthracene in toluene at 293 K (from PCA method; reprinted from Ref. 29; copyright 1992 Elsevier Science.)

eigenvectors and by the LS constraint that all the spectra are normalized to unit area $[\alpha(\lambda_{exc})\mathbf{V}_\alpha + \beta(\lambda_{exc})\mathbf{V}_\beta = 1]$. This constraint, together with those relative to the nonnegativity of the emission intensities and the fact that every mixture spectrum must be a positive combination of the pure spectra, significantly limits the ambiguous regions of the calculated pure spectra. In fact, these LS-constraints define two ranges of potential pure-component spectra that correspond to the combination coefficients of the two outer segments of the normalization line, starting with those of the experimental spectra richest in a specific component (see, as an example, the C and D points of Figure 5.5) and ending at the outer combination coefficients (the A and B points of Figure 5.5), which start to generate some percentage of negative spectral intensities.

In the treatment of a two-component system, if at least one λ_{em} exists for each pure-component spectrum, where the fluorescence intensity of the other is zero,[26a] their unknown spectra are expected to be uniquely defined by the coefficients corresponding to the extremes of the uncertainty regions (the A and B points of Figure 5.5; PCA-LS method). The λ_{exc}-independent emission spectra of the two pure components, $\mathbf{S}_A$ and $\mathbf{S}_B$, will then be given by

$$\mathbf{S}_A = \alpha_A\mathbf{V}_\alpha + \beta_A\mathbf{V}_\beta \tag{14}$$

and similarly for B.

The PCA-LS procedure was applied to the spectrothermal matrix of the UV spectra of *trans*-2-StAn,[28] to the excitation–emission matrix of the *trans* isomers of 2-styrylnaphthalene (2-StN),[27a] 1,2-di-(2-naphthyl)ethene (2,2′-DNE),[27c] 3-styrylphenanthrene (3-StPh),[29] 2-styrylanthracene (2-StAn),[29,30] and 1-(2-naphthyl),2-(6′-quinolyl) ethene (2,6′-NQE).[31]

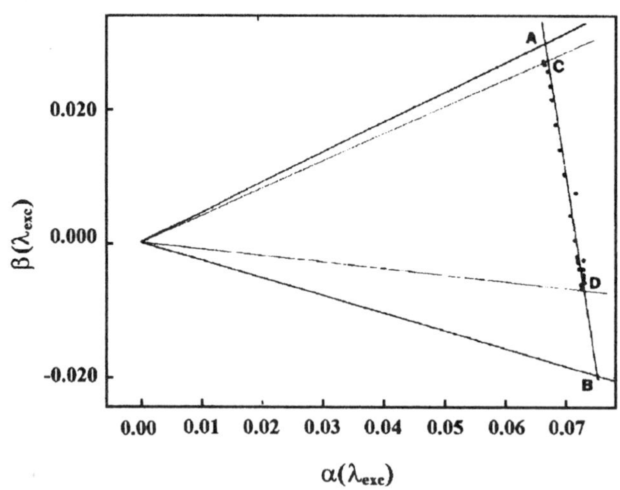

Figure 5.5. Plot of $\beta(\lambda_{exc})$ against $\alpha(\lambda_{exc})$ for *trans*-2-styrylanthracene in toluene at 293 K. The normalization line is drawn through the experimental points, from the PCA-LS method. (Reprinted from Ref. 29; copyright 1992 Elsevier Science.)

Another PCA-SM treatment, which introduces the further constraint that the Stern–Volmer (SV) quenching constant of each rotamer for a specific added quencher (see Section 2.4.5) be independent of λ_{exc} (PCA-SV), was applied to the fluorescence of the *trans* isomers of 2-StN,[32] 2,2′-DNE,[27d] and 2-StAn.[33,34] This different procedure is necessary to obtain unique conformer spectra in the case where the two emission spectra span nearly identical wavelength ranges. In such a situation, the LS constraint cannot be applied.[26a,27] For other details on the complications involved, one should address the specialized literature.[26]

The unit-area normalized experimental fluorescence spectrum, $N(\lambda_{exc})\mathbf{S}(\lambda_{exc})$, can also be given as the sum of the fractional contribution of the two pure spectra, $X_i(\lambda_{exc})\mathbf{S}_i$, by

$$N(\lambda_{exc})\mathbf{S}(\lambda_{exc}) = X_A(\lambda_{exc})\mathbf{S}_A + X_B(\lambda_{exc})\mathbf{S}_B = 1, \tag{15}$$

where $N(\lambda_{exc})$ is the normalization factor and the $X_i(\lambda_{exc})$ values can be derived using Equation (16) from the pure-component coefficients and those generated by the mixture spectra:

$$X_B(\lambda_{exc}) = \frac{\alpha(\lambda_{exc}) - \alpha_A}{\alpha_B - \alpha_A} = \frac{\beta(\lambda_{exc}) - \beta_A}{\beta_B - \beta_A}, \tag{16a}$$

and

$$X_A(\lambda_{exc}) = \frac{\alpha_B - \alpha(\lambda_{exc})}{\alpha_B - \alpha_A} = \frac{\beta_B - \beta(\lambda_{exc})}{\beta_B - \beta_A}. \tag{16b}$$

The unit-area normalization leads to pure-component spectra that are also normalized ($\mathbf{S}_i = 1$). Equation (15) reduces to

$$\mathbf{S}(\lambda_{exc}) = \frac{X_A(\lambda_{exc})}{N(\lambda_{exc})} + \frac{X_B(\lambda_{exc})}{N(\lambda_{exc})}, \tag{17}$$

where the distribution of the two terms on the right displays the shape of the fluorescence excitation spectra. Since the experimental quantum yield $\bar{\phi}_F(\lambda_{exc})$ is proportional to the area of the emission spectrum $\mathbf{S}(\lambda_{exc})$, one can relate the contributions of A and B to the emission to the corresponding fractions of excited molecules of each species:

$$X_i(\lambda_{exc})\bar{\phi}_F(\lambda_{exc}) = f_i(\lambda_{exc})\phi_{F,i}. \tag{18}$$

If there is some wavelength at which only one species (i) absorbs, thus allowing its yield $\phi_{F,i}$ to be measured by selective photoexcitation, $f_i(\lambda_{exc})$ can be obtained by Equation (18), and then the relative absorption spectra and abundances of the two species can be determined.

Provided that $\phi_{F,i}$ values of the pure components are independent of λ_{exc}, the quantitative relationship[32,33] between $\bar{\phi}_F$ and X_A is given by

$$\frac{1}{\bar{\phi}_F(\lambda_{exc})} = \frac{1}{\phi_{F,B}} + \left(\frac{1}{\phi_{F,A}} - \frac{1}{\phi_{F,B}}\right)X_A(\lambda_{exc}). \tag{19}$$

The plot of the extensive set of $\bar{\phi}_F$ as a function of the fractional contribution of one component allows the $\phi_{F,i}$ values of the rotamers to be obtained.

2.3.5 Differential Quenching

When the fluorescence lifetimes of the rotamers are significantly different, one can resort to a selective quenching of the longer-lived component B, combined with selective excitation, to obtain the quasi-pure spectrum of A, which is less quenched by a factor of about $\tau_{F,A}/\tau_{F,B}$. The rate constant ($k_{Q,i}$) for fluorescence quenching of a distinct rotamer by an added concentration of a quencher $Q(c_Q)$ is given by the usual expression,

$$k_{Q,i} = \frac{\tau_{F,i} - \tau_{F,i(Q)}}{\tau_{F,i}\tau_{F,i(Q)}c_Q}. \tag{20}$$

For a strong quencher, like oxygen, it is to be expected that $k_{Q,A} = k_{Q,B}$. The Stern–Volmer constant for the fluorescence yield of each rotamer i is given by

$$K_{Q,i} = \frac{\phi_{F,i} - \phi_{F,i(Q)}}{\phi_{F,i(Q)}c_Q} = k_{Q,i}\tau_{F,i}, \tag{21}$$

while the Stern–Volmer constant of the mixture is

$$\bar{K}_Q(\lambda_{exc}) = \frac{\bar{\phi}_F(\lambda_{exc}) - \bar{\phi}_{F(Q)}(\lambda_{exc})}{\bar{\phi}_{F(Q)}(\lambda_{exc})c_Q}, \tag{22}$$

which is a complex function of the fraction of excited molecules f_i. The method of differential quenching was used extensively by Saltiel and co-workers in the study of naphthyl- and anthryl-styrenes.[32–34]

2.3.6 Frozen-In Nonequilibrium Mixtures of Rotamers

Another method for obtaining information on distinct rotamers consisted in freezing in nonequilibrium rotamer mixtures of *trans*-DAEs by irradiation of the *cis* isomers in rigid matrices.[35] The *trans* → *cis* process is more inhibited than the *cis* → *trans* one in viscous media. Therefore, one can photochemically produce the *trans* isomer at sufficiently low temperatures and high viscosities such that thermal equilibration between rotamers no longer occurs.[35] Since the rotamer composition is then controlled by the conformational equilibrium of the *cis* isomer (which is expected to differ from that of the *trans* isomer), the detection of a particular species can be favored. Comparison of the spectra recorded at low temperature with those obtained after heating the solution (to get equilibration) and refreezing proved to be a useful tool in generating information on the contribution of the different species to the spectra of the mixture. It is important to note that by this method it is possible to compare, under identical conditions, the emission spectra of a mixture of two components with largely different compositions. The freezing-in method may reflect a general phenomenon of preferential enrichment of the mixture in a specific rotamer when the system is included in an organized medium (e.g., by crystallization or incorporation into a host crystal).

2.3.7 Emission in Shpolskii Matrices

Quasi-line fluorescence emission and excitation spectra can be obtained in *n*-alkane frozen (Shpolskii) matrices at liquid-helium temperatures. The high-resolution characteristics of this method may lead to detection of the pure electronic spectra of distinct rotamers and to the assignment of the corresponding structures. The technique was used for the study of DAE rotamerism,[36] giving interesting results, but also evidencing some problems related to the shift of the equilibrium towards the most stable species at low temperatures and to the role played by the matrix.

2.3.8 Structural Information by Comparison with Restricted Analogues

Structural identification of rotamers was often based on the relative positions of their absorption and emission spectral properties or on information obtained by theoretical calculations of the energy gap between their ground and first excited electronic states. Other more specific and very useful methods for the assignment of the resolved spectra of rotamers of DAEs to specific conformations were based on the analysis of their photoproducts or on a comparison with analogous compounds that, because of their structure or for steric reasons, are constrained to a single conformation. When steric hindrance by convenient substituents is introduced, one of the possible conformations will often predominate in the equilibrium mixture, depending on the nature and position of the substituents.

The 3-methyl derivatives of 1,2-di-(2-naphthyl)ethene (DNE) were used by Saltiel et al. as guides to obtaining the resolution of three-component emission spectra of DNE measured as a function of λ_{exc} and $[O_2]$.[27b] These authors successfully applied the PCA method by using a matrix of experimental spectra of DNE and the slightly blue-shifted 3-methyl-DNE (two rotamers expected) and 3,3′-dimethyl-DNE (one rotamer expected; Scheme **2**). The substituted compounds helped Saltiel and his colleagues in the selection of pure-component combination coefficients for the three expected rotamers of DNE.[5]

An interesting example of the helpful use of restricted analogues is also offered by *trans*-2-StN. The structural assignment of the short-lived (A) and longer-lived (B) conformers to the *s-cis* and *s-trans* species, respectively,[4,5,32] was confirmed by comparison of their resolved spectra with those of the 1- and 3-methyl derivatives and of the rigid compound 3-styrylidenebenz[e]indane (3-StBI), which are sterically and structurally constrained analogues, respectively (Scheme **3**).[5,32]

Other examples of the method of constrained analogues are reported in the sections that follow.

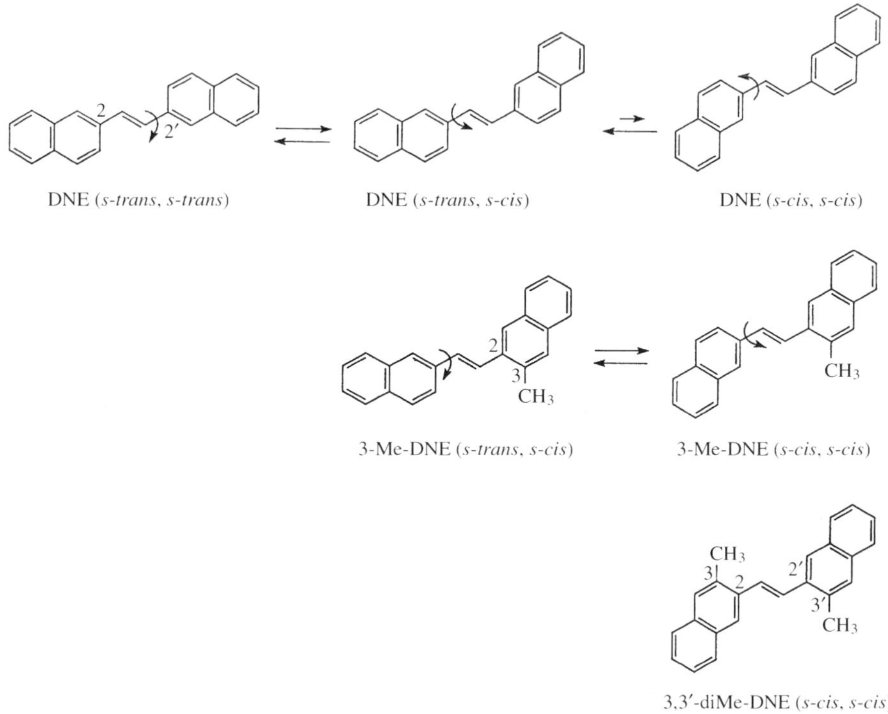

Scheme 2

2.4 Evidences of Rotamer Interconversion in the Excited States (Non-NEER Behavior)

Structural factors reducing the height of the energy barrier to the *s-trans** → *s-cis** interconversion or long excited-state lifetimes can allow the interconversion of excited rotamers against the NEER principle.[37]

In compounds where the excitation energy is largely localized on the aryl group, the bounds at the ethene bridge do not undergo the usual reversal of the single/double-bond order upon excitation in both of the lowest excited states of singlet and triplet multiplicity.[22,38] Therefore, torsional barriers undergo minor changes upon excitation, with the consequence that *trans* → *cis* isomerization around the double bond is inhibited, while rotamer interconversion around the single bond can occur.

Two cases of excited rotamer interconversion in the excited states are possible: (*i*) a two-way process (the equilibrated case), and (*ii*) a one-way process (the non-equilibrated case). In the first case, the basic condition is that the rate constants of rotamer interconversion are about equal to the inverse of the corresponding lifetimes. This requirement is just met in those systems in which the excitation energy, essentially localized in the large aromatic groups, decreases the order of the C–C

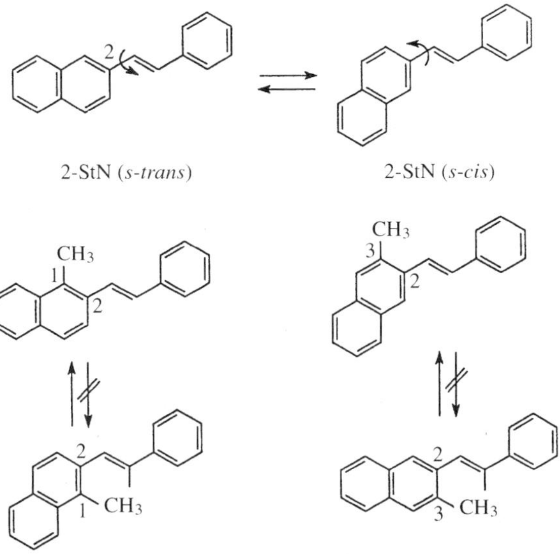

2-StN (*s-trans*) 2-StN (*s-cis*)

1-Me, 2-StN 3-Me, 2-StN

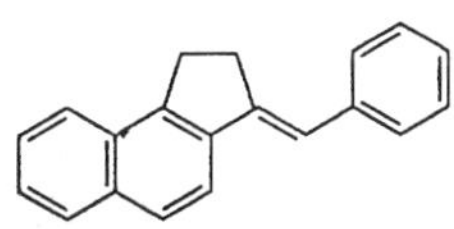

3-StBI

Scheme 3

bond between the aromatic and ethenic moieties in both S_1 and T_1 states. Another opportunity is offered by the longer lifetimes of the triplet states compared with the singlet lifetimes.

The experimental pieces of evidence of two-way interconversion are as follows: (*i*) the shape of the emission spectrum is independent of λ_{exc} and changes with temperature in isoviscosity conditions; (*ii*) the shape of the fluorescence excitation spectrum is independent of λ_{em} and reproduces the absorption spectrum; (*iii*) the absorption spectrum changes with temperature in isoviscosity conditions; (*iv*) the fluorescence decay is monoexponential (the excited equilibrated rotamers decay with the same temperature-dependent average lifetime) and (*v*) the fluorescence quantum yield of the mixture is independent of λ_{exc}, but could depend on temperature if the rotamers have different $\phi_{F,i}$ values or if activated pathways are present among the deactivation processes of their excited states.

In this case, the relative abundance of the excited rotamers in S_1 or T_1 depends on their enthalpy difference. The standard enthalpy change between the A and B rotamers in the S_1 state can be estimated by the equation

$$\Delta H(S_1) = (\bar{\nu}_{00,B} - \bar{\nu}_{00,A}) - \Delta H(S_0), \tag{23}$$

where $\bar{\nu}_{00}$ is the 0–0 transition energy of the $S_1 \rightarrow S_0$ radiative process of the two rotamers and $\Delta H(S_0)$ is the standard enthalpy change in the ground state.

In the nonequilibrated case, the height of the energy barrier of the excited-state interconversion of one rotamer (e.g., A*) into the more stable one (e.g., B*) must be such as to make the rate constant ($k_{A^* \rightarrow B^*}$) about equal to the inverse of the lifetime of the rotamer ($1/\tau_{F,A}$), while that of the opposite process must be much lower, namely, $(k_{B^* \rightarrow A^*}) \ll (1/\tau_{F,B})$.

Some experimental evidence of one-way interconversion parallels the preceding pieces of evidence reported for systems that obey the NEER principle: (*i*) the shape of the excitation and emission spectra depends on λ_{em} and λ_{exc}, respectively, (*ii*) the absorption spectrum changes with temperature in isoviscosity conditions; (*iii*) the fluorescence decay is multiexponential; and (*iv*) the fluorescence quantum yield of the mixture depends on λ_{exc} and the temperature. Furthermore, in this case unlike the systems which obey the NEER principle, one can observe emission from one rotamer by selective excitation of the other one, moreover, the fluorescence decay, observed at wavelengths where the excited rotamer formed by interconversion from the other one is strongly emitting, shows a rise component (a negative preexponential factor). These observations indicate that one rotamer is being converted into the other in the S_1 state. In the absence of other activated processes (e.g., *cis–trans* isomerization around the double bond) in the deactivation pathways of S_1, one can derive the Arrhenius parameters of the conformational isomerism by the equation

$$\frac{1}{\tau_{F,A}} - \frac{1}{\tau_{F,A}^{lim}} = A_{A^* \rightarrow B^*} \exp\left(\frac{-{}^1\Delta E_{A^* \rightarrow B^*}}{RT}\right), \tag{24}$$

where $\tau_{F,A}^{lim}$ indicates the lifetime of the rotamer A*, which turns into B* at temperatures where the conversion process is prevented.[39]

Overall, the results reported in the literature (specific examples are found in Section 3.1.2) indicate that any fluorimetric study of ground-state rotamerism should start with a preliminary verification that the requirements for the validity of the NEER principle are fullfilled.

3 Typical Examples of Rotamerism in Some Selected Series of Olefins

One of the most intensely studied systems concerns molecules wherein one or both phenyl groups of stilbene are replaced by a 2-naphthyl group, namely, the *trans* isomers of 2-styrylnaphthalene and 1,2-di-(2-naphthyl)ethene. Their conformational equilibria, investigated by several techniques and methodologies (some aspects are mentioned in Section 2), have been described in recent review articles[4,5,32] and will not be further treated here.

Given the availability in the literature of articles that offer an exhaustive panorama of rotamerism studies on a large number of DAEs and related molecules, this section illustrates the behavior of a restricted number of compounds only, just occasionally touching other typical cases. The cases of the anthryl derivatives (a two-component system) and 1-(2-naphthyl),2-(6′-quinolyl)ethene (a three-component system) will be described in detail as representative examples of compounds whose behavior has been satisfactorily explained. The choice of a few other cases was based on compounds for which new results are available or whose rotamerism can affect the yield and direction of photochemical relaxation (Section 4).

3.1 Two-Component Systems: The Case of *Trans*-Anthryl Derivatives

The existence of distinct rotamers in DAEs bearing the anthryl group was demonstrated by emission spectroscopy studies. The experimental observations showed the validity of the NEER principle[8] for some systems, while in other cases the interconversion of excited rotamers in the S_1 or T_1 states was established.

3.1.1 Systems that Respect the NEER Principle

For a number of *trans*-2-anthryl derivatives, the Fischers[40] first observed a variation in the emission spectra with λ_{exc}, which was most pronounced when excitation occurred in the tail of the absorption bands (Fig. 5.6). In all cases, the emission spectra could be described by a superposition of two sets of peaks shifted 8–15 nm one from the other. The oxygen-selective quenching of the short-wavelength

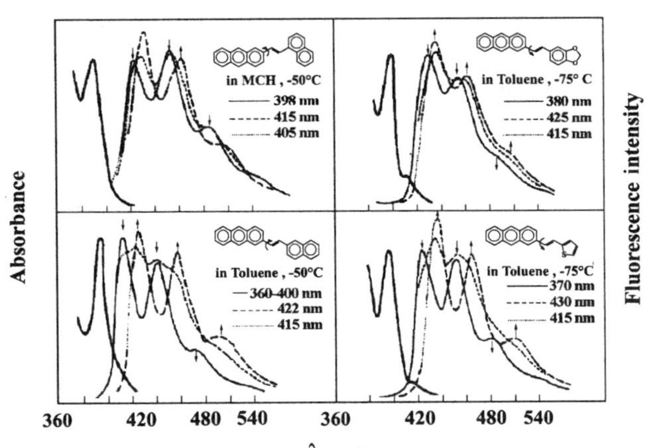

Figure 5.6. Examples of the λ_{exc} effect for some derivatives of *trans*-2-styrylanthracene in nonpolar solvents at different temperatures. (Adapted from Ref. 40; copyright 1981 American Chemical Society.)

absorbing species indicated that the two rotamers have different fluorescence lifetimes.

A subsequent paper[41] on the fluorescence lifetimes and selective quenching of the *trans* geometrical isomers of 2-styrylanthracene (2-StAn) and its 2′-naphthyl (2,2′-AnNE) and 2′-thienyl (2,2′-AnTE) analogues confirmed the previous observations. The good fit of the biexponential decay suggested that two fluorescing species were the predominant contributors in all these compounds, although four structurally distinct rotamers are possible for the naphthyl and thienyl derivatives. Some of these species were suggested either to be very similar in terms of their photophysical properties or to contribute to the emission too weakly to be revealed under low-resolution conditions. The same authors succeeded in obtaining a good separation of the A and B spectra by applying the technique called "phase resolution of spectra,"[42] as shown in Figure 5.7.

The *trans*-2-StAn isomer was also investigated by Ghiggino et al.[28] By time-resolved spectroscopy, the emission spectrum of the two distinct rotamers was detected. The results obtained were in excellent agreement with the earlier ones.[40,41] Since the time-resolved measurements clearly distinguished two separate lifetimes (in toluene at room temperature, $\tau_{F,A} = 9$ ns, and $\tau_{F,B} = 27$ ns) with no observable rise times, no appreciable interconversion of the excited rotamers was supposed to occur in the S_1 state. The temperature dependence of the emission intensities at $\lambda_{em} = 415$ and 452 nm in toluene, assigned to long-lived and short-lived rotamers, respectively, showed that the $\bar{F}_{415}$ signal decreased with an increase in temperature over the 200–294 K range, while $\bar{F}_{452}$ increased, providing evidence for a change in the relative abundances of the ground-state rotamers with temperature.

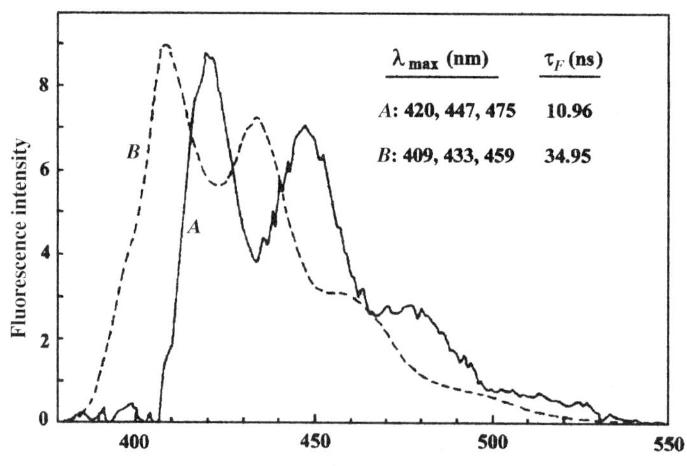

Figure 5.7. Fluorescence spectra of A and B rotamers of *trans*-2-styrylanthracene in toluene ($\lambda_{exc} = 370$ nm), as obtained by the "phase resolution of spectra" technique. (Adapted from Ref. 41; copyright 1984 American Chemical Society.)

More recently, detailed fluorimetric analyses of 2-StAn were carried out as a function of λ_{exc}, temperature, or quencher concentration by the two procedures described, that is KFA[30,43] (Section 2.3.3) and PCA, the latter with both self-modeling methods (PCA-LS[28–30] and PCA-SV[33,34]; Section 2.3.4). These procedures allowed the researchers to obtain the spectral and photophysical properties of the two distinct rotamers from the mixture data and their relative abundance in the ground state at room temperature.

The KFA method was applied to the biexponential decay function of 2-StAn, even though the scarce dependence of the fluorescence quantum yield of the mixture ($\bar{\phi}_F$) on λ_{exc} and of the preexponential factor on temperature limited the extent of its application.[43]

The kinetic method, combined with the statistical procedure (called KSFA version),[29] was used in a refined study of the "anomalous" fluorescence of 2-StAn.[30] Since the KFA procedure[24,25] is independent of the overlap between the rotamer emission spectra (which is an important condition in the statistical method), but dependent on the difference in their lifetimes, it turned out to be a very useful method for deriving the correct excitation spectrum and photophysical properties of the two rotamers, because they have very different fluorescence lifetimes (i.e., in a 9:1-methylcyclohexane:3-methylpentane mixture, MCH/3MP at room temperature; $\tau_{F,A} = 10$ ns and $\tau_{F,B} = 64.6$ ns).[28,30,38,41,43]

The excitation and emission fluorescence spectra of 2-StAn are shown in Figure 5.8 as a function of λ_{em} and λ_{exc}, respectively, in toluene at room temperature.[30] The pronounced λ-dependence of the spectra clearly indicates the existence of different emitting species, which do not interconvert during the S_1 lifetime, in accordance

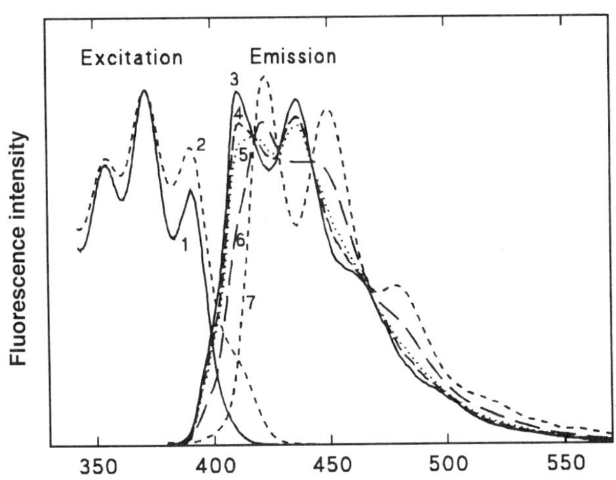

Figure 5.8. Corrected fluorescence excitation spectra at $\lambda_{em} = 390$ (1) and 451 nm (2) and emission spectra at $\lambda_{exc} = 354$ (3), 386 (4), 400 (5), 406 (6), and 420 nm (7) of *trans*-2-styrylanthracene in toluene at 293 K. (Adapted from Ref. 30; copyright 1996 Elsevier Science.)

with the NEER principle. The fluorescence spectra as a function of λ_{exc} were normalized for the same absorbed quanta to determine the isoemissive wavelengths (λ'_{em}). At the isoemissive frequency, the fluorescence excitation spectrum overlaps the absorption spectrum, the fluorescence intensity becomes independent of λ_{exc}, and the contribution of each component to the fluorescence decay, %(i) (λ_{exc}) [namely, the fraction of excited molecules at each excitation wavelength, $f_i(\lambda_{exc})$] can be evaluated (Eq. (10)). Figure 5.9 shows the relative absorption spectra $\varepsilon_A(\lambda_{exc})c_A/\bar{c}$ and $\varepsilon_B(\lambda_{exc})c_B/\bar{c}$ of the two rotamers in MCH/3MP at room temperature, obtained by the %(i) (λ_{exc}) and $\bar{\varepsilon}(\lambda_{exc})$ values, in turn derived from deconvolution of the decay curves and the absorption spectrum of the mixture, respectively, and by Equation (3).[30]

By the PCA-SM mathematical procedure (Section 2.3.4) based on both self-modeling methods (PCA-SL by our group[29,30] and PCA-SV by Saltiel's group[33,34]), the spectral properties (excitation and emission spectra) of the two rotamers were derived.

Figures 5.3, 5.4, and 5.5, respectively, show the map of the excitation–emission fluorescence input matrix (41 × 95), the two significant eigenvectors, which

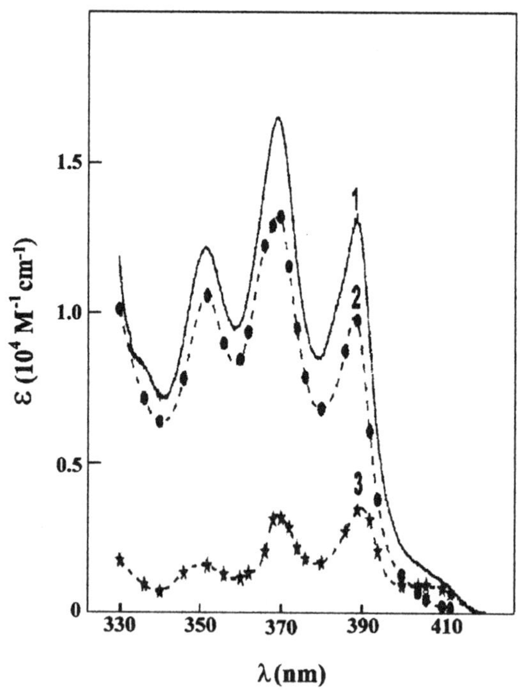

Figure 5.9. *trans*-2-styrylanthracene in MCH/3MP at 293 K: curve 1, absorption spectrum $\bar{\varepsilon}(\lambda_{exc})$; curves 2 and 3, relative absorption spectra of the B, $\varepsilon_B(\lambda_{exc})c_B/\bar{c}$, and A, $\varepsilon_A(\lambda_{exc})c_A/\bar{c}$, rotamers, respectively (from KFA analysis; reprinted from Ref. 30; copyright 1996 Elsevier Science).

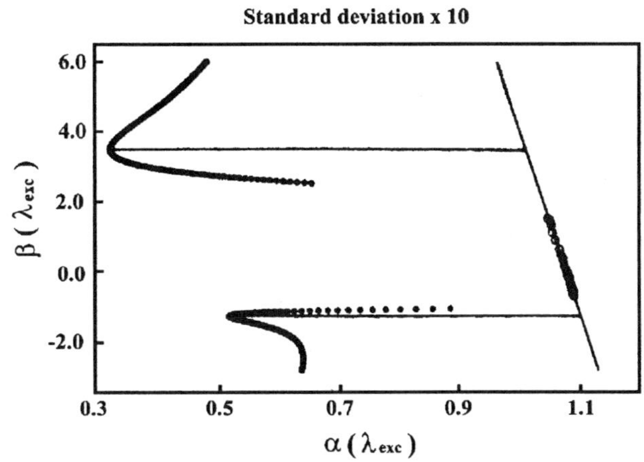

Figure 5.10. Normalization line and combination coefficients for the spectral matrix of *trans*-2-styrylanthracene in toluene at 292 K (PCA-SV method). The curves give standard deviation/slope ratios for the global Stern–Volmer plots. (Reprinted from Ref. 34; copyright 1997 American Chemical Society.)

accounted for 98% (94% and 4%, respectively) of the total variance of the measured spectra, and the plot of the $\beta(\lambda_{exc})$ versus $\alpha(\lambda_{exc})$ values.

Figure 5.10 shows the combination coefficients of the experimental fluorescence spectra of 2-StAn, obtained as a function of both λ_{exc} and the quencher concentration (input matrix 24×221), which adhere closely to the normalization line on which the pure-component coefficients (α_i, β_i) are also shown. These two coefficient pairs were unambiguously identified by well-defined minima on the curves, giving standard-deviation : slope ratios for the global Stern–Volmer plots (PCA-SV).[34]

The spectral properties of the rotamers of 2-StAn in two solvents at room temperature, as derived from the KFA and PCA procedures, are reported in Table 1. The derived spectra of the pure rotamers, obtained by three different methods (the

Table 1 Spectral data (λ/nm) of *trans*-2-styrylanthracene rotamers in two solvents at 293 K, as derived from kinetic (KFA) and statistical (PCA-SL or PCA-SV) procedures. (Adapted from Ref. 30; copyright 1996 Elsevier Science.)

	MCH/3MP		Toluene		
Spectrum	2-StAn(A)	2-StAn(B)	2-StAn(A)	2-StAn(B)	Analysis Method
Excitation	351, 369, 390, 406	352, 369, 389	356^{sh}, 372, 392, 412	356, 372, 392	KFA
	352, 370, 389, 406^{sh}	352, 370, 389	356, 373, 391, 411^{sh}	356, 373, 392	PCA-SL
			368, 387, 407	355, 373, 392	PCA-SV
Emission	416, 443, 473, 506	393^{sh}, 405, 429, 457, 490	423, 450, 479	400^{sh}, 410, 435, 462	PCA-SL
			422, 450, 479	400^{sh}, 411, 436, 460^{sh}	PCA-SV

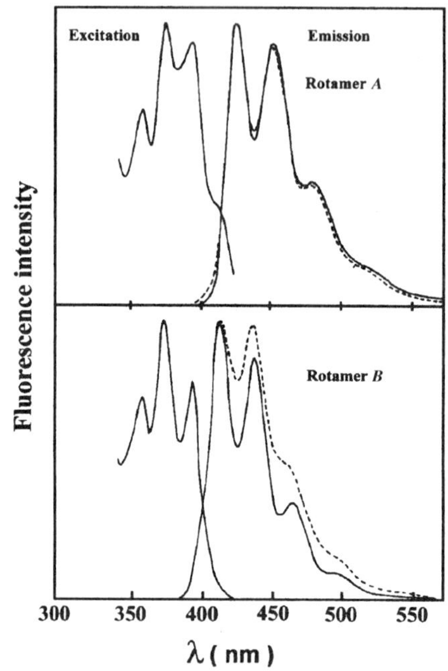

Figure 5.11. Excitation and emission fluorescence spectra of the two rotamers of *trans*-2-styrylanthracene in toluene at 293 K, calculated by the PCA-LS method; the dashed line represents the experimental emission spectrum richer in one rotamer. (Reprinted from Ref. 29; copyright 1992 Elsevier Science.)

spectra obtained by the PCA-LS procedure are shown in Figure 5.11), are practically coincident, except for small differences in the excitation spectrum of the short-lived A rotamer.[30,34]

The emission spectrum of the long-lived B rotamer shows a shoulder at the short λ_{em} onset, which becomes relatively more intense in solvents of lower dielectric constant. This emission, observed at 393 nm in MCH/3MP (at 400 nm in toluene), was considered part of the $S_2 \rightarrow S_0$ transition, which originates from the S_2 state of the B rotamer, thermally populated from S_1 during its lifetime, while the emission at 405 nm (410 nm in toluene) is part of the $S_1 \rightarrow S_0$ transition.[44]

The fluorescence quantum yields, derived by Equations (18) or (19), the lifetimes, and the radiative rate constants ($k_{F,i} = \phi_{F,i}/\tau_{F,i}$) of *trans*-2-StAn rotamers in MCH/3MP and toluene are reported in Table 2.[30] The very different k_F values of the two rotamers point to a different nature of their lowest excited singlet state. On the basis of the k_F values, it was assumed that the S_1 state of the A and B rotamers are of (^1L_a–^1B_u) and (^1L_b) character, respectively, in agreement with the results of theoretical calculations.[38] Moreover, the k_F value of the A rotamer was found to be independent of the solvent dielectric constant, D. By contrast, that of the B rotamer decreased with D, according to its fluorescence lifetime, which became longer in MCH/3MP compared with toluene, the radiative process being by far the most important deactivation pathway of the S_1 state. These experimental results for the B

Table 2 Kinetic and photophysical parameters of *trans*-2-styrylanthracene in two solvents at 293 K. (Reprinted from Ref. 30; copyright 1996 Elsevier Science.)

	MCH/3MP		Toluene	
Parameter	2-StAn(A)	2-StAn(B)	2-StAn(A)	2-StAn(B)
ϕ_F	1.0	0.74	1.0	0.76
τ_F (ns)	10.0	64.6	9.2	27.8
$k_F(10^8\ s^{-1})$	1.0	0.11	1.1	0.27

rotamer were explained on the basis of the mixing between the two lowest excited singlet states S_1 and S_2 of low (^1L_b) and high ($^1L_a-^1B_u$) oscillator strength, respectively. This mixing, which becomes larger with increasing solvent polarity, makes the forbidden $S_1 \rightarrow S_0$ transition become partially allowed.[45]

The different bathochromic shift of the $S_2 \rightarrow S_0$ ($\Delta\bar{\nu}_{0,0} \cong 445\ cm^{-1}$) and $S_1 \rightarrow S_0$ ($\Delta\ \bar{\nu}_{0,0} \cong 300\ cm^{-1}$) transitions of the B rotamer and of the $S_1 \rightarrow S_0$ transitions of the A rotamer ($\Delta\bar{\nu}_{0,0} \cong 400\ cm^{-1}$) with the increase of the solvent dielectric constant led us to confirm the nature of the S_2 state of the B rotamer and the different character of the S_1 state of the two rotamers.[30]

Moreover, the high fluorescence quantum yields in Table 2 imply that the S_1 state decay is mainly radiative, even if a photoreactive process (the geometrical *trans* → *cis* isomerization) is also operative. In fact, a very small, but real, formation of *cis* isomer ($\bar{\phi}_{t\rightarrow c} \cong 0.003$ at $\lambda_{exc} = 302$ nm in MCH/3MP at room temperature) was recently found. The quantum yield depends on λ_{exc} and is drastically reduced (more than halved) if *trans*-2-StAn is excited at 410 nm,[46] where the A rotamer absorbs more than B.[30] On the basis of the λ_{exc}-dependence of $\phi_{t\rightarrow c}$ and of the high ϕ_F value derived for the A rotamer ($\phi_{F,A} \cong 1$),[30] the B rotamer was considered to be the reactive conformation.

By assuming that the Franck–Condon envelopes of the first absorption band system (from S_0 to S_1 and S_2) are similar in shape for the two rotamers,[14,25,29] Equation (11) could be applied.[47]

The ratio $\varepsilon_{00,A}c_A/(\varepsilon_{00,B}c_B)$ was estimated from the relative absorption spectra in toluene as derived from the PCA-LS method (Fig. 5.12). By combination with Equation (12) the relative abundances of rotamers (62% for the B rotamer in the ground state at 293 K) were estimated.[29] From the Boltzmann distribution (neglecting the entropy term), the enthalpy difference between the two ground-state rotamers was obtained ($\Delta H = 300$ cal mol^{-1}). This $\Delta H(S_0)$ value is in excellent agreement with that suggested by the temperature effect on the preexponential factor ($\Delta H(S_0) \leq 400$ cal mol^{-1})[43] and with the values of around 200 cal mol^{-1} obtained by the CS-INDO method[43] and about 300 cal mol^{-1} derived by QCFF/PI and CNDO/S calculations.[38]

The rotameric behavior of the three isomeric *trans*-*n*-StAn ($n = 1, 2, 9$) was well described by the S_0 torsional potential-energy curves using the CS-INDO method, as shown in Figure 5.13.

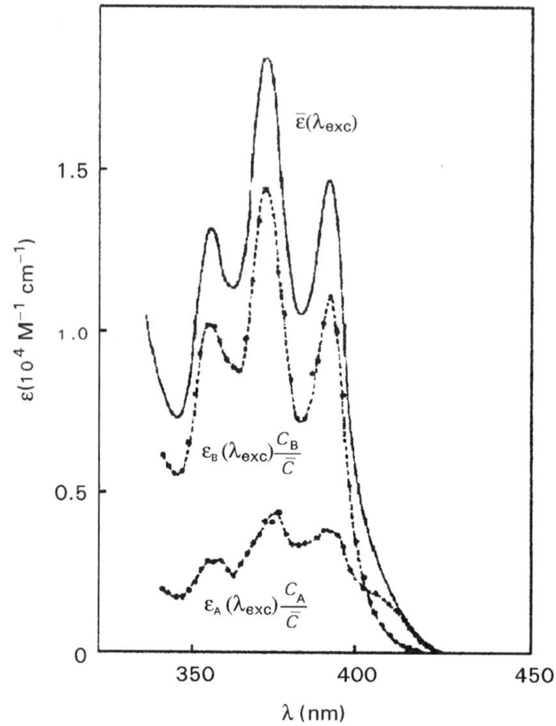

Figure 5.12. Measured absorption spectrum, $\bar{\varepsilon}(\lambda_{exc})$, of *trans*-2-styrylanthracene in toluene at 293 K and relative absorption spectra of the A and B rotamers, $\varepsilon_A(\lambda_{exc})c_A/\bar{c}$ and $\varepsilon_B(\lambda_{exc})c_B/\bar{c}$, respectively, as derived from the PCA-LS method. (Reprinted from Ref. 29; copyright 1992 Elsevier Science.)

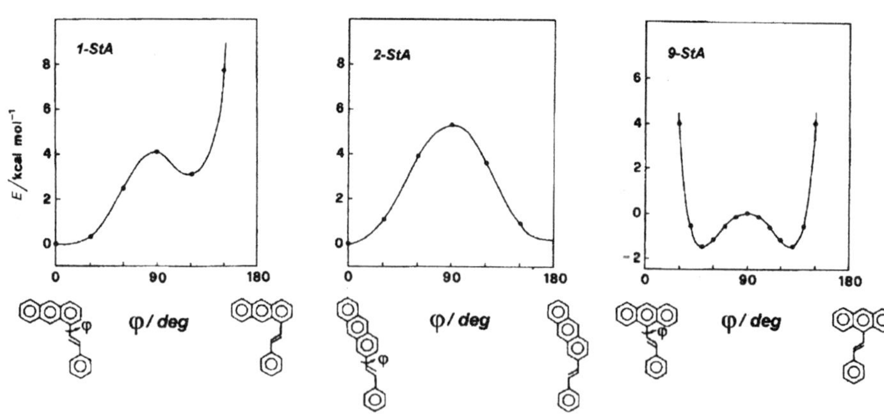

Figure 5.13. Ground-state potential-energy curves for internal rotation of *trans*-*n*-styrylanthracenes ($n = 1, 2, 9$) around the "single" bond connecting the anthracenic group to the ethenic bridge. (Reprinted from Ref. 43; copyright 1989 Elsevier Science.)

As expected by virtue of symmetry, two equivalent minima, markedly nonplanar (52 and 128°), and due to steric strain, were found for 9-StAn. In the case of 1-StAn, on the other hand, CS-INDO calculations predicted an absolute, rather flat, minimum at $\phi = 115-120°$, with an energy difference (ΔH value) of 3.1 kcal mol^{-1}, which implies a negligible abundance (ca. 0.6%) of the less stable rotamer at room temperature.[43] Therefore, for both compounds, the presence of a detectable mixture of rotamers with different spectral and photophysical properties is not expected at room temperature, as was confirmed by the experimental results.[38] The S_0 potential-energy curve for 2-StAn indicated two quasi-isoenergetic and coplanar rotamers separated by a barrier of about 5 kcal mol^{-1}, the more stable form being the more elongated one (the *s-trans* conformation: rotamer B).[43]

3.1.2 Systems that are not Governed by the NEER Principle

In the case of olefins bearing anthryl groups, rotation during the lifetime of the excited state has been observed both in the fluorescent state and, even easier, in the longer-lived triplet state. An example of the first type is 2-vinylanthracene (2-VA) and its β-alkyl derivatives, which undergo almost one-way rotational isomerization around the single bond between the anthryl group and the ethenic carbon.[39,48–50] The fluorescence excitation and emission spectra, obtained by steady-state excitation, are dependent on λ_{em} and λ_{exc}, respectively. The fluorescence quantum yield depends on λ_{exc} (e.g., for 2-VA in benzene at room temperature, it goes from 0.58 to 0.41 in the 316–405-nm range),[39] and the time-resolved fluorescence spectra change with time, as shown in Figure 5.14.[39,48–50]

The fluorescence decay showed two decay components whose lifetimes are independent of λ_{em}. The preexponential factors are, however, strongly dependent on λ_{em}. At wavelengths where the short-lived A component preferentially emits (for example, near 390 nm), the amplitudes (preexponential factors) are both positive. In contrast, at wavelengths where B is strongly emitting (for example, near 430 nm), one decaying exponential (A_B positive) and one rising exponential (A_A negative) were found. (See, for example, Figure 5.15).[39,48–50] The presence of the rise component shows that the $A^* \rightarrow B^*$ (one-way) interconversion occurs in the excited singlet state.[39,50]

The decay curves, monitored at various wavelengths, showed a considerable temperature dependence of the lifetime of the A rotamer, suggesting the involvement of an activated process in S_1 (Table 3). In the case of 2-VA, a ${}^1\Delta E_{A^*\rightarrow B^*}$ value of 3.9 kcal mol^{-1} was derived for rotamer interconversion from the Arrhenius plot according to Equation (24). Assuming an efficiency less than 1/10 for the reverse process (not detected experimentally), and assuming that the preexponential factors for the backward and forward processes are the same, a ${}^1\Delta E_{B^*\rightarrow A^*}$ value larger than 6 kcal mol^{-1} was estimated.[39] Similar values of ${}^1\Delta E_{A^*\rightarrow B^*}$ were found for 2-VA alkyl-substituted at the ethenic carbon [4.3 kcal mol^{-1} for *trans*-1-methyl,2-(2-anthryl)ethene and *trans*-1-(*t*-butyl),2-(2-anthryl)ethene].[50] In the case of *trans*-1-(2-propenyl),2-(2′-anthryl)ethene, the shape of the fluorescence spectrum was insensitive to λ_{exc} at room temperature. However, at lower temperatures (below

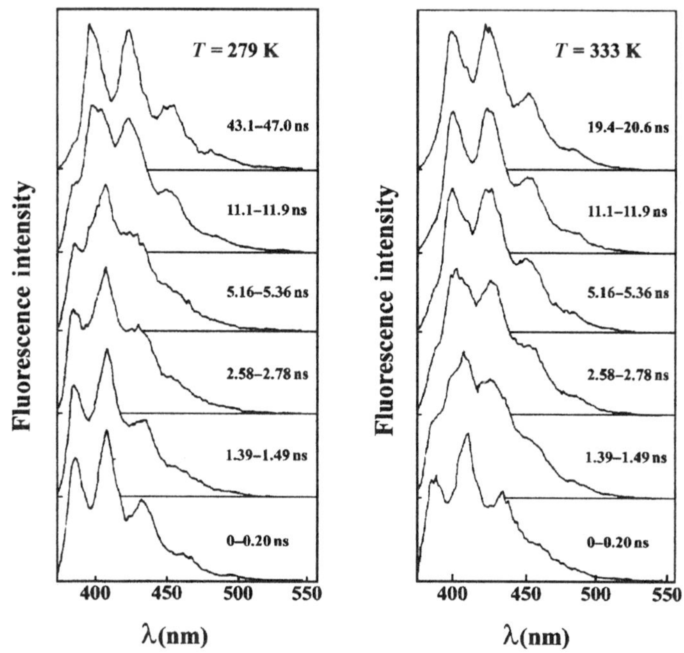

Figure 5.14. Time-resolved fluorescence spectra of 2-vinylanthracene in deaerated benzene at two temperatures. (Adapted from Ref. 39; copyright 1989 Elsevier Science.)

190 K), a huge λ_{exc} effect was observed.[49] The fluorescence kinetics of the propenyl derivative are closely analogous to those previously described for 2-VA. The decay is biexponential with a short (0.55 ns) and a longer (10 ns) lifetime, due to the presence of two rotamers. The fast rise–decay time observed at 421 nm (near the second peak of the B spectrum) reflects the rotamer interconversion.[49] The relative magnitude of the two lifetimes explains why there is no photoselection at room temperature. Since the equilibration time is over 15-times shorter than the average fluorescence lifetime, nearly all of the emission comes from the equilibrium mixture and is thus

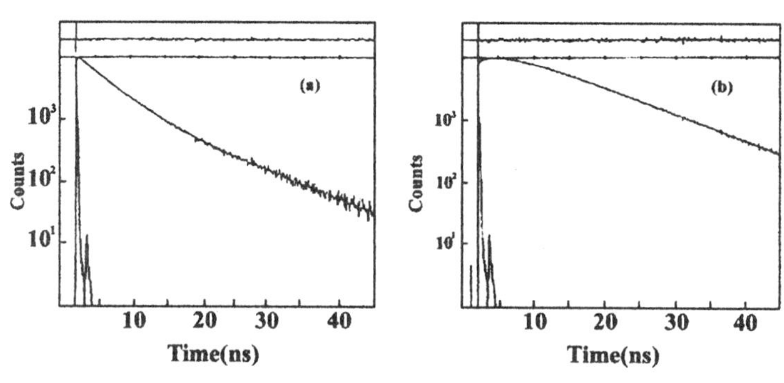

Figure 5.15. Fluorescence decay and rise curves of 2-vinylanthracene monitored at (a) 390 and (b) 430 nm in deaerated benzene at 24°C. (Adapted from Ref. 39; copyright 1989 Elsevier Science.)

Table 3 Fluorescence decay parameters of 2-vinylanthracene in benzene. (Reprinted in part from Ref. 39; copyright 1989 Elsevier Science.)

T (K)	λ_{em} (nm)	τ_A (ns)[a]	τ_B (ns)
279	390	5.37	9.52
	430	5.66 (−)	9.87
309	390	3.11	9.79
	430	3.07 (−)	9.88
333	390	2.03	9.85
	430	1.97 (−)	9.84

[a] The sign (−) indicates the rise component.

independent of λ_{exc} (two-way interconversion). In pentane, a value of about 3.5 kcal mol^{-1} was derived for the activation energy of the $^1A^* \rightarrow {}^1B^*$ interconversion, and a value of 4.1 kcal mol^{-1} was estimated for the reverse rotation, as calculated from

$$^1\Delta E_{B^* \rightarrow A^*} = {}^1\Delta E_{A^* \rightarrow B^*} - \Delta H(S_1). \tag{25}$$

The $\Delta H(S_1)$ value of about -0.6 kcal mol^{-1} between A and B, derived from the differential fluorescence spectra measured in the picosecond region by using Equation (23), indicates a relative abundance of about 73% of the B* rotamer in the equilibrated mixture in S_1. In contrast, the A rotamer was found to be the most stable conformation in the ground state; its relative abundance was about 92% at room temperature, as indicated by a $\Delta H(S_0)$ value of about 1.5 kcal mol^{-1}.[49]

A similar interesting example of the nonvalidity of the NEER principle was reported for *trans-N*-methoxy-1-(2-anthryl)ethanimine (Scheme **4**).[51] The temperature effect on the absorption spectrum of this compound in toluene indicated that two rotamers must exist in the ground state and that the short-lived hypsochromic form (*s-trans*) is the most stable in energy in S_0. Time-resolved fluorimetry showed that the short-lived component was the one that contributed the most to the spectrum, which then converted to that of the most stable species in S_1 (*s-cis*) in the

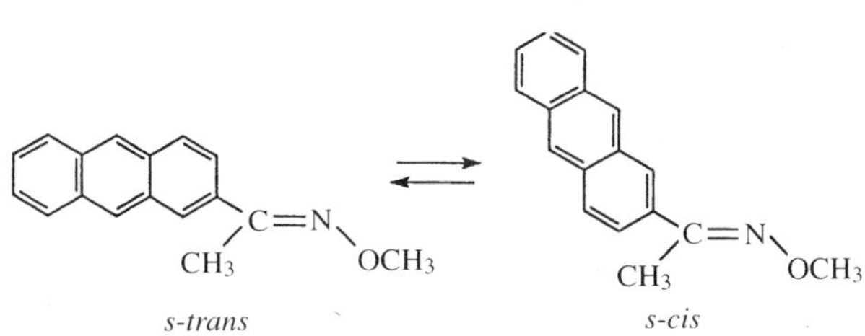

trans-N-methoxy-1-(2-anthryl)ethanimine

Scheme 4

nanosecond timescale at a rate that increased with temperature. Decay curves observed at 390 and 460 nm and at 331 and 183 K are shown in Figure 5.16.

At low temperatures, the decay observed at both wavelengths was monoexponential ($\tau_F = 10.4$ ns). Above 243 K, the decay becomes biexponential, the longer lifetime corresponding to the longer-λ absorbing species. At 331 K, with monitoring carried out at 460 nm, the time profile shows a rise (3.6 ns) and a decay (7.3 ns) component. However, when monitoring was performed at 390 nm, only two decay components were observed (3.6 and 7.7 ns). The agreement between the rise time at 460 nm and the short decay time at 390 nm strongly indicates that rotational isomerization takes place during the S_1 lifetime, in this case from the short-lived hypsochromic rotamer to the longer-lived species. A rotational barrier of about 4 kcal mol^{-1} and a preexponential factor of $1.8 \times 10^{11}\ s^{-1}$ were estimated by the Arrhenius plot for the lifetime of the short-lived rotamer.[51] It has to be noted that these activation parameters are similar to those found for 2-VA and its analogues.

Another example of one-way interconversion was observed for the 9,9′-dianthrylmethanol rotamers. By using a global analysis, the fluorescence spectra were resolved into three components, which were assigned to geometries with very different overlap of chromophores. On the basis of the presence of negative preexponential factors (rise components) in the deconvolution of the decay curves, the existence of three emitting species, which may at least partially interconvert, was considered.[52]

The second typical case where the NEER principle is not satisfied is that of the triplet state of styrylanthracenes and other DAEs bearing polycyclic groups of low T_1 energy. Interesting experimental evidence by both room temperature laser-flash photolysis[21] and conventional flash photolysis at variable temperatures[20b] and convincing theoretical interpretations[22] of the triplet state equilibration of 2-StAn and related compounds were reported. In fact, the triplet state of the *trans* isomers of anthryl-substituted DAEs decays slowly to S_0 without isomerizing to *cis*, thus persisting long enough (ca. 100 μs for *trans*-2-StAn)[20b,21,53] to make rotamer conversion possible in fluid solutions.

The *s-trans* → *s-cis* rotational isomerization of 2-StAn (one-way interconversion in the T_1 state) was studied more recently by the time-dependent change of its T–T

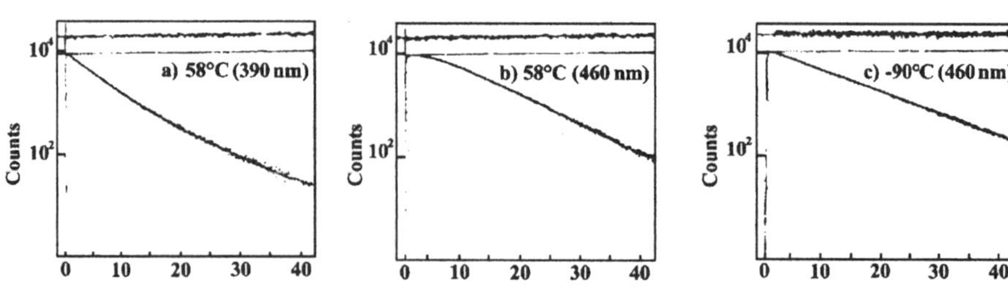

Figure 5.16. Fluorescence decay and rise curves of *trans*-*N*-methoxy-1-(2-anthryl)ethanimine at 390 and 460 nm in deaerated toluene at 58 and −90°C. (Adapted from Ref. 51; copyright 1991 American Chemical Society.)

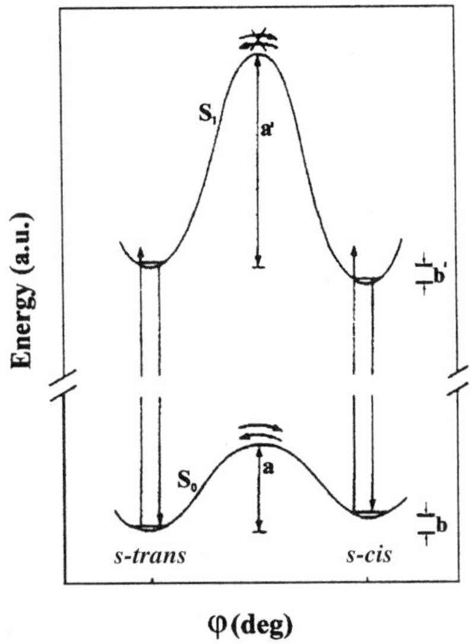

Figure 5.17. Schematic potential-energy curves for the *s-trans* → *s-cis* isomerization of *trans*-diarylethenes in the S_0 and S_1 states. Typical values of the energy barrier and energy difference (kcal mol^{-1}) of rotamers in S_0 are a ≅ 5 and b ≅ 0.5 − 1, respectively. (Reprinted from Ref. 37; copyright 1992 *EPA Newsletter.*)

absorption; an energy barrier of 7.2 kcal mol^{-1} was estimated.[54] CS-INDO S-CI calculations predicted a barrier of 5.3 kcal mol^{-1} in T_1, a value almost halved compared with that of stilbene, so that the combination of both the small barrier and the long lifetime explains why the NEER principle breaks down with these compounds. On the other hand, for the short-lived fluorescent state (tens of ns), smaller internal rotation barriers are required to allow partial equilibration of the rotamers in fluid solutions, so that interconversion in S_1 was found for 1-(*t*-butyl),2-(2-anthryl)ethene,[50] but not for 2-StAn, for which a higher barrier was calculated (Fig. 5.17).[22]

3.2 Three-Component Systems: The Case of *Trans*-1-(2-Naphthyl),2-(6′-Quinolyl)Ethene

The most thoroughly investigated three-component system is that of *trans*-1,2-di(2-naphthyl)ethene (2,2′-DNE).[12,27d,40,55] Here, the behavior of a quinolyl analogue of DNE, *trans*-1-(2-naphthyl),2-(6′-quinolyl)ethene (2,6′-NQE), is reported.[31] The experimental findings in the spectroscopic and photophysical study of 2,6′-NQE (dependence of the emission and excitation fluorescence spectra on λ_{exc} and λ_{em}, respectively, as shown in Figure 5.18, parallel change of ϕ_F with λ_{exc}, and multi-exponential decay) pointed to a system that obeys the NEER principle. Therefore, the study of the fluorescence behavior of NQE by combined kinetic and statistical procedures (KSFA) gave information about the excited singlet state properties of the rotamers and their relative abundances in the ground state.

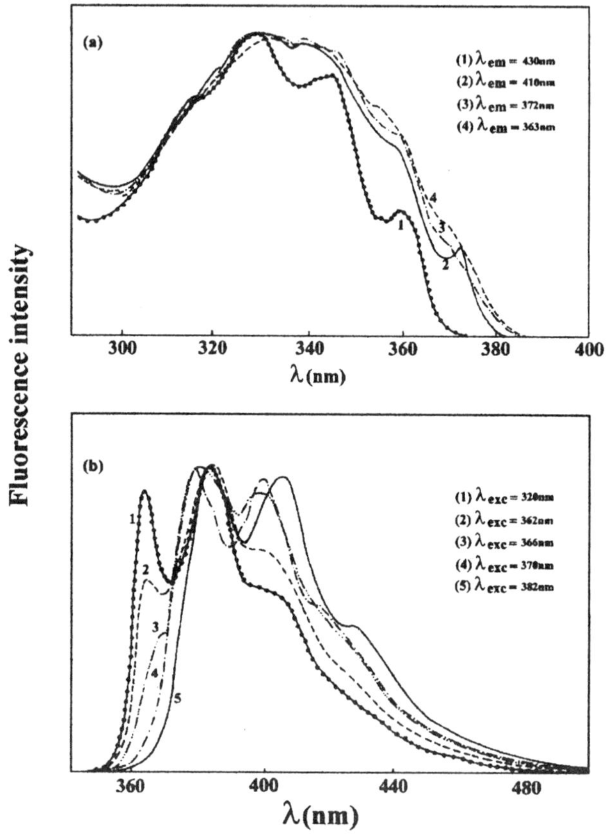

Figure 5.18. Excitation (a) and emission (b) fluorescence spectra of *trans*-2,6′-NQE in *n*-hexane at 293 K, as a function of λ_{em} and λ_{exc}, respectively. (Reprinted from Ref. 31; copyright 1993 Elsevier Science.)

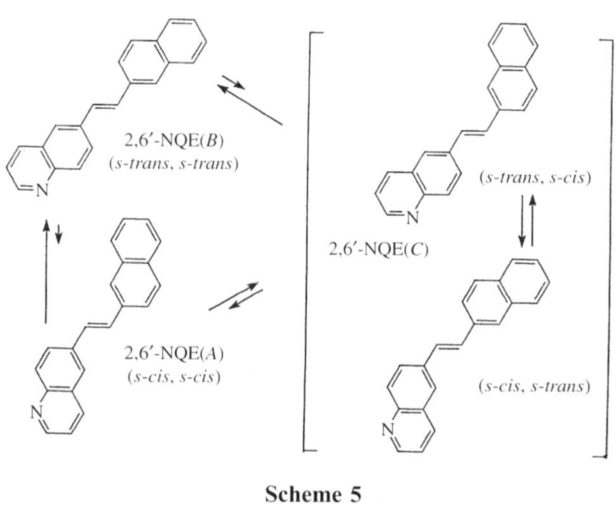

Scheme 5

In this molecule, the rotation of both aryl groups around the essential single bonds with the ethene bridge should give rise to a mixture of four distinct conformers (Scheme **5**).

The application of the PCA-SM statistical analysis (Section 2.3.4) to an emission–excitation matrix of 2,6′-NQE in *n*-hexane at room temperature, using the NIPALS algorithm with cross-validatory estimation,[26] gave only three significant components, as in the case of the symmetrical analogue DNE (Scheme **2**).[27c] In agreement with this result, three significant eigenvectors were found; the shape of the fourth eigenvector clearly showed that it was produced by instrumental noise. The three eigenvectors, shown in Figure 5.19, express 99.5% (93, 5, and 1.5%, respectively) of the total variance of the measured spectra, thus indicating that two of the four expected rotamers of NQE (the *s-cis,s-trans*, and *s-trans,s-cis* rotamers; see Scheme **5**) are expected to have very similar geometries and then practically indistinguishable spectral and photophysical properties, as previously reported for related systems.[14,24,36b]

For a three-component system, the experimental emission spectrum [$\mathbf{S}(\lambda_{exc})$] is reproduced by the linear combination[56]

$$\mathbf{S}(\lambda_{exc}) = \alpha(\lambda_{exc})\mathbf{V}_{\alpha} + \beta(\lambda_{exc})\mathbf{V}_{\beta} + \gamma(\lambda_{exc})\mathbf{V}_{\gamma} \tag{26}$$

of three eigenvectors that define a normalization plane in the α, β, and γ space. All combination coefficients representing these spectra must fall within a triangle in

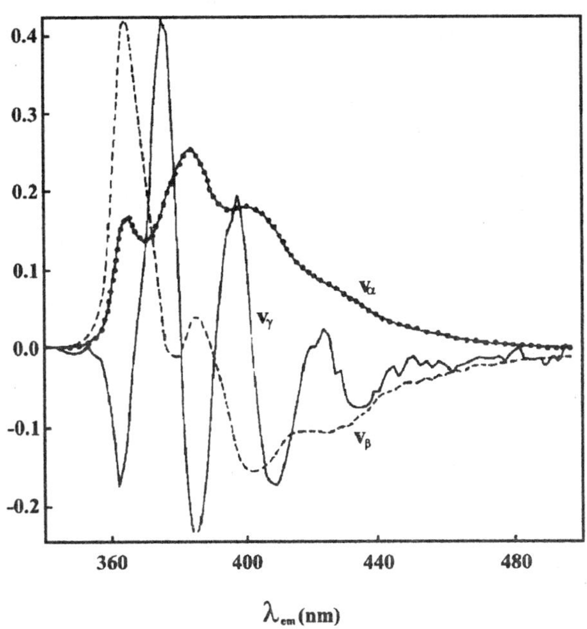

Figure 5.19. $\mathbf{V}_{\alpha}$, $\mathbf{V}_{\beta}$ and $\mathbf{V}_{\gamma}$ eigenvectors of the fluorescence matrix of *trans*-2,6′-NQE in *n*-hexane at 293 K as a function of λ_{em}. (Reprinted from Ref. 31; copyright 1993 Elsevier Science.)

that plane, as shown in Figure 5.20. The procedure in the choice of the coefficient triplets (α_i, β_i, γ_i) corresponding to the pure components was improved by applying an algorithm that maximizes the differences among the spectra calculated at each point of the contour line including all coefficients that generate positive emission spectra[56] (Fig. 5.20). The coefficients for the three pure components are those that produce the most different spectra and then identify the corners of the ABC triangle containing all the experimental coefficients.

The pure emission spectra of the three rotamers, calculated by

$$\mathbf{S}_i = \alpha_i \mathbf{V}_\alpha + \beta_i \mathbf{V}_\beta + \gamma_i \mathbf{V}_\gamma, \tag{27}$$

are shown in Figure 5.21. From the pure-component coefficients (α_i, β_i, and γ_i) and those that are λ_{exc}-dependent, generated by the mixture spectra, the fractional contributions of each rotamer to the unit-area-normalized experimental spectrum (X_i) could be calculated as a function of λ_{exc} by

$$\begin{bmatrix} \alpha(\lambda_{exc}) \\ \beta(\lambda_{exc}) \\ \gamma(\lambda_{exc}) \end{bmatrix} = \begin{bmatrix} \alpha_A & \alpha_B & \alpha_C \\ \beta_A & \beta_B & \beta_C \\ \gamma_A & \gamma_B & \gamma_C \end{bmatrix} \begin{bmatrix} X_A(\lambda_{exc}) \\ X_B(\lambda_{exc}) \\ X_C(\lambda_{exc}) \end{bmatrix}. \tag{28}$$

The excitation spectrum of each 2,6′-NQE rotamer (Fig. 5.21) is then represented by the distribution of the $X_i(\lambda_{exc})/N(\lambda_{exc})$ value as a function of λ_{exc}.

Despite the system complexity, useful information arose from selective analysis (λ_{exc} and λ_{em}) of the fluorescence decay of the mixture (Table 4) that gave the

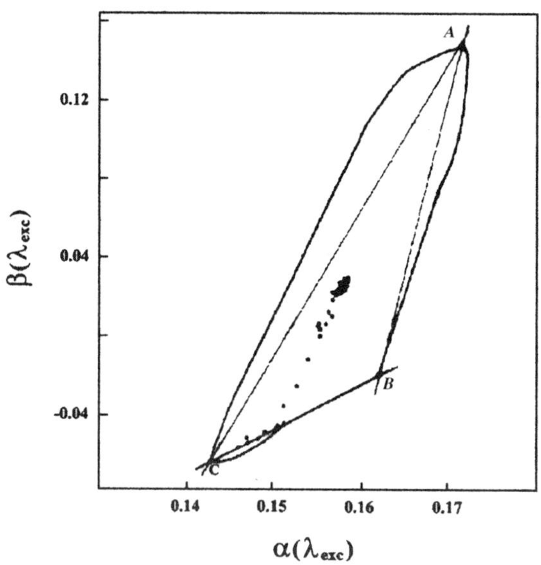

Figure 5.20. Plot of $\beta(\lambda_{exc})$ against $\alpha(\lambda_{exc})$ for the three-component system *trans*-2,6′-NQE. The A, B, and C points identify the pure-component coefficients. (Reprinted from Ref. 31; copyright 1993 Elsevier Science.)

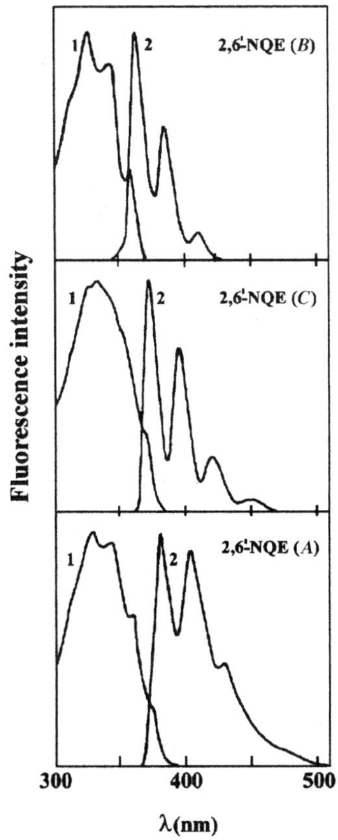

Figure 5.21. Normalized excitation (1) and emission (2) fluorescence spectra of the three rotamers of *trans*-2,6′-NQE in *n*-hexane at 293 K, as derived by the KSFA method. (Adapted from Ref. 31; copyright 1993 Elsevier Science.)

Table 4 Fluorescence decay parameters of *trans*-2,6′-NQE in *n*-hexane at 293 K, as a function of λ_{exc} and λ_{em}. (Reprinted in part from Ref. 31; copyright 1993 Elsevier Science.)

λ_{exc} (nm)	λ_{em} (nm)	$\tau_{F,A}$ (ns)	%(A)	$\tau_{F,B}$ (ns)	%(B)	$\tau_{F,C}$ (ns)	%(C)
300	355			3.9	100		
326	400	1.9[a]	18	3.9	50	3.4	32
372	417	1.7	40			3.1	60
384	417	2.1	79			3.3	21

[a] Imposed value in the triexponential deconvolution.

lifetimes of the three rotamers and the fractional contributions [%(C) and %(A)] of the two bathochromic species in the wavelength range where the system behaves as a two-component mixture ($\lambda_{exc} > 370$ nm).

By using the $f_i(\lambda_{exc})$ (i = C, A) and the corresponding $\bar{\phi}_F(\lambda_{exc})$ values, the $\phi_{F,i}$ of the C and A rotamers were derived from Equation (4). From the $\phi_{F,C}$, $\phi_{F,A}$, and $X_i(\lambda_{exc})$ values, the fractions of excited molecules of C and A rotamers were derived at all the explored λ_{exc}'s by Equation (18).[31] Since $\Sigma f_i(\lambda_{exc}) = 1$, the fluorescence quantum yield of the B rotamer at each λ_{exc} was then calculated by Equation (18). A mean value of 0.45 was thus obtained. The relative absorption spectra of the three rotamers are shown in Figure 5.22.

The standard enthalpy changes among B, C, and A in the ground state were thus obtained by using Equations (11) and (12), where the ratios $\varepsilon_{00,B}c_B/(\varepsilon_{00,c}c_C)$ and $\varepsilon_{00,C}c_C/(\varepsilon_{00,A}c_A)$ were derived from the relative absorption spectra of Figure 5.22 and the $k_{F,i}$ values from the $\phi_{F,i}/\tau_{F,i}$ relationship. All the photophysical and thermodynamic parameters of the 2,6′-NQE rotamers in n-hexane at 293 K are collected in Table 5.

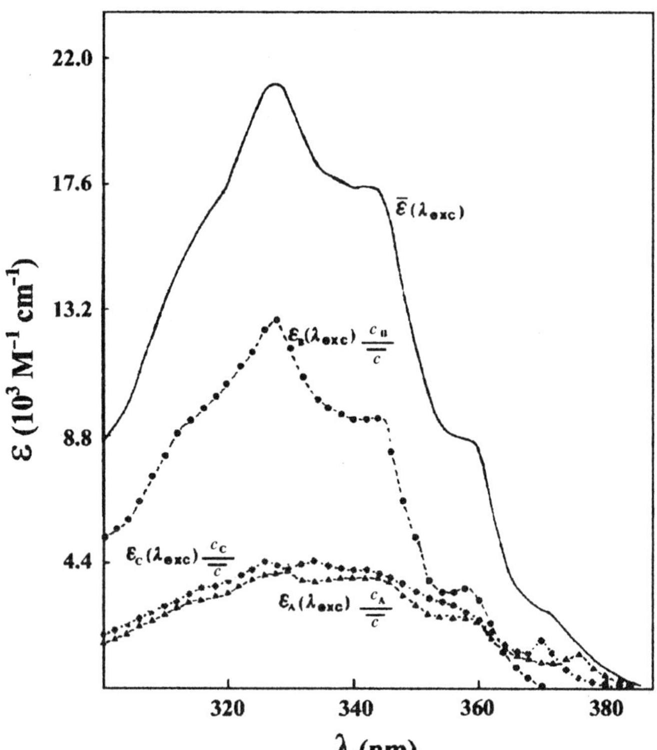

Figure 5.22. Experimental absorption spectrum, $\bar{\varepsilon}(\lambda_{exc})$, of *trans*-2,6′-NQE in n-hexane at 293 K and relative absorption spectra of the A, B, and C rotamers, $\varepsilon_A(\lambda_{exc})c_A/\bar{c}$, $\varepsilon_B(\lambda_{exc})c_B/\bar{c}$ and $\varepsilon_C(\lambda_{exc})c_C/\bar{c}$, respectively, as derived by the KSFA method. (Reprinted from Ref. 31; copyright 1993 Elsevier Science.)

Table 5 Photophysical and thermodynamic parameters of *trans*-2,6′-NQE rotamers in *n*-hexane at 293 K, obtained by the KSFA method. (Reprinted in part from Ref. 31; copyright 1993 Elsevier Science.)

Parameter	2,6′-NQE(B)	2,6′-NQE(C)	2,6′-NQE(A)
ϕ_F	0.45	0.87	0.77
τ_F (ns)	3.9	3.3	1.9
Relative abundance in S_0	0.74	0.18	0.08
$\Delta H(S_0)_{B-C}$ (cal mol^{-1})	790		
$\Delta H(S_0)_{C-A}$ (cal mol^{-1})		500	

The B rotamer was found to be the most stable conformer in the ground state, as shown also by the effect of temperature on the emission spectrum of the mixture. The low enthalpy values obtained are in agreement with the experimental evidence that the emission spectrum of NQE is a mixture of all species down to 170 K.

3.3 Rotamerism in α,ω-Diarylpolyenes

Ground-state single-bond rotational conformers play a role in the excited-state properties of polyenes and diphenylpolyenes (DPhPs). Their structure has received much attention, mainly because of the stereochemical specificity of their reactions and the relationship with carotenoids, vitamin-D derivatives, and related visual pigments.[8] The π-electron system of linear polyenes is particularly suitable for theoretical study,[4,7] therefore, the experimental work was also helped by suitable semiempirical and ab initio calculations.[7,57] In the ground state, short-chain polyenes, such as butadienes and hexatriene, were found, by optical and magnetic spectroscopic studies, to exist as an equilibrium of conformations produced by rotation about the essentially single bonds of the conjugated chain.[58] Following the NEER principle, the rotation about formerly single bonds is inhibited in the excited state, thus leading to mixtures of distinct species that do not interconvert and that may have different lifetimes and reactivities, as in the case of the ethene derivatives. Generally, *s-trans* rotamers are more stable because of only modest steric interactions and then predominate in both the *trans* and *cis* geometrical isomers. Steric effects are responsible for the deviation from planarity of the *s-cis* conformations. A Raman spectroscopy study on 1,3,5-hexatriene (H) and its methyl derivatives has been reported.[15] The time-resolved resonance Raman spectra of the lowest triplet state populated by acetone sensitization showed that identical transient spectra were obtained from both geometrical isomers of H, confirming that a common conformation predominates.

The situation is different in substituted compounds, where steric hindrance may affect the ground-state conformational equilibrium. The equilibrium in the ground state of the *cis* isomer is shifted from the predominantly *s-trans,s-trans* configuration in H to *s-cis,s-trans* and *s-cis,s-cis* in 2-methyl (2-MeH) and 2,5-dimethyl (2,5-

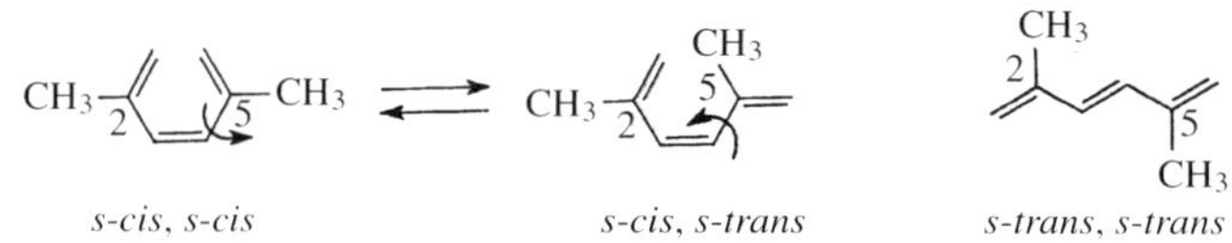

s-cis, s-cis *s-cis, s-trans* *s-trans, s-trans*

cis-2,5-DMeH *trans*-2,5-DMeH

Scheme 6

DMeH) derivatives. By the same time-resolved resonance Raman technique, different triplet species were produced from the *trans* and *cis* isomers of the sterically hindered 2,5-DMeH, in which the *cis* isomer can exist in two different conformations (Scheme **6**).[15]

These rotamers deviate from planarity (the steric strain can be relieved after excitation, by twisting of this weakened bond), and the differences in conformations in the ground state are preserved in T_1. The twisted geometry corresponds to a diallylic structure (in agreement with theoretical predictions) in which the bonds that are formal single bonds in the ground state acquire more of a double-bond character.[15]

These results, which were interpreted in favor of the validity of the NEER principle, not only for singlet excited rotamers, but also in the case of longer-lived triplets (some important exceptions were described earlier), are in agreement with the results obtained from the study of the selective photochemistry of the different conformations of these compounds. (See Section 4.2.1). Analogically to polyenes, the rotamers of α,ω-diphenylpolyenes (DPhP) were found to have different absorption spectra and then could be excited selectively by changing the irradiation wavelength.[59,60] Since they satisfy the NEER principle,[8] their photochemistry depends on the rotamer distribution in the ground state, which follows the Boltzmann law. However, because the *s-cis* absorption spectrum is partly buried under the more intense spectrum of *s-trans*, the contribution of the *s-cis* conformer to fluorescence, increasing with λ_{exc}, was observable during selective excitations in the extreme red onset region of the ground-state absorption profile. Application of the "singular value decomposition (SVD) analysis," a statistical method similar to PCA, to the fluorescence spectra recorded as a function of λ_{exc}, allowed the spectral contribution of the distinct species of *all-trans*-diphenylbutadiene (DPhB) to be resolved (Fig. 5.23). The fluorescence spectra of the rotationally constrained analogues of DPhB, namely, 1,5-diphenyl-2,3,4,6,7,8-hexahydronaphthalene (a) and 1,4-diphenyl-1,3-cyclopentadiene (b) (Scheme **7**) did not show any λ_{exc}-dependence and were found to be very similar to those of the corresponding distinct rotamers, thus confirming their assignment to the *s-trans* and *s-cis* conformations.[59]

A recent paper[60] describes the use of the PCA method for quantitative separation of these fluorescence spectral components. By using λ_{exc} within the main absorption band (280–350 nm), the recorded spectra are due to the *s-trans* rotamer and do not depend on λ_{exc}. On the other hand, when excitation is occurring in the extreme red onset region, the emission spectra become very different from that of the *s-trans*

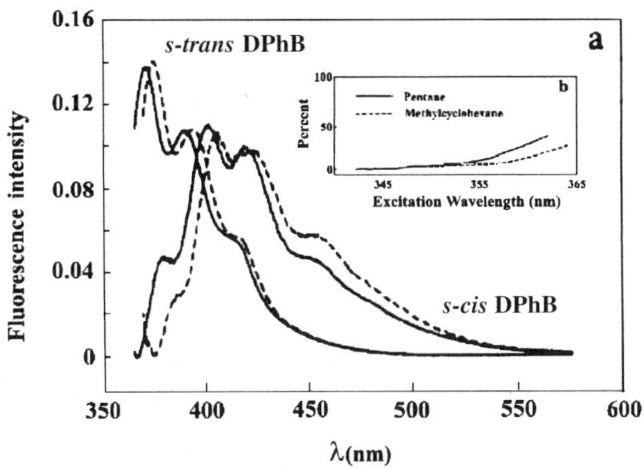

Figure 5.23. (a) The *s-trans* and *s-cis* spectra corresponding to least overlap between rotamers, as determined from the SVD analysis and fitting of the scaled wavelength-dependent *all-trans*-1,4-diphenyl-1,3-butadiene fluorescence emission spectra, and (b) the fractional contribution of the *s-cis* rotamer as a function of λ_{exc}. (Adapted from Ref. 59; copyright 1993 American Chemical Society.)

conformer, depend strongly on λ_{exc}, and reveal spectral mixtures of the two rotamers. The relative fluorescence contributions and fluorescence quantum yields of the two species at two temperatures were thus estimated (Table 6). An enthalpy difference of 1.9 kcal mol^{-1} was obtained for the conformational equilibrium of DPhB in MCH. The latter was also evaluated on the basis of theoretical calculations. Molecular energy and geometry were first optimized by using the AM1 semiempirical method and then further optimized by using an ab initio method. The results showed that both conformers are planar, with the *s-cis* having an energy 2.21 kcal mol^{-1} higher than that of the *s-trans*, in very good agreement with the experimental findings.

A similar, but stronger, red edge effect was observed with diphenylhexatriene (DPhH).[6] Excitation at 355 nm (near the center of the absorption band) and 385 nm (at the red edge) produced noticeably different emission spectra. Application of the

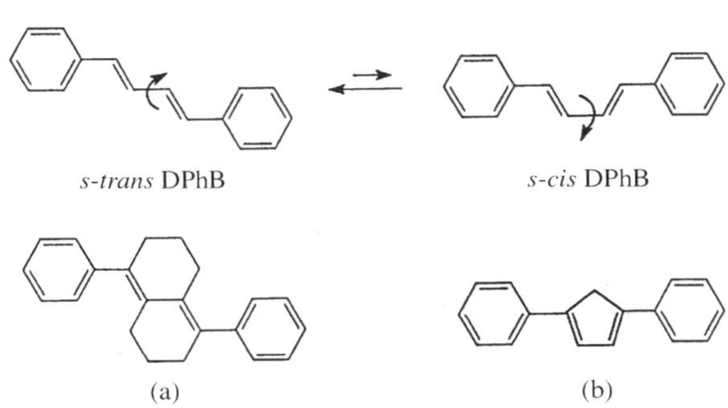

Scheme 7

Table 6 Fluorescence quantum yields of DPhB rotamers in *n*-hexane at room temperature (from Ref. 60)

λ_{exc} (nm)	*s-cis* (%)	$\bar{\phi}_F$	$\phi_{F,s\text{-}trans}$	$\phi_{F,s\text{-}cis}$
328	0	0.37	0.37	
368	57	0.41		0.45

PCA-SV method (see Section 2.4.4) to matrices composed of fluorescence spectra measured at different λ_{exc} and constant temperature allowed their resolution into two components, one of which is the combined emission from the two lowest excited singlet states of A_g and B_u character of the *all-s-trans* conformer and the red-shifted one was assigned to the emission of the *s-trans-s-cis* and/or *s-cis-s-cis* conformers. The fractional contribution of the latter at $\lambda_{exc} = 355$ nm increased strongly with temperature (from 6% at 270 K to 25% at 364 K) and was significantly higher at longer values of λ_{exc}, leading to an estimated enthalpy difference of 3.4 kcal mol^{-1} for conformer equilibration in S_0.[61]

When one or both phenyl rings of DPhPs are substituted by polycyclic groups or asymmetrically substituted phenyl groups, additional conformers may exist in their solution due to rotation of the aryl groups about the single bond with the α,ω carbon atoms of the chain. An indication of rotamerism was recently found in our laboratory for symmetrical di-(2-thienyl) and di-(3-thienyl) polyenes (D2TP and D3TP)[62] and for asymmetrical α-phenyl,ω-(2-anthryl)butadienes (Ph2AnB).[63]

In the case of the *all-trans*-dithienylpolyenes that were investigated (with the chain length $n = 2$, 3, and 4; (see Scheme **8**), only for di-(3-thienyl)hexatriene (D3TH) were the excitation and emission fluorescence spectra clearly dependent on λ_{em} and λ_{exc}, respectively.[62] Even in this case, selective excitation at the tail of the absorption ($\lambda_{exc} = 380$ nm) led to the emission spectrum of the quasi-pure short-lived A rotamer, while at $\lambda_{em} = 367$ nm, where A does not emit, the quasi-pure excitation spectrum of the hypsochromic B rotamer was obtained (Fig. 5.24). At low temperatures, the equilibrium shifts towards the more stable B rotamer, whose fluorescence spectrum was recorded in rigid matrices at liquid-nitrogen temperatures.[62]

In the case of the anthryl derivatives,[63] a behavior similar to that found for the corresponding ethene derivatives, *n*-styrylanthracene (*n*-StAn), was observed; that is, both excitations to the two lowest excited states S_1 and T_1 are largely localized in the anthryl group, and the single bonds do not increase their order much upon excitation. Therefore, *s-trans* → *s-cis* interconversion in the excited states could be operative against the NEER principle. In any case, the fluorescence excitation spectrum was practically coincident with the absorption spectrum, and its shape did not depend on λ_{em}, while the fluorescence emission spectrum and quantum yield did not depend on λ_{exc}. Moreover, the fluorescence decay was monoexponential. The concurrence of these results indicates that the presence of rotamers is hardly detectable, at least in fluid solutions at room temperature. Only by exciting 1-phenyl,2-(2-anthryl)butadiene (Ph2AnB) at the absorption tail with $\lambda_{exc} > 410$ nm does the shape of the emission spectrum change markedly (then remaining practically independent of λ_{exc} at longer wavelengths, above 426 nm; see spectrum

Scheme 8

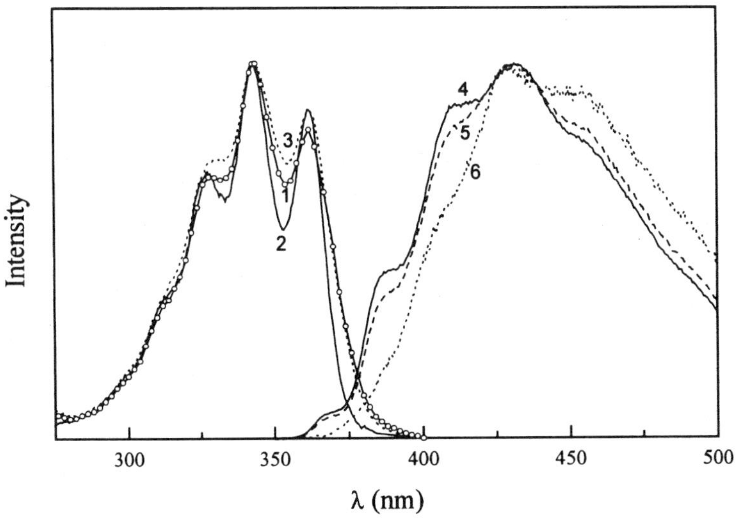

Figure 5.24. Normalized absorption (curve 1), excitation (curves 2 and 3, at $\lambda_{em} = 367$ and 455 nm, respectively), and emission (curves 4, 5, and 6, at $\lambda_{exc} = 344$, 355, and 380 nm, respectively) spectra of *all-trans*-di(3-thienyl)hexatriene in MCH/3MP at room temperature. (Adapted from Ref. 62; copyright 1999 American Chemical Society.)

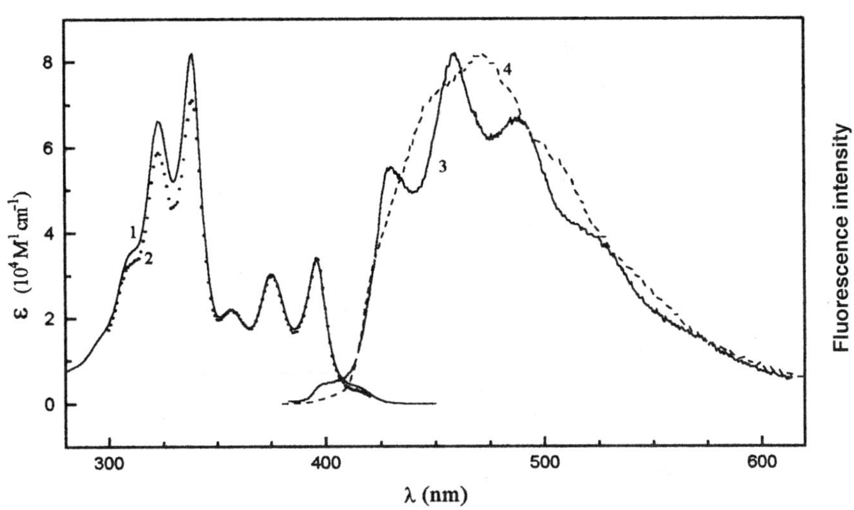

Figure 5.25. Absorption (1), excitation (2), and fluorescence spectra [$\lambda_{exc} < 405$ nm (3) and ≥ 425 nm (4)] of *trans, trans*-Ph2AnB in MCH/3MP at room temperature. (Reprinted from Ref. 63; copyright 1999 American Chemical Society.)

4 of Fig. 5.25) and the fluorescence decay become biexponential. By analogy with the behavior of *trans*-2-StAn, the presence of a short-lived rotamer was then evidenced by measuring the fluorescence decay at $\lambda_{em} = 470$ nm and $\lambda_{exc} = 416$ nm.[63]

4 Conformational Control of Photochemistry

The existence of different conformers in solutions of several series of organic compounds may affect the direction and the yield of their reactions in both the ground and excited states. For the ground-state reactions, when the thermal reactivity of the conformers can lead to different products, the overall reaction rate of the mixture depends on its composition. Also in the case of photochemical reactions, under irradiation the existence of different conformers in solution may affect the direction and the quantum yield of their reactive relaxation. Selective excitation of different ground-state conformers leads to a wavelength-dependent photoreactivity. Generally, the rotamers of the *trans*-arylalkenes can display only a change in the photoisomerization quantum yield, while those of the *cis* isomers can lead to different photoproducts (e.g., by isomerization or cyclization reactions). Some examples are described in the following subsections.

4.1 Effect on the Photoreaction Quantum Yield

In the presence of conformational equilibria, the *trans–cis* photoisomerization quantum yield of the mixture of DAE rotamers, $\bar{\phi}_{t\to c}$, depends on the excitation energy and temperature. These anomalous results may be explained on the basis of different rate constants for the radiative and radiationless deactivation processes of rotamers, due to their different geometry.

The photochemical properties of each rotamer of the fluorescent *trans* isomers can be derived through two methods, both of which use the parameters (f_i and $\phi_{F,i}$) previously derived by kinetic and statistical analysis of fluorescence (see Sections 2.3.3 and 2.3.4). In the first case, the $\bar{\phi}_{t\to c}(\lambda_{exc}, T)$ value can be expressed as the sum of the contributions of each rotamer to the geometrical *trans* → *cis* isomerization, without discriminating the photoreactive (singlet or triplet) mechanism, by

$$\bar{\phi}_{t\to c}(\lambda_{exc}, T) = \Sigma f_i(\lambda_{exc}, T)\phi_{t\to c,i}(T), \tag{29}$$

where $f_i(\lambda_{exc}, T)$ are the fractions of excited molecules in S_1 belonging to each species at any temperature and λ_{exc}, and $\phi_{t\to c,i}(T)$ is the photoisomerization quantum yield of each rotamer. The latter value can be obtained at any temperature by Equation (29), where $\bar{\phi}_{t\to c}(\lambda_{exc})$ and $f_i(\lambda_{exc})$ must be obtained at two values of λ_{exc}s, when they are the most different. If the $S_1 \to S_0$ internal conversion (IC) is a negligible process in the S_1 deactivation, as is generally observed in *trans*-DAEs [i.e., $\bar{\phi}_{t\to c}(\lambda_{exc}, T)/\alpha + \bar{\phi}_F(\lambda_{exc}, T) \cong 1$], the $\phi_{t\to c,i}(T)$ values can also be derived by

$$\phi_{t\to c,i}(T) = \alpha[1 - \phi_{F,i}(T)]. \tag{30}$$

The factor α represents the fraction of molecules that decay from the perpendicular configuration in the lowest excited states S_1 or $T_1(^{1,3}\text{perp})$ to the *cis* ground state. The potential-energy curves for both states, as a function of the torsional angle, have a minimum at 90°. From there, decay to the maximum of the ground-state curve at 90° (photoisomerization by a diabatic mechanism) and about 50 : 50 partitioning ($\alpha \cong 0.5$) to the *trans* and *cis* ground-state molecules takes place.

If the derived $\phi_{t\to c,i}$ and $\tau_{F,i}$ values are temperature dependent, the *trans* → *cis* torsional process is an activated pathway, and its rate constant can be expressed as

$$^1k_{t\to p,i}(T) = {}^1A_{t\to p,i}\exp\left(\frac{-^1\Delta E_{t\to p,i}}{RT}\right). \tag{31}$$

A different, more complicated, procedure allows one to derive the singlet and triplet contributions of each rotamer to *trans* → *cis* photoisomerization ($^1\phi_{t\to c,i}$ and $^3\phi_{t\to c,i}$, respectively) and their reactive kinetic constants on the basis of the results of a detailed analysis of the temperature effect on the fluorescence properties of the rotamer mixture. This study allowed the authors to derive the kinetic rate constants of the $^1trans^* \to {}^1perp^*$ torsional-activated process in S_1 ($^1k_{t\to p,i}$) and of the $^1trans^* \to {}^3trans^*$ intersystem crossing ($k_{ISC,i}$) for each rotamer, together with the $f_i(\lambda_{exc}, T)$ and $\tau_{F,i}(T)$ values. Therefore,

$$\bar{\phi}_{t\to c}(\lambda_{exc}, T) = \Sigma f_i(\lambda_{exc}, T)\tau_{F,i}(T)[\alpha^1k_{t\to p,i}(T) + k_{ISC,i}\phi^{sens}_{t\to c,i}(T)], \tag{32}$$

where $\phi^{sens}_{t\to c,i}$ represents the photoisomerization quantum yield, induced by triplet-energy donors. The $^1k_{t\to p,i}(T)$ and $k_{ISC,i}$ values can be derived by an Arrhenius-type equation, applied to the dependence of $\tau_{F,i}$ on temperature; that is,

$$\frac{1}{\tau_{F,i}(T)} = \frac{1}{\tau^{lim}_{F,i}} + \left[{}^1A_{t\to p,i}\exp\left(\frac{-^1\Delta E_{t\to p,i}}{RT}\right)\right] \tag{33}$$

and

$$k_{ISC,i} = \frac{1}{\tau_{F,i}(T)} - [k_{F,i} + {}^1k_{t\to p,i}(T)], \tag{34}$$

if a negligible IC process is assumed.

4.1.1 Aza-Analogues of trans-Styrylnaphthalenes

The rotamer-induced anomalies in the fluorescence behavior of stilbenelike compounds bearing a naphthyl and a pyridyl group were studied for the 1-(*n*-naphthyl),2-(3′-pyridyl)ethenes (*n*,3′-NPEs).[24,64] Four distinct, almost coplanar, isomers are expected for these asymmetric compounds, due to internal rotation of the two side groups. However, as indicated by previous studies on styrylnaphthalene (*n*-StN)[14,25] and styrylpyridine (*n*-StP)[65] isomers, only the rotation of the naphthyl group is expected to produce substantial effects on the spectral behavior, a λ_{exc}-effect due to the pyridyl group being detectable only in rigid matrices. Since the rotamer behavior is practically controlled by the "naphthyl rotamers," one expects to find evidence of two main conformers for the 2-naphthyl compounds (2,3′-NPEs) and only one for the 1-naphthyl derivatives (1,3′-NPEs), as in the case of the StNs analogues (Scheme **9**).[4,12,14] In fact, only the most different "naphthyl rotamers" could be spectrally resolved, the differences between the possible "pyridyl rotamers" being masked in the spectral envelope. This behavior was confirmed by the good fit of the bi-exponential decay of 2,3′-NPE and by the spectra at 77 K, where only two main transitions were observable in the fluorescence emission.[24,64] This is particularly true for the 2-naphthyl derivatives, since, for the 1-naphthyl derivatives, a monoexponential decay was found, as in the case of 1-StN, where the steric hindrance between the naphthyl group and the ethenic hydrogens in one of the two conformers shifts the equilibrium practically completely towards the less hindered rotamer. The photophysical properties of *n*,3′-NPE mixtures are reported in Table 7.[6]

The photoisomerization quantum yield depends slightly on λ_{exc} for 2,3′-NPE only, as expected. The higher quantum yield measured at 336 nm, where the ratio of the preexponential factors shows an increase in the weight of the short-lived rotamer, indicates that the latter is less fluorescent and more photoreactive.

A more detailed analysis of the rotamer mixture of *trans*-2,3′-NPE[24] indicated the absence of isoemissive points; therefore, a more complicated version of the KFA analysis proposed by Birks et al.[25] was applied.[24] This version allowed for a separation of the distinct photophysical properties of the two conformers ($\phi_F = 0.25$ and 0.85 for the A and B rotamers of 2,3′-NPE, respectively), which have similar radiative rate parameters ($k_{F,i}$ ca. $5 \times 10^7\ s^{-1}$), but quite different nonradiative rate parameters [$k_{NR,i} = (1 - \phi_{F,i})/\tau_{F,i}$]. The latter values were $1.5 \times 10^8\ s^{-1}$ for the more abundant (62%) A rotamer and $0.1 \times 10^8\ s^{-1}$ for the longer lived rotamer B, in agreement with its reduced reactivity, as indicated in Table

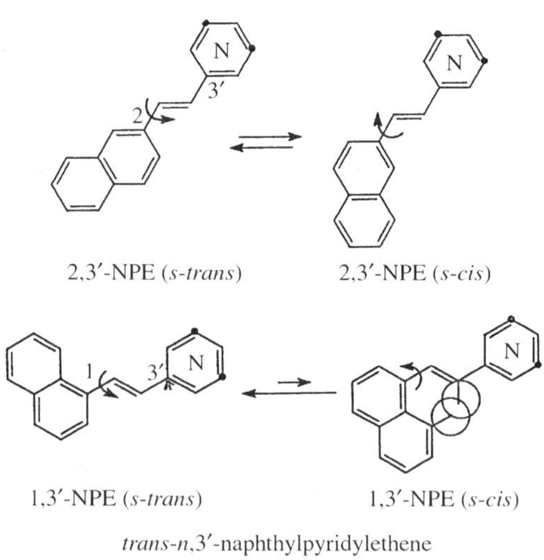

Scheme 9

Table 7 Fluorescence parameters and photoisomerization quantum yields of *trans-n*,3′-NPE in *n*-hexane at room temperature (from Ref. 64)

	$\lambda_{exc} = 315$ nm			$\lambda_{exc} = 336$ nm		
Isomer	$\bar{\phi}_F$	$\tau_F{}^a$ (ns)	$\bar{\phi}_{t\to c}$	$\bar{\phi}_F$	$\tau_F{}^a$ (ns)	$\bar{\phi}_{t\to c}$
1,3′-NPE	0.48	1.8	0.24	0.54	1.8	0.24
2,3′-NPE	0.78	5.6 16.3	0.15	0.64	5.0 14.9	0.19

[a] Measured at $\lambda_{em} = 380$ nm.

7. A study of the effect of temperature on the fluorescence decay of *trans*-2,3′-NPE allowed important information to be obtained on the activated $^1trans^* \rightarrow {}^1perp^*$ internal rotation of the two excited rotamers in the singlet manifold. Using Equation (33), the Arrhenius parameters $^1\Delta E_{t\to p,i}$ and $\log {}^1A_{t\to p,i}$ in MCH (6.7 kcal mol^{-1} and 12.8, respectively, for A and 7.3 kcal mol^{-1} and 12.4, respectively, for B) were derived (Fig. 5.26). Subsequently, the $^1k_{t\to p,i}(T)$ values were derived by Equation (31).[24]

In the presence of a relevant IC process $[\bar{\phi}_F + (\bar{\phi}_{t\to c}/\alpha) < 1]$, as is observed for 2,3′-NPE, an evaluation of the role of the $S_1 \rightarrow S_0$ IC in the deactivation of S_1 can be obtained by

$$k_{IC,i} = \frac{1}{\tau_{F,i}(T)} - [k_{F,i} + {}^1k_{t\to p,i}(T) + k_{ISC,i}]. \tag{35}$$

The $k_{ISC,i}$ values can be obtained from the $\phi_{t\to c,i}$ values (see Equation (29)), and from the equation

$$\phi_{t\to c,i} = {}^1\phi_{t\to c,i} + {}^3\phi_{t\to c,i} = \alpha\tau_{F,i}k_{t\to p,i} + \tau_{F,i}k_{ISC,i}\phi^{sens}_{t\to c,i} \tag{36}$$

assuming an equal value for $\phi^{sens}_{t\to c,i}$ of the two rotamers.

The results thus derived, which give the $\phi_{F,i}$, $\tau_{F,i}$ and $\phi_{t\to c,i}$ values of the rotamers of two aza-analogues of *trans*-2-StN, at room temperature are summarized in Table 8, together with the kinetic parameters of their S_1 deactivation pathways.

4.1.2 cis-2-Styrylanthracene

The *cis* isomer of 1-(2-anthryl),2-(phenyl)ethene (*cis*-2-StAn) belongs to a family of 1,2-diarylethenes that exhibits one-way *cis* → *trans* adiabatic photoisomerization on the lowest singlet or triplet energy surface ($^{1,3}cis \rightarrow {}^{1,3}trans$). Following direct excitation, the *cis* → *trans* photoisomerization was assumed to involve ^1cis-2-StAn → ^3cis-2-StAn intersystem crossing as the initial key step.[66,67] More recent papers indicate that the photoreactive pathway proceeds predominantly through a singlet adiabatic mechanism, at least at temperatures higher than room temperature.[46,68,69]

The shape of the fluorescence spectrum, following excitation of *cis*-2-StAn (Fig. 5.27), is similar to that of the *trans* isomer, obtained at the same λ_{exc} (Fig 5.11). By

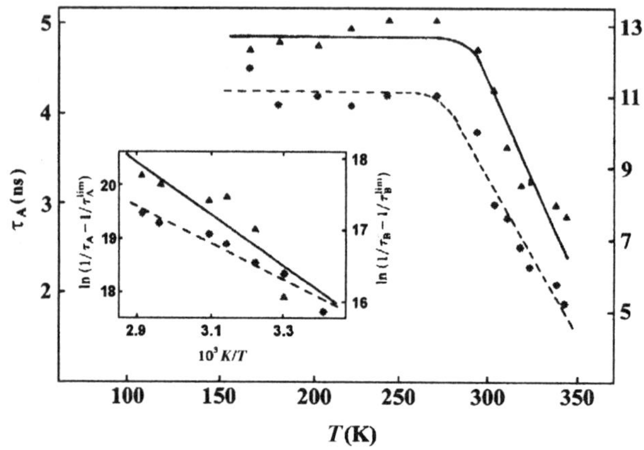

Figure 5.26. Plots of fluorescence lifetimes against temperature for 2,3′-NPE(A) (– – –) and 2,3′-NPE(B) (—) in MCH. The inset shows the corresponding Arrhenius plots. (Reprinted from Ref. 24; copyright 1986 Royal Society of Chemistry.)

contrast, its excitation spectrum is λ_{em}-independent and is a replica of the absorption spectrum, which shows a different shape compared with that of the *trans* isomer.

The quantum yield of the photoisomerization was found to be dependent on λ_{exc}, with values of 0.63 and 0.50 at 366 nm and 290 nm, respectively, at room temperature, in deaerated solutions of MCH/3MP.[46,68] The fluorescence decay was found to be biexponential and its parameters to depend on the analysis wavelength ($\tau_{F,1} = 4.6$ ns and $\tau_{F,2} = 63$ ns, at $\lambda_{em} = 395$ nm; $\tau_{F,1} = 14.6$ ns and $\tau_{F,2} = 63.5$ ns, at $\lambda_{em} = 435$ nm).[46]

On the basis of the previous results (Table 2), the $\tau_{F,i}$ values of 4.6, 14.4, and 63 ns were assigned to the A or B rotamers of $^1cis^*$, ^1trans-A*, and ^1trans-B* conformations, respectively.[46] The latter component is the largest at any experimental conditions, thus pointing up the fact that the ^1cis-B* → ^1trans-B* adiabatic process is the most important reactive pathway.

The results of experiments[46, 69] and calculations[46] indicate that (i) two rotamers (*cis*-A and *cis*-B) are present in the ground state in similar concentrations (about

Table 8 Photophysical and photochemical parameters together with kinetic rate constants (10^8 s^{-1}) of rotamers of *trans*-2-StN and its aza-analogues, 2,3′-NPE and 2,6′-NQE, in *n*-hexane at room temperature ($\lambda_{exc} = 320$ nm; from Refs 14, 24, and 31)

Rotamer	$\bar{\phi}_F$	$\bar{\phi}_{t\to c}$	$\phi_{F,i}$	$\tau_{F,i}$ (ns)	f_i	$k_{F,i}$	$^1k_{t\to p,i}$	$k_{ISC,i}$	$k_{IC,i}$	$\phi_{t\to c,i}$
2-StN(A)	0.67	0.16	0.83	5.2	0.23	1.6	0.03	0.3	negligible	0.08
2-StN(B)			0.50	27.6	0.77	0.18	0.01	0.17	negligible	0.25
2,3′-NPE(A)	0.58	0.15	0.25	4.9	0.45	0.51	0.63	0.56	0.24	0.31
2,3′-NPE(B)			0.85	14.9	0.55	0.57	0.09	0.01	negligible	0.07
2,6′-NQE(B)			0.45	3.9	0.60	1.2	—	—	negligible	0.28
2,6′-NQE(C)	0.57	0.22	0.87	3.3	0.21	2.6	—	—	negligible	0.06
2,6′-NQE(A)			0.77	1.9	0.19	4.0	—	—	negligible	0.12

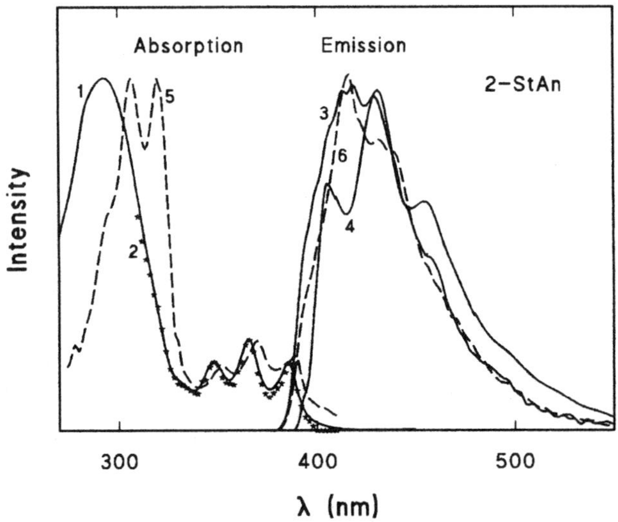

Figure 5.27. Normalized absorption (1) and excitation (2) spectra of the *cis* isomer of 2-styrylanthracene in MCH/3MP at room temperature together with fluorescence spectra obtained by exciting the *cis* isomer ($\lambda_{exc} = 290$ nm) at 298 K (3) and 80 K (4). Absorption (5) and emission (6) spectra of the *trans* isomer are also reported for comparison.

66% for *cis*-B), as derived from the $\Delta H(S_0)$ value,[46] and (*ii*) in agreement with the NEER principle,[8] the *cis* rotamers of 2-StAn do not interconvert in the excited states during their lifetime.

Based on the spectral and fluorescence decay results, the $\bar{\phi}_F$ values (0.53–0.46 in toluene[69] and 0.50–0.42 in MCH/3MP[46] at room temperature in the 290–390-nm λ_{exc} range) can be assumed to be the sum of two main contributions: from $^1cis^*$, directly excited, and from ^1trans-B*, produced by the specific adiabatic singlet isomerization ($^1\phi_{c\text{-}B^* \rightarrow t\text{-}B^*}$). The $^1\phi_{c\text{-}B^* \rightarrow t\text{-}B^*}$ values of about 0.60 in MCH/3MP[46] and 0.45 in toluene[69] were obtained at room temperature from the equation

$$^1\phi_{c\text{-}B^* \rightarrow t\text{-}B^*} = \%(t\text{-}B)\bar{\phi}_F/\phi_{F,t\text{-}B}, \tag{37}$$

where the %(*t-B*) values were 94 in MCH/3MP and 75 in toluene, derived from a treatment of the fluorescence decay curve[46] and the PCA-SV method,[69] respectively. The $^1\phi_{c\text{-}B^* \rightarrow t\text{-}B^*}$ values thus obtained in MCH/3MP at room temperature (0.66 and 0.55 at $\lambda_{exc} = 366$ and 290 nm, respectively) account for the overall *cis* → *trans* quantum yield, measured under the same experimental conditions.[46]

The small contribution from $^1cis^*$ to the fluorescence, following excitation of *cis*-2-StAn, can be roughly evaluated by

$$\phi_{F,c} = \%(cis)\bar{\phi}_F. \tag{38}$$

Values of about 0.03 and 0.025 were derived in MCH/3MP[46] and toluene,[69] respectively.

For the specific adiabatic singlet $^1cis\text{-B}^* \rightarrow {}^1trans\text{-B}^*$ process, a unit $^1\phi_{ad,c\text{-}B}$ value was then derived at room temperature from the relation

$$^1\phi_{\text{ad,c-B}} = {}^1\phi_{\text{c}\rightarrow\text{t-B}^*}/f_{\text{c-B}}, \tag{39}$$

where $f_{c\text{-}B}$ was assumed to be 0.66 at $\lambda_{\text{exc}} = 366$ nm, on the basis of similar molar extinction coefficients for the $S_0 \rightarrow S_1$ transition of the two rotamers of *cis*-2-StAn derived by calculations.[46]

The most important reactive process in the deactivation of $^1cis\text{-A}^*$ is the formation of a dihydrophenanthrene-type product,[70] which converts to both the stable dehydrogenated cyclic compound 1,2-benzotetraphene (BT) ($\phi_{\text{c-A}\rightarrow\text{BT}} = 0.1$) and ^1cis-A in the ground state.[46]

All these results show that, in a nonpolar solvent at room temperature, the most relevant deactivation pathways of the S_1 state of the *cis*-2-StAn rotamers are the $S_1 \rightarrow S_0$ IC and the cyclization process for the A rotamer and the $^1cis\text{-B}^* \rightarrow {}^1trans\text{-B}^*$ adiabatic reactive channel for the B rotamer.[46] The *cis* isomer of 2-StAn is a clear example of how the existence of a conformational equilibrium in fluid solution affects the direction and quantum yield of its reactive processes.

4.2 Effect on the Direction of Photoreaction (Multiway Cyclization of *cis* Isomers)

4.2.1 Dimethylhexatriene

An illustrative example of how rotamerism can affect not only the yield, but also the direction, of photochemical reactions of polyenes concerns the photocyclization of methylated *all-cis*-hexatriene,[8] already mentioned in Section 3.3. The unsubstituted triene cannot cyclize, since it exists mainly in the *s-trans–s-trans* form (with respect to the central double bond) and only *cis* → *trans* photoisomerization could be observed. As mentioned earlier, Raman spectroscopy studies on the triplets of the *cis* isomers of methylated hexatrienes showed that different spectra were obtained from the sterically hindered 2,5-DMeH (Scheme **6**), in which the *cis* isomer can exist in two different conformations.[15] The results are in agreement with the photochemical results on the cyclization products of these compounds, which were found to be controlled by rotamerism.[8] In fact, for the 2,5-dimethyl derivative, in which the *s-cis–s-cis* form was expected to prevail, two main cyclization products were found, dimethyl-vinylcyclobutene (**10d**) and dimethyl-cyclohexadiene (**10e**), while for the 2-methyl derivative, in which the *s-cis–s-trans* form prevails, three photoproducts were obtained: methyl-1,2,4-hexatriene (**10a**), methyl-vinylcyclobutene (**10b**), and methyl-bicyclo[3.1.0]hexene (**10c**) (Scheme **10**).

In more recent works on the same compound and other dialkyl derivatives with substituents of different size, which can affect the relative abundances of the rotamers by steric effects, the photorearrangements of substituted hexatrienes was

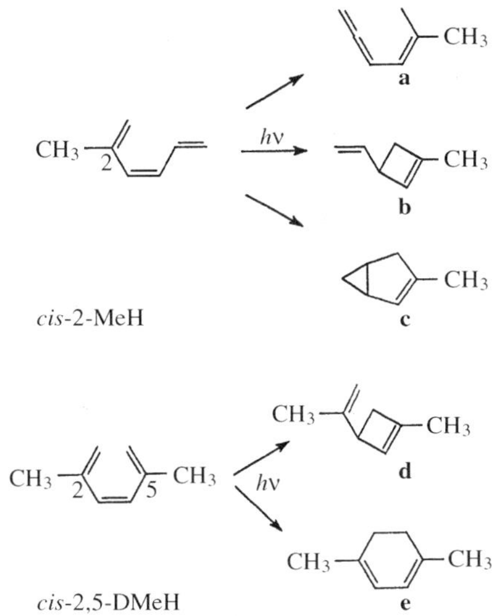

Scheme 10

investigated further.[58] Under the constraints imposed by the NEER principle, it was shown that each rotamer of the equilibrium mixture of the ground state can be associated with its own specific photochemistry.

4.2.2 *Diarylethenes*

Rotamerism of *cis* isomers of DAEs and related compounds has been relatively much less investigated in comparison to the *trans* isomers, mainly because of their very short S_1 lifetime, reflected by the lack of fluorescence emission from their fluid solutions, but probably also because of their more difficult preparation and lower photostability. The relaxation of the lowest excited states of *cis* geometry takes place, apart from possible contributions of nonreactive decay, by isomerization to *trans* and cyclization to 4a,4b-dihydrophenanthrene-type (DHPhe) compounds. The latter can reconvert to *cis* thermally or photochemically or, in the presence of hydrogen acceptors, can form stable (phenanthrene-type) polycyclic oxidation products.[70] When the starting *cis* isomer can exist as several different rotational conformers, more than one photocyclization product can be observed under irradiation. The process has been studied particularly for meta-stilbenes and for DAEs bearing naphthyl groups.

Note that the relative abundances of the cyclized photoproducts cannot be considered to quantitatively reflect the rotamer abundances, since the reactivity and chemical yield of rotamer cyclization depend not only on the starting rotamer concentration, but perhaps also on steric and substituent-induced electronic factors (e.g., the different electron density and free-valence index at the closure centers.).[64,70] In any case, this method can offer important qualitative information on the presence and composition of the rotamer mixture. Some interesting applications to stilbene and its naphthyl analogue derivatives will be described next.

Early examples of the effect of rotamerism on cyclization products were reported for meta-substituted stilbenes,[71] particularly for methylated compounds when comparable amounts of 2- and 4-substituted phenanthrenes were obtained[71b] and for 3-styrylpyridine,[72] which gave a 1:4 mixture of 4- and 2-azaphenanthrene (4-APhe, and 2-APhe, respectively; Scheme **11**). In the latter case, simple calculations of the charge density and localization energies on the reactive position by the Hückel method gave a satisfactory explanation of the higher reactivity of position 4 (*para* to the nitrogen), not excluding cyclization in position 2 (*ortho* to the nitrogen), as was found experimentally.[72]

A series of *cis*-stilbenes bearing various substituents (*t*-butyl, methoxy, and fluorine) at the meta (3 or 3,5) position(s) with respect to the ethene carbons were studied by Fischer's group by conventional stationary and pulsed techniques.[73] Photocyclization of the *cis* isomers of tetramethoxy- and tetrafluoro-stilbenes led to two-colored DHPhes, only one of which was reconverted into the starting *cis* isomer by visible light. In tetramethoxy-stilbene, the primary cyclization product, DHPhe, formed under irradiation, underwent a two-stage decay, providing indirect evidence for the presence of two modifications, which were subsequently generally studied by their absorption spectra at low temperatures (around 133 K). After heating at around 193 K, a thermal equilibrium between the two species was established. Microsecond flash photolysis indicated that the the primary transient decayed fast (its half-life was about 80 μs), being the precursor of a second, longer-lived transient (with a half-life of about 50 s). The fast decay and the growing-in were assigned to partial thermal conversion to attain the equilibrium between the two modifications.

E. Fischer and co-workers have also described interesting studies of *cis*-DAEs bearing naphthyl groups. In their pioneering flash photolytic study on *cis*-1,2-di(2-naphthyl)ethene (*cis*-2,2′-DNE),[74] these researchers got evidence for the presence of two isomeric dibenzodihydrophenanthrenes (DHPhes) (Scheme **12**), formed by photocyclization of two different conformers of the three that in principle, could exist as an equilibrium mixture in solutions of 2,2′-DNE.

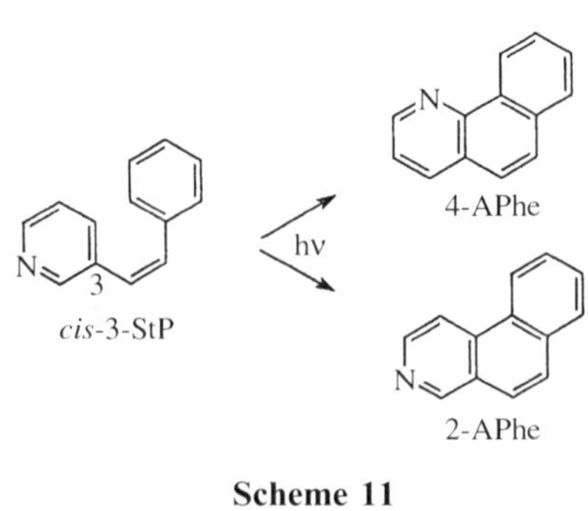

Scheme 11

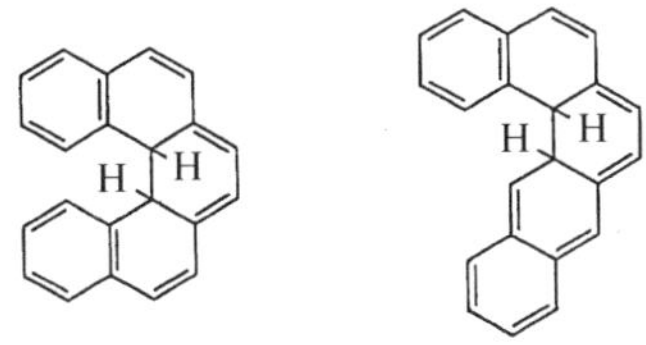

Scheme 12

Two relatively stable DHPhe-type compounds also were formed as cyclization products in 1,2-di-(2-naphthyl)cyclopentene. As expected from the results of CFF–SCF–PI–CI calculations, each of them can exist in two modifications. Experimental evidence for these species was found for one of the two DHPhes.[75]

In later improved experiments, these systems were revisited, with the study extended to other methylo- and fluorine-substituted DNEs.[76,77] Photocyclization products of two rotamers were generally confirmed under appropriate conditions. One of the two DHPhes was found to exist in two colored interconvertible modifications. In fact, a relatively stable DHPhe-like product was formed through a precursor absorbing at longer wavelengths, the two modifications being separated by a substantial potential-energy barrier. As for the unsubstituted DNE, two colored interconvertible modifications were also observed for one of the two rotamers of dimethyl-DNE, substituted at the ethenic carbons.[76]

4.2.3 Diarylethenes Bearing Naphthyl and Pyridyl Groups

Some information on the behavior of the nonfluorescent *cis*-1-(*n*-naphthyl),2-(*n*′-pyridyl)ethenes (*cis-n,n*′-NPEs) was obtained from their photocyclization products.[64] From both the 1-naphthyl and 2-naphthyl derivatives, only two cyclized compounds were produced: azabenzo[a]phenanthrenes (azachrysenes), in the case of *cis*-1,3′-NPE, and azabenzo[c]phenanthrenes (ABPhe), in the case of *cis*-2,3′-NPE (Scheme **13**).

In both cases, the compound with the quinoline structure was found in much smaller yield with respect to the isomer with the isoquinoline structure. No production of anthracene-type compounds (ABAn) was observed. This was explained, in analogy with related compounds,[70] by the inertness towards ring closing when the intermediate (dihydro derivative)—which should lead finally to aza-benzanthracene (ABAn)—has a highly quinoid structure.

4.2.4 Diarylethenes Bearing Phenanthryl Groups

Conformational equilibria of *cis* compounds bearing the phenanthryl or benzologue moieties were studied by Laarhoven and co-workers.[78–80] Information was obtained

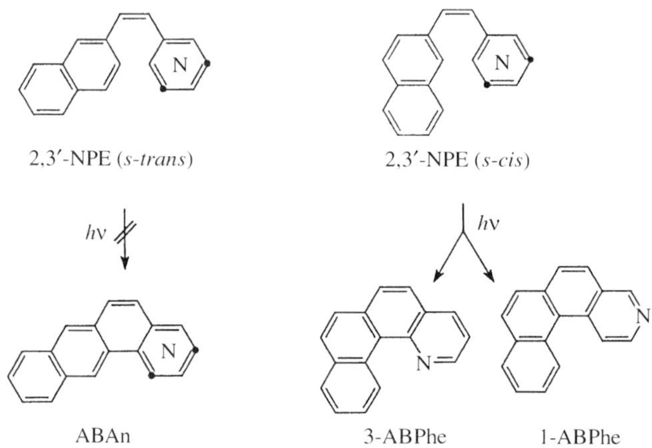

Scheme 13

by NMR spectroscopy and from the structure of these compounds' photocyclization products (Scheme **14**). Interestingly, unlike *cis*-3-styrylphenanthrene (Scheme **14,1**), where the *s-trans* (anti) rotamer (Scheme **14,1b**) was prevailing in the mixture, for *cis*-2-styrylbenzo[c]phenanthrene (Scheme **14,2**), the *s-cis* (syn) conformation (Scheme **14,2a**) was found to prevail in spite of its larger crowding.[79] Other analogues, such as styryl-dibenzo[c,g]triphenylene (Scheme **14,4**) and hexahelicene (Scheme **14,3**) behave as **1**, with the anti rotamer prevailing at varying extents. The conformational preference of **2a** was ascribed to van der Waals interaction between the aryl groups, induced by the higher polarizability of the benzo[c]phenanthryl moiety in comparison with the phenanthryl one. For similar reasons, the 1,1′-diphenyl derivative of *cis*-1,2-di-(9-phenanthryl)ethene was found to prefer conformation **5**, which leads to the photoproduct **6** and in which both phenyl substituents are in the overcrowded region. Moreover, the photochemical formation of the cyclobutane derivative **8** from *trans-trans*-2,2′-distyrylbiphenyl was taken as an indication of the conformation **7** where the styryl groups are strongly crowded. It was shown by NMR spectroscopy that further enlargement of the aryl groups can lead to a helical overcrowded conformation in *cis* isomers of styryl aromatics. Whether this occurs depends on a favorable balance between attractive and repulsive forces in the overcrowded region.

4.2.5 1-Styrylanthracene

In MCH/3MP at room temperature, we found that *cis*-1-styrylanthracene (*cis*-1-StAn) has very small fluorescence and *cis* → *trans* photoisomerization quantum yields (0.054 and 0.04, respectively) and a short fluorescence lifetime (0.45 ns).[81] The most important process in the deactivation of the lowest excited singlet state is the formation of a dihydrophenanthrene-type product (DHPhe), which converts, in

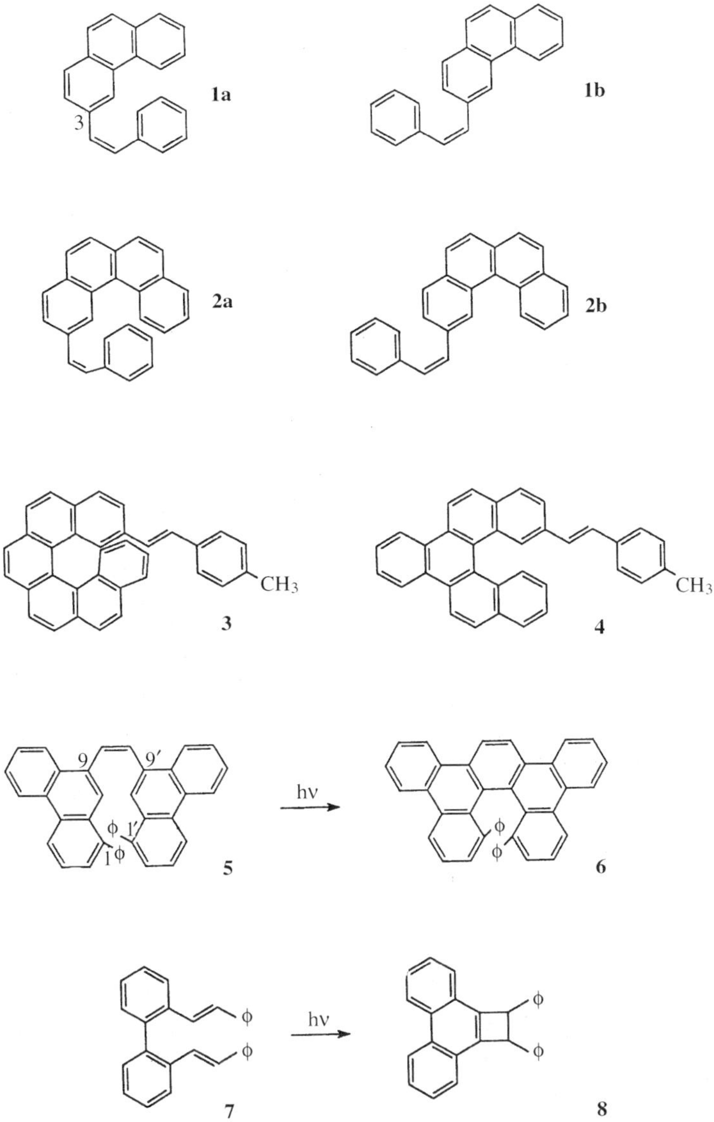

Scheme 14

the presence of oxygen, to benzo[b]chrysene (BC) ($\phi_{c\rightarrow BC} = 0.45$),[81] as previously found in benzene.[82]

Above room temperature, HPLC analysis of diluted solutions irradiated at 366 nm showed the formation of three photoproducts: the *trans* isomer, adiabatically produced by a mixed (singlet and triplet) mechanism; the BC derivative, and another photoproduct of intramolecular cycloaddition photodehydrocyclization (CY) with

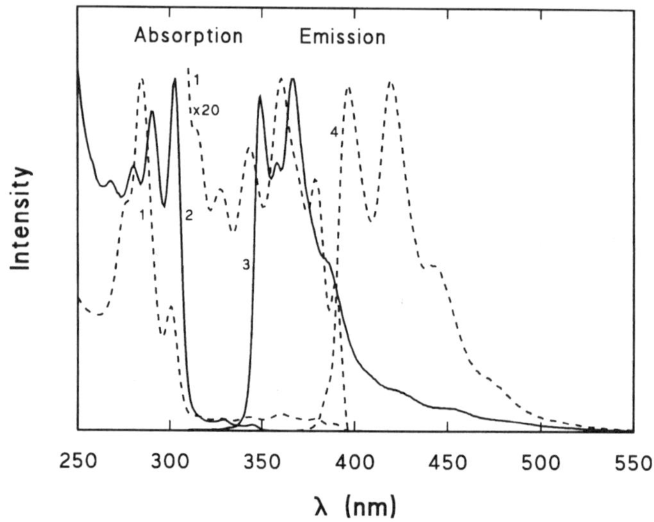

Figure 5.28. Normalized absorption and emission spectra of cyclization products of *cis*-1-StAn in water–acetonitrile at room temperature: 1 and 4 (-----) for benzo[b]chrysene; 2 and 3 (—) for the CY compound. (Reprinted from Ref. 81; copyright 1997 Royal Society of Chemistry.)

the same mass of 278. The latter was characterized by its absorption spectrum, whose properties suggest a naphthalenic nature of the S_1 state (based on the very low oscillator strength and the vibronic structure of the $S_0 \rightarrow S_1$ transition) and, consequently, the presence of a naphthalenic structure in this cyclization product. The $S_0 \rightarrow S_2$ transition shows a much higher oscillator strength and a shape which is

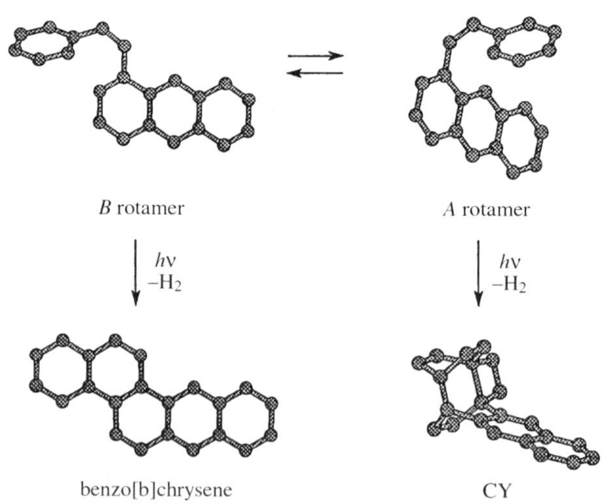

Scheme 15

reminiscent of that of compounds (9,10-dihydroanthracene, lepidopterenes) where the anthracenic aromatic structure is destroyed.[83] The absorption and emission spectra of the products of intramolecular cycloaddition *cis*-1-StAn[81] are shown in Figure 5.28.

With the use of a modeling program based on ZINDO/S-CI and PM3-CI calculations,[81] the nature of CY was tentatively assigned as originating from the A rotamer (see Scheme **15**). This cyclization structure should involve the C(4) carbon atom of the anthracene group and the C(4′) of the phenyl group with loss of the anthracenic structure. The computed spectral parameters were in good agreement with those obtained from the experimental absorption spectrum. The formation of cyclization products was not observed when *cis*-1-StAn was irradiated in the presence of triplet-energy donors, thus showing that the T_1 state is not involved in these photoreactions.[81]

5 Concluding Remarks

This chapter presents an overview of the researches—mainly experimental work—carried out in the last few decades on the phenomenon of rotational isomerism in organic molecules, with emphasis on a specific series of olefin molecules, the geometrical *trans* and *cis* isomers of diarylethenes. The description of conformational equilibria between *s-cis* and *s-trans* rotamers in the ground and excited states is based mainly on the observation of spectral anomalies in solution (e.g., the dependence of fluorescence properties on the excitation energy). Other techniques (flash photolysis and infrared and NMR spectroscopy), as well as quantum mechanical calculations, gave significant, but less extensive, contributions. The type of information afforded by the fluorimetric technique depends on whether the basic NEER assumption that no equilibration occurs between the excited rotamers during the lifetime of the emitting state applies to the situation at hand. In the first case, the fluorescence behavior reflects the ground-state equilibrium composition. On the other hand, when the NEER principle is not applicable, pulsed fluorimetry can give information on the equilibration process in the S_1 state. Two quantitative methods of fluorescence analysis, KFA (a kinetic method based on the λ_{exc} effect on the fluorescence quantum yields and radiative decay curves) and PCA-SM (a mathematical method based on statistical treatment of spectra pertaining to multicomponent systems), are described in detail in Sections 2.3.3 and 2.3.4, and several applications are reported in Section 3 to show how the properties of each rotamer (abundance, spectra, and photophysical and thermodynamic parameters) are resolved from those of their mixtures. The application of both methods is delicate and meets with limiting difficulties: the KFA method gives both spectral and kinetic information, but is practically limited to mixtures of rotamers having different lifetimes and is hardly applicable to systems of more than two components, while the PCA method can be applied to a larger number of components (if they have at least slightly different spectra), but gives only spectral information. The rotamerism of two systems, investigated in depth by both methods of analysis and representing examples of two-component (*trans*-2-styrylanthracene) and three-component [*trans*-1-(2-naphthyl),2-(6′-quinolyl)ethene] systems, has been described in particular, but several other examples are reported for ethene and polyene derivatives. Section 4 deals with the effect of rotamerism on photochemistry. Several examples are shown in which the quantum yield or the direction of photoreaction is affected by the existence of different rotamers in solution.

The study of conformational equilibria has reached a good level of understanding, particularly for DAEs, but further work is needed in order to obtain a clear picture of the intrinsic spectral and kinetic properties of the separate rotamers and a more general interpretation of the breakdown of the NEER principle. The use of fast spectroscopy and ultrafast spectroscopy and of other techniques different from fluorimetry (such as NMR, ENDOR, IR and Raman, particularly at low temperatures) should be more extensively investigated. Important contributions are expected from spectroscopic investigations into isolated-molecule conditions (e.g., Shpolskii matrices, already successfully used in a few cases, and supersonic jets). The experimental studies should be accompanied by parallel quantum mechanical calculations, which have proven to be of great help in the interpretation of the rotamer behavior. Finally, the role of the solvent on the intrinsic spectral and kinetic properties of each rotamer and, particularly, its effect on the interconversion barriers and on the enthalpy difference should be more extensively investigated as well.

Acknowledgments

The authors would like to dedicate this chapter to two unforgettable late friends who convinced us to face this interesting research field: Ernst Fischer, who carried out an important pioneering work with many important observations on the rotamerism of several series of diarylethenes, and John B. Birks, who proposed the elegant KFA method to us. The financial support of the Italian Consiglio Nazionale delle Ricerche and Ministero dell'Università e della Ricerca Scientifica e Tecnologica are gratefully acknowledged. Thanks are also due to M. Pelliccia for help in the graphics.

References

1. Eliel, E.L.; Wilen, S.H.; Mander, L.N. *Stereochemistry of Organic Compounds*, Wiley: New York, **1994;** Testa, B. *Principles of Organic Stereochemistry*, Dekker: New York, **1979;** Tamm, C. *Stereochemistry, New Comprehensive Biochemistry, Vol. 3*, Elsevier Biomedical Press: Amsterdam, **1982;** Marples, B.A. *Elementary Organic Stereochemistry and Conformational Analysis, Vol. 34*, Roy. Soc. Chem., Monographs for Teachers, London, **1981.**

2. Orville-Thomas, W.J. *Internal Rotation in Molecules*, Wiley: London, **1974.**

3. (a) Alfimov, M.V.; Sheck, Y.B.; Kovalenko, N.P. *Chem. Phys. Lett.*, **1976**, *43*, 154–156; Sheck, Y.B.; Kovalenko, N.P.; Alfimov, M.V. *J. Lumin.*, **1977**, 15, 157–168; (b) Fischer, E. *J. Photochem.*, **1981**, *17*, 331–340; *J. Mol. Struct.*, **1982**, *84*, 219–226; (c) Mazzucato, U. *Pure Appl. Chem.*, **1982**, *54*, 1705–1721; (d) Ghiggino, K.P., *J. Photochem.*, **1980**, *12*, 173–177; Matthews, A.C.; Sakurov, R.; Ghiggino, K.P. *J. Photochem.*, **1982**, *19*, 235–244.

4. Mazzucato, U.; Momicchioli, F. *Chem. Rev.*, **1991**, *91*, 1679–1719 and references therein.

5. Saltiel, J.; Sun, Y.-P. In *Photochromism. Molecules and Systems*, Dürr H. and Bouas-Laurent, H. eds., Elsevier: Amsterdam, **1990;** pp 143–152 and references therein.

6. Mulliken, R.S. *J. Chem. Phys.*, **1939**, *7*, 121–135.

7. Klessinger, M.; Michl, J. *Excited States and Photochemistry of Organic Molecules*, VCH Publishers: New York, **1995**, pp 63–70.

8. Havinga, E. *Experientia*, **1973**, *29*, 1181–1193; Vroegop, P.J.; Lugtenburg, J.; Havinga, E. *Tetrahedron*, **1973**, *29*, 1393–1398; Jacobs, H.J.C.; Havinga, E. *Adv. Photochem.*, **1979**, *11*, 305–373.

9. Cherkasov, A.S. *Dokl. Acad. Sci. USSR*, **1962**, *146*, 852–855.

10. Fletcher, A.N. *J. Phys. Chem.*, **1968**, *72*, 2742–2749.

11. Hammond, G.S.; Shim, S.C.; Van, S.P. *Mol. Photochem.*, **1969**, *1*, 89–106.

12. Fischer, G.; Fischer, E. *Mol. Photochem.*, **1974**, *6*, 463–472; Haas, E.; Fischer, G.; Fischer, E. *J. Phys. Chem.*, **1978**, *82*, 1638–1643.

13. Scoponi, M.; Gallinella, E.; Momicchioli, F. *J. Chem. Soc. Faraday Trans. 2*, **1988**, *84*, 95–103.

14. Bartocci, G.; Masetti, F.; Mazzucato, U.; Marconi, G. *J. Chem. Soc. Faraday Trans. 2*, **1984**, *80*, 1093–1105.

15. Langkilde, F.W.; Jensen, N.-H.; Wilbrandt, R.; Brouwer, A.M.; Jacobs, H.J.C. *J. Phys. Chem.* **1987**, *91*, 1029–1040.

16. (a) Dezube, B.; Dobson, C. M.; Teague, C. E. *J. Chem. Soc. Perkin II*, **1981**, 730–735; (b) Yu, H.-T.; Vela, M.A.; Sattler, M.C.; Fronczek, F.R.; McLaughlin, M.L.; Barkleyy, M.D. *J. Am. Chem. Soc.*, **1995**, *117*, 348–357; (c) McMahon, L.P.; Yu, H.-T.; Vela, M.A.; Morales, G.A.; Shui, L.; Fronczek, F.R.; McLaughlin, M.L.; Barkley, M.D. *J. Phys. Chem.*, **1997**, *101*, 3269–3280.

17. Fisichella, S.; Mineri, G.; Scarlata, G.; Sciotto, D. *Tetrahedron*, **1975**, *31*, 2445–2447; Bottino, F.A.; Mineri, G.; Sciotto, D. *Tetrahedron*, **1978**, *34*, 1557–1559; Bottino, F.A.; Scarlata, G.; Sciotto, D.; Torre, M. *Spectrochim. Acta*, **1980**, *36A, 205–206.*

18. Higuchi, J.; Ishizu, K.; Nemoto, F.; Tajima, K.; Suzuchi, H.; Ogawa, K. *J. Am. Chem. Soc.*, **1984**, *106*, 5403–5406.

19. Sorriso, S.; Lumbroso, H. *Bull. Soc. Chim. France*, **1973**, 1583–1586; Jakusek, E.; Freundlich, P.; Kolodziej, H.A.; Koll, A.; Sorriso, S. *Chem. Phys.* **1984**, *84*, 441–451.

20. (a) Fischer, E.; Castel, N. *Mol. Struct.*, **1986**, *145*, 367–376; (b) Krongrauz, V.; Castel, N.; Fischer, E. *J. Photochem.*, **1987**, *39*, 285–300.

21. Wismontski-Knittel, T.; Das, P.K. *J. Phys. Chem.*, **1984**, *88*, 1168–1173.

22. Momicchioli, F.; Baraldi, I.; Fischer, E. *J. Photochem. Photobiol. A: Chem.*, **1989**, *48A*, 95–107.

23. Aloisi, G.G.; Elisei, F.; Mazzucato, U.; Latterini, L.; Rodgers, M.A.J. in *Fast Elementary Processes in Chemical and Biological Systems*, A. Tramer Ed., Am. Inst. Phys. Conf. Proc., **1996**, *364*, 139–146.

24. Bartocci, G.; Masetti, F.; Mazzucato, U.; Spalletti, A.; Bruni, M.C. *J. Chem. Soc., Faraday Trans. 2*, **1986**, *82*, 775–788.

25. Birks, J.B.; Bartocci, G.; Aloisi, G.G.; Dellonte, S.; Barigelletti, F. *Chem. Phys.*, **1980**, *51*, 113–120.

26. (a) Lawton, W.H.; Sylvestre, E.A. *Technometrics*, **1971**, *13*, 617–633; (b) Wold, S. In *Evaluation and Optimization of Laboratory Methods and Analitical Procedures*, Massart, D.L.; Dijkstra, A.;

Kaufman, L. eds., Elsevier: Amsterdam, **1978;** Otha, N. *Anal. Chem.*, **1973**, *45*, 553–557; Malinowski, E.R. *Anal. Chem.*, **1977**, *49*, 612–617; (c) Vandeginste, B.: Essers, R.; Bosmann, T.; Reijnen, J.: Kateman, G. *Anal. Chem.*, **1985**, *57*, 971–985; (d) Wold, S. *Tecnometrics*, **1978**, *20*, 397–405 and references therein.

27. (a) Saltiel, J.; Eaker, D.W. *J. Am. Chem. Soc.*, **1984**, *106*, 7624–7626; (b) Saltiel, J.; Sears Jr., D.F.; Mallory, F.B.; Mallory, C.W.; Buser, C.A. *J. Am. Chem. Soc.*, **1986**, *108*, 1688–1689; (c) Sun, Y.-P.; Sears, Jr., D.F.; Saltiel, J. *Anal. Chem.*, **1987**, *59*, 2515–2519; (d) Sun, Y.-P.; Sears, Yr., D.F.; Saltiel, J.; Mallory, F.B.; Mallory, C.W.; Buser, C.A. *J. Am. Chem. Soc.*, **1988**, *110*, 6974–6984.

28. Ghiggino, K.P.; Skilton, P.F.; Fischer, E. *J. Am. Chem. Soc.*, **1986**, *108*, 1146–1149.

29. Spalletti, A.; Bartocci, G.; Masetti, F.; Mazzucato, U.; Cruciani, G. *Chem. Phys.*, **1992**, *160*, 131–144.

30. Bartocci, G.; Mazzucato, U.; Spalletti, A. *Chem. Phys.*, **1996**, *202*, 367–376.

31. Bartocci, G.; Spalletti, A.; Masetti, F.; Cruciani, G. *J. Mol. Struct.*, **1993**, *298*, 165–175.

32. Saltiel, J.; Sears, Jr., D.F.; Choi, J.-O.; Sun, Y.-P.; Eaker, D.W. *J. Phys. Chem.*, **1994**, *98*, 35–46.

33. Saltiel, J.; Zhang, Y.; Sears, Jr., D.F. Choi, J.-O. *Res. Chem. Interm.*, **1995**, *21*, 899–921.

34. Saltiel, J.; Zhang, Y.; Sears, Jr., D.F. *J. Phys. Chem.: A*, **1997**, *101*, 7053–7060.

35. Alfimov, M.V.; Razumov, V.F.; Rachinsky, A.G.; Listvan, V.N.; Sheck, Y.B. *Chem. Phys. Lett.*, **1983**, *101*, 593–597; Castel, N.; Fischer, E. *J. Mol. Struct.*, **1985**, *127*, 159–166.

36. (a) Muszkat, K.A.; Wismontski-Knittel, T. *Chem. Phys. Lett.*, **1981**, *83*, 87–90; *J. Phys. Chem.*, **1981**, *85*, 3427–3431; (b) Lamotte, M.; Morgan, F.J.; Muszkat, K.A.; Wismontski-Knittel, T. *J. Phys. Chem.*, **1990**, *94*, 1302–1309.

37. Mazzucato, U.; Momicchioli, F. *EPA-Newsletter*, **1992**, *44*, 31–39.

38. Bartocci, G.; Masetti, F.; Mazzucato, U.; Spalletti, A.; Orlandi, G.; Poggi, G. *J. Chem. Soc. Faraday Trans. 2*, **1988**, *84*, 385–399.

39. Arai, T.; Karatsu, T.; Sakuragi, H.; Tokumaru, K.; Tamai, N.; Yamazaki, I. *Chem. Phys. Lett.*, **1989**, *158*, 429–434.

40. Fischer, G.; Fischer, E. *J. Phys. Chem.*, **1981**, *85*, 2611–2613.

41. Wismontski-Knittel, T.; Das, P.K.; Fischer, E. *J. Phys. Chem.*, **1984**, *88*, 1163–1168.

42. Mc Gown, L. B.; Bright, F. V. *Anal. Chem.*, **1984**, *56*, 2195–2199.

43. Bartocci, G.; Masetti, F.; Mazzucato, U.; Baraldi, I.; Fischer, E. *J. Mol. Struct.*, **1989**, *193*, 173–183.

44. Bartocci, G.; Mazzucato, U.; Spalletti, A.; Elisei, F. *Spectrochim. Acta*, **1990**, *46 A*, 413–418.

45. Hug, G.; Becker, R.S. *J. Chem. Phys.*, **1976**, *65*, 55–63.

46. Spalletti, A.; Bartocci, G. *Phys. Chem. Chem. Phys.*, **1999**, *1*, 5623–5634.

47. Birks, J. B.; Dyson, D. J. *Proc. Roy. Soc.: A*, **1963**, *275*, 135–148.

48. Flom, S. R.; Nagarajan, V.; Barbara, P. F. *J. Phys. Chem.*, **1986**, *90*, 2085–2092.

49. Brearley, A. M.; Flom, S. R.; Nagarajan, V.; Barbara, P. F. *J. Phys. Chem.*, **1986**, *90*, 2092–2099.

50. Arai, T.; Karatsu, T.; Sakuragi, H.; Tokumaru, K.; Tamai, N.; Yamazaki, I. *J. Photochem. Photobiol. A: Chem.*, **1992**, *65*, 41–51.

51. Furuuchi, H.; Arai, T.; Sakuragi, H.; Tokumaru, K.; Nishimura, Y.; Yamazaki, I. *J. Phys. Chem.*, **1991**, *95*, 10322–10325.

52. CorneliBen-Gude, C.; Rettig, W. *Chem. Phys.*, **1998**, *229*, 325–334.

53. Galiazzo, G.; Spalletti, A.; Elisei, F.; Gennari, G. *Gazz. Chim. Ital.*, **1989**, *119*, 277–280.

54. Karatsu, T.; Yoshikawa, N.; Kitamura, A.; Tokumaru, K. *Chem. Lett.*, **1994**, 381–384.

55. Castel, N.; Fischer, E. *J. Mol. Struct.*, **1985**, *127*, 159–166.

56. Cruciani G.; Spalletti, A.; Bartocci, G. Z. *Phys. Chem.*, **1991**, *172*, 227–252 and references therein.

57. Orlandi, G.; Zerbetto, F.; Zgierski, M.Z. *Chem. Rev.*, **1991**, 91, 867–891 and references therein.

58. Brouwer, A.M.; Cornelisse, J.; Jacobs, H.J.C. *Tetrahedron*, **1987**, *43*, 435–438; Brouwer, A. M.; Bezemer, L.; Cornelisse, J.; Jacobs, H.J.C. *Recl. Trav. Chim. Pays-Bas*, **1987**, *106*, 613–624.

59. Wallace-Williams, S.E.; Moller, S.; Goldbeck, R.A.; Hanson, K.M.; Lewis, J.W.; Yee, W.A.; Kliger, D.S. *J. Phys. Chem.*, **1993**, *97*, 9587–9592.

60. Bunker C.E., Lytle C.A., Rollins H.W.; Sun Y.-P. *J. Phys. Chem.*, **1997**, *101*, 3214–3221 and references cited therein.

61. Saltiel, J.; Sears, D.F.; Sun, J.-P.; Choi, J.-O. *J. Am. Chem. Soc.*, **1992**, *114*, 3607–3612.

62. Bartocci, G.; Spalletti, A.; Becker, R.S.; Elisei, F.; Floridi, S.; Mazzucato, U. *J. Am. Chem. Soc.*, **1999**, *121*, 1065–1075.

63. Spalletti, A; Bartocci, G.; Galiazzo, G.; Macchioni, A.; Mazzucato, U. *J. Phys. Chem.*, **1999**, *103*, 8994–9002

64. Masetti, F.; Bartocci, G.; Mazzucato, U. *Gazz. Chim. Ital.*, **1982**, *112*, 255–260.

65. Bartocci, G.; Masetti, F.; Mazzucato, U.; Dellonte, S.; Orlandi, G. *Spectrochim. Acta*, **1982**, *38A*, 729–735.

66. Karatsu, T.; Arai, T.; Sakuragi, H.; Tokumaru, K. *Chem. Phys. Lett.*, **1985**, *115*, 9–15.

67. Arai, T.; Karatsu, T.; Tsuchiya, M.; Sakuragi, H.; Tokumaru, K. *Chem. Phys. Lett.*, **1988**, *149*, 161–166.

68. Mazzucato, U.; Spalletti, A.; Bartocci, G.; Galiazzo, G. *Coord. Chem. Rev.*, **1993**, *125*, 251–260.

69. Saltiel, J.; Zhang, Y.; Sears, Jr., D.F. *J. Am. Chem. Soc.*, **1996**, *118*, 2811–2817.

70. Muszkat, K.A. *Top. Curr. Chem.*, **1980**, *88*, 89–143.

71. (a) Hugelshofer, P.; Kalvoda, J.; Schaffner, K. *Helv. Chim. Acta*, **1960**, *43*, 1322–1332; (b) Mallory, F.B.: Wood, C.S.; Gordon, J.T., *J. Am. Chem. Soc.*, **1964**, *86*, 3094–3102.

72. Bortolus, P.; Cauzzo, G.; Galiazzo, G. *Tetrahedron Letters*, **1966**, 239–242, 3717–3721; Bortolus, P.; Cauzzo, G.; Mazzucato, U.; Galiazzo, G. *Z. Phys. Chem. N.F.*, **1969**, *63*, 29–38.

73. Castel, N.; Fischer, E.; Rauch, K.; Jabalameli, M.; Luettke, W. *J. Photochem. Photobiol. A: Chem.*, **1989**, *50*, 221–237.

74. Wismontski-Knittel, T.; Bercovici, T.; Fischer, E. *J. Chem. Soc. Chem. Comm.*, **1974**, 716–717.

75. Wismontski-Knittel, T.; Kaganowitch, M.; Seger, G.; Fischer, E. *Recl. Trav. Chim. Pays-Bas*, **1979**, *98*, 114–117.

76. Muszkat, K.A.; Jakob, A.; Castel, N.; Fischer, E.; Rauch, K.; Luettke, W. *J. Photochem. Photobiol. A: Chem.*, **1991**, *60*, 193–205.

77. Ittah, Y.; Jakob, A.; Muszkat, K.A.; Castel, N.; Fischer, E. *J. Photochem. Photobiol. A: Chem.*, **1991**, *56*, 239–247.

78. De Jong, M.H.; Laarhoven, W.H. *Recl. Trav. Chim. Pays Bas.*, **1973**, *92*, 673–682.

79. Somers, J.B.M.; Laarhoven, W.H. *J. Roy. Neth. Chem. Soc.*, **1980**, *99*, 160–166 and references therein.

80. Laarhoven, W.H. *Recl. Trav. Chim. Pays Bas.*, **1983**, *102*, 185–204, 241–254.

81. Spalletti, A.; Bartocci, G.; Elisei, F.; Mancini, V.; Mazzucato, U. *J. Chem. Soc. Faraday Trans.*, **1997**, *93*, 211–219.

82. Karatsu, T.; Kitamura, A.; Zeng, H.; Arai, T.; Sakuragi, H.; Tokumaru, K. *Bull. Chem. Soc. Jpn.*, **1995**, *68*, 920–928.

83. Becker, H.-D. *Adv. Photochem.*, **1990**, *15*, 139–227

CHAPTER

6

Calculations of Excited-State Conformational Properties

Massimo Olivucci

Istituto di Chimica Organica, Università degli Studi di Siena, Siena, Italy

Michael A. Robb

Department of Chemistry, King's College, London Strand, London, United Kingdom

Fernando Bernardi

Dipartimento di Chimica "G. Ciamician" Università di Bologna, Bologna, Italy

1 Introduction

During the last decade or so, the speed of computers has grown considerably and the computational investigation of realistic models of organic compounds is rapidly becoming a standard practice.[1,2] Standard quantum chemical methods[1] (see Section 3 for details) provide a complete description of what happens at the molecular level during the processes that characterize a complex conformational change or a chemical reaction. In particular, it is possible to compute[3] the equilibrium geometry of two interconverting conformers (or of a reactant and a product), the transition structure that connects them, and the associated energy barrier with almost chemical accuracy (ca. 0.01 Å and 1.0 kcal mol^{-1} errors). Furthermore, the "reaction path" (i.e., the detailed progression of the reactant's molecular structure towards the transition state and the product), can be determined in a completely unbiased way by computing the minimum-energy path (MEP)[3b,4] connecting the reactant to the product on the $3N-6$-dimensional potential-energy surface of the system (where N

Conformational Analysis of Molecules in Excited States, Edited by Jacek Waluk
ISBN 0-471-29707-0 © 2000 Wiley-VCH

is the number of nuclei of the molecular system under investigation). Finally, quantum chemical and experimental data may be employed to develop computationally inexpensive, but accurate, molecular mechanics force fields that allow the automatic mapping of the vast multidimensional conformational space of complex molecules, the determination of the preferred conformers, and the simulation of the conformational equilibria using Monte Carlo or molecular dynamics methods.[1]

A detailed understanding of the conformational properties of electronically excited-state species would increase our ability to control known photochemical processes and to design new photochemical systems. In ground-state molecules, the investigation of conformational problems is facilitated by the fact that, usually, the molecule has a well-defined connectivity (or constitution), including the double-bond position, and the possible conformers are related by rotations about *single* bonds. In contrast, the conformational analysis of electronically excited molecules appears to be far more complex, due to the *variable and less well understood nature of the chemical bonding*. For instance, the photoexcitation of common organic chromophores, such as a conjugated hydrocarbon, leads to an electronic reorganization that changes the pattern of the single or double bonds, generating a different connectivity and, thus, a different set of conformers.

An additional type of excited-state conformational problem arises when we deal with the selectivity of photochemical reactions. Selectivity (either regioselectivity or stereoselectivity) is not a property linked to the structure of the reactant or product, but it is a specific property of the reaction mechanism. In fact, in a selective reaction, the molecular structure of each "element" of the *favored* reaction mechanism (i.e., the structures of all the intermediates and transition structures along the reaction path), must be consistent with the structure of the *favored* product. In ground-state chemistry, the selectivity of the reaction is often controlled by the conformational properties of the relevant transition-state structure such as those in bimolecular substitution—S_N2—or elimination—$E2$—reactions. For instance, the existence of a single energetically favored transition-state conformer, such as the *anti* or *syn* transition-state conformer in an addition or elimination reaction, leads to a stereospecific product. Despite the fact that a high degree of stereospecificity has been observed in photochemical reactions (e.g., in photochemical pericyclic reactions)[5], the mechanism that makes possible the production of a stereochemically rigid product from a highly energetic excited-state species appears to be difficult to unveil. In fact, the photoproduct formation process involves the decay from the excited state to the ground state, where the primary products accumulate. Thus, *a stereochemically relevant mechanistic element in a photochemical reaction (i.e., the analog of the transition structure) appears to be the molecular structure associated with the decay channel*. In conclusion, the conformational properties of such a structure must provide the basis for the rationalization of the observed stereospecificity.

In principle, the problem of the variable nature of the excited-state bonding, or that of the selectivity of photochemical reactions, can be successfully investigated by using ab initio quantum chemical computations. With this methodology, one can determine and characterize the molecular structure and relative stability of an arbitrary conformer irrespective of the complexity of its bonding status. However, in practice, *excited-state quantum chemical computations require nonconventional methodologies and strategies.* For this reason, excited-state computations are not routinely used by chemists. Thus, one target of this chapter is to convince the reader that it is now technically feasible to describe the conformational properties, and their relationship with photochemical reactivity, by using methods that are available in standard quantum chemistry packages (e.g., *Gaussian*,[6] *Molcas*,[7] *Columbus*,[8] *Gamess*,[9] and others[10]). We hope that the "practical" discussion of these problems and the examples given throughout the chapter can provide the information necessary for the reader to carry out his or her own computations.

In Section 2, we focus on a fairly complete case history that serves to define the most relevant aspects of the excited-state structure and bonding in a common organic chromophore: buta-1,3-diene. Through this specific example, we shall provide a general overview of the excited-state conformational structure of a π-system, the conformational structure of a decay channel, its relationship with photochemical reactivity, and its influence on the stereochemistry of the observed photoproducts.

The application of modern computational tools to the investigation of the conformational properties of an excited-state species is not a trivial task. In practice, one needs to deal with two general problems. The first problem arises from the fact that the equilibrium values of the excited-state length and energy of a covalent bond (e.g., of a π-bond) differ from the corresponding ground-state values and from the values of a different excited state. For instance, entities such as *Z*- and *E*-diastereoisomers, which are not usually interconverted in ground-state conformational equilibria, may be interconverted when the molecule is electronically excited. Often, in these situations, the electronic configuration of an excited-state molecule cannot be described in terms of pairs of coupled electrons. In other words, the electronic structure may present an *open-shell* (e.g., *biradical*) *character.* A correct treatment of such systems requires a multireference wave function. As a consequence, the type of ab initio quantum chemical methods required for excited-state computations belong to the so-called post-SCF methods,[11] rather than the standard single-reference SCF wave function. These methodologies are based upon the use of multireference wave functions capable of describing states that must be represented as a mixture of different electronic configurations.

The second "practical" problem concerns the general structure (i.e., the shape) of excited-state potential-energy surfaces. Ground-state potential-energy surfaces contain only two chemically relevant features: energy minima, which correspond to stable species such as reactants, products, and intermediates; and saddle points, which correspond to transition structures. While minima and saddle points are also found in excited-state potential-energy surfaces, a novel feature appears to be of fundamental importance for excited states: an excited-state potential-energy surface may cross the energy surface of the ground state or that of a different excited state, leading to a *conical intersection* or to a *singlet–triplet crossing.*[12] Notice that these are unavoided crossings that provide the major decay channel from the excited state to the ground state (or to a lower excited state). Since the structure and location of these crossings play a fundamental role in determining certain excited-state properties (e.g., the excited-state lifetime and, as mentioned, the photoproduct distribution), the search for and characterization of these entities represents an important step in the mapping of the energy surface. However, while the computation of minima and saddle points can be carried out with standard gradient methods, the determination of crossing points requires special computational tools and strategies.

In Section 3, we deal with both the problem of the accurate excited-state energy evaluation and that of the excited-state surface mapping, giving a practical overview of the tools required for exploring the shapes of the potential-energy surfaces of excited-state species. We will also try to illustrate the general strategy presently used in our group for dealing with an excited-state problem.

Finally, in Section 4, we examine a series of different organic chromophores, which have recently been investigated via ab initio quantum chemical computations, by focusing on the relationship between conformational properties, the reaction mechanism, and stereochemistry. These molecules comprise linear and cyclic conjugated hydrocarbons, protonated Schiff bases, α,β-enones, and silanes.

2 The Case of Buta-1,3-diene

Conformers are stereoisomers that can be interconverted by overcoming a rotational barrier. Usually, when the rotation takes place about a single bond, the interconversion between conformers is possible at room temperature, and the chemical nature of the species remains unchanged. On the other hand, conformers that differ by/in a rotation about a double bond cannot be easily converted, and the *Z*- and *E*-diastereoisomers of a disubstituted ethylene correspond to different, isolable chemical species: The *Z*- to *E*-interconversion corresponds to a true chemical change characterized by double-bond breaking and reconstitution.

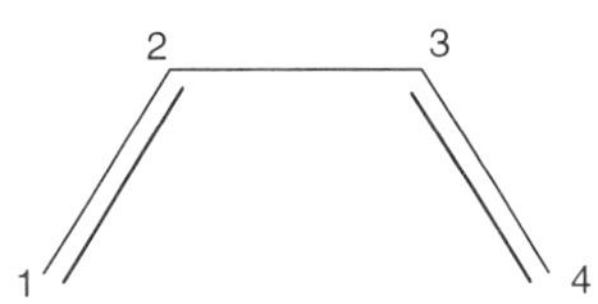

1

The difference between real conformers and *Z*- and *E*-diastereoisomers is usually very clear for ground-state unsaturated hydrocarbons. In fact, *Z*- and *E*-but-2-ene will not rotate about the central C_2–C_3 double bond (bond length 1.35 Å), and the only possible interconverting conformers are related to the methyl group rotation about the terminal C_1–C_2 and C_3–C_4 single bonds. This behavior also holds for ground-state linear conjugated hydrocarbons (i.e., polyenes). Despite the changes in character of single and double bonds due to the delocalization of the π-system (the double bonds become longer and weaker, and the single bonds become shorter and stronger), the bonds do not lose their identity. Thus, as shown in Figure 6.1 (top), *s-cis* buta-1,3-diene (**1**) can easily rotate about the central bond (bond length 1.48 Å and rotational barrier ca. 5 kcal mol^{-1}), but cannot rotate about the two lateral

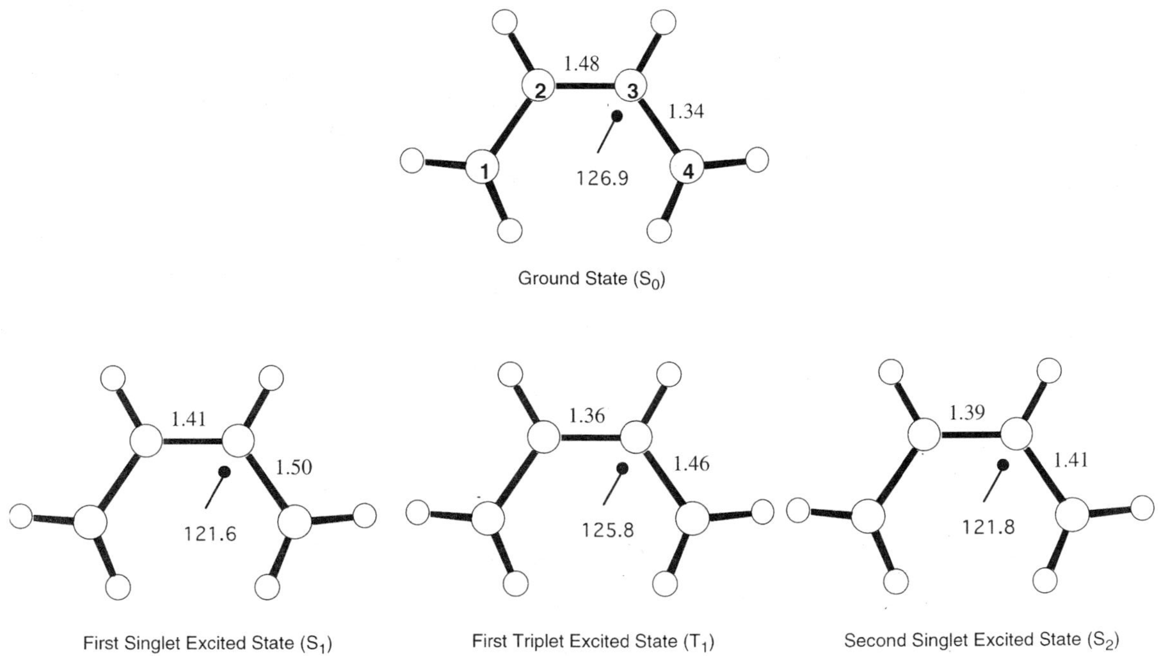

Figure 6.1. Relevant geometrical parameters (Å and degrees) for the S_0, S_1, T_1, and S_2 relaxed *planar* molecular structures of *s-cis* buta-1,3-diene. These parameters have been computed at the CAS(4,4)/6-31G* level of theory (S_0, S_1, T_1) and CIS/6-31G* level of theory (S_2). A planarity constraint was applied during the geometry optimization. The acronyms CAS (complete active space) and CIS (configuration interaction singles) refer to the quality of the wave function used in the computation. This point is discussed in Section 3.1. The acronym 6-31G* indicates the type of Gaussian basis set used to describe the molecular orbitals. (see Ref. 1 or any good quantum chemistry book for a detailed explanation).

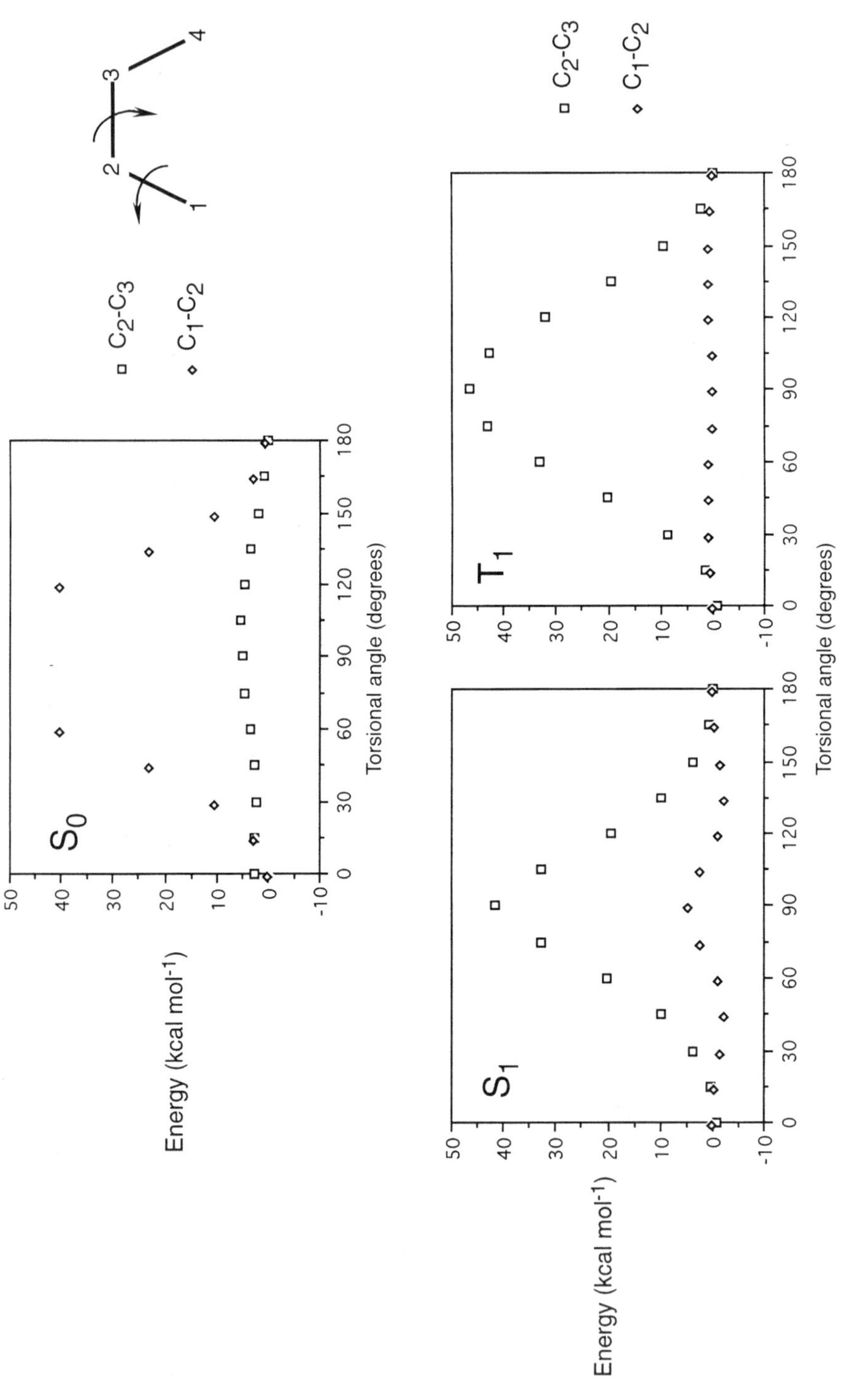

1
2
3
4
S_0
S_1
T_1
C_2-C_3
C_1-C_2
C_2-C_3
C_1-C_2
Energy (kcal mol^{-1})
Energy (kcal mol^{-1})
Torsional angle (degrees)
Torsional angle (degrees)

double bonds (bond length 1.34 Å and rotational barrier ca. 65 kcal mol^{-1}). Notice that the *s-cis* buta-1,3-diene conformation is actually a transition structure between two gauche conformers[13] located in two very shallow energy minima (see also the energy minima along the C_2–C_3 curve at the top of Figure 6.2). Nevertheless, throughout this chapter we will refer to the *s-cis* buta-1,3-diene "conformer" only.

If we now consider the first singlet excited state (S_1) of **1** we find that the equilibrium geometry is very different from that of the corresponding ground-state species (compare the S_0 and S_1 structures in Figure 6.1). In particular, one has an inversion in the bonding character along the chain. In fact, as an effect of the change in the electronic structure, the central bond acquires a double-bond character (bond length 1.41 Å), while the terminal bonds become single bonds (bond lengths 1.50 Å). The strength of these bonds can be evaluated by looking at the energy profile along the corresponding torsional coordinate. In the first singlet excited state, the rotation about the original double bond is allowed (the C_1–C_2 rotational barrier in S_1 is ca. 5 kcal mol^{-1}), while the rotation about the original single bond is substantially forbidden (the C_2–C_3 rotational barrier in S_1 is ca. 40 kcal mol^{-1}) at room temperature. In conclusion, promotion to S_1 has the effect of changing the position of the conformationally active bond.

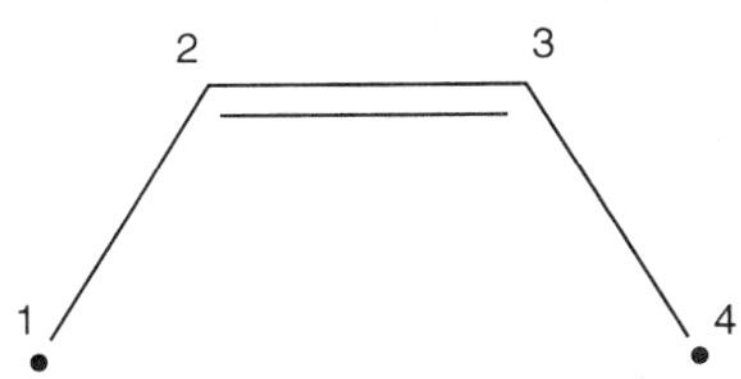

2

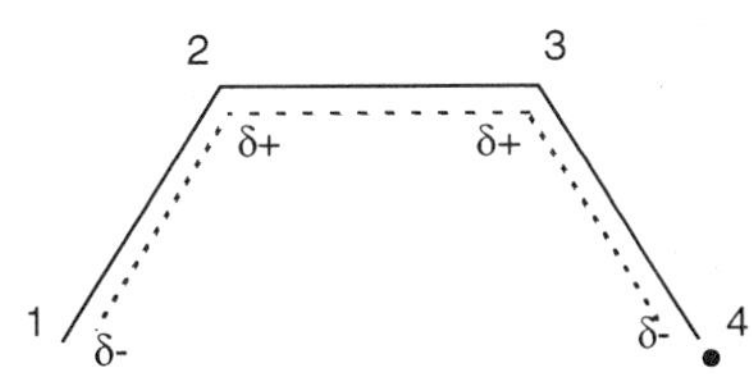

3

Similarly, promotion of buta-1,3-diene to the first triplet state (indicated with T_1) leads to an equilibrium structure (Fig. 6.1) with the central bond featuring an even shorter bond length (1.36 Å) and terminal bonds that are almost single bonds (1.46 Å). Similar to the case of the S_1 state, these molecules are conformationally impaired with respect to the rotation about the original single bond (the rotational barrier in T_1 is ca. 50 kcal mol^{-1}), but undergo facile rotation about the original double bonds (the rotational barrier in T_1 is ca. 1 kcal mol^{-1}). Both the S_1 and T_1 species are *diradicals* with two unpaired electrons localized on the two terminal methylenes. Of course, in S_1 the unpaired electrons are singlet spin coupled, while in T_1 the electrons are triplet spin coupled. Globally, both the S_1 and T_1 species are apolar, with a fairly homogeneous electronic distribution along the carbon skeleton (i.e., each π electron is localized approximately at a different carbon center). In fact, the terminal carbon atoms (C_1 and C_4) and the central carbon atoms (C_2 and C_3)

←

Figure 6.2. Energy profiles along the torsional coordinates C_1–C_2 and C_2–C_3 of the S_0, S_1, and T_1 optimized structures of Figure 6.1. The computations were carried out with the use of the CAS(4,4)/6-31G* level of theory (see caption to Figure 6.1 for the acronyms). The values of the remaining $3N - 7$ internal coordinates are kept fixed at their initial (optimized) values.

have −0.013 and 0.013 charge, respectively. (These values express the charge of a carbon center with the attached hydrogen charges summed on to it.) Thus, in these species, the bonding can be described by structure **2**.

Promotion to the second excited singlet state (S_2) leads to a further change in the nature of the bonding. The relaxed planar structure of *s-cis* buta-1,3-diene promoted to its S_2 state is shown in Figure 6.1. In this structure, the three C–C bond lengths have approximately the same values (1.39 and 1.41 Å for the central and terminal bond distances, respectively), and thus, we have a very different bonding pattern with respect to both the ground and the S_1 state. Furthermore, the electronic structure of S_2 buta-1,3-diene is different from that of the other states and corresponds to that of a more polar, closed-shell species in which the four π electrons are less evenly distributed along the carbon skeleton. Indeed, the terminal carbon atoms are more negatively charged (−0.024), while the central carbons are more positively charged (0.024) than in the S_1 structure. The corresponding excited-state species can be represented by structure **3** in Formula **2**.

Experimental evidence for a change in the conformationally active bonds in the first excited state of conjugated hydrocarbons comes from classic photochemical studies. Let us briefly discuss a result from the work of Jacobs and Havinga[14] on the photochemistry of short polyenes, mostly hexa-1,3,5-trienes. These authors have established the "principle of nonequilibration of excited-state rotamers" (the NEER principle), which states that during the short (singlet) excited-state lifetime, the excited-state conformers (i.e, the rotamers) do not equilibrate. The mechanistic hypothesis behind the NEER principle is that electronic excitation enhances the torsional barriers about the single bonds (as seen in Figure 6.2 for the torsional energy profiles of *s-cis* buta-1,3-diene), inducing a slow or fully impaired equilibration process of the conformers. However, this principle is empirical and was established on the basis of the observation of a correlation between the equilibrium populations of the conformers in the ground state and the structure of the major photoproducts. In particular, if each type of conformer yields its own specific assortment of photoproducts, then the NEER principle implies that the outcome of the reaction specifically depends on the most populated ground-state conformer. In other words, the composition of the primary products that originates from irradiation should form a qualitative and quantitative reflection of the composition of the ground-state conformational equilibrium.

The NEER principle has been validated by analyzing several experimental results, especially in the field of vitamin-D photochemistry[14] and other hexatriene derivatives. A particularly clear illustration of the principle is given in the work of Vroegop et al.,[15] who have investigated the relationship between the ground-state conformational equilibrium and the photoproduct decomposition in hexatriene and methyl-substituted hexatrienes. Irradiation of unsubstituted *Z*-hexa-1,3,5-triene was found to yield the *E*-isomer as the only photoproduct. On the other hand, irradiation of *Z*-2,5-dimethylhexa-1,3,5-triene yielded photocyclization (i.e., the formation of cyclohexadiene) with a fair quantum yield. Finally, irradiation of *Z*-2-methylhexa-1,3,5-triene yielded 3-methylbicyclo[3.1.0]hex-2-ene, 1-methyl-3-isopropenyl-cyclobut-1-ene, and 5-methylhexa-1,2,4-triene. The different photoproduct distributions obtained by irradiation of *Z*-hexa-1,3,5-triene, *Z*-2,5-dimethylhexa-1,3,5-triene, and *Z*-2-methylhexa-1,3,5-triene were explained on the basis of the dominant ground-state conformer, as illustrated in Scheme 6.1. In fact, in the unsubstituted trienes, the *tZt*- (*s-trans, Z, s-trans*) conformer is expected to dominate. After substitution with one methyl group (*Z*-2-methylhexa-1,3,5-triene), the dominant ground-state conformer happens to be the *tZc*- (*s-trans, Z, s-cis*), and after substitution with two methyl groups, the *cZc*- (*s-cis, Z, s-cis*) conformer dominates. Notice that the geometrical structure of the observed photoproducts can be easily related to that of the dominating conformer.

Going back to the diene **1**, we see that the analysis of the S_1 energy profile in Figure 6.2 and of the optimized planar structure in Figure 6.1 suggests that, due to impaired rotation about the central C_2–C_3 double bond, the *s-cis* form cannot be interconverted into the *s-trans* in the S_1 state. Thus, according to the NEER principle, we expect that photolysis of *s-cis* buta-1,2-diene would only lead to production of the ring-closure photoproduct cyclobutene and, in substituted dienes, to Z/E isomerization of the terminal double bonds (i.e., these two photoproducts can be related to the *s-cis* ground-state conformer in the way suggested by Scheme **1**). Similarly, from irradiation of the *s-trans* conformer, we expect the production of bicyclo[1.1.0]butane and, again, Z/E double-bond isomerization, but not a *s-trans*→*s-cis* conformational change. In conclusion, both the computational results of Figure 6.1 and Figure 6.2 and the observed conformer-specific photoproduct distributions suggest that the mechanistic hypothesis underlying the NEER principle is correct. Indeed, it is known[16] that photolysis of the only *s-cis* conformer gives cyclobutene and Z/E double-bond isomerization (in substituted or deuterated

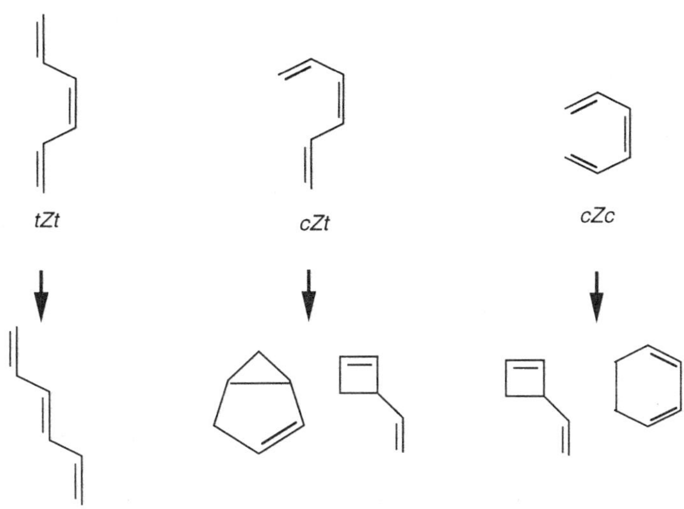

Scheme 6.1

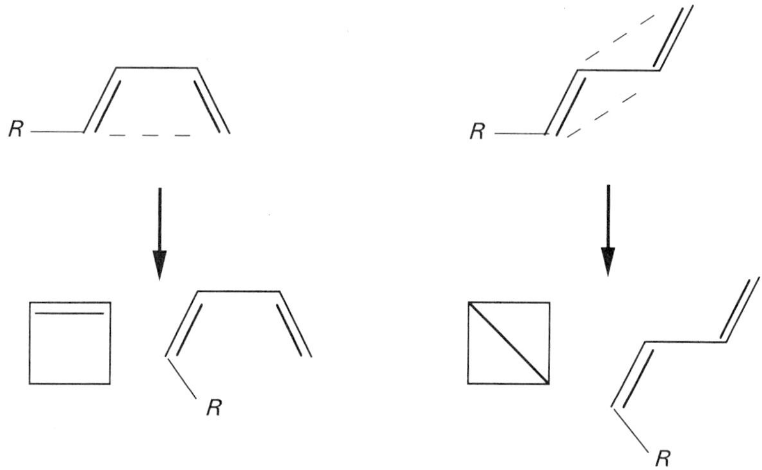

Scheme 6.2

dienes) and that the *s-trans* conformer yields only Z/E double-bond isomerization plus a small amount of bicyclo[1.1.0]butane (see Scheme 6.2).

The photoinduced change in conformation in buta-1,3-diene has been investigated in a more direct fashion in matrix isolation studies by Squillacote et al. and Sheridan et al.[17] It was demonstrated that irradiation of *s-cis* buta-1,3-diene in an argon matrix at 20 K (i.e., under conditions in which the *thermal s-cis→s-trans* conformational conversion is fully blocked) leads to a facile conversion of the *s-cis* form to the *s-trans* form, which is substantially faster than the ring-closure reaction to cyclobutane. A similar behavior was observed for the substituted diene 2,3-dimethylbuta-1,3-diene,[18] although in this case the ring-closure reaction was shown to be 50 times faster than *s-cis→s-trans* conformational interconversion. These more direct experimental results challenge the idea of a large excited-state conformational barrier and therefore the basic mechanistic interpretation of the NEER principle. Most important, in the present context, the experimental results contradict the prediction based upon the S_1 energy profile given in Figure 6.2, which shows about a 40-kcal-mol^{-1} *s-cis→s-trans* rotational barrier. The failure of this prediction indicates severe limitations in the method used to compute the rotational barrier. We now demonstrate that such limitations are due to the arbitrary choice of the reaction coordinate. In fact, this coordinate has not been computed (e.g., by locating the MEP connecting the two excited-state conformers), but it has been assumed to correspond to the rigid torsional deformation about the C_2–C_3 bond. However, buta-1,3-diene is a system with 24 vibrational degrees of freedom, and its potential-energy surface has 24 dimensions. Thus, the correct *s-cis→s-trans* conversion coordinate may be complicated, involving coupling of the torsional deformation with other molecular deformations, including C–C stretching, C–C–C bendings, or other torsions.

2.1 The Structure of the S_1 Buta-1,3-diene Energy Surface

Modern computational tools allow the computation of excited-state intermediates, transition states, and reaction coordinates in an unbiased way. As we shall see in the next section, this is achieved by exploiting the capability of quantum chemical methods to provide the accurate energy, slope (i.e., the energy gradient), and curvature (i.e., the Hessian matrix) at any point of the potential-energy surface and taking into account the full set of the molecular degrees of freedom. In recent years, the technology has been applied to the investigation of the interconversion of the *s-trans* and *s-cis* buta-1,3-diene in the first singlet excited state.[19–21]

The computation of the conformational interconversion coordinate requires, as a preliminary step, the calculation of the molecular structures of the reactant and product. The computed equilibrium structures of *s-cis* and *s-trans* buta-1,3-diene are shown in Figure 6.3. These correspond to a nonplanar structure with partially twisted CH_2 groups. The relaxed S_1 buta-1,3-diene has two distinct conformers (C_s-MIN and C_2-MIN), which differ in the relative direction of twisting of the two terminal methylenes. The most stable conformer corresponds to C_s-MIN, where the terminal methylenes are twisted in a disrotatory fashion at about 50°. The 2.3-kcal-mol^{-1} higher conformer C_2-MIN has C_2 symmetry, and the two terminal methylenes are twisted in a conrotatory fashion. Both these conformers have a planar or nearly planar *s-cis* carbon skeleton and may interconvert via the rotation of a single CH_2 group. The structure of the corresponding transition state (C_2/C_s TS) is given in Figure 6.3. Notice that the S_1C_{2v} structure of Figure 6.1, computed by enforcing the system planarity, is about 10 kcal mol^{-1} higher in energy than that of C_s-MIN and relaxes to either C_s-MIN or C_2-MIN when the planarity constraint is removed.

We have stressed the importance of performing fully unconstrained structure optimizations and of computing the conformational interconversion path in a completely unbiased way. An *s-cis*→*s-trans* interconversion path on the S_1 energy surface can be computed starting from the C_s-MIN conformer and constructing the MEP leading towards the *s-trans* form. The procedure required to perform this computation will be discussed in the next section. An *s-trans*→*s-cis* path can be computed in a similar way starting from the *s-trans* form. This corresponds to the "virtual" conformer of S_2 symmetry given in Figure 6.3. The resulting energy profiles are shown in Figure 6.4 (top) as a function of the C_2–C_3 torsional angle. Figure 6.4 shows at least three remarkable features:

1. The path is substantially barrierless for both the *s-cis*→*s-trans* and *s-trans*→*s-cis* processes. In the *s-cis*→*s-trans* path (crosses), there would be about a 1-kcal-mol^{-1} barrier, but an improved theoretical treatment (open circles) suggests a fully barrierless process.
2. The conformational interconversion is not completed on the excited-state potential-energy surface, since about half-way along the path the S_1 energy surface crosses the ground-state surface (S_0). A more detailed description of this remarkable feature is given in Figure 6.4 (bottom), where the energy profiles of both the S_1 and S_0 states are plotted as a function of the S_1 MEP coordinate (see

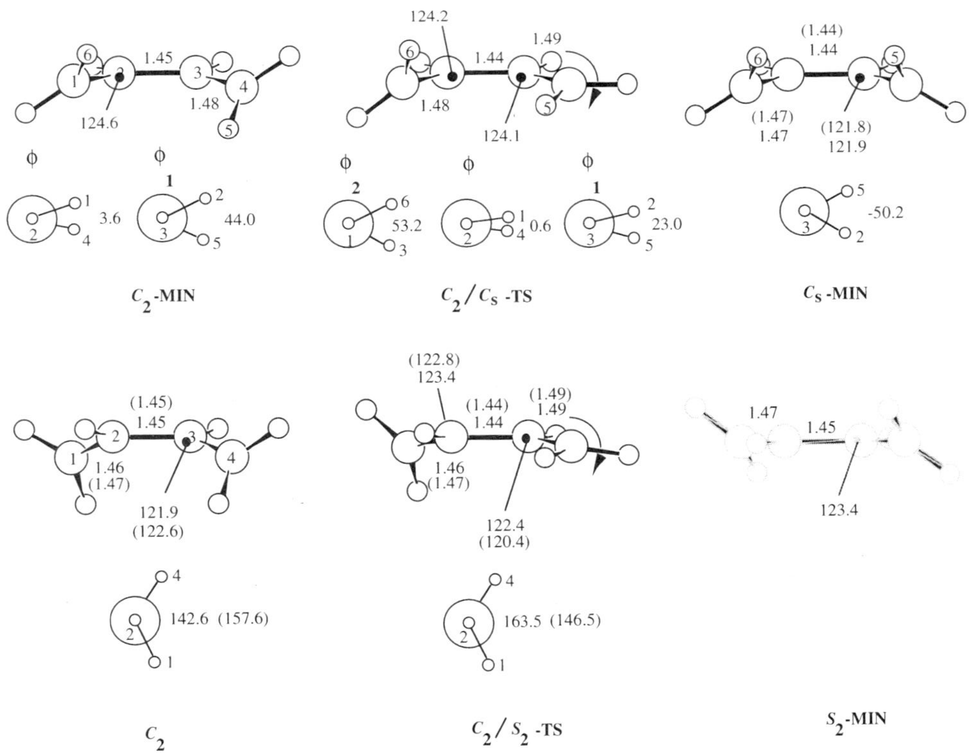

Figure 6.3. Optimized S_1 conformations of *s-cis* buta-1,3-diene (top) and *s-trans* buta-1,3-diene (bottom). (CAS(4,4)/DZ + d level of theory. See caption to Figure 6.1 for the acronyms.) The optimized CAS(4,4)/DZ + spd structure of the C_s minimum is also reported (values in parentheses). The acronyms DZ + d and DZ + spd indicate two different Gaussian basis sets. The optimization was carried out without geometrical constraints in the full space of the $3N - 6$ degrees of freedom of the molecule. The values of the relevant structural parameters are given in Å and degrees. The labels are defined in Section 2.1. Adapted from Figure 1 in Ref. 20 and Figure 3 in Ref. 21.

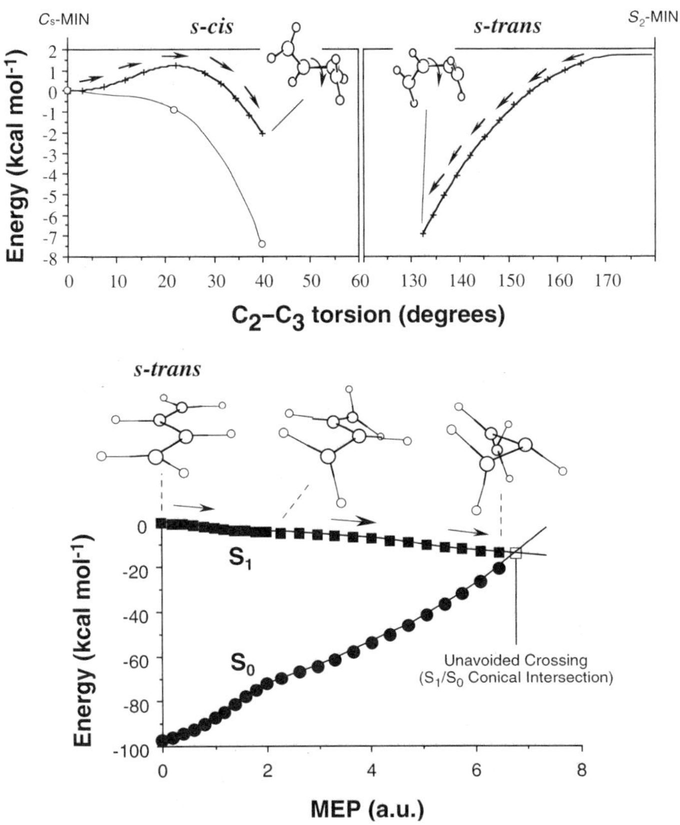

Figure 6.4. *s-cis* and *s-trans* buta-1,3-diene energy profiles along the S_1 excited-state relaxation path leading to an *s-cisoid* and an *s-transoid* conical intersection, respectively. Top: S_1 energy profiles computed at the CAS(4,4)/4-31G (crosses) and CAS(4,4)/DZ + d (circles) level of theory. The acronyms DZ + d and 4-31G indicate two different Gaussian basis sets. Bottom: S_1 (full squares) and S_0 (full circles) energy profiles of *s-trans* buta-1,3-diene along the S_1 relaxation path leading to the *s-transoid* conical intersection, computed with the use of the DZ + d basis set. Adapted from Figure 3 in Ref. 20 and Figure 1 in Ref. 47.

also figure legend) for the *s-trans*→*s-cis* path. Notice that as the *s-trans* form proceeds towards the *s-cis* form, the S_1 and S_0 energies become degenerate at an S_1–S_0 conical intersection[22–24]—an unavoided crossing between potential-energy surfaces of the same spin multiplicity. As mentioned, this crossing provides the decay channel that delivers the excited-state species to the ground-state directly.[23]

3. The conical intersection points lie about 8 kcal mol^{-1} lower in energy than the initial excited-state intermediates. Thus, to the best of our knowledge, the lowest energy structure on the S_1 energy surface of buta-1,3-diene corresponds to a conical intersection point (i.e., to the decay channel).

The two paths just described are the lowest lying conformational interconversion paths in the S_1 potential-energy surface of buta-1,3-diene. These paths are *nonadiabatic*, because the interconversion is not completed on the excited state, but involves decay to the ground state via the crossing. An alternative adiabatic path for the excited state *s-cis*→*s-trans* interconversion has not been located in this system.

As has been pointed out, the structure of the hypothetical *s-cis*→*s-trans* reaction coordinate of Figure 6.2 (i.e., a single torsional deformation) leads to a reaction path that cannot explain the experimental results reported by Squillacote et al.[17] An analysis of the computed reaction coordinate reveals that the evolution along the *s-cis* and *s-trans* paths involves at least two torsional coordinates and one bending coordinate (see Figure 6.5). The geometrical deformation that characterizes the conical intersection structures can be understood on the basis of Scheme 6.3, which illustrates the *s-cis* conical intersection (we will call it *s-cisoid*) in a more detailed way. This structure features an almost isolated CH_2-radical centre (at C_1) connected to a –CH–CH–CH_2 allyl fragment featuring a small C_2–C_3–C_4 bending angle, and a pyramidalized C_3 carbon. In Sections 4.1 and 4.2, we shall see that this peculiar structure is a common feature of the S_1 state of linear polyenes and of certain cyclic aromatic and antiaromatic compounds. Notice that, despite the fact that the conical intersections entered along both the *s-cis*→*s-trans* and *s-trans*→*s-cis* paths satisfy the common structural features previously described, the conical intersection located at the end of the *s-trans* path (we shall call it *s-transoid*) has a torsional angle greater than 90°, while the corresponding *s-cis* structure has a torsional angle smaller than 90°, (compare the two bottom structures in Figure 6.5). In other words, the *s-cisoid* and *s-transoid* structures are two distinct conical intersection conformers that "remember" the initial conformations of the excited-state reactants.

The *s-cisoid* and *s-transoid* conical intersection conformations are not "isolated," but are connected by a "line" of infinite conical intersection points (i.e., points at which the S_0 and S_1 potential-energy surfaces are degenerate). It is possible to carry out a "conformational analysis" of the buta-1,3-diene conical intersection by plotting the change in energy along the path connecting the *s-cisoid* to the *s-transoid* structures. When this is done, IT is found that these two conformers are connected via a conformer of higher symmetry (i.e., of C_2-symmetry) characterized by a 90°-twisted central C–C bond and two 90°-twisted terminal methylenes.

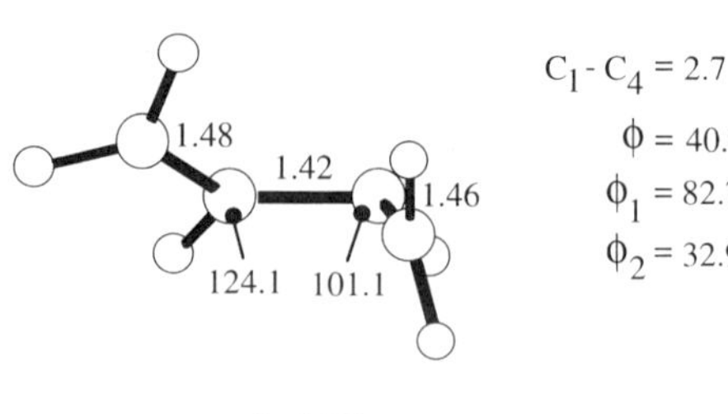

Scheme 6.3

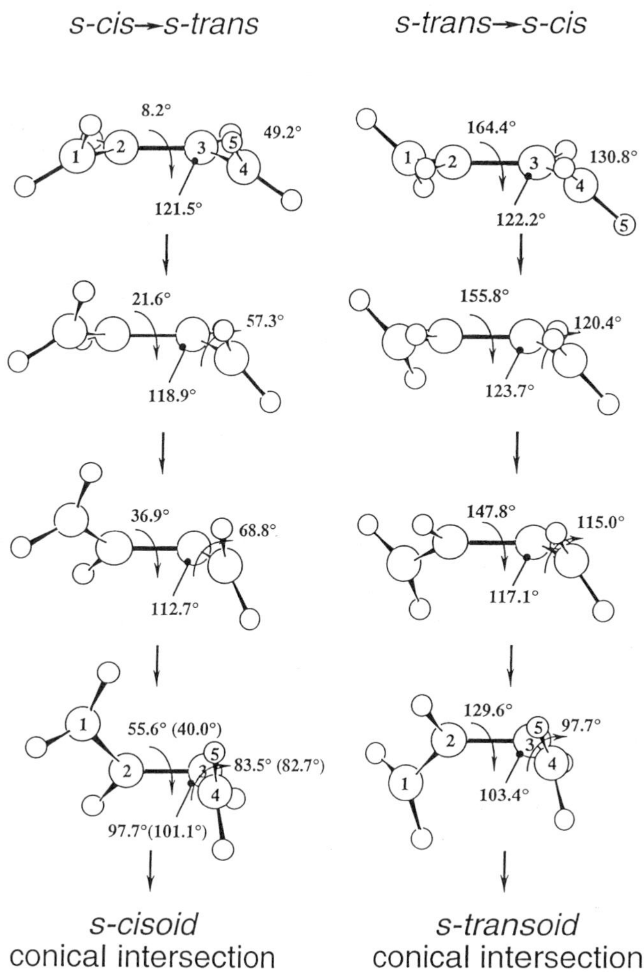

Figure 6.5. Molecular structure evolution along the energy profiles given in Figure 6.4 (top). The data were computed at the CAS(4,4)/4-31G level of theory. For *s-cis* butadiene, the starting geometry corresponds to the more stable S_1 conformer (C_s-MIN) seen in Figure 6.3. Adapted from Figures 6 and 7 in Ref. 20.

However this *all-twisted* conformation is located at a slightly higher energy than that of the *s-cisois* and *s-transoid* conformers.

2.2 Conical Intersection Conformers and Reaction Selectivity

The computational results just discussed indicate that for buta-1,3-diene, the original hypothesis behind the NEER principle is not confirmed by rigorous quantum chemical computations, in agreement with modern experimental data.[17,18] The same results indicate the novel mechanistic hypothesis that *the lack of excited-state equilibration of the excited-state conformers is due to the presence of ultrafast radiationless decay channels located along the conformational interconversion path*. In other words, it is suggested that the excited-state conformers do not equilibrate because they decay to the ground state at the conical intersection. A direct consequence of this hypothesis is that *the photoproduct distribution observed after irradiation of each conformer must reflect the geometrical structure of the associated decay channel*. In fact, as pointed out earlier, it is at this point that the system starts to relax along the ground-state energy surface and initiate the product-formation process.

As reported in Scheme 6.2, the photolysis of *s-cis* butadiene yields cyclobutene and Z/E double-bond isomerization products (in properly substituted systems); however, the photolysis of *s-trans* butadiene yields Z/E double-bond isomerization products and small amounts of bicyclo[1.1.0]butane. If the new hypothesis is correct, then the structure of the *s-cisoid* and *s-transoid* conical intersections must be related to these photoproducts. Inspection of the bottom structures of Figure 6.5 suggests the following:

1. Both structures feature only *one* of the two terminal CH_2 groups fully twisted (84° in the *s-cisoid* and 98° in the *s-transoid*). The near-90° value of this angle suggests an efficient production of Z/E double-bond isomers. Furthermore, the fact that there is only one (not two) fully twisted double bond suggests that there will be only a single Z/E isomerization.
2. In the *s-cisoid* structure, the two terminal methylenes are relatively close to each other (ca. 2.5 Å apart) and face each other. In contrast, in the *s-transoid* structure, the terminal methylenes are far apart (3.3 Å). This suggests that the cyclobutene ring is formed from the *s-cisoid* decay point only.
3. In both structures, the C_2–C_3–C_4 angle is close to 100° and thus there is a relatively short distance between one terminal methylene and the central carbon atom not directly attached to it. The chain arrangement suggests the formation of the cyclopropylcarbinyl diradical as a precursor of bicyclo[1.1.0]butane. This sterically demanding process (i.e., the formation of a cyclopropyl ring) cannot compete with the facile cyclobutene formation during relaxation from the *s-cisoid* structure. However, it may be populated during relaxation from the *s-transoid* structure, where the other cyclization processes cannot compete.

The conclusions given in points 1 to 3 are based on analogies between the molecular structure of the conical intersection (an intrinsically transient entity) and that of the photoproducts and thus are merely qualitative. A quantitative validation of the new mechanistic hypothesis requires the computation of all the possible product-formation paths. These are a set of ground-state MEPs that start in the vicinity of a conical intersection point (i.e., of the $S_1 \rightarrow S_0$ decay channel) and end at different equilibrium structures on the S_0 potential-energy surface, establishing a link between a conical intersection structure and a set of primary photoproducts. The procedure required to perform such a computation is called the IRD method[25] and will be discussed in Section 3.2.

In Figure 6.6, we show the result of product-formation path computations starting at the *s-cisoid* conical intersection. (Notice that C_s-MIN is the S_1 structure given in

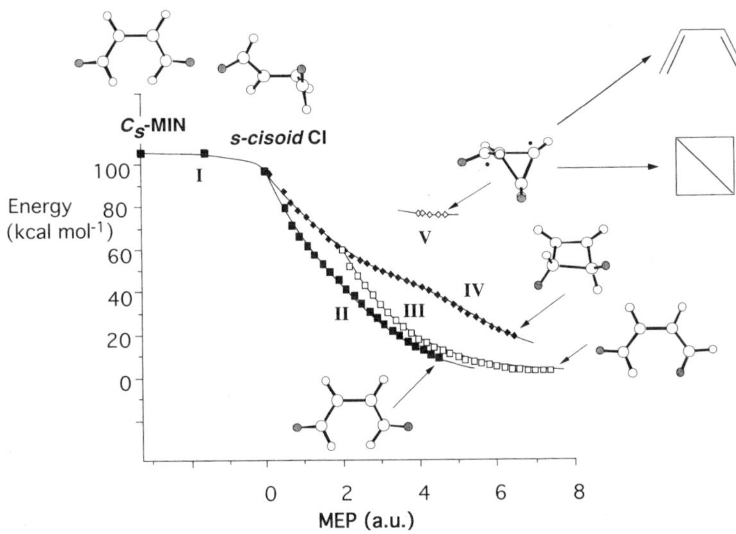

Figure 6.6. *s-cis* Buta-1,3-diene ground-state relaxation paths departing from the *s-cisoid* conical intersection conformer. Path **II** (full squares) leads to reactant back-formation. Path **III** (open square) leads to double-bond E/Z isomerization. Path **IV** (full diamonds) leads to ring closure and, finally, Path **V** (open diamonds) leads to a cyclopropyl diradical. The S_1 relaxation path seen in Figure 6.4 and connecting C_s-MIN and the conical intersection is also reported (Path **I**, full squares). All energy profiles are computed using the CAS(4,4)/DZ + d level of theory. Adapted from Figure 6 of Ref. 22a.

Figure 6.1 and Path **I** is the S_1 barrierless *s-cis* path given in Figure 6.2). Path **II** describes the reactant back-formation process. Paths **III, IV**, and **V** are photochemical paths, since they all describe reactive processes. In particular, Path **III** describes a (single) double-bond isomerization, Path **IV** illustrates cyclobutene formation, and Path **V** describes the *stepwise* formation of bicyclobutane via a cyclopropylcarbinyl diradical.

No path describing photoinduced *s-cis*→*s-trans* conformational isomerization has been located. Therefore, the experimentally observed facile *s-cis*→*s-trans* interconversion cannot be explained on the basis of this computation. At present, we can only provide a qualitative—although reasonable—argument, which is awaiting a more direct validation. Because of the barrierless nature of the S_1 branch of the *s-cis*→*s-trans* path (Path **I**) of Figure 6.6, the system will accelerate towards the *s-cisoid* conical intersection. After the decay, this direction of motion will be initially conserved due to inertial factors. The MEP coordinates given in Figure 6.4 show that the motion involves a large contribution from the central torsion and, therefore, along the *s-cis*→*s-trans* direction. The production of ground-state *s-trans* buta-1,3-diene would then be due mainly to *dynamical factors* and not only to the structure of the ground-state potential-energy surface.

Several attempts to locate a valley pointing in the direction of Path **V** (i.e., leading to a cyclopropylcarbinyl diradical) in the *vicinity* of the S_1/S_0 crossing (i.e., the decay channel) have failed. Rather, this valley starts to exist only at a much larger (4.0 a.u.) distance than the valley defining Paths **II, III**, and **IV**. Hence, one expects that a system moving in the direction of the tip of the cone will hop to S_0 and populate the *closer* double-bond isomerization and cyclobutene formation valleys. Another problem that arises with the cylopropylcarbinyl diradical intermediate is that it may revert to buta-1,3-diene rather than evolve towards bicyclobutane.[26] Thus, only a small number of S_1 molecules are expected to reach the bicyclobutane valley. This seems to be consistent with the undetectable or extremely low[16] production of bicyclobutane from *s-cis* butadiene.

Another piece of evidence in favor of the conical intersection mechanism for the buta-1,3-diene photolysis comes from a series of experimental studies on locked dienes. This situation arises when carbons C_2 and C_3 are connected by short hydrocarbon chains (one or two CH_2 units) or by a ring system that results in small C_2–C_3-chain (α_2) and chain-C_2–C_3(α_1) bending angles,[27–29] as illustrated in Scheme 6.4. (The reported values of α_1 have been computed by using MM2 molecular mechanics calculations.) The observations of Aue[28] in the series 1,2-bismethylenecyclobutane, 2,3-bisdimethylenebicyclo[2.2.1]heptane, 1,2-bisdimethylenecyclopentane, 1,2-bisdimethylenecyclohexane are summarized in Scheme 6.4 (adapted from Scheme **VII** of Ref. 20). The quantum yield for the production of cyclobutene decreases as the values of α_1 and α_2 are decreased in the corresponding reactant. The same trend has been observed by Leigh[27,29] in the photochemistry of 1,2-bisdimethylidenecycloalkanes. In the case of 1,2-bismethylenecyclobutane[28] and 1,2-bismethylidenecyclobutane,[27] no product has been detected. This behavior can be explained by the computed S_1 reaction coordinate (Figure 6.4). In fact, it can be seen that along the pathway, the value of one of the bending angles (ω in Scheme 6.4) must change from the sp^2 equilibrium value (120°) at C_s MIN to about 90.0° at the *s-cisoid* intersection. If the adjacent bending angle (α_1 in Scheme 6.4) is already highly strained, the energy must rise very rapidly on distorting the system towards

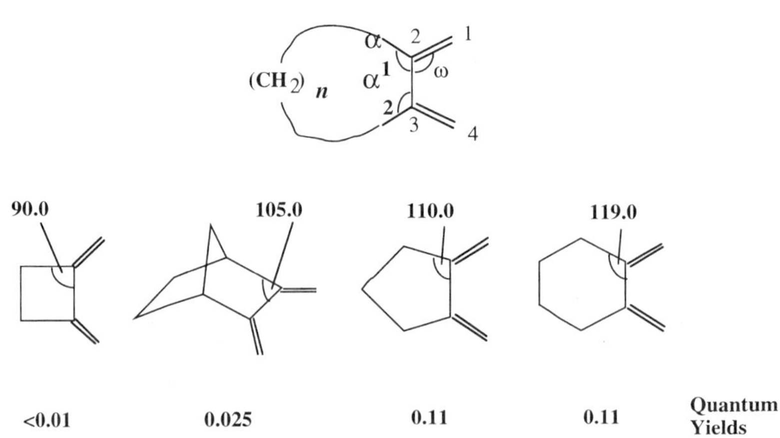

Scheme 6.4

the conical intersection, which will increase both the barrier to the conical intersection and the lifetime of the excited-state intermediate. This theoretical interpretation suggests a regular increase in product quantum yields by gradual relaxation of the α_1 and α_2 angles in accordance with the trend[28] shown in Scheme 6.4 and with the work of Leigh.[27,29]

2.3 Excited-State Conformations and Stereoselectivity of the Photochemical *s-cis* Buta-1,3-diene Ring Closure

In thermal chemistry, the stereoselectivity of a reaction is usually determined by the preferred (i.e., more stable) conformation of the relevant transition-state structure. One classic example is that of a pericyclic reaction, such as *s-cis* buta-1,3-diene ring closure. In this case, the transition-state conformation is defined with respect to the relative direction of rotation of the two terminal methylenes of the diene, which can be conrotatory or disrotatory, with the conrotatory favored. A high degree of stereoselectivity is also observed in the photochemical counterpart of the same reaction, which yields the disrotatory stereoisomer of cyclobutene. However, due to the nonadiabatic nature of a photochemical reaction path, the explanation of the observed stereoselectivity does not appear to be straightforward. On the other hand, earlier we showed that the cyclobutene formation arises via decay at the *s-cisoid* conformer of an S_1/S_0 conical intersection. Therefore, it appears natural to assume that, ultimately, it must be the preferred *con*rotatory or *dis*rotatory orientation of the two terminal methylenes in this structure that determines the stereoselectivity of the reaction.

In order to assign the *con* or *dis* conformation to the computed *s-cisoid* conical intersection conformer, it is necessary to examine in detail the way in which this structure may be generated in a typical photochemical experiment. The absorption of a 254-nm photon promotes *s-cis* buta-1,3-diene to the S_2 excited state and not to the S_1 state directly. In fact, the S_2 state corresponds to the spectroscopically allowed $1B_2$ state of polyenes, while the S_1 state corresponds to a dark, spectroscopically forbidden state ($2A_1$) of this class of molecules. The way in which the S_1 potential-energy surface is entered has been documented by mapping[19–21] the S_2 and S_1 energy surfaces in the **FC** region (i.e., the region near the vertical excitation point) by means of geometric optimization and MEP computations. These results are pictorially summarized, together with those of the previously discussed nonadiabatic *s-cis*→*s-trans* path computation, in the diagram of Figure 6.7.

The initial relaxation of the reactant on the S_2 energy surface (Path **VII**) leads to an energy minimum of C_s symmetry (C_s-ION) that features two 32° twisted CH_2 groups shown in Figure 6.7 (top). This structure is lower in energy than the C_{2v} planar S_2 structure given in Figure 6.1, which was computed under a planarity constraint. Thus, along Path **VII**, one has a disrotatory motion of the two methylenes. Remarkably, at the bottom of this energy minimum, the S_1 energy surface crosses the S_2 energy surface, thus promoting a fully efficient $S_2 \rightarrow S_1$ decay.

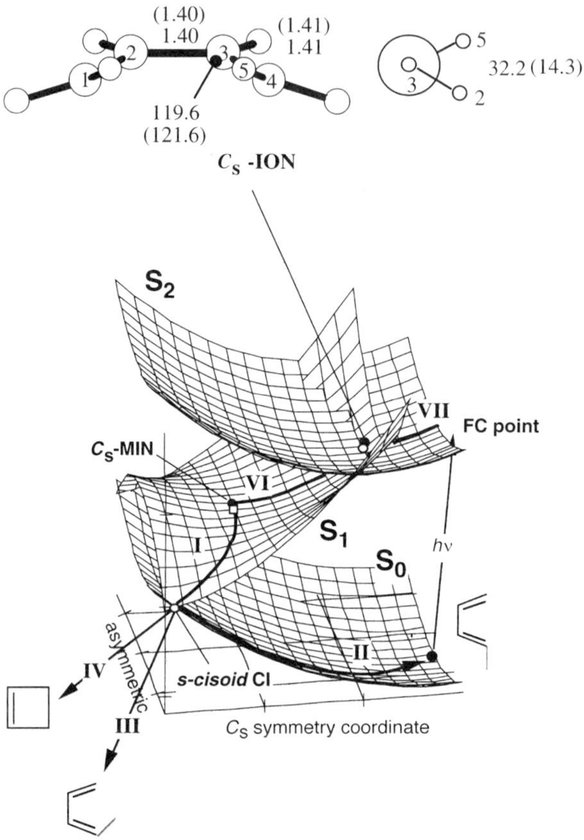

Figure 6.7. Three-dimensional representation of the photochemical $S_2 \rightarrow S_1 \rightarrow S_0$ reaction path for *s-cis* buta-1,3-diene from the reactant ground-state equilibrium structure (the Franck–Condon (**FC**) point) to the photoproducts. Paths **VI** and **VII** indicate the relaxation from the spectroscopic state S_2. Paths **I, II, III**, and **IV** correspond to the paths illustrated in Figure 6.6. Full circles indicate equilibrium structures, and open circles indicate decay channels (conical intersections). The relaxation paths of Figure 6.6 are reported.

A relaxation path (Path **VI**) starting at the S_2/S_1 conical intersection ends in the region of the C_s-MIN structure already seen in Figure 6.2.

According to Figure 6.7, the initial relaxation of the reactant along the symmetric S_2 and S_1 composite relaxation path (i.e., Path **VII** + **VI**) yields the C_s-MIN structure through a motion involving a disrotatory rotation of the two terminal methylenes. The subsequent relaxation of C_s-MIN occurs along Path **I**, whose coordinate has already been described in Figure 6.4. It is clear from this figure that, while the path is characterized by a fully asymmetric deformation of the structure, the rotation of the two terminal methylenes is disrotatory. In conclusion, the *s-cisoid* conical intersection generated by relaxation of the C_s-MIN structure has a *dis* conformation. While this feature alone suggests that the experimentally observed cyclobutene stereoisomer is produced, prediction can be confirmed by examining the CH_2 motion along the product-formation paths given in Figure 6.6. It is clear from the figure that the path leading to cyclobutene production (Path **IV**) describes a disrotatory motion of the two terminal methylenes.

The alternative conrotatory path on the S_1 energy surface has also been located. This path originates from the highest C_2-MIN conformer of S_1 buta-1,3-diene already shown in Figure 6.3. The S_1 disrotatory and conrotatory paths are compared in Figure 6.8, which represents, on a two-dimensional diagram, Paths **VII, VI**, and **I** energy profiles seen in Figure 6.7 (full line). Notice that, in contrast with the disrotatory path, the conrotatory path (dashed line) is not barrierless, and its transition state lies about 7 kcal mol^{-1} higher in energy than the C_s-MIN conformer. This energy difference must be, at least in part, responsible for the fact that the conrotatory route is not populated. On the other hand, the structure of the potential-energy surface of the spectroscopic state S_2 suggests that the C_2-MIN would not be populated upon $S_2 \rightarrow S_1$ decay. In fact, no conformer generated via conrotatory motion (and therefore showing C_2 symmetry) has been located on this surface.

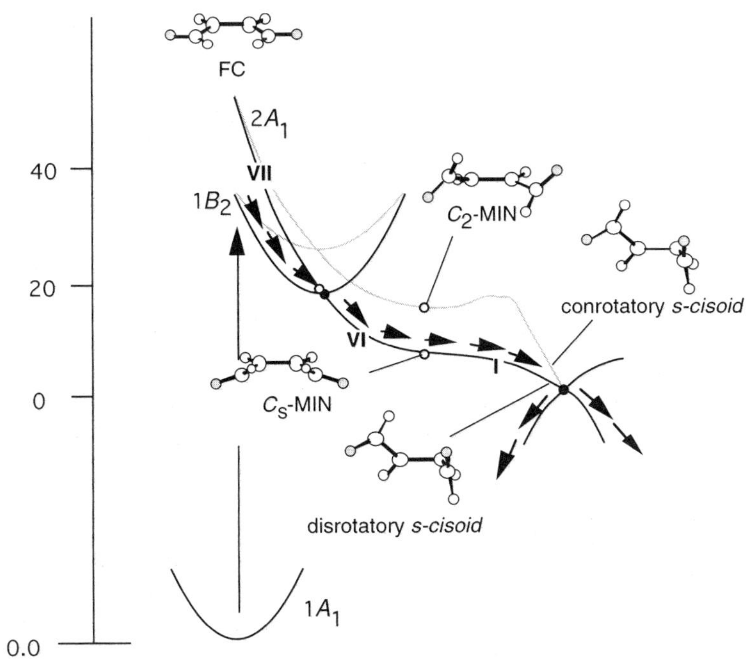

Figure 6.8. Detailed structure and energetics of the S_2 and S_1 relaxation paths for *s-cis* buta-1,3-diene. *Dis*rotatory (full line) and *con*rotatory (light line) energy profiles along the corresponding paths. Paths **VI** and **VII** correspond to the paths illustrated in Figure 6.7. The conrotatory and disrotatory *s-cisoid* structures represent two stereochemically different conical intersections. The labels are defined in Section 2.1. Adapted from Figure 5b of Ref. 22a.

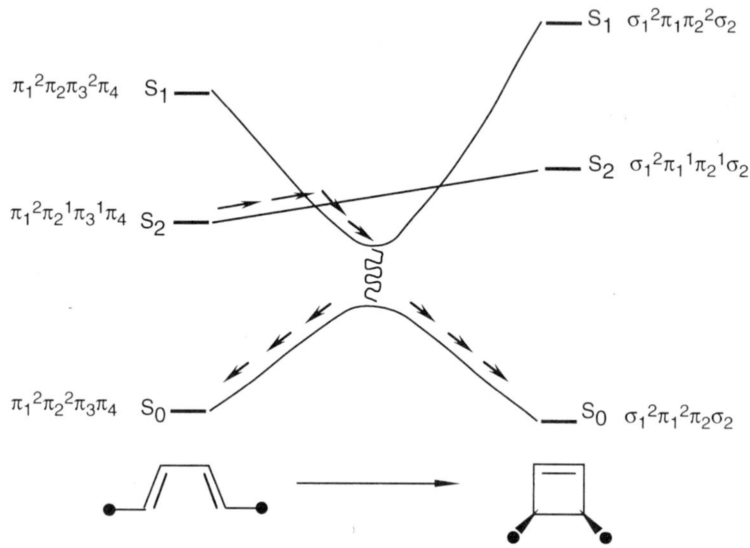

Scheme 6.5

Finally, notice that there is a correspondence between the diagram in Figure 6.8 and the well-known state correlation diagrams originally suggested by Woodward and Hoffmann[30a] (WH) and subsequently modified by Van der Lugt and Oosteroff[30b] (Scheme 6.5). As is well known, the WH rules successfully predict the observed stereochemistry by analyzing the motion in the $1B_2$ (HOMO-LUMO singly excited) spectroscopic state.

However, the diagram in Figure 6.8 contains certain flaws. First, the WH treatment seems to suggest that allowed photochemical processes yield products in an excited state. Experimentally, however, relaxation of the excited products to the ground state is not observed. Furthermore, it has been noted that the first singlet excited state of cyclobutene is some 50–60 kcal mol^{-1} higher in energy than the corresponding singlet state for butadiene, suggesting that the formation of the spectroscopic excited singlet state of cyclobutene from the corresponding singlet state of butadiene is unlikely.[30c]

The understanding of these problems was advanced by Van der Lugt and Oosteroff,[30b] who proposed (on the basis of computational results) that along distinct disrotatory and conrotatory ring-closure coordinates, the $2A_1$ state has two deep minima. The observed disrotatory stereochemical evidence was then explained on the basis of a higher rate of internal conversion at the disrotatory $2A_1$ minimum, owing to a substantially smaller excited-state–ground-state energy gap at this point.

The recent computational reinvestigation of the interplay between the $2A_1$, the $1B_2$ and the $1A_1$ states of *s-cis* butadiene rationalizes the photostereochemistry completely. In fact, as shown in Scheme 6.5, the $S_2(1B_2)$ state (i.e., the spectroscopic state) corresponds to the singly excited state $\pi_1^2\pi_2^1\pi_3^1\pi_4^0$. ($\pi_1$, π_2, π_3, and π_4 are the four π orbitals of buta-1,3-diene). At the geometry of the reactant (the left end of the diagram), this is the first excited state in the WH state correlation diagram, but it becomes the second excited state at the transition-state geometry. On the other hand, the $S_1(2A_1)$ state (the dark state) is a doubly excited state ($\pi_1^2\pi_3^2$) and thus corresponds to the second excited state at the geometry of the reactant, but to the first excited state at the transition-state geometry. Notice that the WH diagram suggests that a symmetry-imposed real crossing will exist between S_2 and S_1, in full agreement with the computed diagram. However, it also suggests the existence of an avoided crossing at the transition-state geometry. Thus, the only qualitative difference between the recently computed diagram and those previously predicted is that the avoided crossing needs to be replaced by a conical intersection (see also the similar discussion in Ref. 22b). (This point will be further discussed at the end of Section 3.2.)

2.4 The Nonstereospecific Ring Opening of Cyclobutene

As discussed earlier, the computed nonadiabatic reaction path for ring closure of *s-cis* buta-1,3-diene suggests a stereochemical behavior that appears to be consistent with both the experimental data and the prediction of the WH rules[30] (given in the form of the state correlation diagram). On the other hand, the reverse photochemical process (i.e., the ring opening of cyclobutene to buta-1,3-diene) is one of the few examples of a violation of the WH rules. In fact, this pericyclic reaction is not stereospecific.[31] We now show that the reaction paths illustrated in Figure 6.6 and the conical intersection structure of Scheme 6.3 provide an explanation for the observed lack of stereoselectivity.[32] The experimental observations can be explained by assuming that the cyclobutene ring opens on the S_2 state. In this way, cyclobutene photolysis would yield S_2 *s-cis* buta-1,3-diene, which decays to the dark S_1 state following the pathway illustrated in Figure 6.6. The final photoproducts must then originate via decay at the same S_1/S_0 *s-cisoid* conical intersection and will follow the relaxation paths shown in Figure 6.7. It is then obvious that, since decay at this point involves concurrent butadiene formation and double-bond E/Z isomerization, the production of a stereospecific (disrotatory) ring-opening product is impossible. Thus, the loss of stereochemistry is due to the unavoidable concurrent *cis–trans* isomerization and butadiene formation.

2.5 Ultrafast Photochemistry of Buta-1,3-diene

Resonance Raman studies on isoprene (2-methyl-buta-1,3-diene)[33] indicate that the absorbing S_2 state is depopulated on a timescale of 10 fs due to fast internal conversion to the nearby S_1 state. A very short S_2 excited-state lifetime is also consistent with the large bandwidths observed in the low-temperature absorption spectra of parent and substituted butadienes.[9b,34] Thus, geometric relaxation and photochemical reaction must occur mainly on the S_1 potential-energy surface, and the femtosecond S_2 state dynamics will have only a minor influence. On the other hand, it is well known that, in contrast with longer polyenes (such as octa-1,3,5,7-tetraene),[35] short polyenes (such as buta-1,3-diene and hexa-1,3,5-triene) do not fluoresce.[36] This important photophysical feature implies that the photochemical reaction of buta-1,3-diene must occur on a timescale shorter than the typical fluorescence lifetime. Such an ultrafast reaction cannot be compatible either with the existence of a stable excited-state intermediate or with nonnegligible excited-state energy barriers. Thus, both the available spectroscopic data and the missing fluorescence of buta-1,3-diene are consistent with the substantially barrierless excited-state relaxation–reaction path seen in Figures 6.6 and 6.8. In particular, the ultrafast decay from S_2 to S_1 is due to the existence of a conical intersection at the bottom of the S_2 surface (i.e., at the C_s-ION structure). Similarly, the lack of fluorescence from the S_1 state must be due to the barrierless path that connects the C_s-MIN structure to the *s-cisoid* conical intersection.

2.6 Buta-1,3-diene Triplet Conformers and Triplet Photochemistry

The difference in the decay timescale of the singlet and triplet states forms the basis of the explanation of the observed differences in single-bond c/t and double-bond Z/E photoproduct distribution obtained from these states. Saltiel showed that the lowest singlet excited state of polyenes does not usually decay to the lower triplet excited state.[37] Thus, the only way to access the lowest triplet state is via sensitization with suitable sensitizers. In a series of experiments by Saltiel et al., it was demonstrated that 1,3-dienes (such as the hexa-3,5-diene) undergo, upon triplet sensitization, a two-bond Z/E double-bond isomerization, in contrast with direct irradiation, which leads to one-bond isomerization exclusively.[38] Earlier we saw that the structure of the S_1 reaction path and that of the *s-cisoid* conical intersection satisfactorily account for one-bond isomerization. In order to rationalize the difference between the singlet and triplet double-bond isomerization photochemistry, one has to look at potential-energy structure for the lowest triplet state (T_1), which is shown in Figure 6.9.

There are two different conformers: a C_{2v} planar conformer (C_{2v}-MIN) and a C_s twisted conformer (twisted-MIN), which is a few kcal mol^{-1} lower in energy. Nevertheless, these conformers are separated by a small energy barrier and can easily interconvert at room temperature. The most important point is that near the twisted-MIN structure the S_1 surface crosses the T_1 surface, thus providing a region of decay. Of course, decay from the twisted structure will result in one-bond isomerization. However, triplet states of dienes cannot be short lived, as the spin–orbit coupling in pure diradical species is usually very small.[12] Thus, the T_1 excited-state conformer will have enough time to undergo double-bond isomerization on the excited-state surface prior to decay. As a consequence, both one-bond and two-bond isomerized species will be released in the ground state.

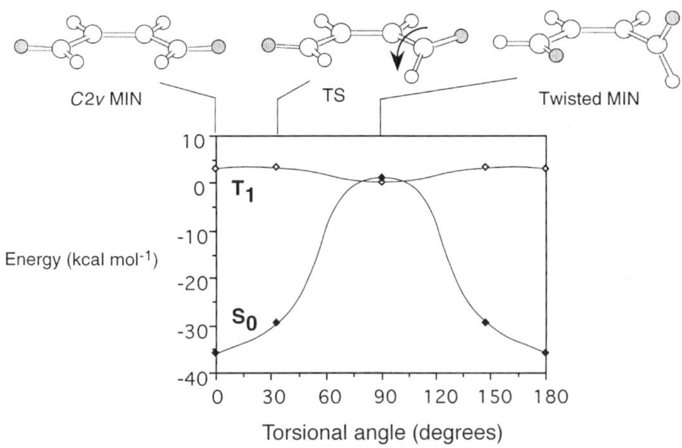

Figure 6.9. Energy profiles along the torsional coordinate C_1–C_2 of the T_1 optimized structure of Figure 6.1. The computations have been carried out using the CAS(4,4)/6-31G* level of theory. The values of the remaining $3N - 7$ internal coordinates are kept fixed at their initial values. The labels are defined in Section 2.6.

3 Methods for Excited-State Computations

As mentioned in the introduction to the chapter, the use of computational chemistry for investigating the properties of electronically excited species is still a nonstandard practice. However, in recent years new computational tools for the calculation of excited-state energy surfaces have been developed and tested. Most importantly, practical computational strategies for tackling a photochemical or a photophysical problem have been proposed. Thus, in this section we discuss the strategy and the tools that recently have been employed to investigate a series of excited-state conformational processes and related reaction mechanisms. These make up the photochemistry of buta-1,3-diene discussed in the previous section and all the examples discussed in Section 6.4. We do not present a comprehensive overview of the computational methods available for excited-state investigation.

The computational investigation of conformational processes and reaction mechanisms is based on the mapping of the potential-energy surface of the electronic states of interest. Thus, in ground-state chemistry, one maps the region of the ground-state energy surface comprising the reactant and product structures and the transition states and intermediates describing their mutual transformation.[3b] Since the potential-energy surface has $3N - 6$ dimensions, one cannot afford to map the energy surface along all the internal coordinates of the molecules. On the other hand, *it is often dangerous to derive mechanistic information by scanning the energy surface along a few selected internal coordinates and keeping the others at fixed values* (see in Section 3.2). As we have already seen in the previous section, an approximate, but effective, solution to this problem is provided by computing the reaction path[4] of the process of interest. Next, we show that while a reaction path is, substantially, a one-dimensional cross section of the multidimensional energy surface, it still conveys the basic information for understanding the reaction mechanism and the molecular motion characterizing ground- and excited-state processes.

In general, there are two steps involved in the computation of a reaction path for an excited-state system. The first step involves the evaluation, at a fixed molecular geometry, of the potential energy, its first derivatives (i.e., the *gradient* vector), and its second derivatives (i.e., the *Hessian* matrix), with respect to all the $3N - 6$ internal coordinates of the molecule (i.e., bond lengths, bending angles, and torsional angles). This step requires the use of quantum chemical methods capable of describing electronically excited states. In the second step, one evaluates the different "elements" forming a reaction path by using a series of complementary computational tools. These elements are energy minima and saddle points, minimum-energy paths, and potential-energy surface crossings.

3.1 The Potential-Energy Surface Computation: The CAS–SCF Method

In recent years the CAS–SCF method[39] has been used extensively for mapping excited-state potential-energy surfaces of organic compounds.[22a] This ab initio quantum chemical method also provides the "natural" basis for the accurate evaluation of the excited-state energetics. In fact, vertical and 0–0 excitation energies and ground- and excited-state energy barriers can be evaluated to just a few kcal mol^{-1} error by employing multireference methods that include the evaluation of the dynamical correlation energy. These methods, which require the multireference CAS–SCF wave function as the starting point, are the multireference configuration interaction method (MR–CI[40]) and multireference perturbation theory methods (e.g., MR–MP2[41] and CAS–PT2[42]). In what follows, we give a conceptual introduction to the CAS–SCF method.

The CAS–SCF method is essentially a combination of the standard ab initio Hartree–Fock (HF) and configuration interaction (CI) methods. In the HF method, the wave function used to describe the electronic structure of the molecule is a single reference wave function Ψ_{HF}, which means that it is constructed by using a single Slater determinant (D_i). The "structure" of a Slater determinant is illustrated in Scheme 6.6. Each molecular orbital ϕ_{MOi} is constructed as a linear combination of atomic orbitals (ϕ_{AOi}) by using a suitable atomic basis set. The shape of each orbital is "optimized" in a variational process by changing the values of the orbital coefficients $c_1, c_2, \ldots, c_i, \ldots, c_n$ (n is the number of atomic orbitals), to get the lowest possible value for the energy. A wave function of this type is certainly capable of describing the electronic configuration of systems in which all the molecular orbitals are doubly occupied. These are the so-called closed-shell systems. Thus, the HF method is usually applied to the investigation of the equilibrium structure of ground-state species and to the study of ground-state conformational problems. However, this method cannot be applied to chemical processes that involve singly or

$$\phi_{\mathrm{MOi}} = c_1\,\phi_{\mathrm{AO1}} + c_2\,\phi_{\mathrm{AO2}} + \ldots + c_i\,\phi_{\mathrm{AOi}} + \ldots + c_n\,\phi_{\mathrm{AON}}$$

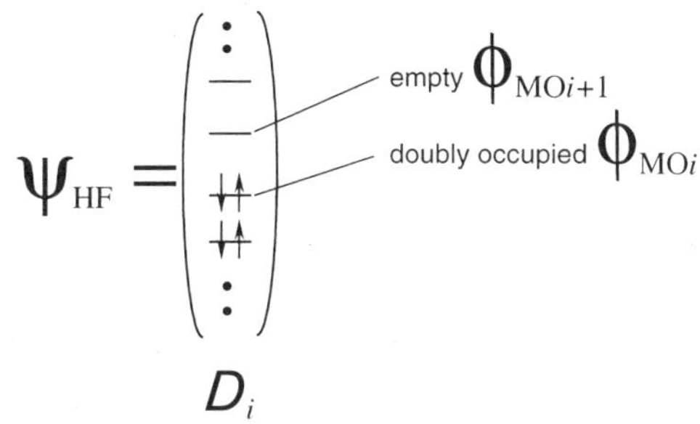

Scheme 6.6

partially occupied orbitals. These chemical processes correspond to ground-state reactions involving radical pairs of diradical intermediates or excited-state processes in which a diradical character develops after photoexcitation. To deal with such problems, one may try to use the configuration interaction (CI) method.

In the CI method, the molecular orbitals produced by an HF computation are employed to generate a multireference wave function. As shown in Scheme 6.7, the CI wave function is defined as a linear combination of Slater determinants describing different electronic configurations, including both closed-shell and diradical configurations. Most important, each configuration corresponds to a different electronic excitation of the reference closed-shell electronic configuration. The coefficients $C_1, C_2, \ldots, C_i, \ldots, C_m$ (where m is the number of Slater determinants) are determined by computing the eigenvectors of an $m \times m$ CI Hamiltonian matrix whose elements have the form $\langle D_i|H|D_j\rangle$, where D_i is a Slater determinant and H is the Hamiltonian operator of the molecule under investigation. The eigenvalues of the CI Hamiltonian provide the potential energies associated with the ground state and $m - 1$ different excited states. Each eigenvector describes the corresponding electronic structure of these states. However, in order to get accurate results with the CI method, the wave function must contain an extremely large number of Slater determinants, which often makes the calculation very difficult if not impossible.

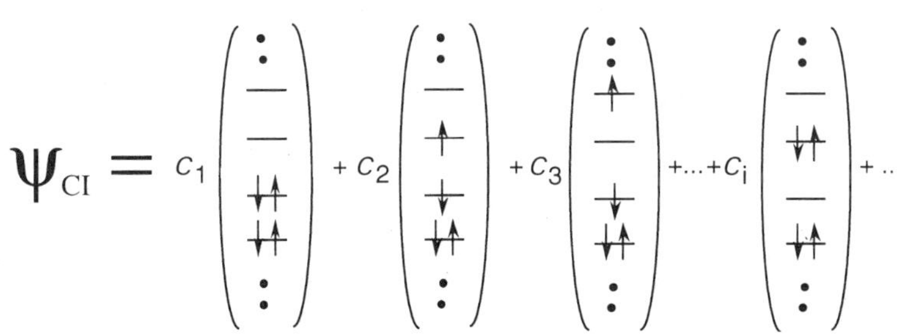

Scheme 6.7

In order to solve this problem, one can think of using a wave function containing a limited number of Slater determinants (i.e., a so-called truncated CI wave function), but rather than "optimize" only the coefficients of the CI wave function, $C_1, C_2, \ldots, C_i, \ldots, C_m$, one simultaneously optimizes these coefficients and the atomic orbital coefficients $c_1, c_2, \ldots, c_i, \ldots, c_n$ of the molecular orbitals that are used to construct the Slater determinants. Thus, one has $n + m$ variational coefficients. This type of wave function defines the multireference–self-consistent-field (MC–SCF) method. Thus, the MC–SCF wave function is a CI wave function that contains only a very limited number of Slater determinants that describe only a limited number of electronic excitations. *The choice of the subset of the electronic excitations that must be used to build the MC–SCF wave function is critical.* In order to define this subset, the molecular orbitals are grouped into three different sets: the core orbitals, the valence orbitals, and the virtual orbitals. The core orbitals are the orbitals that remain doubly occupied in the entire MC–SCF Slater determinant set. The virtual orbitals are the orbitals that remain empty. Finally, the valence orbitals (or active orbitals) are the orbitals that change occupancy. The electrons filling the valence orbital set are the valence electrons (or active electrons). One variety of the MC–SCF method is the CAS–SCF method. This method is defined by a wave function whose Slater determinants describe *all* possible electronic excitations within the chosen set of active electrons and orbitals. Thus, CAS stands for complete active space.

The choice of orbitals that must be included in the valence space is a function of the molecular system and, most importantly, the type of chemical or physical process that one must investigate. For instance, if we are interested in the behavior of an excited state of a chromophore containing a π-system, the valence orbitals must contain the orbitals involved in the photoexcitation process. Similarly, if we want to investigate a chemical reaction, then all the orbitals taking part in the bond-breaking and bond-forming processes must be enclosed in the valence space. A simple example of a CAS–SCF wave function can be described for the most basic π system: the ethylene molecule. By selecting the π and π^* orbitals as the active orbitals and the two π electrons as the active electrons, one can construct a CAS–SCF wave function containing four different Slater determinants (see Scheme 6.8).

The determinant corresponding to the electronic configuration with the doubly-occupied π orbital (C_1 coefficient) describes a closed-shell system, while the others describe all the configurations that can be generated by moving one or both of the electrons to the higher energy orbital. For the equilibrium geometry of ethylene in its ground state, the configuration that dominates the wave function (i.e., the configuration with the largest weight) is expected to be the closed-shell configuration. In fact, this configuration describes the π bonding in an MO treatment. Nevertheless, if the ethylene molecule is highly twisted (e.g., along the reaction path describing a Z/E isomerization), then other configurations start to make a relevant contribution. For instance, near a 90° rotation, the C_1 and C_4 coefficients (C_4 is the coefficient of a doubly excited configuration) have similar magnitude, since the molecule will posseses a pair of quasi-degenerate orbitals. Of course, the weights of the configurations are expected to change when the molecule is promoted to an excited state

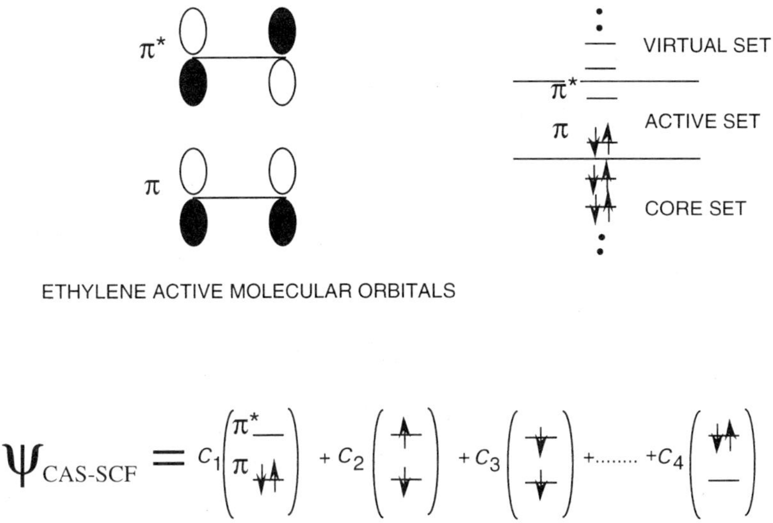

Scheme 8

via the absorption of a photon. In this case, the configuration that will be representative of the singlet excited state will be the singly excited configuration C_2 (rather then the doubly excited configuration C_4). On the other hand, the configuration corresponding to coefficient C_3 is the triplet-state configuration, which (by symmetry) does not mix with the other singlet configurations.

Once the CAS–SCF wave function has been constructed by choosing the proper set of orbitals and electrons, the energy of the molecule can be computed as an eigenvalue of the CI matrix of $m \times m$ dimensions constructed by using simultaneously optimized orbitals. Thus, there will be m eigenvalues and m corresponding eigenvectors of the CI matrix. The lowest eigenvalue provides a value for the ground-state energy, while the others will be excited-state energies. The corresponding eigenvector gives a description of the electronic structure of the state. One of the advantages of the CAS–SCF method is that, for each state of interest (ground or excited), one gets the best set of orbitals, thus improving the description with respect to a simple CI treatment. The success of the CAS–SCF method in excited-state potential-energy surface mapping is due to two specific features. The first is that the method is accurate enough to be able to model the correct shape of the energy surface and the positions (i.e., the molecular structure) of the energy minima, transition states, and reaction coordinates. The second is that the first and second derivatives of the potential energy (i.e., the gradient vector and Hessian matrix, respectively) can be computed analytically. This very important technical feature that allows the use of different powerful computational tools for the exploration and mapping of the energy surface is described next.

The CAS–SCF energetics are usually not accurate enough to reproduce excitation energies and reaction barriers with chemical accuracy. For that reason, one needs to go beyond the CAS–SCF method. Indeed, the CAS–SCF wave function needs to be further improved, since it cannot satisfactorily describe the "dynamic correlation" effects. By that term, we mean the instantaneous mutual repulsion of two electrons with opposite spins. Methods that provide a wave function capable of describing the dynamic correlation are the MR–CI and multireference–MP2 methods. On the other hand, these methods are computationally very expensive and presently currently do not allow the evaluation of the energy derivatives. Therefore, a common strategy is that of evaluating the reaction coordinate at the CAS–SCF level of theory and then correcting the energy profile by computing a few relevant points, using one of the "correlated" methods. Recently, the multireference–MP2 method has been extensively employed to evaluate the excitation energies and reaction barriers of medium-sized organic molecules. In this method, the dynamic correlation is "added" to the CAS–SCF wave function that acts as the zeroth–order wave function, using perturbation theory. In practice, the basic CAS–SCF wave function is enriched by adding to the set of Slater determinants enclosed in the CAS other Slater determinants representing single and double excitations from the set of the core orbitals to the virtual orbitals, from that of the core to that of the valence, and from the valence to the virtual. In our work, we have used the CASSCF–MP2[41] and the CAS–PT2[42] multireference–MP2 methods.

3.2 Search and Computation of Excited-State Reaction Paths

The structure of a "stable" molecular species, such as a conformer, can be determined by using the standard tools for *geometry-optimization*.[1,3] These tools exploit the knowledge of the energy, gradient and Hessian to locate a local energy minimum on the potential-energy surface of interest. As illustrated in Figure 6.10, for a model energy surface, the computation starts at a suitable guess structure (i.e., a structure in the proximity of the conformer). At this point, the energy, gradient and Hessian are evaluated by using a quantum chemical method (such as the CAS–SCF method previously described) and then are used to predict a structure that is closer to the true energy minimum (e.g., $\mathbf{Min_1}$ in Figure 6.10). This procedure is repeated until the energy minimum is reached (see Figure 6.10.) The equilibrium structure of a stable species (e.g., that of an intermediate or of one of its conformers) corresponds to the structure of the energy minimum. In a similar way, it is possible to optimize the geometry of a transition structure between two different conformers (or between a reactant and a product); however, in this case, the optimization procedure stops at a saddle point (**TS** in Figure 6.10).

Energy minima, saddle points, and energy maxima (**Max** in Figure 6.10) are characterized by their gradient and Hessian.[1,2] For all these points, the gradient vector must have zero length. This means that the intensity of the forces acting on the atoms at the optimized structure is zero; hence, such a point is also known as a *stationary point*. The type of stationary point (minimum, saddle point, or maximum) can be determined from the second derivatives, which form a $3N - 6 \times 3N - 6$ matrix called a *Hessian*. In a multidimensional energy surface, the eigenvalues of the

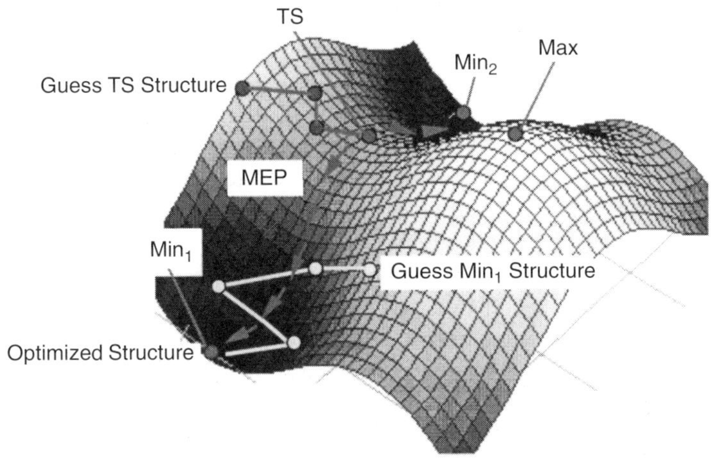

Figure 6.10. Model potential-energy surface showing a maximum or peak (**Max**), and two minima ($\mathbf{Min_1}$, $\mathbf{Min_2}$) connected by a transition state **TS**. The MEP describing the reaction path is indicated by a stream of arrows. The open circles and full circles indicate the "iterations" required to locate a minimum ($\mathbf{Min_1}$) and a transition structure (**TS**) respectively.

diagonalized Hessian determine the characteristics of the stationary point. [In fact, these eigenvalues determine the local curvature (convex or concave) of the energy surface.] Furthermore, the vibrational frequencies of the molecule are proportional to the square root of the Hessian eigenvalues. Thus, if all the eigenvalues are positive or, equivalently, all frequencies are real, the point is a local minimum, and it is located at the bottom of a valley of the potential surface. From such a point, motion in any direction leads to a higher energy. A (first-order) saddle point has one negative eigenvalue and all the rest positive, which means that there will be one imaginary frequency, a maximum in one direction and a minimum in all the other directions. Transition structures are first-order saddle points. An nth-order saddle point has n negative eigenvalues (i.e., it is a maximum with respect to n mutually perpendicular directions). If all the eigenvalues are negative (i.e., if all frequencies are imaginary), the point is a local *maximum*. These points, anyway, have no chemical meaning. Figure 6.10 illustrates all these features for a model energy surface (where only two degrees of freedom are considered).

In Figure 6.10, we show the so-called MEP—that is, the minimum-energy path, which is a line of steepest descent in mass-weighted coordinates that connects a first-order saddle point (**TS**) to two minima ($\mathbf{Min_1}$ and $\mathbf{Min_2}$). Thus, a MEP provides a rigorous and unambiguous definition of the reaction path (for a one-step reaction). A MEP can be computed by using standard computational tools. One of these exploits the so-called intrinsic reaction coordinate (IRC) method.[43] In this method, the reaction coordinate and the energy profile along it are simultaneously determined by constructing two steepest descent lines that start at the transition structure **TS** (i.e., the first-order saddle-point structure) and end at the "reactant" and "product" minima, respectively.

In the previous section, we saw that excited-state potential-energy surfaces have few special features that are not present in ground-state energy surfaces. As shown in Figures 6.11 and 6.12, an excited-state energy surface has energy minima (which correspond to excited-state intermediates) and first-order saddle points (i.e., transition structures) that may connect different excited-state conformers or intermediates. Nevertheless, the most relevant feature in these two-dimensional model surfaces are the surface crossings between two singlet states (e.g., the S_1 and S_0 states in Figure 6.11) and between one singlet state and one triplet state (e.g., the T_1 and S_0 states in Figure 6.12). The presence of a surface crossing corresponding to a local energy minimum on the excited-state energy surface allows for the existence of a peculiar type of reaction path that, however, has been demonstrated to be a common feature in different classes of organic compounds. (See also Section 4). This path connects an excited-state intermediate $\mathbf{M}^*$ to a crossing point via a standard transition structure $\mathbf{TS}^*$. The crossing point between two energy surfaces of the same spin multiplicity (S_1 and S_0, T_1 and T_0, etc.) is called a conical intersection, and indeed, it forms a "conical funnel." [A conical intersection is an $n-2$-dimensional object (with n being the number of internal coordinates), and therefore, it appears as a

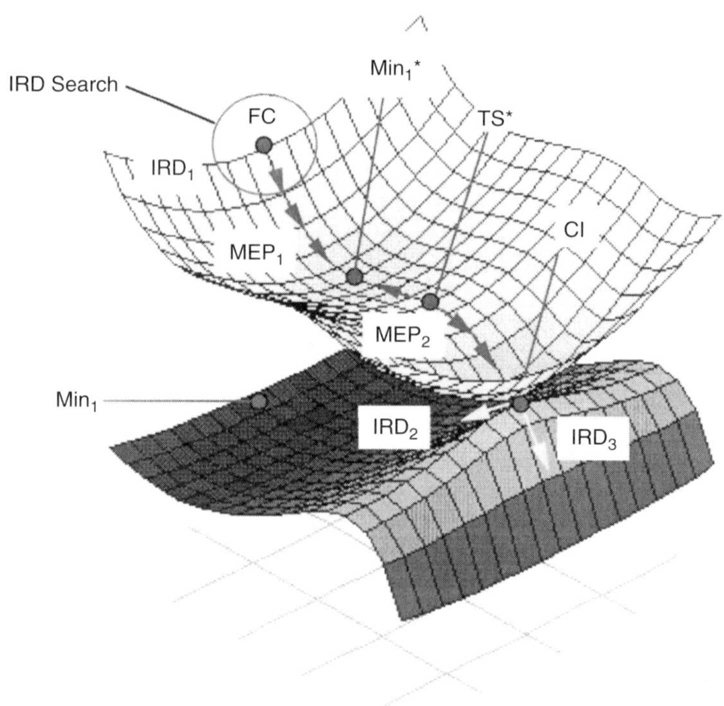

Figure 6.11. Model intersecting potential-energy surfaces showing an excited-state minimum ($\mathbf{Min}_1^*$), a transition state ($\mathbf{TS}^*$), and a conical intersection (**CI**). The composite excited-state MEP (i.e., $\mathbf{MEP_1}$, $\mathbf{MEP_2}$) and the initial direction of relaxation from the FC point ($\mathbf{IRD_1}$) are indicated by the stream of arrows. The ground-state initial directions of relaxation ($\mathbf{IRD_2}$ and $\mathbf{IRD_3}$) starting at **CI** are indicated by arrows.

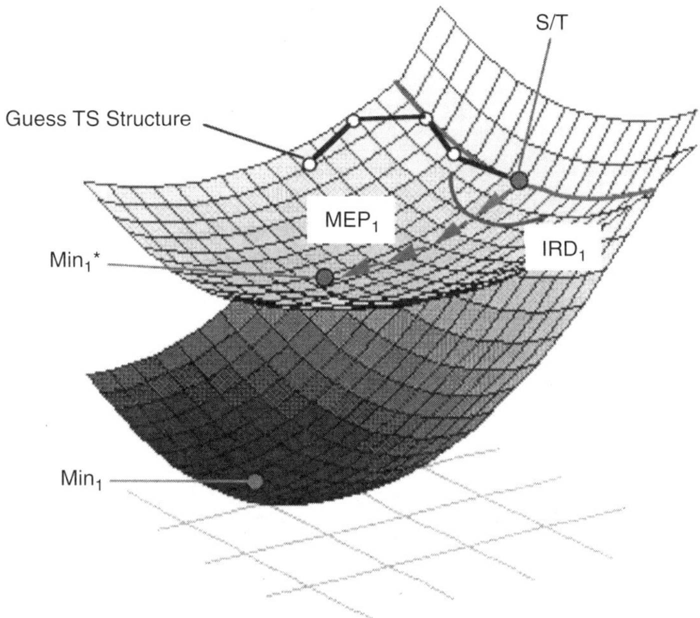

Figure 6.12. Model intersecting potential-energy surfaces showing an excited-state minimum (**Min$_1^*$**) and singlet–triplet crossing (**S/T**). The excited-state MEP (i.e., **MEP$_1$**) and the initial direction of relaxation from the minimum-energy **S/T** point (**IRD$_1$**) are indicated by the stream of arrows. The open circles indicate the "iterations" required to locate the crossing minimum **S/T**.

single point in a two-dimensional energy surface]. This is the *nonadiabatic* path seen in the previous section, which forms a radiationless decay channel whereby the system decays and continues its relaxation motion along the ground-state energy surface and, therefore, along the energy surface of a different state.

An additional specific feature of certain excited-state paths is that they start, not at an energy minimum, but at a nonstationary point (e.g., point **FC** in Figure 6.11). For instance, this happens when one is interested in investigating the initial relaxation process occurring soon after the photoexcitation. According to the Franck–Condon approximation, the photoexcitation delivers the ground-state species to the excited state vertically (i.e., the molecular geometry of the initial ground-state species remains unchanged). Since the ground- and excited-state equilibrium geometries are usually quite different, the excited-state relaxation process will start at a point of the excited-state surface that corresponds to the ground-state equilibrium structure. The reaction path describing the relaxation process may be determined by computing the so-called initial relaxation direction (IRD) and then using this direction to start a steepest descent path computation that will terminate at an energy minimum (**M$_1^*$** in Figure 6.11). The initial relaxation direction is determined via the IRD optimization procedure. In practice, this procedure locates a local steepest descent direction among the directions of a rotating vector (see circle centered at **FC** in Figure 6.11) of a fixed length and centered on the initial point (e.g., the **FC** point). In the case of the two-dimensional model surface of Figure 6.11, there is only one steepest descent direction at the **FC** point, which corresponds with the one pointing towards the energy minimum (see arrow). In general, the search can be extended to *all* the possible relaxation paths starting from a *nonstationary point* on a $3N - 6$-dimensional energy surface. However, the same methodology can be applied to search for all the *ground-state relaxation paths* starting from a conical intersection point (which appears as a cusp on the ground-state energy surface). In this case, there is usually more than one relaxation direction. In the model surface of Figure 6.11, we show that there are two ground-state relaxation directions. One leads towards the original reactant and the other leads towards a photoproduct.

In Figure 6.12, we show a two-dimensional model surface with a crossing that lies, not at the bottom of an excited-state energy valley, but above it. While this is illustrated for the case of a singlet–triplet crossing,* a similar feature has also been documented for the intersection of energy surfaces of the same spin multiplicity (i.e., when the crossing corresponds to a conical intersection). In this case, one is interested in the path that connects the lowest energy point located on the crossing seam to the excited-state equilibrium structure. In order to map this path, one has to apply two different tools. The first tool is required to locate and optimize a point of minimum energy along the conical intersection (or singlet–triplet) crossing seam. This point will be referred to here as the conical intersection optimizer (CIO)[44,45]. As illustrated in Figure 6.12, a CIO calculation starts at a suitable guess geometry and, after a few optimization steps, locates a close crossing point. After a point of crossing has been located, the optimization proceeds by following, roughly, the crossing seam in the downhill direction until an energy minimum has been reached. This minimum represents the point of the crossing seam that can be more easily accessed through a displacement of the molecular structure of the excited-state intermediate. Hence, it is of interest to map the MEP that connects M_1^* to the seam minimum. In order to achieve this aim, we have to use a different methodology. Since the point at the seam minimum is nonstationary, we can use the IRD method to locate the direction of the preferred relaxation path. After the initial relaxation path has been determined, one can compute the steepest descent path from the crossing point to the excited-state intermediate.

Finally, it is important to point out that the reaction path controlling an excited-state conformational or reactive process must be evaluated by computing the MEP and *not* by scanning the excited-state energy surface along a hypothetical reaction coordinate. To better illustrate this point, we consider the seminal work by Van der Lugt and Oosteroff[30b] and by Devaquet et al.[46] on the mechanism of the photochemical electrocyclic ring closure of *s-cis* buta-1,3-diene (or ring opening of

---

* In contrast to a conical intersection, a singlet–triplet crossing is an $n - 1$-dimensional object and therefore appears as a line on a two-dimensional energy surface. This is because the off-diagonal matrix element of the Hamiltonian between a singlet and a triplet state is zero by symmetry, thus, there is only one trivial condition to be satisfied at a singlet–triplet crossing point: The energy of the singlet state must be equal to the energy of the triplet state.[12]

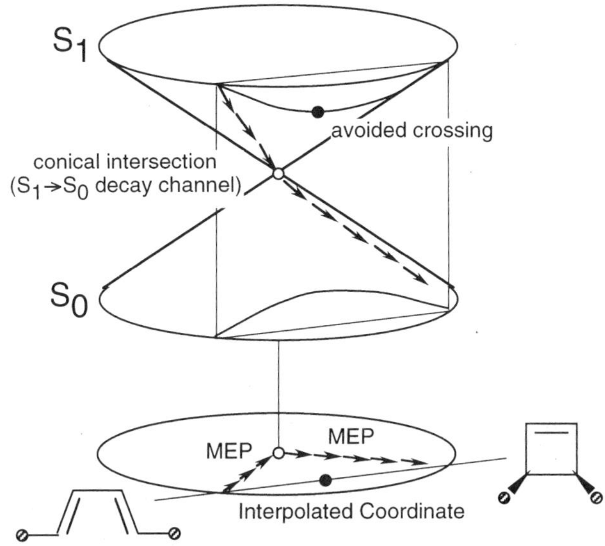

Scheme 6.9

cyclobutene). In the original papers, the excited-state reaction pathway for this reactive process was evaluated by scanning the potential-energy surface along a hypothetical reaction coordinate derived by interpolating the molecular structure of the ground-state reactant and product. As illustrated in Scheme 6.9, the procedure yields an energy profile featuring an avoided crossing (full circle) between the ground- and the excited-state potential-energy surfaces. On the other hand, if one computes the S_1 MEP starting from the reactant structure on the energy surface, the procedure will reach the conical intersection point (discussed in the previous section), which corresponds, in this case, to the S_1 minimum. Thus, despite the fact that Van der Lugt and Oosteroff's and Devaquet's computational work contributed to establish the general mechanistic picture of a photochemical pericyclic reaction, the computational strategy used (the only strategy available at that time) missed the real nature of the decay channel in this system.

The strategy we have illustrated is summarized in the following flowchart (see top of p. 332).

4 Excited-State Conformations of Organic Chromophores

In the introduction to the chapter and in Section 2, we learned that there are two important aspects of the excited-state conformational properties. The first is related to the variable nature of the excited-state bonding, which makes the distinction between conformational and reactive processes less straightforward than in the ground state. The second deals with the conformational structure of intrinsically transient entities, such as the conical intersections and singlet–triplet crossings. The

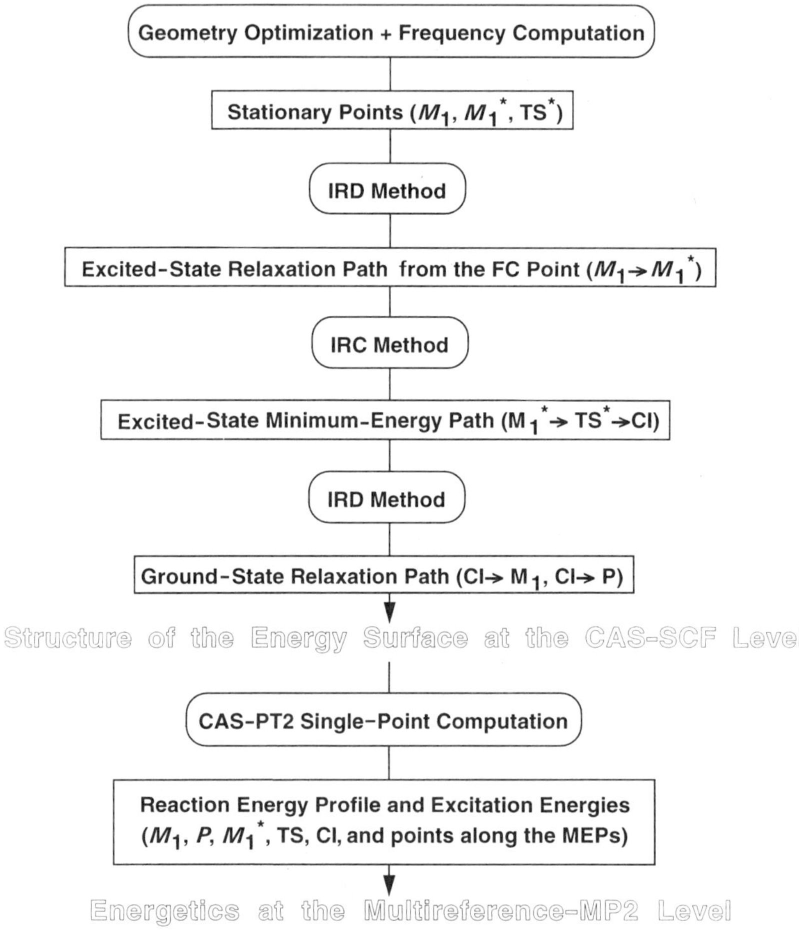

Flowchart

favored conformation of these entities may determine, for instance, the stereoselectivity of photochemical reactions in the same way in which the favored conformation of a transition state determines the stereoselectivity in a ground-state reaction. Shortly we shall examine a series of selected computational results that are related to these two aspects. The term "favored conformation" of an intrinsically transient entity requires some further remarks. In ground-state chemistry, the favored conformation of a transition state can be rigorously and unambiguously defined as its lowest energy conformation (consider, as an example, the *anti* versus *syn* conformation of a β-elimination transition state), as depicted in Scheme 6.10a, where TS and TS′ represent two different transition state conformations. Thus, the conformer TS′ determines, under kinetic control, the most populated reaction path and the final reaction stereochemistry. On the other hand, in photochemistry, the lowest energy decay channel (i.e., the lowest energy conical intersection conformation—**CI** in Scheme 6.10b) may not be the channel that determines

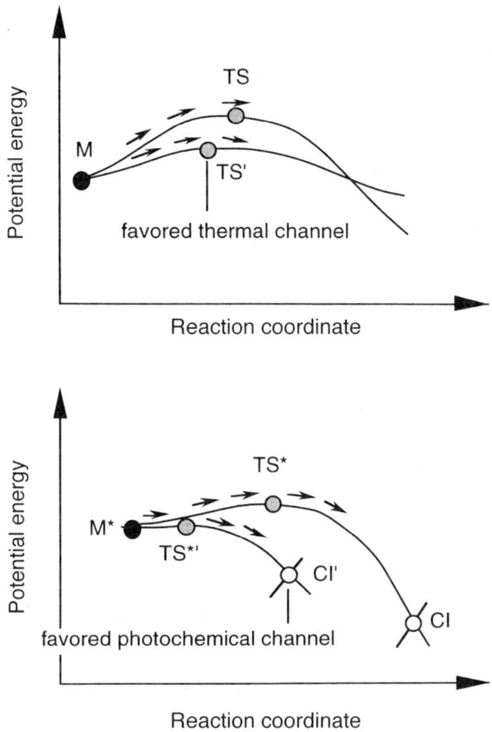

Scheme 6.10

the observed stereochemistry. In fact, the population of the decay channel depends on the accessibility of the channel itself, which can be controlled by the lowest excited-state transition-state barrier (i.e., the barrier corresponding to TS*′ in Scheme 6.10b) or by more complex dynamical factors, especially when the competing paths are both nearly barrierless—as we documented earlier in the case of *s-cis*→*s-trans* photochemical isomerization (see Section 2.2).

The reaction *stereoselectivity* represents only one aspect of the problem of reaction selectivity in organic chemistry. The *regioselectivity* (for instance, the orientation—that is, the position—of the initial attack in a thermal electrophilic addition reaction) and the *chemoselectivity* (i.e., the preference for one type of reaction over another one) represent the other two important aspects of this fundamental problem. These three selectivity levels will be discussed later in the context of photochemical processes.

4.1 Linear Conjugated Hydrocarbons (Polyenes and Polyenyl Radicals)

In Section 2, we saw that a conformational process in a short polyene (e.g., the *s-cis*→*s-trans* isomerization of buta-1,3-diene) triggers the radiationless decay from the singlet excited state to the ground state. Thus, the conformational interconversion and the photophysics of electronically excited conjugated molecules are intertwined. Furthermore, the radiationless decay occurring upon *s-cis*→*s-trans* isomerization initiates the process of photoproduct formation on the ground-state energy surface. We have also seen that the excited-state conformational processes are connected, at least in polyenes, to the excited-state reactivity and, consequently, to the reaction selectivity for the Z/E isomerization and cyclization reactions. In what follows, we shall see that these concepts are fully general in linear polyenes.

The S_1 equilibrium molecular structure of the *all-trans* conformer of short linear polyenes is basically planar. As we have seen for buta-1,3-diene, the bonding is such that the single-bond–double-bond distribution is almost reversed with respect to the ground state, and the molecule has diradical character. This is illustrated in Figure 6.13 for the case of *all-trans* octa-1,3,5,7-tetraene (*all-trans*-OT). The molecular geometry at the $S_1 \rightarrow S_0$ radiationless decay channel (an S_1/S_0 conical intersection) shows, invariably, a sharp "kink" located at a $-(CH)_3-$ segment in the middle of the molecule. In Figure 6.14, we compare the structure of this segment for *all-trans*-OT with the *all-trans* heptatrienyl radical (*all-trans*-ER). The comparison of these structures with that of the buta-1,3-dienes seen in Figure 6.5 reveals that each structure contains the same triangular arrangement of the carbon skeleton that we have already discussed in Section 8.2.1. The $-(CH)_3-$ kink electronic structure corresponds to three weakly interacting $p\pi$ electrons set in a triangular arrangement that are loosely coupled to an isolated $p\pi$ radical moiety and a polyenic fragment (an allyl radical and a vinyl group in *all-trans* OT and an H atom and a CH_2 group in buta-1,3-diene) in polyenes and to two polyenic fragments in polyenyl radicals (two vinyl groups in *all-trans* ER).

The $-(CH)_3-$ conical intersection has proven to be general for a series of linear polyenes and polyene radicals[47] with the formula C_nH_{n+2}. This is related to the fact that in these molecules the first excited state has a doubly excited $\pi-\pi^*$ orbital with a diradical nature. Invariably, those intersections are located along a reaction path

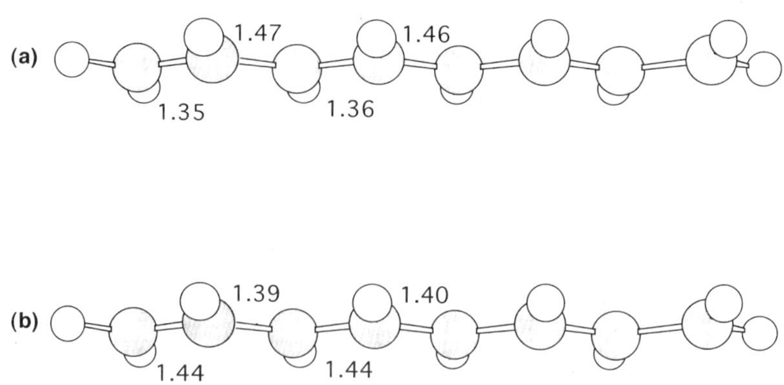

Figure 6.13. Ground-state (a) and excited-state (b) equilibrium structures of *all-trans* octa-1,3,5,7-tetraene. The data are from Ref. 50.

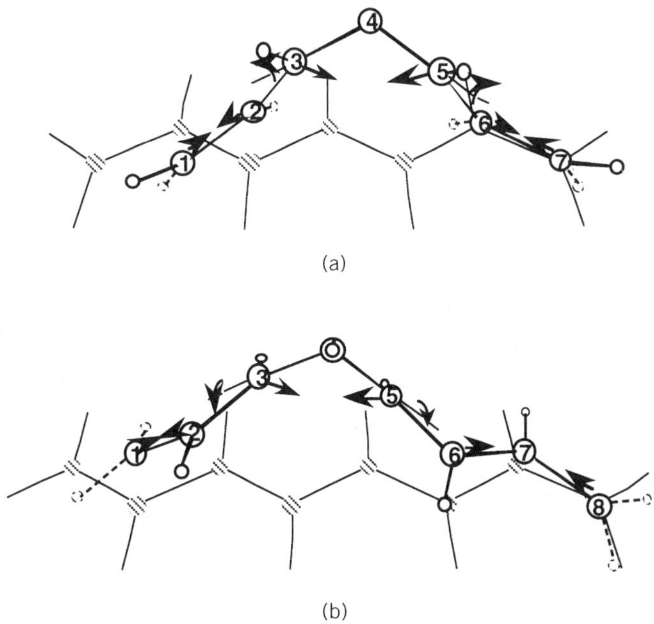

Figure 6.14. (a) Structure of the conical intersection decay channel of the *all-trans* hepta-1,3,5-trienyl radical and (b) *all-trans* octa-1,3,5,7-tetraene. In both structures, the central–$(CH)_3$–kink corresponds to the C_3–C_4–C_5–segment. The data are from Ref. 47.

describing a conformational change of the molecule involving the simultaneous rotation about two C–C bonds and a decrease in one of the C–C–C angles (see angle C_3–C_4–C_5 in Figure 6.14 and angle C_2–C_3–C_4 in Figure 6.5). Thus, as already discussed in Section 2.2 for buta-1,3-diene, the conformational process cannot be completed on the excited-state surface. The potential-energy surfaces of six ($n = 3, \ldots, 8$) *all-trans* conjugated hydrocarbons have been computed by using the ab initio CAS–SCF method with the DZ + d basis set and an active space including all π and π^* orbitals and electrons. The lowest excited-state surface starting at the optimized planar S_1 or D_1 equilibrium structures has the general structure illustrated in Figure 6.11. This surface is composed of a low-lying transition structure (**TS***), separating the excited-state minimum (**Min**$_1^*$) from the conical intersection (**CI**). Such paths have been computed by using the IRC method implemented in the *Gaussian94*[6] software package (see Section 3.2). The barrier heights have been corrected for dynamical correlation effects by single-point computations, using second-order multireference Møller–Plesset perturbation methods (see Section 3.1). For $n = 3$ and 4, the existence of a conical intersection at the end of the computed reaction path was also demonstrated by a direct search for and optimization of the conical intersection point by means of the CIO methodology (see Section 3.2) implemented in the same package.

As shown in Table 6.1, the computed barrier heights are slightly larger than their experimental counterparts and are correct within 1 kcal mol^{-1} wherever the compar-

Table 6.1 Energy barriers computed with second-order multireference Møller–Plesset perturbation methods (PT2F). (See Section 3.1. Adapted from Table 1 in Ref. 47.)

Structure (C_nH_{n+2})		Calculated (kcal mol^{-1})	Experimental (kcal mol^{-1})
0	$n = 3$	0.0[a]	—[c]
0	$n = 4$	0.0[a]	—[c]
	$n = 5$	8.7	—
0	$n = 6$	0.0	0.0–0.3
	$n = 6$ (E/Z)[b]	1.5	0.2–0.5
0	$n = 6$ (c/t)[b]	1.0	~0.0
0	$n = 7$	16.2	—
0	$n = 8$	7.6	6.0

[a] For $n = 3$ and 4, the reaction path is barrierless.

[b] In order to extend the comparison between experimental and computed values of the energy barriers, we have also recomputed the energy barriers for the two *tZt*-hexa-1,3,50-triene decay paths reported in Ref. 19.

[c] Fluorescence not detected.

ison with experimental results is possible. However, one has confidence that the computed paths correspond to the paths populated at the range of the low vibrational excess energies used in the jet experiments.[35,48] As discussed elsewhere,[22a] the quantum mechanical origin of this feature can be understood by comparison with H_3, in which any equilateral triangle configuration corresponds to a point on the S_0/S_1 conical intersection in which the three H electrons have identical pairwise interactions. The same feature also characterizes the $-(CH)_3$ kink, in which three singly occupied p orbitals interact through two π interactions and one σ interaction.

4.2 Adiabatic versus Nonadiabatic Paths and Photochemical Stereoselectivity

In the previous subsection, we described the rather complex torsional deformation characterizing the nonadiabatic pathway ending at the $-(CH)_3$–kink—an S_1/S_0 conical intersection. As discussed in Section 2, this intersection provides the "photochemical" decay channel prompting Z to E or, probably, *s-cis* to *s-trans* isomerization in buta-1,3-diene. However, in longer linear polyenes, such as *all-trans* hexa-1,3,5-triene (*all-trans* HT) or *all-trans* OT there are, in principle, multiple distinct S_1 paths leading to distinct conical intersections. Each path is characterized by the "position" of the kink along the hydrocarbon chain. In other words, the kink may develop at the center or near the end of the chain, and thus, the conical intersections are, substantially, different diastereoisomeric conformers. As a consequence, we expect that decay from each intersection will lead to a different Z/E diastereoisomer. Furthermore, since the excited-state equilibrium structure has C–C bonds whose lengths are close to those of single bonds, one may expect the existence of conventional adiabatic paths leading to different excited-state Z/E diastereoisomers. The computed adiabatic and nonadiabatic S_1 paths of *all-trans* HT[49] and *all-trans* OT[50] were recently reported. Each path is characterized by a transition structure and, consequently, by a barrier height. The molecular structure of the various transition structures is given in Figure 6.15.

In *all-trans* HT, all paths start at a planar C_{2h} energy minimum. There are two nonadiabatic and two adiabatic Z/E isomerization paths defined by four transition structures. The nonadiabatic transition structures of Figure 6.15a and Figure 6.15b are characterized, as expected, by the presence of a $-(CH)_3-$ kink that involves the C_2–C_3–C_4 and C_1–C_2–C_3 angles, respectively. The first transition state is associated with a *nonadiabatic* $E \rightarrow Z$ (because of the large, approximately 50° torsion about the central double bond) isomerization and the second with a *nonadiabatic* single-bond (i.e. $t \rightarrow c$) isomerization (because of the 43° torsion about the single bond). The adiabatic transition structures of Figure 6.15c and Figure 6.15d define two isomerization paths on the excited state, connecting the *all-trans* S_1 equilibrium structure to itself (via a 180° rotation of one terminal CH_2 group) and to the Z–HT structure (via a 180° $E \rightarrow Z$ rotation about the central C=C bond), respectively. These structures show an approximately 90° twisted double bond and hence lie halfway along the S_1 isomerization path.

The energy barriers controlling access to the four isomerization paths of *all-trans*-HT previously described are given in Table 6.2. It is remarkable that *the adiabatic paths have larger barriers*. Indeed, the lowest adiabatic reactive process, which corresponds to CH_2 rotation, has about a 7 kcal-mol^{-1} barrier. Both nonadiabatic paths (i.e., those defined by the transition structures of Figures 6.15a and 6.15b) are energetically favored, and the path associated with $E \rightarrow Z$ motion (i.e., the one with the most "central" kink) is nearly barrierless.

The same type of potential-energy surface topology was recently documented for the S_1 excited state of *all-trans* TO,[50] in which there are five reactive channels on the S_1 energy surface. The barriers and geometries of the transition structures that separate these channels from the S_1 minimum are collected in Table 6.3 and Figures 6.15e–h. The lowest energy channel (ca. 7 kcal mol^{-1}) is a nonadiabatic channel characterized, as in *all-trans* HT, by an $E \rightarrow Z$ rotation about one of the two equivalent central double bonds (Fig. 6.15e) and leading to the conical intersection shown in Figure 6.15i. The remaining four channels lie at higher energies. One nonadiabatic channel with a barrier of 10 kcal mol^{-1} corresponds to a motion similar to that described in Figure 6.5; however, the $-(CH)_3-$ kink is located at the end of the molecule (Fig. 6.15f). There is an adiabatic channel at 11 kcal mol^{-1} with an approximately 90° twisted CH_2 group (Fig. 6.15g). Finally, a fully adiabatic $E \rightarrow Z$ isomerization ($\mathbf{TS}_{E \rightarrow Z\ \mathbf{adiab}}$) channel lies approximately 7 kcal mol^{-1} above the channel for nonadiabatic $E \rightarrow Z$ isomerization (Fig. 6.15h).

The *all-trans* HT and *all-trans* TO data revised in the previous paragraphs show that:

1. The adiabatic paths have larger barriers than the nonadiabatic paths. As a consequence, the nonadiabatic paths are expected to control the photochemical product distribution.
2. The photochemical Z/E isomerization of these molecules must be diastereoselective. In particular, the major Z/E diastereoisomer produced via photolysis of these species is expected to originate via decay at the most easily accessible diastereomeric decay channel.
3. The molecular structure of the diastereomeric decay channel (and the associated S_1 transition-state structure) indicates that the nonadiabatic Z/E isomerization may lead to a simultaneous double-bond and (adjacent) single-bond conformational change.

Table 6.2 Absolute (E) and relative (ΔE) energies computed for *trans*-HT with second-order multireference Møller–Plesset perturbation theory (PT2F) and CAS–SCF (in parentheses). See Section 3.1. The relative energies describe the energetics of the S_1 energy surface. Adapted from Table I in Ref. 49.

Structure	State	E (a.u.)	ΔE(kcal mol^{-1})
S_0 **MIN (FC)**			
	S_0	−232.57108[a]	
		(−231.89131)	
		−232.74169[b]	
		(−231.97080)	
	S_1	−232.37632[a]	25.9[a]
		(−231.68227)	
		−232.55181[b]	
		(−231.76782)	
	S_2 **MIN**	(−231.61265[a]	
		−232.55547[b]	
		(−231.70289)	
S_2 **MIN**			
	S_1	−232.57848[b]	
		(−231.79882)	
	S_2	−232.55520[b]	
		(−231.71236)	
S_1C_{2h}-**MIN**			
	S_0	−232.54399[a]	−79.4[a]
		(−231.85774)	
	S_1	−232.41752[a]	0.0[a]
		(−231.72735)	
		−232.51938[c]	0.0[c]
		(−231.75701)	
S_1C_2-**MIN**			
	S_1	(−231.72740[c])	
S_1 **TS**$_{E\to Z}$			
	S_0	−232.47574[c]	−36.5[c]
		(−231.78621)	

(*Continued next page*)

Table 6.2 (*Continued*)

Structure	State	E (a.u.)	ΔE(kcal mol^{-1})
	S_1	-232.41711^a	0.3^a
		(−231.71797)	
		-232.51944^a	0.0^a
		(−231.74849)	
S_2/S_1 **CI**			
	S_1	-232.55664^b	
		(−231.77367)	
	S_2	-232.55484^b	
		(−231.70511)	
S_1 **LP**[d]			
	S_0	-232.43771^a	-12.7^a
		(−231.74480)	
		$(-231.73975)^{a,e}$	
	S_1	-232.41930^a	-1.1^a
		(−231.71957)	
		$(-231.71518^{a,e})$	
S_1/S_0**CI**[f]			
	S_0	$(-231.71683^{a,e})$	
	S_1	$(-231.71582^{a,e})$	
S_1 $\mathbf{TS}_{t\to c}$			
	S_1	-232.51653^c	1.8^c
		(−231.74997)	
S_1 $\mathbf{TS}_{E\to Z\ adia.}$			
	S_1	-232.49055^c	18.1^c
		(−231.73063)	
S_1 $\mathbf{TS}_{CH_2\ adia.}$			
	S_1	232.50773^c	7.3^c
		(−231.74692)	

[a] Structure and energy computed by using a CAS(6,6)/6-31G* level of theory.
[b] Structure and energy computed by using a CAS(6,8) level of theory and energy computed by using symmetry-adapted wave functions, state-averaged orbitals, and the ANO basis set C[6*s*, 3*p*, 1*d*], H[2*s*, 1*p*] with a set of diffuse functions C[2*p*] on each carbon atom (see Ref. 49 for details).
[c] CAS(6,6)/D95* level of theory.
[d] Last optimized point along the S_1 **MEP** (path **III**).
[e] State-averaged energy. The state averaging includes the first two states (S_0 and S_1) with 0.5 weight (see Ref. 49 for details).
[f] Optimized (lowest energy) crossing point. It is found that this point corresponds to the absolute energy minimum of the S_1 state. The corresponding structure is only slightly different from structure **LP** and <3 kcal mol^{-1} more stable.

These results can be used to validate the mechanistic picture presented below, with respect to the experimental observations. Points 1 and 2 imply that the lowest nonadiabatic Z/E isomerization barrier determines the observed decay kinetics and, in turn, excited-state lifetime. In fact, by inspection of Tables 6.1, 6.2, and 6.3, we see that the computational results reproduce the relative and absolute values of the experimentally determined excited-state barriers within a 1–2 kcal-mol^{-1} error. Furthermore, the computed vertical and 0–0 excitation energies appear to be

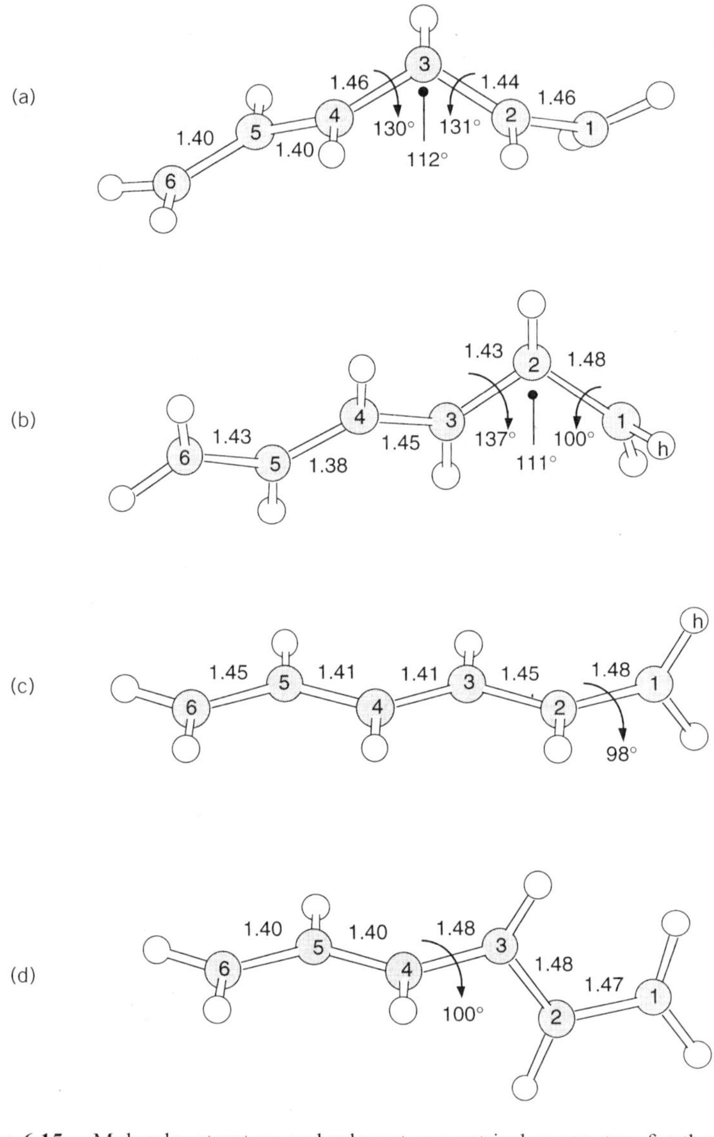

Figure 6.15. Molecular structure and relevant geometrical parameters for the S_1 low-lying transition structures, in order of increasing energy: (a) $TS_{Z/E}$, (b) $TS_{t \rightarrow c}$, (c) $TS_{CH2\ adia}$, (d) $TS_{Z/E\ adia}$.

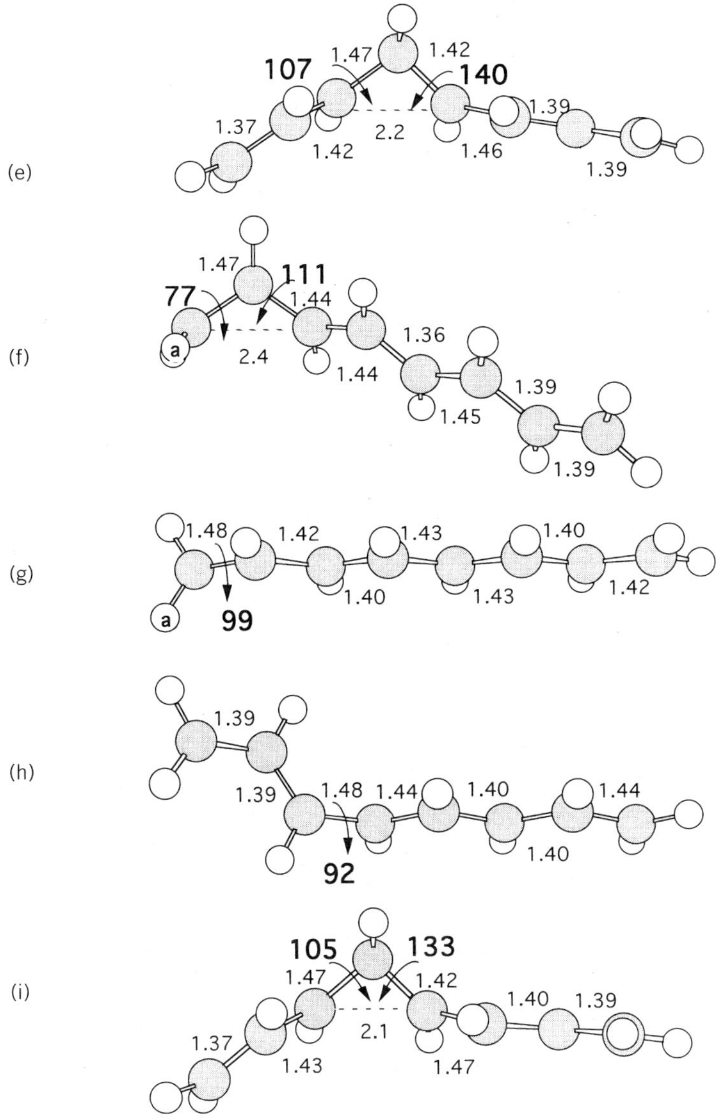

Figure 6.15 (continued). Molecular structure and relevant geometrical parameters for the S_1 low-lying transition structures, in order of increasing energy: Structures corresponding to (e) $TS_{E \rightarrow Z \text{ (nonadiab.)}}$, (f) $TS_{CH_2 \text{ (nonadiab.)}}$, (g) $TS_{CH_2\text{-twist (adiab.)}}$, (h) $TS_{E \rightarrow Z \text{ (adiab.)}}$, and (i) the S_1/S_0 intersection structure (CI). The relevant geometrical parameters are given in Å and degrees (H_a–C–C–C and C–C–C–C torsions are in boldface.) The data are from Ref. 50 (*all-trans* OT) and Ref. 49 (*all-trans* HT).

Table 6.3 Multireference-MP2 (see Section 3.1) absolute (E) and relative (ΔE) energies. Adapted from Table I in Ref. 50.

Structure	State	E (a.u.) S_1[a]	ΔE (kcal mol^{-1})[a]	Experimental
S_0 Min	S_0	−309.84471	−76.7	−82.3
		(−309.75183)	(−78.4)	
S_1 Min	S_1	−309.72241	0.0	0.0
		(−309.62694)	0.0	
$TS_{E \rightarrow Z\ (nonadiab.)}$	S_1	−309.71034	7.6	6.0
		(−309.61422)	(8.0)(7.4[b])	
$TS_{CH_2\ (nonadiab.)}$	S_1	−309.70645	10.0	
$TS_{CH_2\text{-twist}(adiab.)}$	S_1	−309.70428	11.4	
$TS_{E \rightarrow Z\ (adiab.)}$	S_1	−309.69877	14.8	

[a] Values computed using the D95* and 6-31G* (values in parentheses) basis sets.
[b] Vibrational zero-point-energy-corrected value.

consistent with the experimental quantities.[49,50] The result labeled 3 has been the object of a more direct investigation of the validity of the computed mechanistic scheme. Fuss et al.[51] have reported a matrix isolation investigation of the photochemical conversion of 7-dehydrocholesterol (7-DHC) to tachysterol. 7-DHC (a precursor of vitamin D) is a cyclohexa-1,3-diene derivative that produces, by irradiation, pre-vitamin D (P_0), a Z-hexa-1,3,5-triene derivative. In turn, the irradiation of P_0 leads to the production of tachysterol (T), which is an E-hexa-1,3,5-triene derivative. In a cold matrix, the irradiation of 7-DHC produces the *cZc* (*s-cis, Z, s-cis*) conformation of P_0 exclusively, since thermal interconversion to more stable conformers (such as the *cZt* conformer) is blocked. Thus, under these conditions, the primary photoproduct of the irradiation of P_0 must derive from the *cZc* conformer of pre-vitamin D. A model system that can be used to investigate the photochemistry of P_0 is *cZc*-hexa-1,3,5-triene (*cZc*-HT).

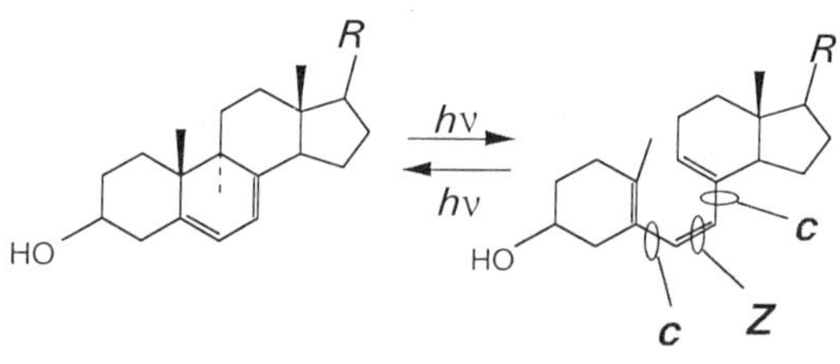

7-DHC **P_0**

The possible excited-state reaction paths for the evolution of photoexcited hexa-1,3,5-triene in its *cZc* conformation have been computationally investigated.[53] The results, summarized in Figure 6.16, suggest that the irradiation of *cZc*–HT leads to the production of a very short-lived S_1 state intermediate (top structure) that can decay according to three different nonadiabatic paths (Paths **I**, **II**, and **III**) leading to three structurally different conical intersections. Despite the apparent difference, it is easy to recognize certain common features in the structure of the three decay

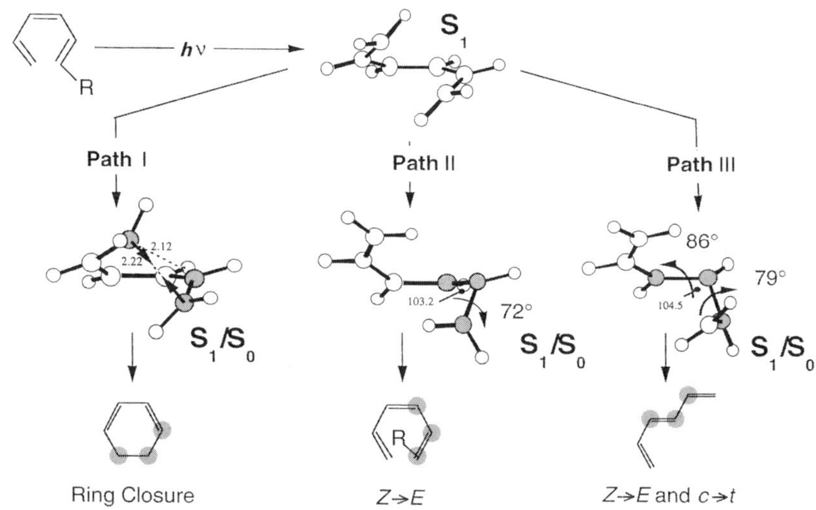

Figure 6.16. Competing S_1 relaxation paths and molecular structure of the corresponding decay channels of S_1 *cZc*-hexa-1,3,5-triene. The geometrical parameters are in degrees and Å. The shaded circles in the photoproduct structures indicate the position of the $-(CH)_3-$ kink of the conical intersections.

channels. In particular, the right-end channel structure features a $-(CH)_3-$ kink located at the center of the hydrocarbon chain ($-C_2-C_3-C_4-$), and the central channel structure features a $-(CH)_3-$ kink located at the end of the hydrocarbon chain ($C_1-C_2-C_3-$). Decay via these two conical intersections must lead (according to the new mechanistic view of the NEER principle discussed in Sections 1 and 2) to different Z/E diastereoisomers. The channel structure suggests that the right-end channel (Path **III**) leads to Z/E double-bond isomerization at the central double bond, while the central (Path **II**) may lead to Z/E double bond isomerization only at the terminal methylenes. (This isomerization can be detected in a properly substituted reactant.) Notice, however, that the structure of these channels suggests that the double bonds may not be the only bonds that undergo a conformational change. Indeed, *the right-end structure has two highly twisted adjacent bonds*, and thus decay at this channel may lead to a conformer that has both the central double bond and the adjacent single bond isomerized with respect to the original S_1 hexa-1,3,5-triene conformation. In other words, one has a simultaneous $Z \rightarrow E$ and $c \rightarrow t$ isomerization. In contrast, the structure of the central decay channel suggests only a single $Z \rightarrow E$ isomerization.

The result of the experiment by Fuss et al.[51] is illustrated in Figure 6.17. The irradiation of 7-DHC in a 90 K matrix leads to the production, via a two-photon process, of the tachysterol conformer T. Notice that this conformer is the *primary photoproduct* of the irradiation of the pre-vitamin D intermediate P_0 generated via photochemical ring opening. The structure of T (which undergoes a conformational change to T_0 upon heating) clearly indicates that the photochemical primary event results in a two-bond isomerization, including the central double bond and one

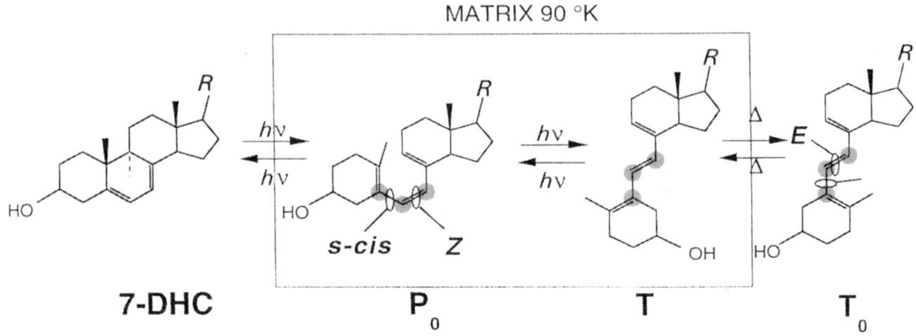

Figure 6.17. 90-K pre-vitamin D photochemistry. 7-dehydrocholesterol (**7-DHC**) is converted to the *cZc* conformer of precalciferol ($\mathbf{P_0}$). In turn, $\mathbf{P_0}$ can be converted to either **7-DHC** or the *cEt* conformer of tachysterol (**T**). The more stable *cZc* conformer is produced thermally. The shaded circles in the photoproduct structures indicate the position of the $-(CH)_3-$ kink of the conical intersections.

adjacent single bond. Thus, the observed path corresponds to Path **III** in our hexa-1,3,5-triene model. The simultaneous $Z \rightarrow E$ and $c \rightarrow t$ isomerization is a typical feature of the $-(CH)_3-$ structure, which has, in general, two highly twisted adjacent bonds. Notice that the efficiency of the right-end decay channel in 7-DHC is due to the fact that the two terminal double bonds in P_0 are embedded in two six-membered rings and the central decay channel cannot be fully active in this system. The only diastereoisomer produced during the photolysis is the one just described.

4.3 Photochemical Chemoselectivity and Regioselectivity: The Case of *Z*-hexa-1,3,5-triene

Paths **II** and **III** in Figure 6.16 differ in the position and magnitude of the *torsional deformation* along the hydrocarbon chain. Thus, these nonadiabatic channels must lead, in principle, to different ground-state Z/E diastereoisomers. If Path **III** is more populated than Path **II**, due, for instance, to a lower excited-state barrier, the reaction will be diastereoselective, and the major product will have an isomerized central double bond. On the other hand, the molecular structure of the decay channel corresponding to Path **I** (i.e., to the left-hand decay channel) is qualitatively different from the other two. As shown in Figure 6.16, the kink appears to be localized, not *along* the hydrocarbon chain, but *between* the two ends of the chain. Thus, the typical "triangular" arrangement is formed by the interaction of $-CHCH_2$ and CH_2- moieties and not by the torsional deformation of double and single bonds (accompanied by bending deformation). The conical intersection of Path **I** cannot thus be considered a different conformer (or diastereoisomer) of the conical intersections of Paths **II** and **III**, but it must rather be regarded as a truly different constitutional isomer. This idea is also consistent with the application of the NEER principle to the decay channel of Path **I**. The predicted photoproducts are the ring-closure photoproduct cyclohexa-1,3-diene and the cyclopropylcarbinyl diradical (a precursor of bicyclo[3.1.0]hex-2-ene), not a different Z/E diastereoisomer. The fact that Path **I** can compete with Path **II** and Path **III** implies that the photochemical reaction of a hexa-1,3,5-triene derivative may present a certain level of *chemoselectivity*. In fact, if, for instance, the excited-state barrier along Path **I** is lower than those placed along Paths **II** and **III**, then Path **I** will be the most populated path, and decay at the corresponding conical intersection will lead to the selective production of cyclization products. If, on the other hand, this barrier is larger than those of Paths **II** and **III**, the only observed photoproducts will be Z/E isomerization products. The computed S_1 barrier for the ring-closure reaction and Z/E isomerization indicates that ring closure is highly favored in unsubstituted *cZc*-hexa-1,3,5-triene.[52] However, in substituted and polycyclic hexa-1,3,5-trienes, the competition between the three different channels may lead to different photoproduct populations. In conclusion, in hexa-1,3,5-trienes, the stereoselectivity and chemoselectivity originate by the competition of conformationally or constitutionally different excited-state paths.

It is also important to recognize that there is a second level of selectivity. This level operates after $S_1 \rightarrow S_0$ decay, during the ground-state relaxation process, starting from the decay channel and leading to the photoproduct S_0 energy wells. This process was discussed in Section 2 when we dealt with the selectivity of the buta-1,3-diene photoreactions. As mentioned in the previous paragraph, the structure of the decay channel of Path **I** suggests that two different cyclic photoproducts may be formed upon ground-state relaxation together with the original reactant (Fig. 6.18.) The first product (cyclohexa-1,3-diene) is formed by the attack of a terminal CH_2 radical center (see the earlier discussion of the electronic structure of the kink) on the terminal CH_2 group at the other end of the chain (i.e., the attack of carbon 1 on carbon 6). The alternative product (a cyclic biradical) is instead formed by the attack of the same CH_2 group on the carbon center adjacent to the CH_2 at the other end of the molecule. These relaxation paths lead to constitutional isomers of the

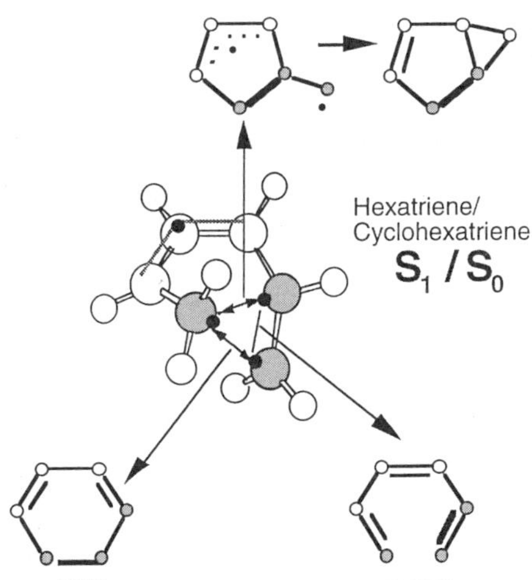

Figure 6.18. Structure of the lowest lying conical intersection of *cZc*-hexa-1,3,5-triene and associated ground-state relaxation processes. The shaded atoms indicate the position of the $-(CH)_3-$ kink of the conical intersections. Data from Ref. 53.

same functional class that are formed through the same chemical reaction, but on two different centers.[53] The computed relaxation paths suggest that the ring closure to cyclohexa-1,3-diene is much more facile than the ring closure to the cyclic biradical, which indicates that the reaction must be regioselective.

4.4 Cyclic Conjugated Hydrocarbons

In the previous subsections, we dealt exclusively with excited-state conformations of linear conjugated hydrocarbons and with the relationship between these conformations and the photochemical reactivity and selectivity. In this short subsection, we show that the same kind of excited-state conformational properties are found in cyclic polyenes, including aromatic and antiaromatic hydrocarbons.

In Figures 6.19a and 6.19b, we show the S_1 equilibrium structure and the S_1/S_0 conical intersection channel of benzene, respectively.[54] It is clear that the deformation required to access the conical intersection and decay to the ground state involves rotation about two adjacent C–C bonds and shortening of the C–C–C angle between these two bonds. Upon inspection of Figure 6.19b, one readily sees that the benzene conical intersection structure shows the same type of $-(CH)_3-$ kink and electronic structure as is seen in linear polyenes. In particular, notice that in the case of a cyclic polyene with an even number of $p\pi$-electrons, there are three weakly interacting $p\pi$-electrons localized on the that kink, and the other $p\pi$-electron is delocalized on the remaining allyl fragment (shaded atoms). Again, the computed conical intersection corresponds to a very efficient ultrafast radiationless decay channel. This channel is placed a few kcal mol^{-1} above the S_1 benzene equilibrium structure excited state and corresponds to the experimentally detected "channel 3" that opens up when the excited-state system has around $3000\,cm^{-1}$ of vibrational excess energy (see citations in Ref. 54). The conical intersection structure of benzene suggests prompt formation of the prefulvene diradical, which then rearranges to fulvene. According to the experiment, fulvene is the major photoproduct of benzene photolysis at 254 nm irradiation and populates ($1A_{1g} \rightarrow 1B_{2u}$) the S_1 state exclusively. This reaction is thus chemoselective.

Benzene is an aromatic molecule, and its S_1 excited state must be antiaromatic. In contrast, cycloocta-1,3,5,7-tetraene (COT) is an antiaromatic molecule, and its S_1 state is aromatic. In fact, while the favored conformation of ground-state COT is a tub-shaped molecule (see Figure 6.19c), excited-state COT is planar and symmetric. Nevertheless, torsional deformation of this molecule leads to two different conical intersections[55a] located above the planar S_1 minimum. Again, one of these intersection points (Fig. 6.19d) features the usual $-(CH)_3-$ kink distortion, which therefore appears to be a general feature in both aromatic ($4n + 2\pi$-electrons) and antiaromatic ($4n$ π-electrons) cyclic polyenes (where $n = 1, 2, 3, \ldots$). However, COT also features a new type of S_1/S_0 intersection with a higher C_{2v} symmetry (Fig. 6.19e) and that lies at lower energy with respect to the $-(CH)_3-$ kink. This different type of conical intersection structure also has a new electronic distribution, which is given by four weakly interacting electrons. The first two are localized on pyramidalized carbon centers in 1,4 positions. The remaining two are delocalized on

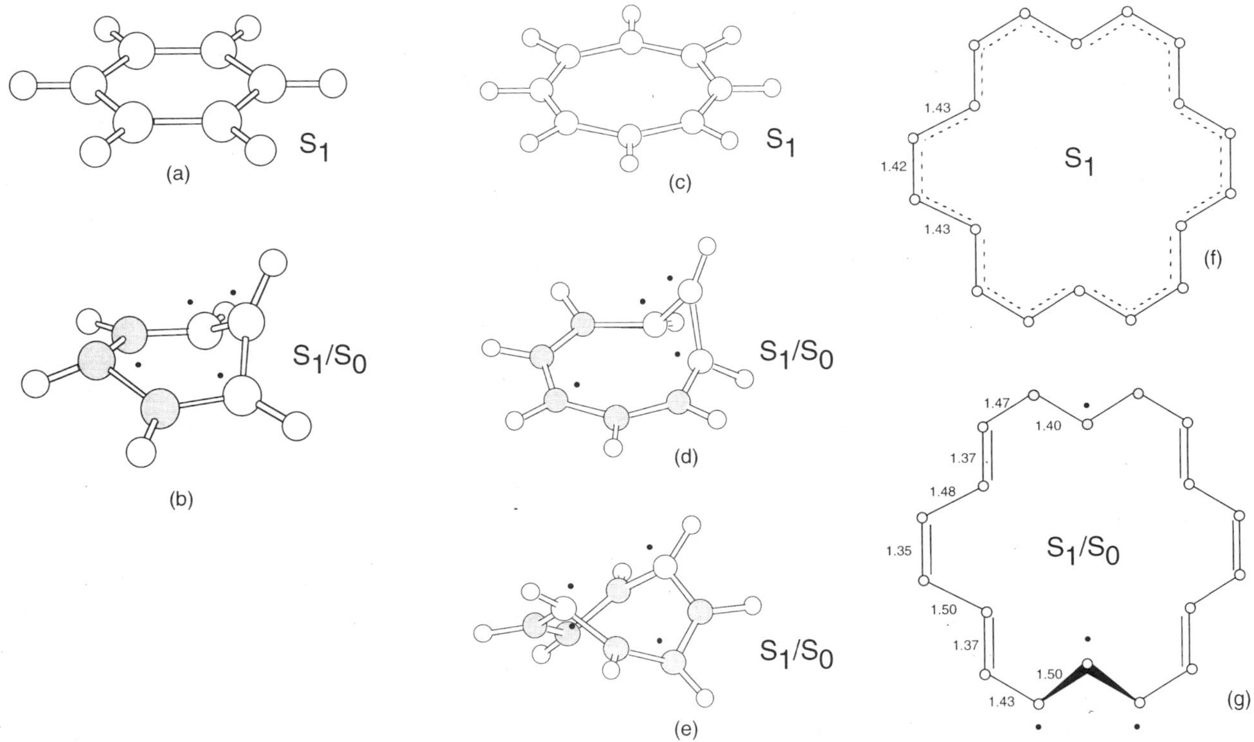

Figure 6.19. Molecular structures for the (a) S_1 benzene, (b) S_1/S_0 benzene conical intersection, (c) S_1 cycloocta-1,3,5,7-tetraene (d) $-(CH)_3-$ kink (benzenoid) type of cycloocta-1,3,5,7-tetraene S_1/S_0 conical intersection, (e) lowest lying cycloocta-1,3,5,7-tetraene S_1/S_0 conical intersection, (f) S_1 18-annulene, and (g) $-(CH)_3-$ kink (benzenoid) type of 18-annulene. The relevant geometrical parameters are given in Å. See also Ref. 54 (for benzene) and Ref. 56a (for 18-annulene). The large 18-annulene system has been investigated with the MM–VB hybrid method (Ref. 56b), which uses a combination of molecular mechanics and a parametrized quantum mechanical force field to lower the computational cost.

two allyl fragments (see shaded atoms.) The lower energy conical intersection structure of COT suggests facile formation of the tricyclic product semibullvalene **4**. Indeed, this product has been experimentally detected.[55b]

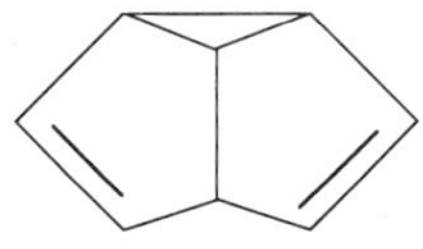

4

The increase in the size of the polyene ring does not seem to alter the general conformational structure of the S_1 energy surface. As shown in Figure 6.19f, the aromatic compound [18]-annulene has a planar excited-state structure[56] with high D_{6h} symmetry. Again, at higher energy, this molecule features the same conical intersection structure (Fig. 6.19g) already seen for benzene and COT. At present this is the only type of conical intersection structure located in this large aromatic system.

4.5 Conformational Effects of Conjugated Heteroatoms: Protonated Schiff Bases

While the photochemistry of linear conjugated hydrocarbons involves a low π-to-π^* doubly excited singlet state, conjugated systems containing a heteroatom may feature a low-lying excited state of a different nature. Thus, in this and in the next subsection, we revise the results of a set of recent computational studies that provide information on the S_1 conformational properties of protonated Schiff bases (PSBs) (consisting of a conjugated hydrocarbon containing a terminal $NH_2(+)$ group) and α,β-enones (consisting of a conjugated hydrocarbon containing a terminal carbonyl group). In these two species, the first singlet excited states are a π-to-π^* singly excited state (i.e., one electron is promoted from the π-HOMO orbital to the π-LUMO orbital) and an n-to-π^* excited state (i.e., one electron is promoted from the nitrogen or oxygen lone pair to the π-LUMO orbital), respectively.

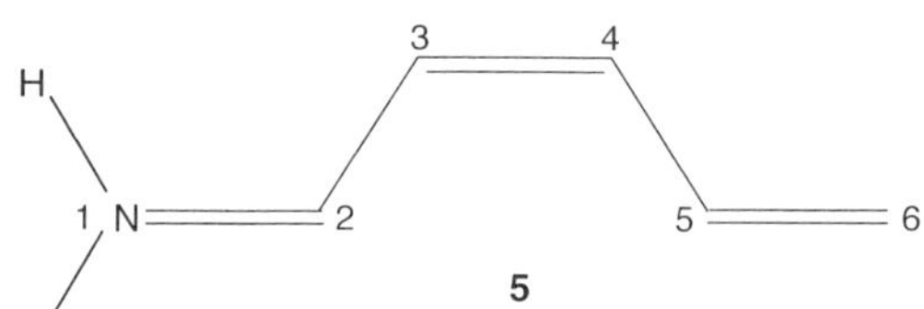

5

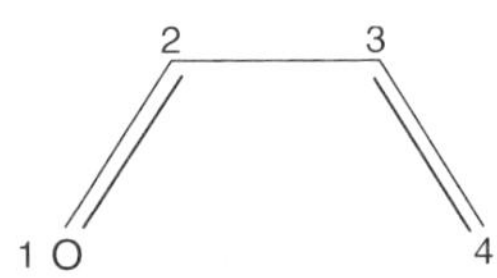

6

In Figure 6.20, we illustrate the structure of the S_1 energy surface of *tZt*-penta-3,5-dieniminium cation[57,58] (*tZt*–PSB) **5**, a PSB that is isoelectronic to the short polyene hexa-1,3,5-triene. It has been found[57] that the first singlet excited state of PSBs is the $1B_2$-like spectroscopic state (corresponding to a *single* HOMO→LUMO excitation), in agreement with previous computational results.[59] In contrast, in polyenes, this state crosses the covalent $2A_1$ (corresponding to a *double* HOMO→LUMO excitation) state near the Franck–Condon region and becomes the second singlet excited state,[21,60] as previously seen for *s-cis* buta-1,3-diene.

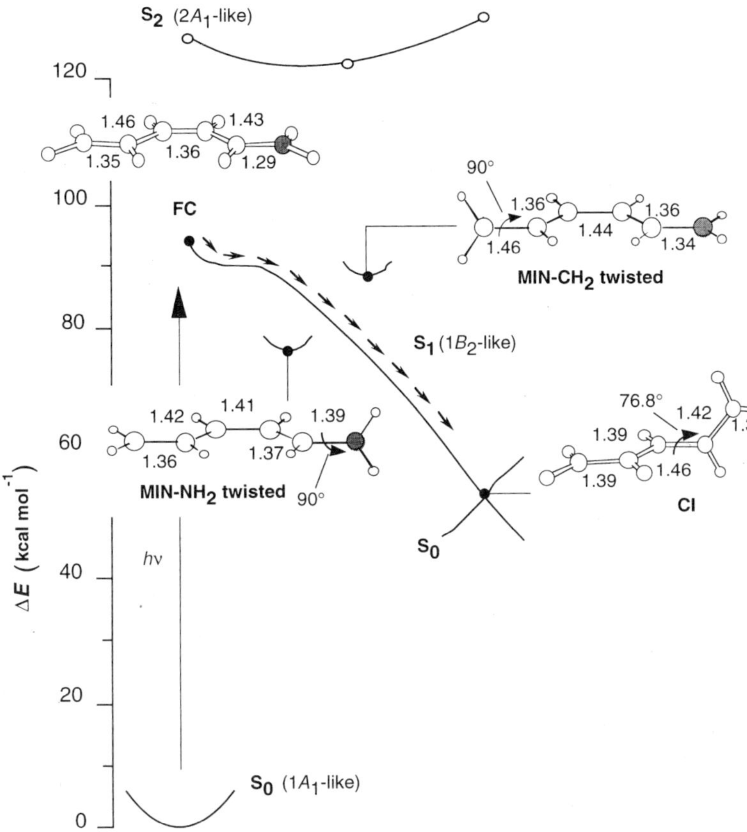

Figure 6.20. Detailed structures and energetics of the S_2 and S_1 energy profiles along the S_1 relaxation path for *cZc* penta-3,5-dieniminium cation. The stream of arrows indicates the S_1 relaxation path from the photoexcited reactant (**FC**) towards the conical intersection decay channel (**CI**). **MIN-CH$_2$** and **MIN-NH$_2$** correspond to two S_1 excited-state twisted stationary points. The values of the relevant structural parameters are given in Å and degrees.

Therefore, in PSBs, the important excited-state conformational structures and processes are those which develop along the ionic $1B_2$-like state (S_1). In these systems, the $2A_1$-like doubly excited state is S_2 (notice that in polyenes the S_1 and S_2 labels have the opposite assignment.)

The S_1 potential-energy surface of **5** comprises three *possible* conformational interconversion paths, which develop through the relaxation out from the FC region. Two adiabatic conformational interconversion paths lead to two excited-state minima, the first of which features a 90° twisted terminal methylene group and the other a 90° twisted amino group (Fig. 6.20). In both minima, the S_0–S_1 energy gap is large (ca. 28 kcal mol^{-1}). As a consequence, these Z/E diastereoisomers *cannot* provide efficient radiationless decay channels in this species. The third path is nonadiabatic and is dominated by a Z-to-E conversion of the central double bond of the molecule. This path has been fully documented[58] via a MEP computation and corresponds to a barrierless path that leads to a 90° twisted conformation wherein the S_1 energy surface crosses the ground-state energy surface at an S_1/S_0 conical intersection. Of course, this conical intersection provides an efficient decay channel, and the *E* isomerization photoproduct accumulates on S_0. The 90° twisted conical intersection is the point of lowest energy on the S_1 energy surface, and it is several kcal mol^{-1} lower than the two S_1 twisted minima previously described. The structure of the MEP coordinate and of the conical intersection (CI in Figure 6.20) indicates that the excited-state relaxation path in this system involves a torsional change at one bond (i.e., at the central double bond). Thus, this conformational process is very different from the one computed for hexa-1,3,5-triene, in which the change involves two adjacent bonds.

The application of the NEER principle to the decay channel described for the *tZt* conformer suggests that one should get, upon photolysis, the corresponding *tEt* conformer *only.* While there are no experimental data available on this short *tZt*-PSB, it provides a model for the 11-*cis* retinal PSB chromophore of rhodopsin: the human retina's visual pigment. The results discussed suggest that in the *isolated* PSB11 there is no driving force for the previously proposed "volume-saving" isomerization mechanisms, which involve the simultaneous rotation about different bonds.[61] Indeed, the so called Hula-twist model[61] for the photoisomerization of rhodopsin, which involves the photoisomerization of two adjacent bonds along the conjugated chain, is, as pointed out by Fuss,[51] very close to the computed conformational path of polyenic hydrocarbons, but not to that of PSBs. In conclusion, the nature of the first excited state in conjugated systems determines the type of conformational changes that lead to efficient decay and, eventually, to photoproduct formation. *In doubly-excited states (such as those of conjugated hydrocarbons), a two-bond conformational distortion occurs involving two adjacent bonds. In contrast, when the first excited state corresponds to a singly excited state, a single-bond conformational change occurs at a double bond.* Furthermore, we have seen that in these systems, the photoinduced double-bond *Z*/*E* isomerization must be diastereoselective. In fact, efficient decay to the ground state must occur at the conical intersection (isomerization about the central double bond), but not at the two minima (isomerization about the two terminal double bonds). In longer PSBs, there are more double bonds in the central portion of the molecule, and thus, one must have more than one efficient conical intersection decay channel. In these molecules, a less strict diatereoselectivity is expected due to the competition between the decay channels, which have different diastereomeric structures. These expectations are supported by both computational results on the *all-trans* hepta-3,5,7-trieniminium cation[62] and experimental observations on the *all-trans* retinal PBS.[63]

4.6 Conformational Effects of Conjugated Heteroatoms: α,β-Enones

The basic difference between conjugated carbonyl compounds, polyenes, and protonated Schiff bases is the presence, in the carbonyl C=O group, of an oxygen lone pair, which gives rise to a different kind of electronic excitation: the n–π^*

excitation. Actually, this is, in general, the lowest energy allowed excitation incarbonyl compounds, including α, β-enones. Thus, the electronic structure of the S_1 state of these molecules has n–π* character, wherein two unpaired electrons reside in mutually orthogonal orbitals. One important property of the n–π* states is that they can give rise to extremely fast intersystem crossing to triplet states with a π–π* electronic structure, which in turn leads, even without the need of a sensitizer, to the efficient production of triplet-state intermediates that undergo an independent evolution with respect to the singlet excited state. The efficiency of singlet–triplet crossing in this system is due to the large value of singlet–triplet spin–orbit coupling (SOC) due to the different relative orientations of the two singly occupied orbitals of this system (i.e., the singly occupied orbital n is "changed" into the orthogonal p orbital in passing from the n–π* to the π–π* configuration).[23,24]

We will next show that recent computational results[64] indicate that large excited-state conformational changes are involved in the radiationless decay of *s-cis* (see structure **6**) and *s-trans* acrolein, the simplest α,β-enone. These changes involve the Z/E rotation of the terminal CH_2 group of acrolein, but not the *s-cis* to *s-trans* conformational change about the central single bond.

In Scheme 6.11a, we give the general mechanistic picture for the evolution of $^1(n–\pi^*)$ *s-cis*-acrolein (structure *s-cis* **F**). Along the S_1 path, the molecule undergoes a conformational change involving rotation of the terminal methylene until an S_0/S_1 conical intersection is reached and the decay to S_0 takes place. While this kind of S_1 evolution has also been described for polyenes and PSBs (see earlier), the molecular structure of the conical intersection in α,β-enones is different. The structure of this conical intersection is shown in Figure 6.21 for a *s-cis* acrolein (structure *s-cis* **J**). Notice that the terminal CH_2 group is fully twisted and faces the carbonyl oxygen atom. Furthermore, in contrast with the conical intersection structure of *s-cis* buta-1,3-diene[19–21] seen earlier (for instance, Fig. 6.6) and that of the corresponding PSB (*s-cis* prop-3-eniminium cation),[58] in which there is more than one twisted bond (however, notice that, as mentioned earlier, in longer PSBs there is only one twisted double bond), the rotation of the terminal methylene is the only torsional displacement documented in acrolein. This result is consistent with the fact that in the S_1 state the central double bond is, substantially, a fully formed double bond (ca. 1.35 Å) and therefore cannot easily rotate.

s-cis **J** *s-cis* **B**

Figure 6.21. Schematic electronic and molecular structures of the S_1/S_0 conical intersection (***s-cis J***) and the T_1/S_0 single–triplet crossing (***s-cis B***) in *s-cis* acrolein. Data from Ref. 64.

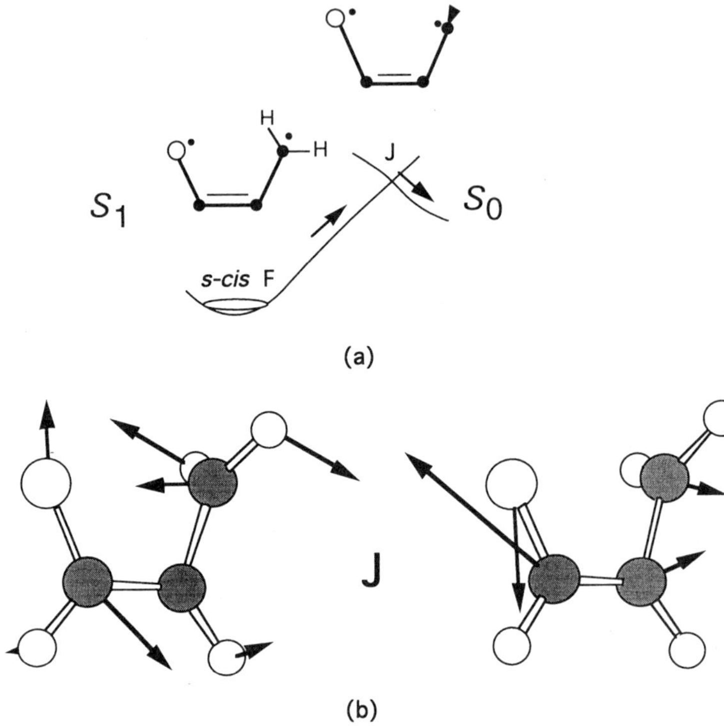

Scheme 11

The application of the NEER principle to this S_0/S_1 decay channel leads to the prediction of a simultaneous production of cyclization products (oxetenes) and Z/E isomerization products resulting from CH_2 rotation. This prediction is supported by the shape of the computed vectors, which describe two independent molecular deformations causing the S_1/S_0 energy splitting at structure **J** (i.e., moving the system geometry out from the tip of the conical intersection). According to the theory of conical intersections, there are always *two* such molecular displacements (See also Section 3.2). Their shapes, shown in Scheme 6.11b, provide qualitative information on the initial S_0 relaxation motion (see Figure 6.11 and the related discussion). The left-hand structure in Scheme 6.11b clearly describes a distortion that suggests ring-closure motion. In contrast, the right-hand structure suggests the formation of a ground-state acrolein molecule. However, since the CH_2 group is 90° twisted, the final product will display a Z or E conformation (when a substituent is present). Therefore, both Z or E conformational diastereoisomers and constitutional isomers (i.e., cyclic systems) can be generated simultaneously from this decay channel, as in the photolysis of *s-cis* buta-1,3-diene. However, as previously mentioned, and in contrast with buta-1,3-diene, the photoinduced *s-cis*-to-*s-trans* conformational change appears to be completely impaired in S_1 acrolein due to the fully formed central double bond (1.34 Å). Thus, the system starts its relaxation on

S_0 at a geometry that is unlikely to initiate a motion about the central C–C bond. On the other hand, the reason for facile oxetane production is shown in Figure 6.21. In fact, in the $^1(n\text{–}\pi^*)$ state, the oxygen has an unpaired electron that lies in the n orbital in the molecular plane. Due to the terminal methylene rotation along the S_1 path, the interaction of the two singly occupied p orbitals (the oxygen n orbital and the CH_2 p orbital), which are initially orthogonal, is strengthened and reaches a maximum at the decay point.

While it is well known that double-bond Z/E isomerization in β-substituted α,β-enones occurs at both the 313- and 254-nm wavelengths, oxetene forms in low quantum yields from 3,4-dimethyl-3-penten-2-one only when *short* wavelengths (obtained with a Vycor filter) are used.[65a,b] It has also been reported that the oxetene production is insensitive to the use of butadiene as a triplet quencher[65b] and that the reaction must therefore involve the singlet manifold only. Accordingly, the reactive singlet channel must open up only when wavelengths around 250 nm are used. According to our computational results, this channel could well correspond to point **J** in Scheme 6.11 and Figure 6.21. The intersection lies approximately 10–20 kcal mol^{-1} above the S_1 minimum. Thus, irradiation of acyclic α,β-enones with the short wavelength would provide the necessary excess energy to reach this singlet decay channel.

Butadiene irradiation at 254 nm also produces two ring-closure photoproducts: cyclobutene and bicyclobutane (see Section 2). Cyclobutene has been proved to originate exclusively from *s-cis* butadiene, whereas bicyclobutane is thought to originate from the *s-trans* conformer. While acrolein photoexcitation leads to oxetene (i.e., the four-membered ring analog of cyclobutene), Friedrich et al.[65] have never been able to detect the formation of oxabicyclobutane (i.e., the analog of bicyclobutane) during short-wavelength (Vycor filter) irradiation of different acyclic α,β-enones. This behavior can be promptly explained by comparing the conical intersection structures of *s-cis* and *s-trans* butadiene (Fig. 6.5) and acrolein (Fig. 6.21). In fact, as previously discussed, the rigid structure of the *s-cis* acrolein conical intersection strongly favors the ring-closure process to oxetene. On the other hand, the planar backbones of the *s-trans* or even *s-cis* acrolein conical intersections suggest that the formation of bicyclic products would be basically impossible. (The production of bicyclic products involves the formation of an S_0 three-membered ring diradical that then closes to the bicyclic compound.) However, this process may be possible in butadiene (see Section 2), in which the intersection structure is not planar and the reaction is less selective.

After 254-nm irradiation, only a fraction of the photoexcited α,β-enones will decay via the singlet S_1/S_0 conical intersection channel, whereas the majority of the photoreactants will follow an alternative *triplet* decay route. The lifetime of a triplet intermediate measured for a series of acyclic and cyclic α,β-enones does not show any change when the system is photoexcited with 353 or 265 nm wavelength.[66] This means that the triplet decay channel remains operative (and is probably the main decay channel) even at short-wavelength irradiation. The same trend has been detected in the double-bond Z/E isomerization quantum yield, measured by using 254 and 313 nm photoexcitation.[67]

The evolution of the initial S_1 intermediate **F** (in either the *s-cis* acrolein or *s-trans* equilibrium conformation) towards the S_1/T_2 crossing (structure **K**), where the decay to the triplet state T_2 occurs, does not involve torsional deformation. In fact, this motion is characterized by a very facile expansion of the C=O bond and compression of the C–C bond. Along this path, the C–O bond shrinks by 0.03 Å and the $C{-}CH_2$ bond increases by 0.07 Å. The T_2 state is populated by fast intersystem crossing from the initial $S_1^1(n{-}\pi^*)$ excited-state molecule to a $T_2^3(\pi{-}\pi^*)$ diradical intermediate that evolves immediately towards a T_2/T_1 conical intersection (**G**) between the $^3(\pi{-}\pi^*)$ and $^3(n{-}\pi^*)$ electronic configurations (See Scheme 6.12 for a substituted *s-trans* acrolein; this scheme is adapted from Scheme **V** in Ref. 64). Near point **G**, the system reaches the lowest $T_1^3(n{-}\pi^*)$ minimum (point **C**), which is planar and structurally similar to the S_1 minimum at point **F**. Nevertheless, this is not the lowest energy T_1 minimum. In fact, relaxation from the planar conical intersection point **G** results in the production of the lowest energy T_1 intermediate **B**, which has a $^3(n{-}\pi^*)$ electronic structure. This intermediate has a fully twisted CH_2 group and corresponds to a T_1/S_0 singlet–triplet crossing that represents the major decay channel from triplet acrolein. The top left structure in Scheme 6.12 displays the vector describing the molecular deformation required to lift the degeneracy (i.e., to enter the S_0 state). It is clear that the torsion of the terminal CH_2 group represents one important component of this mode, implying Z/E isomerization. Notice that in this structure the C–O bond corresponds to a fully formed double bond and the two uncoupled electrons lie in two orthogonal *p* orbitals of the terminal vinyl moiety. The fact that the T_1 and the S_0 states have the same electronic structure implies a small magnitude and efficiency of the SOC. The singlet–triplet crossing rate at **B** is therefore predicted to be slow.

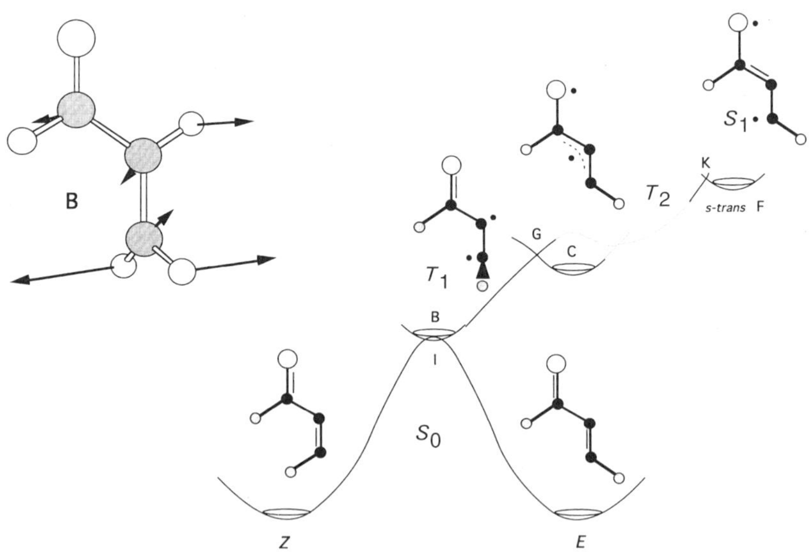

Scheme 6.12

The existence of a twisted excited-state minimum $^3(\pi-\pi^*)$ **B** is of considerable importance for interpreting the very weak phosphorescence and short triplet lifetimes (from transient absorption spectroscopy and time-resolved photoacoustic calorimetry data)[66,67] observed in acyclic and cyclic α,β-enones. However, in order to assign the observed phosphorescence to the twisted triplet intermediate, one must assume a small **C**-to-**B** energy barrier on the T_1 reaction path and therefore a rapid rotation from planar to 90° twisted structure. The experimental data seem to correlate well with this idea. For example, the triplet transient lifetime of rigid versus nonrigid α,β-enones, obtained along a series of molecules wherein the acrolein moiety is embedded in rings of decreasing size, increases when the rigidity of the system is increased. This fact can be explained on the basis of Scheme 6.13. adapted from Scheme **III** in Ref. 64.)

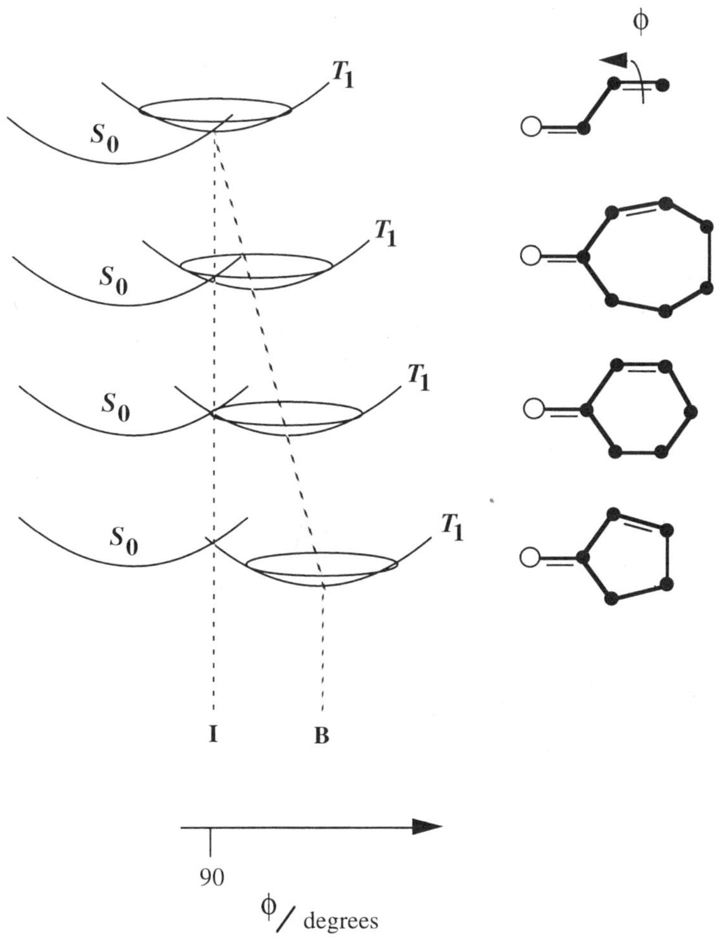

Scheme 6.13

Because of the increase in the rigidity of the system, the **B** equilibrium geometry must shift toward less twisted structures ($< 90°$). In this way, the bottom of the **B** well is removed from the efficient intersystem crossing region coinciding with point **I** that occurs at a fully twisted geometry ($f = 90°$). As a consequence, the lifetime of the (twisted) triplet intermediate corresponding to **B** increases from top to bottom for the four cases illustrated in Scheme 6.13. Thus, the observed short-lived triplet transient absorption at 280–310 nm should be assigned to the twisted $^3(\pi\text{–}\pi^*)$ structure corresponding to **B**, in agreement with the hypothesis of Bonneau[66] and Schuster[67]. However, our computational data also suggest that the system should initially decay to the $^3(n\text{–}\pi^*)$ equilibrium structure **C**, which then interconverts to **B**. Since it is likely that the barrier to this interconversion increases in rigid cyclic α,β-enones, where twisting motion is hindered, we might expect an increase in lifetime for a planar $^3(n\text{–}\pi^*)$ intermediate and, in the limit of very rigid systems, the most stable T_1 minimum. This prediction is in agreement with the hypothesis advanced by Schuster[67b] that the $^3(n\text{–}\pi^*)$ might be the lowest energy triplet state in bicyclo[4.3.0]non-1(6)-en-2-one.

The twisted **B** minimum is thus the most important triplet intermediate along the triplet decay route that has been described. Once the photoexcited system has reached **B**, it can decay to the ground state, going back to the planar form and thus, in β-substituted α,β-enones, leading to double-bond isomerization products exclusively.

4.7 π versus σ Conjugation: Oligosilanes

In Sections 4.1–4.4, we dealt with polyenes and other compounds that exhibit an extensive π delocalization. The direct consequence of the π delocalization is that the photoexcitation leads to an electronically excited state that displays the properties we have discussed including the presence of easily accessible radiationless decay channels, the special $-(CH)_3-$ kink structure of the decay channels, and the fact that the major photochemical events are the reconstitution of the π-system in the form of *s-cis–s-trans* and Z/E conformational stereoisomers or the depletion of the π-system, resulting in cyclizations (e.g., the electrocyclization in *s-cis* buta-1,3-diene and *cZc*-hexa-1,3,5-triene).

In recent years, much progress has been made in the investigation of saturated systems exhibiting σ delocalization. Probably the best known example of saturated structures with extensive σ delocalization are the polysilanes (i.e., saturated polymers with a silicon backbone). Remarkably, because of the weakness of the σ Si–Si bonding in these molecules, photoexcitation to the $\sigma\text{–}\sigma^*$ excited state requires near-UV radiation, similarly to that employed in the π-system of conjugated hydrocarbons.

A series of detailed computational and experimental results on the chemistry and photochemistry of permethylated oligosilanes (Si_nMe_{2n+2}, $n = 3, 7$) was recently reported by Michl and co-workers.[68] The studies indicate that the ground state of these molecules can be described by a mixture of conformers that differ in their torsional angle about one or more Si–Si bonds. In Si_4Me_{10}, there are three basic

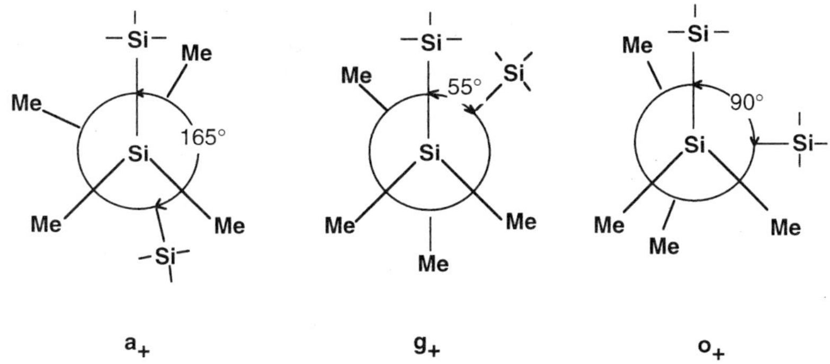

Scheme 6.14

conformers. As illustrated in Scheme 6.14, these are, in order of increasing energy, an *anti* conformer (a_+) a *gauche* conformer (g_+), and an *ortho* conformer (o_+). Each conformer is enantiomeric and exists as a racemic mixture with its corresponding enantiomer (a_-, g_-, o_-). Of course, in longer oligosilanes, one has conformers that are combinations of these three basic structures. For instance, the most stable conformer in Si_6Me_{14} (structure **7**) has an *all-trans* ($a_+a_+a_+$, $a_-a_-a_-$) geometry, as supported by the analysis of its absorption spectra.

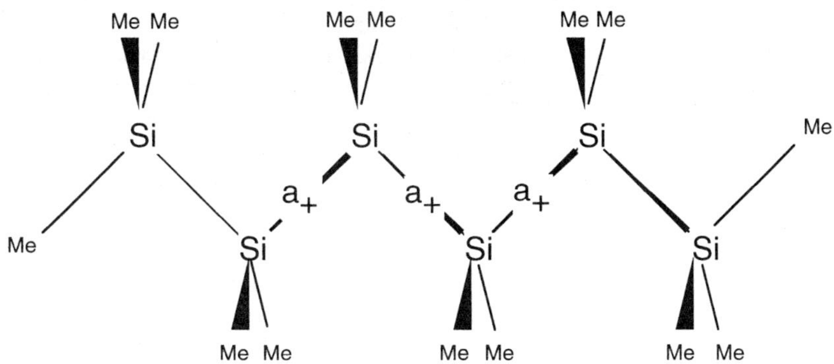

7

The shorter oligosilanes Si_2Me_6 and Si_3Me_8 do not fluoresce; thus, they must have an extremely short excited-state lifetime. This fact suggests, as in the short polyenes buta-1,3-diene and hexa-1,3,5-triene (which show negligible or extremely weak fluorescence), the existence of a low-lying conical intersection decay channel located along a barrierless excited-state path (Figs. 6.4 and 6.8). The longer oligosilanes Si_4Me_{10}, Si_5Me_{12}, Si_6Me_{14} and Si_7Me_{16} fluoresce, and this suggests, again, a behavior similar to that of longer polyenes such as octa-1,3,5,7-tetraene, where the decay is controlled by an energy barrier (Fig. 6.11). Since the short oligosilanes undergo a σ–σ^* excitation, rather than a π–π^* excitation, their relaxed S_1 structures are expected to show expanded Si–Si bonds instead of expanded double bonds. Indeed, a computational investigation of the excited-state structure in Si_4Me_{10} has predicted a central bond that is 0.2 Å longer than the usual Si–Si bond (2.35 Å).[69] In other words, photoexcitation of short oligosilanes leads to a *localized* electronic excitation. Since the expanded Si–Si bond may be located in different positions along the oligosilane chain, one has the possibility of "excitation localization isomerism." It has been proposed that this form of isomerism is at the basis of the dual fluorescence observed[68c] in Si_6Me_{14}. Michl et al.[68] have provided evidence that in longer oligosilanes, such as Si_8Me_{18}, $Si_{10}Me_{22}$ and $Si_{16}Me_{34}$, the character of the electronic excitation changes. In these systems, one has a *delocalized* excitation, and the corresponding excited-state equilibrium structure does not seem to differ much from the ground-state equilibrium structure. The short Si_7Me_{16} system provides the borderline between the two different behaviors and displays both "localized" and "delocalized" excited-state structures.[68d]

Information on the excited-state evolution of short oligosilanes with a localized excitation comes from the investigation of the temperature dependence of their dual-fluorescence quantum yield. For Si_6Me_{14}, a qualitative mechanism has been proposed that accommodates the experimental data.[68c] As mentioned earlier, the dual fluorescence would originate from two fluorescent species corresponding to two distinct locally excited isomers [i.e., two excited-state intermediates—say, Is_1 and Is_2—that differ in the location of the stretched bond (e.g., either a terminal or a central bond)]. At low temperatures, one of the two isomers would fluoresce with a high quantum yield. However, as the temperature is raised, the fluorescence quantum yield decreases. This fact has been attributed to the opening of an efficient radiationless decay channel that is controlled by an energy barrier. As mentioned earlier, this decay channel would correspond to a conical intersection between the first singlet excited state and the ground state of the system. As the temperature is raised further, fluorescence from the alternative conformer (Is_2) is observed, and the fluorescence of the first isomer (Is_1) drops dramatically as a consequence of irreversible population of the Is_2 excited-state energy minimum, possibly a more stable isomer. At a temperature of about 100 K, the fluorescence from Is_2 peaks while fluorescence from the Is_1 stops. Finally, a further increase in the temperature leads to a decrease in the fluorescence from the second species (which cannot be measured accurately at a temperature above 180 K) due to the opening of a second, activated, radiationless decay channel. Again, this channel would correspond to an isomeric conical intersection. This mechanistic scheme proposed for short oligosilanes resembles, qualitatively, the one previously illustrated for *cZc*-hexa-1,3,5-trienes (Fig. 6.16), with the only exception that the generally low fluorescence observed in *cZc*-hexa-1,3,5-trienes is due to the small barriers controlling the competing conical intersection channels. Furthermore, in *cZc*-hexa-1,3,5-trienes, the conical intersections corresponding to Path **II** and Path **III**, in Figure 6.16 are conformational stereoisomers, while in silanes there are two isomers that differ in the location of one stretched bond.

Chain abridgement by 1,1-elimination

-SiRR'-SiRR'-SiRR'- ⟶ -SiRR'-SiRR'- + $\ddot{\mathrm{S}}$iRR'

Homolytic chain scission

-SiRR'-SiRR'-SiRR'- ⟶ -$\dot{\mathrm{S}}$iRR' + $\dot{\mathrm{S}}$iRR'-SiRR'-

The nature of the conical intersection channels operative in oligosilanes has been clarified by a computational investigation[70] on 2,2′-dimethyltrisilane ($SiH_3SiMe_2SiH_3$), a simplified model of Si_3Me_8. A "line" of infinite conical intersection points (i.e., points where the S_0 and S_1 potential-energy surfaces are

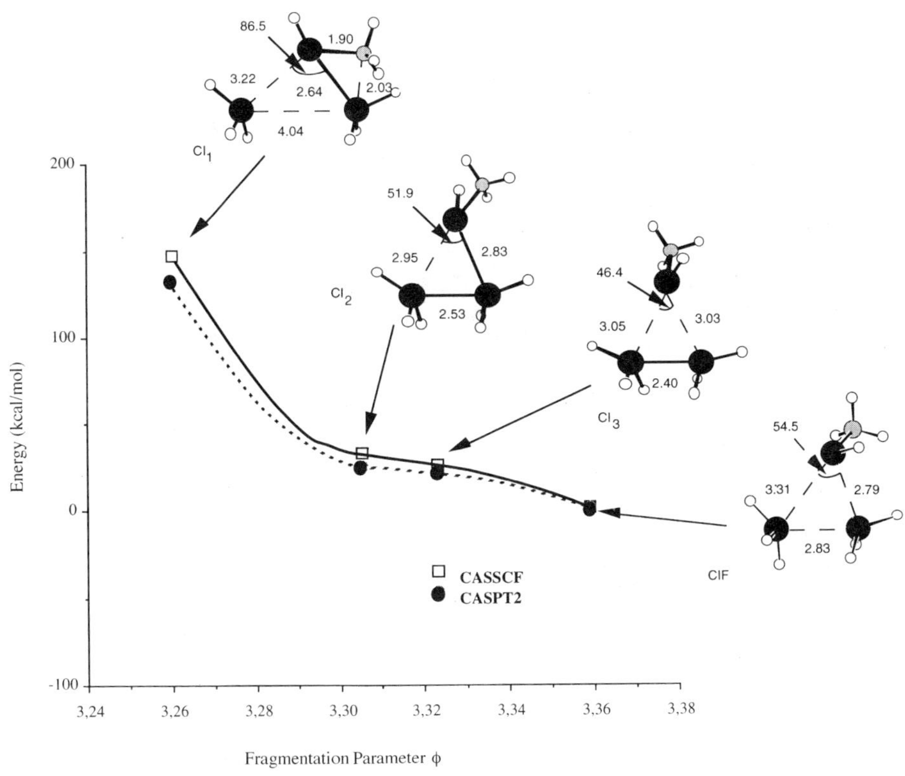

Figure 6.22. CAS-SCF/6-31G* and multirefence MP2 (PT2) (dashed line) energies versus the fragmentation parameter ϕ for the four optimized conical intersections CI_1, CI_2, CI_3 and CIF located on the first singlet excited state of 2-methyl-trisilane. (The main structural parameters are reported in Å and degrees.) The fragmentation parameter is the weighted average of the heavy-atom distances (Si and C) (See Ref. 70 for a detailed definition.) Adapted from Figure 1 of Ref. 70.

degenerate) has been located at the bottom of the S_1 potential-energy surface. A portion of that line, shown in Figure 6.22, corresponds to a region of the intersection space already mentioned in Section 3.2. The conical intersection structures located along the line suggest that the radiationless decay channel in oligosilanes is actually a photofragmentation channel (i.e., a photochemical channel). No excited-state equilibrium structure seems to exist for this species, consistent with the lack of fluorescence observed for Si_3Me_8. The type of photoproducts that can be obtained by ground-state relaxation from the low-energy conical intersection structures (see the low-energy portion of the intersection line shown in Figure 6.22) are fragmented products such as silyl radical pairs. A more complex and experimentally known process, consisting of the chain abridgement of two chain moieties and the release of a silylene particle (the silicon analog of a methylene) can also be predicted by the lowest energy conical intersection structure (see Scheme 6.15). According to this structure, the chain abridgement may occur with "retention of configuration."

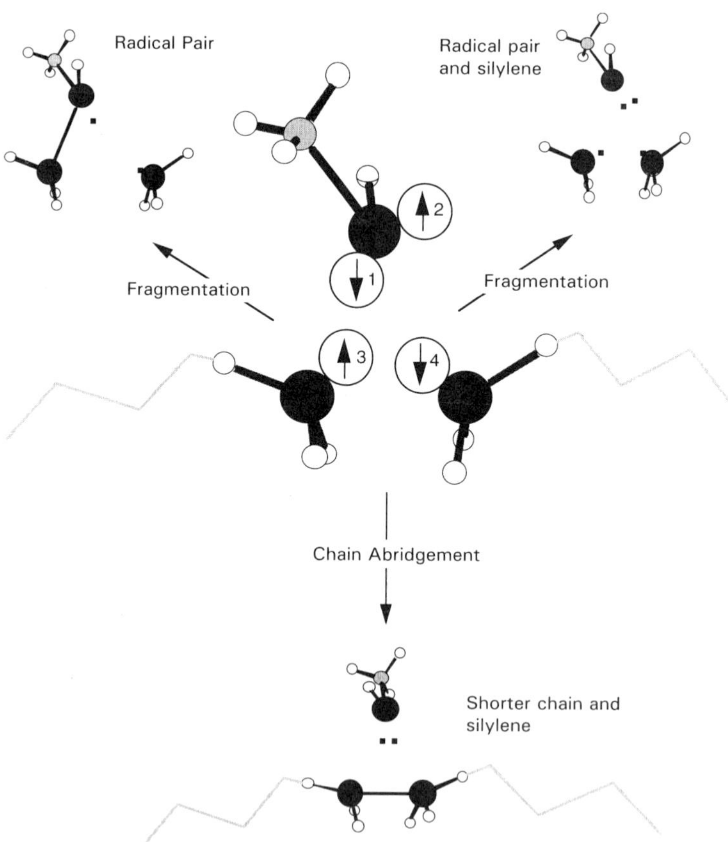

Scheme 15

As illustrated in Scheme 6.15, the lowest lying conical intersection of 2-methyltrisilane shows four weakly interacting electrons. Such an electronic structure suggests an analogy with the four weakly unpaired electrons of polyene conical intersections (see Sections 4.1–4.4, which are located on the $-(CH)_3-$ kink and on the associated delocalized radical fragment. It is thus clear that the silane chain photofragmentation may be seen as the analog of polyene photoisomerizations, with the difference that in the first case no molecular skeleton holds the radical fragments together. Thus, the electron-recoupling process occurring during S_0 relaxation is the source of radical fragments (generated via a recoupling of electrons $1 \leftrightarrow 3$ only), silylene intermediates and a shorter chain (via a recoupling of electrons $3 \leftrightarrow 4$ and $1 \leftrightarrow 2$), or the silylene intermediate and a radical pair (via a recoupling of electrons $1 \leftrightarrow 2$ only).

Finally, notice that the character and structure of the 2-methyltrisilane conical intersection change upon passing from lower to higher energy regions of the intersection space given in Figure 6.22. In particular, at high energy, one has a conical intersection structure, which suggests Si–C bond breaking with the production of methyl radicals. This type of process has also been observed experimentally (see citations in Ref. 70).

5 Conclusions

The computational study of the mechanisms of organic photochemical reactions presents a considerable challenge to computational chemistry. The objective is a complete description of the reaction path from the photoexcited ground-state equilibrium structure (the so-called Franck–Condon point) to the ground-state products. Within this scope, the investigation of the relationship between the excited-state conformational properties and the photochemical outcome acquires a special importance: The chemoselectivity, the regioselectivity and, finally, the stereoselectivity of a photochemical process are highly affected by the photochemically relevant excited-state conformation. In this chapter, *we have provided evidence that this causal relationship corresponds to the conformation of the accessible decay channel (i.e., a conical intersection or a singlet–triplet crossing)*. Remarkably, the shape of this photochemically relevant conformation is determined, in several cases, by the reactant's initial ground-state conformation and by the following relaxation along the excited-state energy surface. In fact, we have also shown that the empirical idea of the nonequilibration of the excited-state rotamers—the NEER principle—formulated by Jacobs and Havinga (see Section 2.1) can now be reformulated on the basis of state-of-the-art quantum chemical computations. We have seen that the computed structure and conformation of the ground-state reactant are related to the structure and conformation of certain specific decay channels (i.e., an intrinsically transient excited-state species). In turn, the geometry of these decay channels can be easily related to the structure of the observed primary photoproducts.

It seems useful to give a qualitative overview of the origin of the selectivity in photochemical reactions. *The dominant ground-state conformation of a reactant determines the structure and conformation of the major decay channel.* Thus, the photolysis of the *s-cis* buta-1,3-diene conformer leads to the *s-cisoid* conical intersection, while the photolysis of the corresponding *s-trans* conformer leads to the *s-transoid* conical intersection. Similarly, the photolysis of *all-trans* hexa-1,3,5-triene leads to the open-chain conical intersection seen in Figure 6.15a, while the photolysis of the curly *cZc* hexa-1,3,5-triene leads to the almost cyclic conical intersection of Figure 6.18. For α,β-enones, the irradiation of *s-cis* acrolein leads to the *s-cis* conical intersection **J** of Figure 6.21 or to the *s-cis* singlet–triplet crossing **B** of the same figure. The corresponding *s-trans* decay channels can be populated only by starting from the corresponding ground-state conformer. Thus, in general, the interconversion of *competing* decay channel conformers is impossible, due to their transient nature. On the other hand, the interconversion of the distinct excited-state conformers that evolve towards the competing decay channels is impaired by large energy barriers. In this manner, the conformationally diverse (photochemical) decay paths are isolated.

The "photochemical" selectivity operates on three levels. The first level is related to the dominant ground-state conformation. For instance, cyclobutene can be produced only by the irradiation of *s-cis* buta-1,3-diene, and cyclohexa-1,3-diene can be produced only from *cZc* hexa-1,3,5-triene irradiation. This discriminates two different chemical reactions (cyclization versus Z/E isomerization). A second level of selectivity operates during the excited-state evolution of the photoexcited conformer. In fact, *different, competing decay channels, often controlled by different excited-state energy barriers, may be accessible after photoexcitation of a specific conformer of the reactant.* Examples are given by the conrotatory and disrotatory decay channels of *s-cis* buta-1,3-diene (Fig. 6.8), by the three decay channels seen in Figure 6.16 for *cZc* hexa-1,3,5-triene, by the two decay channels of *all-trans* hexa-1,3,5-triene given in Figures 6.15a and 6.15b, and by the three decay channels of Figures 6.15e, 6.15f, and 6.15i for *all-trans* octa-1,3,5,7-tetraene. Two different examples are that of the $^1(n\text{–}\pi^*)$ acrolein, which may evolve towards a conical intersection (leading to cyclization and isomerization) or a singlet–triplet decay channel (leading, exclusively, to isomerization) and that of the competing S_1 Z/E isomerization channels of longer PSBs mentioned in Section 4.5. Thus, this second level of selectivity discriminates between different diastereoisomers (Z/E isomerization) and constitutional isomers (cyclization versus Z/E isomerization).

A third level of selectivity operates soon after the decay from a specific decay channel. In fact, *different photoproducts can be generated via relaxation along different relaxation paths originated by the same conical intersection conformer.* This type of relaxation path is a function of the structure of the specific decay channel conformer. Some of these paths have been mapped and are shown in Figure 6.6 for the buta-1,3-diene *s-cisoid* conical intersection conformer and in Figure 6.18 for the curly conical intersection conformer of *cZc*-hexa-1,3,5-triene. These relaxation processes include competing cyclization processes that differ in the position of the first σ bond formed after decay: the cyclobutene–cyclopropylcarbinyl diradical formation processes in *s-cis* buta-1,3-diene, and the cyclohexa-1,3-diene–cyclopentenylcarbinyl diradical formation processes in *cZc*-hexa-1,3,5-triene. Hence, the third level of selectivity seems to control the regioselectivity of certain photochemical primary events.

As a concluding remark, we point out that the majority of the information given in this chapter is based upon the mapping of the ground- and excited-state energy surfaces and on the interpretation of the chemical event in terms of reaction paths. Of course, this represents a *static* view of the reactive process. A more realistic view would be provided by the investigation of the actual molecular motion of the photoexcited conformer. One example is afforded by the failure of the static picture to explain the efficiency of the photochemical *s-cis*→*s-trans* conversion in buta-1,3-diene mentioned in Section 2.2. This type of study, which requires semiclassical or quantum dynamics computations on sizeable systems, represents the modern frontier of computational quantum chemistry applied to organic and bioorganic thermal and photochemical reactivity. While several interesting results are being achieved in this field,[71] a major research effort needs to be undertaken.

References

1. Leach, A.R., *Molecular Modelling* Longman, Singapore, **1996**.
2. Kendall, R.A., Harrison, R.J., Littlefield, R.J., Guest, M.F. In *Reviews in Computational Chemistry*, Vol. VI, Lipkowitz, K.J. and Boyd, D.B., eds., Wiley–VCH: New York, **1995**, 209–301.
3. (a) Schlegel, H.B. In *Adv. Chem. Phys. Volume 67 (Ab Initio Methods in Quantum Chemistry*—I), Lawley, K.P. ed., John Wiley & Sons, Ltd.: New York, **1987**, 249–286. (b) Foresman, J.B.; Frisch, A. In *Exploring Chemistry with Electronic Structure Methods*, Gaussian, Inc.: Pittsburgh, **1996**, 39–60 and references cited therein. (c) Schlick, T. In *Reviews in Computational Chemistry*, Vol III, Lipkowiz, K.B. and Boyd, D.B., eds., Wiley–VCH: New York, **1993**, 1–71.
4. (a) McKee, M.L.; Page M. In *Reviews in Computational Chemistry*, Vol IV, Lipkowiz, K.B. and Boyd, D.B., eds., Wiley–VCH: New York, **1993**, 35–65. (b) Truhlar, D.G.; Steckler, R.; Gordon, M.S. *Chem. Rev.*, **1987**, *1*, 217; (c) Truhlar, D.G.; Gordon M.S. *Science*, **1990**, *249*, 491.
5. Gilbert, A.; Baggott, J. *Essentials of Molecular Photochemistry*, Blackwell Scientific Publications: Oxford, **1991**.
6. Frisch, M.J.; Trucks, G.W.; Schlegel, H.B.; Gill, P.M.W.; Johnson, B.G.; Robb, M.A.; Cheeseman, J.R.; Keith, T.; Petersson, G.A.; Montgomery, J.A.; Raghavachari, K.; Al-Laham, M.A.; Zakrzewski, V.G.; Ortiz, J.V.; Foresman, J.B.; Cioslowski, J.; Stefanov, B.B.; Nanayakkara, A.; Challacombe, M.; Peng, C.Y.; Ayala, P.Y.; Chen, W.; Wong, M.W.; Andres, J.L.; Replogle, E.S.; Gomperts, R.; Martin, R.L.; Fox, D.J.; Binkley, J.S.; Defrees, D.J.; Baker, J.; Stewart, J.P.; Head-Gordon, M.; Gonzalez, C.; Pople, J.A. *Gaussian 94* Gaussian, Inc.: Pittsburgh, PA, **1995**.
7. Andersson, K.; Blomberg, M.R.A.; Fülscher, M.P.; Karlström, G.; Lindh, R.; Malmqvist, P.-Å.; Neogrády, P.; Olsen, J.; Roos, B.O.; Sadlej, A.J.; Schütz, M.; Seijo, L.; Serrano-Andrés, L.; Siegbahn, P.E.M., and Widmark, P.O. *Molcas, version 4*, Lund University: Sweden, **1997**.
8. Dachsel, H.; Shepard, R.; Nieplocha, J.; Harrison, R.J. *J. Comput. Chem.* **1997**, *18*, 430. (A massively parallel multireference configuration interaction program: The parallel Columbus program.)

9. Schmidt, M.W.; Baldridge, K.K.; Boatz, J.A.; Elbert, S.T.; Gordon, M.S.; Jensen, J.J.; Koseki, S.; Matsunaga, N.; Nguyen, K.A.; Su, S.; Windus, T.L.; Dupuis, M.; Montgomery, J.A. *J. Comput. Chem.* **1993**. *14*, 1347. (General atomic and molecular electronic-structure system.)

10. Boyd, D.B. In *Reviews in Computational Chemistry*, Vol. XI, Lipkowitz, K.B. and Boyd, D.B., eds., Wiley–VCH: New York, **1997**, pp. 373–399. (Appendix: Compendium of Software and Internet Tools for Computational Chemistry.)

11. Bartlett, R.J.; Stanton, J.F. In *Reviews in Computational Chemistry*, Vol. V, Lipkowitz K.B., and Boyd, D.B., eds., Wiley–VCH: New York, **1997**, pp. 65–169.

12. Klessinger, M.; Michl, J. *Excited States and Photochemistry of Organic Molecules*, Wiley–VCH: New York, **1994**.

13. Engeln, R.; Consalvo, D.; Reuss, J. *Chem. Phys.*, **1992**, *160*, 427–433.

14. Jacobs, H.J.C.; Havinga, E. *Adv. Photochem.* **1979**, *11*, 305.

15. Vroegop, P.J.; Lugtenburg, J.; Havinga, E. *Tetrahedron*, **1973**, *29*, 1393.

16. Leigh, W.J. In *CRC Handbook of Organic Photochemistry and Photobiology*, Horspool, W.M., and Song, P.-S., ed., SCRC Press: Boca Raton, FJ, **1994**, 123–142.

17. (a) Squillacote, M.E.; Semple T.C.; Mui, P.W. *J. Am. Chem. Soc.*, **1985**, *107*, 6842; (b) Squillacote, M.E.; Sheridan, R.S.; Chapman O.L.; Anet, F.A.L. *J. Am. Chem. Soc.*, **1979**, *101*, 3657.

18. Squillacote, M.; Semple T.C. *J. Am. Chem. Soc.*, **1990**, *112*, 5546.

19. Olivucci, M.; Ragazos, I.N.; Bernardi, F.; Robb, M.A. *J. Am. Chem. Soc.*, **1993**, *115*, 3710–3721.

20. Olivucci, M.; Bernardi, F.; Ottani, S.; Robb, M.A. *J. Am. Chem. Soc.*, **1994**, *116*, 2034.

21. Celani, P.; Bernardi, F.; Olivucci, M.; Robb, M.A. *J. Chem. Phys.*, **1995**, *102*, 5733–5742.

22. (a) Bernardi, F.; Olivucci, M.; Robb, M.A. *Chem. Soc. Rev.*, **1996**, *25*, 321–328. (b) Bernardi, F.; Olivucci, M.; Michl, J.; Robb, M.A. *The Spectrum*, **1996**, *9*, 1–6.

23. Klessinger, M. *Angew. Chem. Int. Ed. Engl.*, **1995**, *34*, 549–551.

24. Michl, J.; Bonačić-Koutecký, V. *Electronic Aspects of Organic Photochemistry*, Wiley:New York, **1990**.

25. Celani, P.; Robb, M.A.; Garavelli, M.: Bernardi, F.; Olivucci, M. *Chem. Phys. Lett.*, **1995**, *243*, 1–8.

26. Nguyen, K.A.; Gordon, M.S. *J. Am. Chem. Soc.*, **1995**, *117*, 3835.

27. Leigh, W.J.; Postigo, J.A. *Chem Comm.*, **1993**, 1836.

28. Aue, D.H.; Reynolds, R.N. *J. Am. Chem. Soc.*, **1973**, *95*, 2027.

29. Leigh, W.J.; Zeng, K. *J. Am. Chem. Soc.*, **1991**, *113*, 4019; Errata: Ibid. **1992**, *114*, 796.

30. (a) Woodward, R.B.; and Hoffmann, R. *Angew. Chem., Int. Ed. Engl.*, **1969**, *8*, 781. (b) Van der Lugt, W.T.A.M.; Oosteroff, L.J. *J. Am. Chem. Soc.*, **1969**, *91*, 6042. (c) Dauben, W.G.; Salem, L.; Turro, N.J. *Acc. Chem. Res.*, **1975**, *8*, 41.

31. Leigh, W.J.; Zeng, K. *J. Am. Chem. Soc.*, **1991**, *113*, 2163.

32. Bernardi, F.; Olivucci, M.; Robb, M.A. *J. Am. Chem. Soc.*, **1992**, *114*, 2752–2754.

33. Trulson, M.O.; Mathies, R.A. *J. Phys. Chem.*, **1990**, *94*, 5741.

34. (a) Vaida, V. *Acc. Chem. Res.*, **1986**, *19*, 114. (b) Leopold, D.G.; Pendley, R.D.; Roebber, J.L.; Hemley R.J.; Vaida, V. *J. Chem. Phys.*, **1984**, *81*, 4218.

35. (a) Petek, H.; Bell, A.J.; Christensen, R.L.; Yoshiara, K. *SPIE*, **1992**, *1638*, 345–356. (b) Petek, H.; Bell, A.J.; Choi, Y.S.; Yoshiara, K.; Tounge, B.A.; Christensen, R.L. *J. Chem. Phys.*, **1993**, *98*, 3777–3794.

36. Orlandi, G.; Zerbetto, F.; Zgierski, M.Z. *Chem. Rev.*, **1991**, *91*, 867–891.

37. Saltiel, J. *J. Am. Chem. Soc.*, **1967**, *89*, 1036.

38. (a) Saltiel, J.; Metts, L.; Wrighton, M. *J. Am. Chem. Soc.*, **1969**, *91*, 5684. (b) Saltiel, J.; Metts, L.; Sykes, A.; Wrighton, M. *J. Am. Chem. Soc.*, **1971**, *93*, 5302.

39. Roos, B.O. *Adv. Chem. Phys.*, **1987**, *69*, 399–446.

40. Kozlowski, P.M.; Dupuis, M.; Davidson, E.R. *J. Am. Chem. Soc.*, **1995**, *117*, 774.

41. McDouall, J.J.W.; Peasley, K.; Robb, M.A. *Chem. Phys. Lett.*, **1988**, *148*, 183.

42. Andersson, K., Malmqvist, P.-A.; Ross, B.O. *J. Chem. Phys.*, **1992**, *96*, 1218. [The CAS-PT2F method is implemented in *Molcas*. (See Ref. 7.)]

43. Gonzalez, C.; Schlegel, H.B. *J. Phys. Chem.*, **1990**, *94*, 5523–5527.

44. Ragazos, I.N.; Robb, M.A.; Bernardi, F.; Olivucci, M. *Chem. Phys. Lett.*, **1992**, *197*, 217.

45. Bearpark, M.J.; Robb, M.A.; Schlegel, H.B. *Chem. Phys. Lett.*, **1994**, *223*, 269.

46. Grimbert, D.; Segal, G.; Devaquet, A. *J. Am. Chem. Soc.*, **1975**, *97*, 6629.

47. Celani, P.; Garavelli, M.; Ottani, S.; Bernardi, F.; Robb, M. A.; Olivucci, M. *J. Am. Chem. Soc.*, **1995**, *117*, 11584–11585.

48. Hayden, C.C.; Chandler, D.W. *J. Phys. Chem.* **1995**, *99*, 7897–7903.

49. Garavelli, M.; Celani, P.; Bernardi, F.; Robb, M.A.; Olivucci, M. *J. Am. Chem. Soc.*, **1997**, *119*, 11487–11454.

50. Garavelli, M.; Celani, P.; Yamamoto, N.; Bernardi, F.; Robb, M.A.; Olivucci, M. *J. Am. Chem. Soc.*, **1996**, *118*, 11656–11657.

51. Muller, A.M.; Lochbrunner, S.; Schmid, W.E.; Fuss, W. *Ang. Chem. Int. Ed. Eng.*, **1998**, *37*, 505–507.

52. unpublished data.

53. Garavelli, M.; Celani, P.; Fato, M.; Olivucci, M.; Robb, M.A. *J. Phys. Chem.:* A, **1997**, *101*, 2023.

54. Palmer, I.J.; Ragazos, I.N.; Bernardi, F.; Olivucci, M.; Robb, M.A. *J. Am. Chem. Soc.*, **1993**, *115*, 673–682.

55. (a) unpublished data. (b) Turro, N.J.; Liu, J.-M.; Zimmerman, H.E.; Factor, R. E. *J. Org. Chem.*, **1980**, *45*, 3511–3512.

56. (a) Bearpark, M.J.; Bernardi, F.; Clifford, S.; Olivucci, M.; Robb, M.A.; Vreven, T. *Mol. Phys.*, **1996**, *89*, 37. (b) Bernardi, F.; Olivucci, M.; Robb, M.A. *J. Am. Chem. Soc.*, **1992**, *114*, 1606–1616.

57. Garavelli, M.; Celani, P.; Bernardi, F.; Robb, M.A.; Olivucci, M. *J. Am. Chem. Soc.*, **1997**, *119*, 6891–6901.

58. Garavelli, M.; Bernardi, F.; Celani, P.; Robb, M.A.; Olivucci, M. *J. Photochem. Photobiol. A: Chem.*, **1998**, *114*, 109–116.

59. (a) Du, P.; Davidson, E.R. *J. Phys. Chem.*, **1990**, *94*, 7013–7020. (b) Du, P.; Racine, C.; Davidson, E.R. *J. Phys. Chem.*, **1990**, *94*, 3944–3951. (c) Martin, C.H. *J. Phys. Chem.*, **1996**, *100*, 14310–14315. (d) Martin, C.H. *Chem. Phys. Lett.*, **1996**, *257*, 229–237.

60. Serrano-Andrés, L. Roos, B.O.; Merchán, M. *Theor. Chim. Acta*, **1994**, *87*, 387.

61. This is the hula-twist (Liu, R.S.H.; Browne, D.T. *Acc. Chem. Res.*, **1986**, *19*, 42–48) mechanism.

62. Garavelli, M.; Bernardi, F. Olivucci, M.; Vreven, T.; Klein, S.; Celani, P.; Robb, M.A. *Faraday Discussion*, **1998**, *110*, 51.

63. Freedman, K.A.; Becker, R.S. *J. Am. Chem. Soc.*, **1986**, *108*, 1245–1251.

64. Reguero, M.; Olivucci, M.; Bernardi, F.; Robb, M.A. *J. Am. Chem. Soc.*, **1994**, *116*, 2103.

65. (a) Friedrich L.E.; Schuster G.B. *J. Am. Chem. Soc.*, **1969**, *91*, 7204–7205. (b) Friedrich L.E.; Schuster G.B. *J. Am. Chem. Soc.*, **1972**, *94*, 1193–1199.

66. Bonneau R. *J. Am. Chem. Soc.*, **1980**, *102*, 3816–3822.

67. (a) Schuster D.I., Dunn D.A., Heibel G.E., Brown P.B., Rao J.M., Woning J.; Bonneau R. *J. Am. Chem. Soc.*, **1991**, *113*, 6245–6255. (b) Schuster, D.I.; Woning, J.; Kaprinidis, N.A.; Pan, Y.; Cai, B.; Barra, M.: Rhodes C.A. *J. Am. Chem. Soc.*, **1992**, *114*, 7029–7034.

68. (a) Albinsson, B.; Teramae, H.; Downing, J.W.; Michl, J. *Chem. Eur. J.* **1996**, *2*, 529. (b) Albinsson, B.; Antic, D.; Neumann, F.; Ottosson, C.H.; Michl, J. unpublished results. (c) Mazières, S.; Raymond, M.K.; Raabe, G.; Prodi, A.; Michl, J. *J. Am. Chem. Soc.*, **1997**, *119*, 6682. (d) Raymond, M.K.; Magnera, T.F.; Zharov, I.; West, R.; Dreczewski, B.; Nozik, A.J.; Sprague, J.R.; Ellingson, R.J.; Michl, J. In *Applied Fluorescence in Chemistry, Biology, and Medicine*, Rettig, W.; Strehmel, B.; Schrader, S., eds, Springer Verlag: Heidelberg, **1999**, p. 79.

69. Teramae, H.; Michl, J. *Chem. Phys. Lett.*, **1997**, *276*, 127.

70. Venturini, A.; Vreven, T.; Bernardi, F.; Olivucci, M.; Robb, M.A. *Organometallics*, **1995**, *14*, 4953–4956.

71. (a) Ben-Nun, M.; Molnar, F.; Lu, H.; James, C.P.; Martinez, T.J.; Schulten, K. *Faraday Discussion*, **1998**, *110*, 447. (b) Vreven, T; Bernardi, F.; Garavelli, M.; Olivucci, M.; Robb, M.A.; Schlegel, H.B. *J. Am. Chem. Soc.*, **1997**, *119*, 12687–12688. (c) Klein, S.; Bearpark, M.; Smith, B.R.; Robb, M.A.; Olivucci, M.; Bernardi, F. *Chem. Phys. Lett.*, **1998**, *292*, 259–266.

Index